Undergraduate Texts in Mathematics

Undergraduate Texts in Mathematics

Undergraduate Texts in Mathematics are generally aimed at third- and fourth-year undergraduate mathematics students at North American universities. These texts strive to provide students and teachers with new perspectives and novel approaches. The books include motivation that guides the reader to an appreciation of interrelations among different aspects of the subject. They feature examples that illustrate key concepts as well as exercises that strengthen understanding.

For further volumes:
http://www.springer.com/series/666

Andrew McInerney

First Steps in Differential Geometry

Riemannian, Contact, Symplectic

Springer

Andrew McInerney
Department of Mathematics
and Computer Science
Bronx Community College
City University of New York
Bronx, NY, USA

ISSN 0172-6056
ISBN 978-1-4614-7731-0 ISBN 978-1-4614-7732-7 (eBook)
DOI 10.1007/978-1-4614-7732-7
Springer New York Heidelberg Dordrecht London

Library of Congress Control Number: 2013941656

Mathematics Subject Classification: 53-01, 53B20, 53D05, 53D10, 53A35, 53A45, 58A10, 58-01, 37J05, 37C05, 37C10, 34A26, 78A05

Printed on acid-free paper

Springer is part of Springer Science+Business Media (www.springer.com)

To Arlen

Preface

This text grew out of notes designed to prepare second-year undergraduate students, primarily mathematics majors, to work on introductory research projects in mathematics. Most of these projects were under the auspices of the Louis Stokes Alliance for Minority Participation at Bronx Community College, City University of New York.

As such, the text has two distinct parts. The first three chapters are introductory, and should be seen as "everything one would need to know" to understand a modern approach to differential geometry. The second part of the book is the core of the text, and showcases three geometric structures that are all prominent areas of current mathematical research. It is a hallmark of this text to present the three together in an introductory way.

There are several conceivable approaches to this text, depending on the level of the class. To cover the entire text in detail, presupposing only minimal exposure to matrix algebra and multivariable calculus, could take two semesters. For that reason, in a one-semester differential geometry class, I recommend a quick review of Chap. 2 (focused on Sects. 2.8–2.10) and Chap. 3, followed by a more careful treatment of Chap. 4, especially Sect. 4.7. This leaves time for approximately three weeks for each of Chaps. 5–7. The student who completes this regimen should have a good sense of differential geometry as the study of smoothly varying tensor structures on the tangent bundle. Alternatively, an instructor might choose to spend more time on Chap. 3, supplementing the text with a more rigorous treatment of the inverse and implicit function theorems, and then focusing on just one or two of the geometric structures later in the text.

The price of the early introduction to geometric structures presented here is that it is purely "local." As mentioned throughout the text, the text stops short of introducing manifolds, and has downplayed the role of topology significantly. I hope that the treatment here provides firm preparation to take the steps in that direction.

I thank the colleagues who have helped me in proofreading, with the usual caveat that all remaining errors are mine alone: Quanlei Fang, Michael Harrison, Alexander Kheyfits, Mohamed Messaoudene, Cormac O'Sullivan, Philippe Rukimbira, and Anthony Weaver. Prof. Fang also provided helpful guidance in creating figures in LaTeX. Prof. Augustin Banyaga and Prof. Dusa McDuff provided welcome encouragement (not to mention inspiration!) at various stages of the project.

I thank the editorial team at Springer, especially Kaitlin Leach and the reviewers, for their patience and helpful guidance in seeing this text through.

Finally, I would like to give special thanks to the many students who have in one way or another encouraged me to write this text: Christian Castillo, Aliou Diop, Linus Mensah, Rosario Tate, Stephany Soria, Raysa Martinez, Aida Wade, Feraz Mohamed, Frances Villar, Mandie Solo, Jean Yao, and Keysi Peralta, to mention only a few. My MTH 35 class in Fall 2012 helped me reshape parts of the first three chapters; Dorian Whyte was especially helpful in proofreading. Without these students, this text would not have come to be.

Bronx, NY Andrew McInerney

Contents

Introduction

Differential geometry can be described as the application of the tools of calculus to questions of geometry. Beginning with the verification of age-old geometric measurements such as the circumference and area of a circle, the new techniques of calculus showed their power by quickly dispensing with questions that had long engaged thinkers from antiquity through the seventeenth and eighteenth centuries.

In the nineteenth century, Gauss displayed the extent to which calculus (in particular, the first two derivatives) determines basic properties of curves and surfaces, at least locally. These results can properly be called the beginning of "classical" differential geometry.

Gauss's student Riemann introduced the notion of a manifold, which brought differential geometry into its own. In particular, the first notions of the tangent space would allow the techniques of linearization and ultimately the tools of linear algebra to be brought to bear on geometric questions. Later and perhaps most prominently, Einstein used manifolds to frame the general theory of relativity.

Another conceptual milestone accomplished with the idea of a manifold was allowing a common framework for the new non-Euclidean geometries that were struggling to gain currency in the nineteenth century. Both of these developments occurred by generalizing the notion of distance and imposing more general metric structures on a manifold. This is what is today known as *Riemannian geometry*.

To lesser fanfare, manifolds allowed a new way of generalizing geometry. Sophus Lie, under the influence of Felix Klein, introduced the notion of transformation groups in the course of investigating differential equations. Later, into the early twentieth century through the continued work of F. Engel and E. Cartan, these notions led to what are today known as *symplectic* and *contact* structures on manifolds.

It is impossible to overstate the importance of the notion of the tangent space for the field of differential geometry. In fact, today it is possible to describe differential geometry as "the study of smoothly varying structures on the tangent bundle." It is the aim of this text to develop this point of view.

The development here is somewhat different from those of classical introductory texts in differential geometry such as the works of Struik [39], do Carmo [12],

O'Neill [33], and Kühnel [25]. Those texts are aimed at introducing Riemannian geometry, and especially the metric tensor and its derived concepts such as curvature.

Here, our goal is to develop the architecture necessary to introduce contact and symplectic geometry alongside their Riemannian cousin. After presenting some preliminary material needed from linear algebra, we spend more time than usual in presenting the definition of the tangent space and notions immediately connected to it, such as vector fields. We then present a chapter on differential forms and tensors, which are the "structures" on the tangent space referred to above.

The first three chapters are really a prelude to the core of the book, which is an exposition of the differential geometry of a symmetric, positive definite $(0,2)$-tensor (Riemannian geometry), a nondegenerate one-form (contact geometry), and a closed, nondegenerate two-form (symplectic geometry). There will be no attempt to give an exhaustive treatment of any of these vast areas of current mathematical research. Rather, the goal is to introduce students early in their mathematical careers to this broader view of geometry.

It is unusual to present these three geometric structures side by side as we do here. We do so to emphasize one of the text's major themes: differential geometry as the study of tensor structures on the tangent bundle. In each case, we will show how a tensor structure not only determines certain key geometric objects, but also singles out special functions or transformations that "preserve the structure."

Differential geometry offers a smooth transition from the standard university mathematics sequence of the first four semesters—calculus through differential equations and linear algebra—to the higher levels of abstraction and proof encountered at the upper division by mathematics majors. Topics introduced or hinted at in calculus and linear algebra are used concretely, but in a new setting. Granted, the simplicity and ab initio nature of first courses in abstract algebra or number theory make them ideal settings for students to learn the practice of proofs. Elementary differential geometry sacrifices these in favor of the familiar ground of derivatives and linear transformations, emphasizing instead the importance of proper definition and generality in mathematics.

Indeed, here lies another main goal of this book: to bring the student who has finished two years with a solid foundation in the standard mathematics curriculum into contact with the beauty of "higher" mathematics. In particular, the presentation here emphasizes the consequences of a definition and the careful use of examples and constructions in order to explore those consequences.

This goal places certain limitations on the presentation. The notion of a manifold, which is the basic setting for modern differential geometry, implies a significant role for topology. A manifold is "locally Euclidean," in the same sense that the surface of the Earth was historically believed to be flat.

This text, however, will steer clear of topology as much as possible in order to give center stage to the role of calculus. For the more advanced reader, this will mean that virtually the entire text is "local." Important theorems about global Riemannian geometry such as the Gauss–Bonnet theorem are thus missing from this presentation. This is an even more severe limitation in the cases of contact and symplectic geometry, as we will discuss in their respective chapters.

For this reason, we will avoid the use of the term "manifold" altogether. This will have some unfortunate consequences in terminology. For example, we will refer to submanifolds as "geometric sets."

As a text aimed at a "transitional" audience, students who have completed the traditional calculus and linear algebra sequence but who have not necessarily been exposed to the more abstract formulations of pure mathematics, we should say a word about the role of proofs. The body of the text is structured to include proofs of most of the main statements and results. This is done not just for the sake of mathematical rigor. Rather, it is premised on the perspective that proofs provide more than mere deductive logical justifications. They also provide a showcase in which the main techniques and concepts can be put on display. In addition, in some cases the reader will be asked to supply proofs or details of a proof as a way to exercise this vital mathematical skill.

Most exercises, however, will be designed to present and explore examples of the mathematical constructions involved. This is based on the point of view that mathematics, and geometry in particular, is not merely a deductive undertaking. There is a rich content to mathematics, the appreciation of which requires intuition and familiarity.

Chapter 1
Basic Objects and Notation

Most of modern mathematics is expressed using the language of sets and functions. This can be a significant hurdle for the student whose mathematical experience, possibly through the entire calculus sequence, has not included any emphasis on sets or set operations. For that reason, we review these basic ideas in this chapter with the goal of both establishing the notation and providing a quick reference that the student can consult when proceeding through the main part of the text.

1.1 Sets

The basic concept that the notion of a set is meant to capture is that of inclusion or exclusion. Unfortunately, there are inherent logical difficulties in writing a formal definition of a set. We can resort to one standard "definition": "A set is a collection of objects." This gives a sense of both the distinction between the objects under consideration and the collection itself, as well as the sense of being "included" in the collection or "not included." Unfortunately, it leaves undefined what is meant exactly by the terms "collection" and "object," and so leaves much to be desired from the perspective of a mathematical definition.

For that reason, we will not try to be too precise in defining a set. Rather, we will call the objects under consideration *elements*, and we think of a *set* as "something that contains elements."

We will generally use uppercase letters to denote sets: A, B, S, etc. We will write elements using lowercase letters: a, b, x, y, etc. We express the relation that "x is an element of A" by writing

$$x \in A.$$

We will sometimes more casually say, "x is in A." If, on the contrary, an element y is not an element of the set A, we write $y \notin A$.

A. McInerney, *First Steps in Differential Geometry: Riemannian, Contact, Symplectic*, Undergraduate Texts in Mathematics, DOI 10.1007/978-1-4614-7732-7_1,

The basic assumption that we make about sets is that they are *well defined*: For every element x and every set A, the statement, "Either $x \in A$ or $x \notin A$" is true, and so the statement "$x \in A$ and $x \notin A$" is false.

In any particular problem, the context will imply a *universal set*, which is the set of all objects under consideration. For example, statisticians might be concerned with a data set of measurements of heights in a given population of people. Geometers might be concerned with properties of the set of points in three-dimensional space. When the statisticians ask whether an element is in a particular set, they will consider only elements in their "universe," and so in particular will not even ask the question whether the geometers' points are in the statisticians' sets.

While the universal set for a given discussion or problem may or may not be explicitly stated, it should always be able to be established from the context. From a logical point of view, in fact, the universal set *is* the context.

There are several standard ways of describing sets. The most basic way is by listing the elements, written using curly braces to enclose the elements of the set. For example, a set A with three elements a, b, and c is written

$$A = \{a, b, c\}.$$

Note that the order in which the elements are listed is not important, so that, for example, $\{a, b, c\}$ is the same set as $\{b, a, c\}$.

Describing a set by means of a list is also possible in the case of (countably) infinite sets, at least when there is a pattern involved. For example, the set of even natural numbers can be expressed as

$$E = \{2, 4, 6, \ldots\}.$$

Here, the ellipsis $(\ldots)$ expresses a pattern that should be obvious to the reader in context.

Most often, however, we will describe sets using what is known as *set-builder notation*. In this notation, a set is described as all elements (of the universal set) having a certain property or properties. These properties are generally given in the form of a logical statement about an element x, which we can write as $P(x)$. In other words, $P(x)$ is true if x has property P and $P(x)$ is false if x does not have property P. Hence we write

$$\{x \in X \mid P(x)\}$$

to represent the set of all x in the universal set X for which the statement $P(x)$ is true. When the universal set is clear from the context, we often simply write $\{x \mid P(x)\}$. In this notation, for example, the set of even natural numbers can be written

$$E = \{n \mid \text{there exists a natural number } k \text{ such that } n = 2k\}.$$

Here the statement $P(n)$ is, "there exists a natural number k such that $n = 2k$." Then $P(4)$ is a true statement, since $4 = 2(2)$, and so $4 \in E$. On the other hand, $P(5)$ is false, since there is no natural number k such that $5 = 2k$, and so $5 \notin E$.

We will also encounter sets of elements described by several properties. For example, the set

$$\{x \mid P_i(x),\ i = 1, \dots, r\}$$

means the set of all elements x for which the r distinct statements

$$P_1(x), P_2(x), \dots, P_r(x)$$

are all true.

Set-builder notation has a number of advantages. First, it gives a way to effectively describe very large or infinite sets without having to resort to lists or cumbersome patterns. For example, the set of rational numbers can be described as

$$\left\{x \in \mathbf{R} \;\middle|\; \text{there are integers } p \text{ and } q \text{ such that } x = \tfrac{p}{q}\right\}.$$

More important, the notation makes explicit the logical structure that is implicit in the language of sets. Since this structure underlies the entire development of mathematics as statements that can be proven according to the rigors of logic, we will emphasize this here.

One special set deserves mention. The *empty set*, denoted by $\emptyset$, is the set with no elements. There is only one such set, although it may appear in many forms. For example, if for some set X, a statement $P(x)$ is *false* for all $x \in X$, then $\emptyset = \{x \mid P(x)\}$.

We now consider relations and operations among sets. For this purpose, we suppose that we are given two sets A and B with the universal set X. We will suppose that both A and B are described in set-builder notation

$$A = \{x \mid P_A(x)\}, \quad B = \{x \mid P_B(x)\},$$

where P_A and P_B are properties describing the sets A and B respectively.

We will say that A *is a subset of* B, written $A \subset B$, if for all $a \in A$, we also have $a \in B$. Hence,

to prove $A \subset B$, show that for all $x \in X$, if $P_A(x)$ is true, then $P_B(x)$ is true.

In fact, there are two standard ways of proving that $A \subset B$. In the direct method, one assumes that $P_A(x)$ is true for all $x \in A$ and then deduces, by definitions and previously proved statements, that $P_B(x)$ must also be true. In the indirect method, by contrast, one assumes that there exists an $x \in A$ such that $P_A(x)$ is true and that $P_B(x)$ is false, and then attempts to derive a contradiction.

The empty set is a subset of all sets: For every set X, $\emptyset \subset X$.

Set equality is defined in terms of the preceding inclusion relationship: Two sets A and B are said to be equal, written $A = B$, if $A \subset B$ and $B \subset A$. Logically, this condition is expressed as an "if and only if" relationship:

To prove $A = B$, show that for all $x \in X$, $P_A(x)$ is true if and only if $P_B(x)$ is true.

There are several basic set operations corresponding to the basic logical connectives "and," "or," and "not." The *intersection of A and B*, written $A \cap B$, is the set

$$A \cap B = \{x \mid x \in A \text{ and } x \in B\}.$$

We have the following:

To prove $x \in A \cap B$, show that $P_A(x)$ is true AND $P_B(x)$ is true.

Also, the *union of A and B* is the set

$$A \cup B = \{x \mid x \in A \text{ or } x \in B\},$$

and

To prove $x \in A \cup B$, show that either $P_A(x)$ is true OR $P_B(x)$ is true.

The *difference of A and B* is the set.

If $P_A(x)$ is true and $P_B(x)$ is true, then the compound statement $P_A(x)$ OR $P_B(x)$ is also true

$$A \backslash B = \{x \mid x \in A \text{ and } x \notin B\},$$

so

To prove $x \in A \backslash B$, show that $P_A(x)$ is true AND $P_B(x)$ is false.

In particular, the *complement of A* is the set $A^c = X \backslash A$. So

To prove $x \in A^c$, show that $P_A(x)$ is false.

There is yet another set operation, one that is of a somewhat different nature from the previous operations. The *Cartesian product of sets A and B* is the set $A \times B$ whose elements are ordered pairs of the form (a, b). More precisely,

$$A \times B = \{(x, y) \mid x \in A \text{ and } y \in B\}, \text{ i.e.,}$$

To prove $(x, y) \in A \times B$, show that $P_A(x)$ is true AND $P_B(y)$ is true.

In this case, the separate entries in the ordered pair are referred to as *components*.

Finally, we list several common sets of numbers along with the standard notation:

$\mathbf{N}$, the set of natural numbers,

$\mathbf{Z}$, the set of integers,

$\mathbf{Q}$, the set of rational numbers,

$\mathbf{R}$, the set of real numbers, and

$\mathbf{C}$, the set of complex numbers.

We will be especially concerned with the set $\mathbf{R}^n = \mathbf{R} \times \cdots \times \mathbf{R}$:

$$\mathbf{R}^n = \{(x_1, \dots, x_n) \mid x_i \in \mathbf{R} \text{ for } i = 1, \dots, n\}.$$

Many times we will encounter special subsets of the real numbers that are defined by the order relations $<$, $\leq$, $>$, and $\geq$. These are the *intervals*, and we use the standard notation

$$\begin{aligned}
[a, b] &= \{r \in \mathbf{R} \mid a \leq r \leq b\}, \\
(a, b) &= \{r \in \mathbf{R} \mid a < r < b\}, \\
(-\infty, b) &= \{r \in \mathbf{R} \mid r < b\}, \\
(-\infty, b] &= \{r \in \mathbf{R} \mid r \leq b\}, \\
(a, \infty) &= \{r \in \mathbf{R} \mid r > a\}, \\
[a, \infty) &= \{r \in \mathbf{R} \mid r \geq a\}.
\end{aligned}$$

We can similarly define the half-open intervals $(a, b]$, etc.

1.2 Functions

Most readers will recall the definition of a function that is typically presented, for example, in a precalculus course. A function is defined to be a rule assigning to each element of one set exactly one element of another set. The advantage of this definition is that it emphasizes the relationship established between elements of the domain and those of the range by means of the rule. It has the disadvantage, however, of lacking mathematical precision, especially by relying on the imprecise term "rule."

In order to be more precise, mathematicians in the 1920s established the following definition of a function. Given two sets A and B, a function f is defined to be a subset $f \subset A \times B$ with the following two properties: First, for all $a \in A$ there

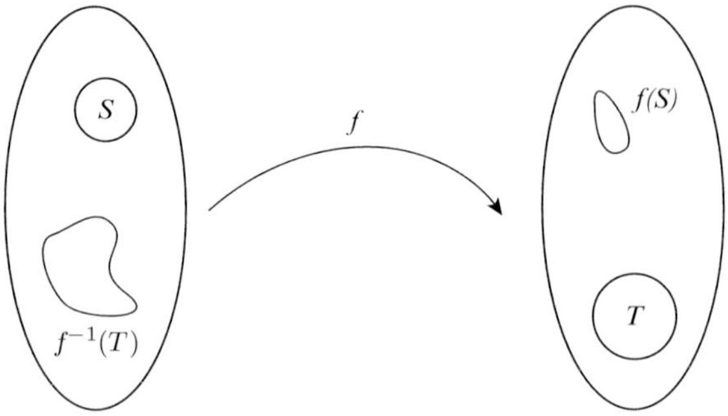

Fig. 1.1 A representation of the image and preimage of a function f.

is $b \in B$ such that $(a,b) \in f$, and second, if $(a_1,b_1),(a_2,b_2) \in f$ and $a_1 = a_2$, then $b_1 = b_2$. (Note that in this formal context, we are faced with competing notational conventions. On the one hand, sets should be denoted with an uppercase letter, while historical usage prefers the lowercase f to denote a function.)

The logical structure of this definition implies a clear distinction between the sets A and B, with a statement about elements of A ($a_1 = a_2$) implying a statement about elements of B ($b_1 = b_2$). The set A is called the *domain*, and the set B is called the *codomain*. (We will reserve the term "range" for a special subset of B defined below.)

Despite this formal definition, throughout this text we will rely on the standard notation for functions. In particular, for a function f with domain A and codomain B, we use the notation $f : A \to B$ to represent the function; the arrow points from the domain to the codomain. We write $f(a)$ to represent the unique element of B such that $(a, f(a)) \in f$. We will occasionally use the notation $a \mapsto f(a)$.

As a central concept in mathematics, a number of different terms have emerged to describe a function. We will use the words *function*, *map*, and *transformation* interchangeably.

Given a function $f : A \to B$ and a set $S \subset A$, the *image of S (under f)* is defined to be the set

$$f(S) = \{y \in B \mid \text{there is } x \in S \text{ such that } f(x) = y\} \subset B.$$

The *range of f* is defined to be the set $f(A)$. For a subset $T \subset B$, the *preimage of T (under f)* is the (possibly empty) set (Fig. 1.1)

$$f^{-1}(T) = \{x \in A \mid f(x) \in T\} \subset A.$$

A function $f : A \to B$ is *onto* if $f(A) = B$, i.e., if the range of f coincides with the codomain of f. To demonstrate that a function is onto, it is necessary then to show that $B \subset f(A)$ (since $f(A) \subset B$ by definition). In other words, it is necessary to show for that for every element $b \in B$, there is $a \in A$ such that $f(a) = b$; this is often done by writing a explicitly in terms of b.

A function $f : A \to B$ is *one-to-one* if for every two elements $a_1, a_2 \in A$, the condition $f(a_1) = f(a_2)$ implies that $a_1 = a_2$. There are two basic approaches to showing that a function is one-to-one. The direct method is to assume that there are two elements a_1, a_2 with the property that $f(a_1) = f(a_2)$ and to show that this implies that $a_1 = a_2$. The indirect method is to show that given any two different elements $a_1, a_2 \in A$ ($a_1 \neq a_2$), this implies that $f(a_1) \neq f(a_2)$.

Given any set A, there is always a distinguished function $\mathrm{Id}_A : A \to A$ defined by $\mathrm{Id}_A(a) = a$ for all $a \in A$. This function is called the *identity map of* A.

Suppose there are two functions $f : A \to B$ and $g : C \to D$ with the property that $f(A) \subset C$. Then it is possible to define a new function $g \circ f : A \to D$ by $(g \circ f)(a) = g(f(a))$ for all $a \in A$. This function is called the *composition* of g with f. This operation on functions is associative, i.e.,

$$(h \circ g) \circ f = h \circ (g \circ f),$$

assuming that all compositions are defined. A function $f : A \to B$ is one-to-one if and only if it has an *inverse* defined on the image of A, i.e., there is a function $g : f(A) \to A$ such that $f \circ g = \mathrm{Id}_{f(A)}$ and $g \circ f = \mathrm{Id}_A$. We normally write f^{-1} to denote the inverse of f.

In addition, for *real-valued* functions $f : A \to \mathbf{R}$, i.e., functions whose codomain is a subset of the set of real numbers, there are a number of operations inherited from the familiar operations on real numbers. If f and g are real-valued functions, then $f+g$, $f-g$, $f \cdot g$, and f/g are defined *pointwise*. For example, $f+g$ is the function whose domain is defined to be the intersection of the domains of f and g and whose value $(f+g)(a)$ for $a \in A$ is given by $f(a) + g(a)$.

Chapter 2
Linear Algebra Essentials

When elementary school students first leave the solid ground of arithmetic for the more abstract world of algebra, the first objects they encounter are generally linear expressions. Algebraically, linear equations can be solved using elementary field properties, namely the existence of additive and multiplicative inverses. Geometrically, a nonvertical line in the plane through the origin can be described completely by one number—the slope. Linear functions $f : \mathbf{R} \to \mathbf{R}$ enjoy other nice properties: They are (in general) invertible, and the composition of linear functions is again linear.

Yet marching through the progression of more complicated functions and expressions—polynomial, algebraic, transcendental—many of these basic properties of linearity can become taken for granted. In the standard calculus sequence, sophisticated techniques are developed that seem to yield little new information about linear functions. Linear algebra is generally introduced after the basic calculus sequence has been nearly completed, and is presented in a self-contained manner, with little reference to what has been seen before. A fundamental insight is lost or obscured: that differential calculus is the study of nonlinear phenomena by "linearization."

The main goal of this chapter is to present the basic elements of linear algebra needed to understand this insight of differential calculus. We also present some geometric applications of linear algebra with an eye toward later constructions in differential geometry.

While this chapter is written for readers who have already been exposed to a first course in linear algebra, it is self-contained enough that the only essential prerequisites will be a working knowledge of matrix algebra, Gaussian elimination, and determinants.

A. McInerney, *First Steps in Differential Geometry: Riemannian, Contact, Symplectic*, Undergraduate Texts in Mathematics, DOI 10.1007/978-1-4614-7732-7_2,

2.1 Vector Spaces

Modern mathematics can be described as the study of sets with some extra associated "structure." In linear algebra, the sets under consideration have enough structure to allow elements to be added and multiplied by scalars. These two operations should behave and interact in familiar ways.

Definition 2.1.1. A (real) *vector space* consists of a set V together with two operations, addition and scalar multiplication.[1] Scalars are understood here as real numbers. Elements of V are called *vectors* and will often be written in bold type, as $\mathbf{v} \in V$. Addition is written using the conventional symbolism $\mathbf{v} + \mathbf{w}$. Scalar multiplication is denoted by $s\mathbf{v}$ or $s \cdot \mathbf{v}$.

The triple $(V, +, \cdot)$ must satisfy the following axioms:

(V1) For all $\mathbf{v}, \mathbf{w} \in V$, $\mathbf{v} + \mathbf{w} \in V$.
(V2) For all $\mathbf{u}, \mathbf{v}, \mathbf{w} \in V$, $(\mathbf{u} + \mathbf{v}) + \mathbf{w} = \mathbf{u} + (\mathbf{v} + \mathbf{w})$.
(V3) For all $\mathbf{v}, \mathbf{w} \in V$, $\mathbf{v} + \mathbf{w} = \mathbf{w} + \mathbf{v}$.
(V4) There exists a distinguished element of V, called the *zero vector* and denoted by $\mathbf{0}$, with the property that for all $\mathbf{v} \in V$, $\mathbf{0} + \mathbf{v} = \mathbf{v}$.
(V5) For all $\mathbf{v} \in V$, there exists an element called the *additive inverse of* $\mathbf{v}$ and denoted $-\mathbf{v}$, with the property that $(-\mathbf{v}) + \mathbf{v} = \mathbf{0}$.
(V6) For all $s \in \mathbf{R}$ and $\mathbf{v} \in V$, $s\mathbf{v} \in V$.
(V7) For all $s, t \in \mathbf{R}$ and $\mathbf{v} \in V$, $s(t\mathbf{v}) = (st)\mathbf{v}$.
(V8) For all $s, t \in \mathbf{R}$ and $\mathbf{v} \in V$, $(s + t)\mathbf{v} = s\mathbf{v} + t\mathbf{v}$.
(V9) For all $s \in \mathbf{R}$ and $\mathbf{v}, \mathbf{w} \in V$, $s(\mathbf{v} + \mathbf{w}) = s\mathbf{v} + s\mathbf{w}$.
(V10) For all $\mathbf{v} \in V$, $1\mathbf{v} = \mathbf{v}$.

We will often suppress the explicit ordered triple notation $(V, +, \cdot)$ and simply refer to "the vector space V."

In an elementary linear algebra course, a number of familiar properties of vector spaces are derived as consequences of the 10 axioms. We list several of them here.

Theorem 2.1.2. *Let V be a vector space. Then:*

1. *The zero vector* $\mathbf{0}$ *is unique.*
2. *For all* $\mathbf{v} \in V$*, the additive inverse* $-\mathbf{v}$ *of* $\mathbf{v}$ *is unique.*
3. *For all* $\mathbf{v} \in V$, $0 \cdot \mathbf{v} = \mathbf{0}$.
4. *For all* $\mathbf{v} \in V$, $(-1) \cdot \mathbf{v} = -\mathbf{v}$.

Proof. Exercise. □

Physics texts often discuss vectors in terms of the two properties of magnitude and direction. These are not in any way related to the vector space axioms. Both of

[1] More formally, addition can be described as a function $V \times V \to V$ and scalar multiplication as a function $\mathbf{R} \times V \to V$.

these concepts arise naturally in the context of *inner product spaces*, which we treat in Sect. 2.9.

In a first course in linear algebra, a student is exposed to a number of examples of vector spaces, familiar and not-so-familiar, in order to gain better acquaintance with the axioms. Here we introduce just two examples.

Example 2.1.3. For any positive integer n, define the set $\mathbf{R}^n$ to be the set of all n-tuples of real numbers:

$$\mathbf{R}^n = \{(a_1, \ldots, a_n) \mid a_i \in \mathbf{R} \text{ for } i = 1, \ldots, n\}$$

Define vector addition componentwise by

$$(a_1, \ldots, a_n) + (b_1, \ldots, b_n) = (a_1 + b_1, \ldots, a_n + b_n),$$

and likewise define scalar multiplication by

$$s(a_1, \ldots, a_n) = (sa_1, \ldots, sa_n).$$

It is a straightforward exercise to show that $\mathbf{R}^n$ with these operations satisfies the vector space axioms. These vector spaces (one for each natural number n) will be called *Euclidean spaces*.

The Euclidean spaces can be thought of as the "model" finite-dimensional vector spaces in at least two senses. First, they are the most familiar examples, generalizing the set $\mathbf{R}^2$ that is the setting for the most elementary analytic geometry that most students first encounter in high school. Second, we show later that every finite-dimensional vector space is "equivalent" (in a sense we will make precise) to $\mathbf{R}^n$ for some n.

Much of the work in later chapters will concern $\mathbf{R}^3$, $\mathbf{R}^4$, and other Euclidean spaces. We will be relying on additional structures of these sets that go beyond the bounds of linear algebra. Nevertheless, the vector space structure remains essential to the tools of calculus that we will employ later.

The following example gives a class of vector spaces that are in general not equivalent to Euclidean spaces.

Example 2.1.4 (Vector spaces of functions). For any set X, let $\mathcal{F}(X)$ be the set of all real-valued functions $f : X \to \mathbf{R}$. For every two such $f, g \in \mathcal{F}(X)$, define the sum $f + g$ pointwise as $(f + g)(x) = f(x) + g(x)$. Likewise, define scalar multiplication $(sf)(x) = s(f(x))$. The set $\mathcal{F}(X)$ equipped with these operations is a vector space. The zero vector is the function $O : X \to \mathbf{R}$ that is identically zero: $O(x) = 0$ for all $x \in X$. Confirmation of the axioms depends on the corresponding field properties in the codomain, the set of real numbers.

We will return to this class of vector spaces in the next section.

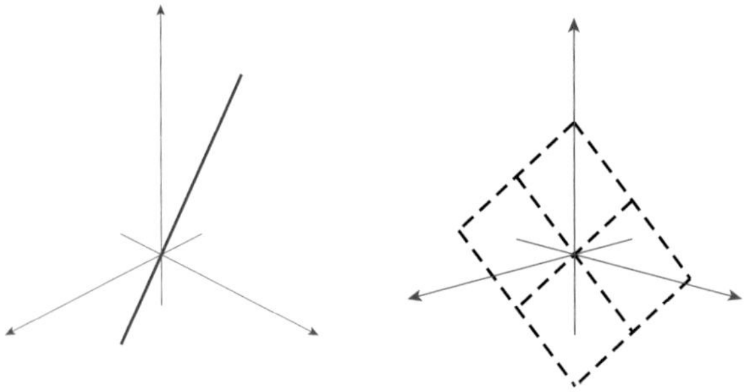

Fig. 2.1 Subspaces in $\mathbf{R}^3$.

2.2 Subspaces

A mathematical structure on a set distinguishes certain subsets of special significance. In the case of a set with the structural axioms of a vector space, the distinguished subsets are those that are themselves vector spaces under the same operations of vector addition and scalar multiplication as in the larger set.

Definition 2.2.1. Let W be a subset of a vector space $(V,+,\cdot)$. Then W is a *vector subspace* (or just *subspace*) of V if $(W,+,\cdot)$ satisfies the vector space axioms (V1)–(V10).

A subspace can be pictured as a vector space "within" a larger vector space. See Fig. 2.1.

Before illustrating examples of subspaces, we immediately state a theorem that ensures that most of the vector space axioms are in fact inherited from the larger ambient vector space.

Theorem 2.2.2. *Suppose $W \subset V$ is a nonempty subset of a vector space V satisfying the following two properties:*

(W1) For all $\mathbf{v},\mathbf{w} \in W$, $\mathbf{v}+\mathbf{w} \in W$.
(W2) For all $\mathbf{w} \in W$ and $s \in \mathbf{R}$, $s\mathbf{w} \in W$.

Then W is a subspace of V.

Proof. Exercise. □

We note that for every vector space V, the set $\{\mathbf{0}\}$ is a subspace of V, known as the *trivial subspace*. Similarly, V is a subspace of itself, which is known as the *improper subspace*.

We now illustrate some nontrivial, proper subspaces of the vector space $\mathbf{R}^3$. We leave the verifications that they are in fact subspaces to the reader.

Example 2.2.3. Let $W_1 = \{(s,0,0) \mid s \in \mathbf{R}\}$. Then W_1 is a subspace of $\mathbf{R}^3$.

Example 2.2.4. Let $\mathbf{v} = (a, b, c) \neq \mathbf{0}$ and let $W_2 = \{s\mathbf{v} \mid s \in \mathbf{R}\}$. Then W_2 is a subspace of $\mathbf{R}^3$. Note that Example 2.2.3 is a special case of this example when $\mathbf{v} = (1, 0, 0)$.

Example 2.2.5. Let $W_3 = \{(s, t, 0) \mid s, t \in \mathbf{R}\}$. Then W_3 is a subspace of $\mathbf{R}^3$.

Example 2.2.6. As in Example 2.2.4, let $\mathbf{v} = (a, b, c) \neq \mathbf{0}$. Relying on the usual "dot product" in $\mathbf{R}^3$, define

$$\begin{aligned} W_4 &= \{\mathbf{x} \in \mathbf{R}^3 \mid \mathbf{v} \cdot \mathbf{x} = 0\} \\ &= \{(x_1, x_2, x_3) \mid ax_1 + bx_2 + cx_3 = 0\}. \end{aligned}$$

Then W_4 is a subspace of $\mathbf{R}^3$. Note that Example 2.2.5 is a special case of this example when $\mathbf{v} = (0, 0, 1)$.

We will show at the end of Sect. 2.4 that all proper, nontrivial subspaces of $\mathbf{R}^3$ can be realized either in the form of W_2 or W_4.

Example 2.2.7 (Subspaces of $\mathcal{F}(\mathbf{R})$). We list here a number of vector subspaces of $\mathcal{F}(\mathbf{R})$, the space of real-valued functions $f : \mathbf{R} \to \mathbf{R}$. The verifications that they are in fact subspaces are straightforward exercises using the basic facts of algebra and calculus.

- $P_n(\mathbf{R})$, the subspace of polynomial functions of degree n or less;
- $P(\mathbf{R})$, the subspace of all polynomial functions (of any degree);
- $C(\mathbf{R})$, the subspace of functions that are continuous at each point in their domain;
- $C^r(\mathbf{R})$, the subspace of functions whose first r derivatives exist and are continuous at each point in their domain;
- $C^\infty(\mathbf{R})$, the subspace of functions all of whose derivatives exist and are continuous at each point in their domain.

Our goal in the next section will be to exhibit a method for constructing vector subspaces of any vector space V.

2.3 Constructing Subspaces I: Spanning Sets

The two vector space operations give a way to produce new vectors from a given set of vectors. This, in turn, gives a basic method for constructing subspaces. We mention here that for the remainder of the chapter, when we specify that a set is *finite* as an assumption, we will also assume that the set is nonempty.

Definition 2.3.1. Suppose $S = \{\mathbf{v}_1, \mathbf{v}_2, \ldots, \mathbf{v}_n\}$ is a finite set of vectors in a vector space V. A vector $\mathbf{w}$ is a *linear combination* of S if there are scalars $c_1, \ldots, c_n$ such that

$$\mathbf{w} = c_1\mathbf{v}_1 + \cdots + c_n\mathbf{v}_n.$$

A basic question in a first course in linear algebra is this: For a vector $\mathbf{w}$ and a set S as in Definition 2.3.1, decide whether $\mathbf{w}$ is a linear combination of S. In practice, this can be answered using the tools of matrix algebra.

Example 2.3.2. Let $S = \{\mathbf{v}_1, \mathbf{v}_2\} \subset \mathbf{R}^3$, where $\mathbf{v}_1 = (1,2,3)$ and $\mathbf{v}_2 = (-1,4,2)$. Let us decide whether $\mathbf{w} = (29,-14,27)$ is a linear combination of S. To do this means solving the vector equation $\mathbf{w} = s_1\mathbf{v}_1 + s_2\mathbf{v}_2$ for the two scalars s_1, s_2, which in turn amounts to solving the system of linear equations

$$\begin{cases} s_1(1) + s_2(-1) = 29, \\ s_1(2) + s_2(4) = -14, \\ s_1(3) + s_2(2) = 27. \end{cases}$$

Gaussian elimination of the corresponding augmented matrix yields

$$\left[\begin{array}{cc|c} 1 & 0 & 17 \\ 0 & 1 & -12 \\ 0 & 0 & 0 \end{array}\right],$$

corresponding to the unique solution $s_1 = 17$, $s_2 = -12$. Hence, $\mathbf{w}$ is a linear combination of S.

The reader will notice from this example that deciding whether a vector is a linear combination of a given set ultimately amounts to deciding whether the corresponding system of linear equations is consistent.

We will now use Definition 2.3.1 to obtain a method for constructing subspaces.

Definition 2.3.3. Let V be a vector space and let $S = \{\mathbf{v}_1, \ldots, \mathbf{v}_n\} \subset V$ be a finite set of vectors. The *span of* S, denoted by $\mathrm{Span}(S)$, is defined to be the set of all linear combinations of S:

$$\mathrm{Span}(S) = \{s_1\mathbf{v}_1 + \cdots + s_n\mathbf{v}_n \mid s_1, \ldots, s_n \in \mathbf{R}\}.$$

We note immediately the utility of this construction.

Theorem 2.3.4. *Let $S \subset V$ be a finite set of vectors. Then $W = \mathrm{Span}(S)$ is a subspace of V.*

Proof. The proof is an immediate application of Theorem 2.2.2. □

We will say that S *spans* the subspace W, or that S is a *spanning set* for the subspace W.

Example 2.3.5. Let $S = \{\mathbf{v}_1\} \subset \mathbf{R}^3$, where $\mathbf{v}_1 = (1,0,0)$. Then $\mathrm{Span}(S) = \{s(1,0,0) \mid s \in \mathbf{R}\} = \{(s,0,0) \mid s \in \mathbf{R}\}$. Compare to Example 2.2.3.

Example 2.3.6. Let $S = \{\mathbf{v}_1, \mathbf{v}_2\} \subset \mathbf{R}^4$, where $\mathbf{v}_1 = (1,0,0,0)$ and $\mathbf{v}_2 = (0,0,1,0)$. Then

$$\mathrm{Span}(S) = \{s(1,0,0,0) + t(0,0,1,0) \mid s,t \in \mathbf{R}\} = \{(s,0,t,0) \mid s,t \in \mathbf{R}\}.$$

Example 2.3.7. Let $S = \{\mathbf{v}_1, \mathbf{v}_2, \mathbf{v}_3\} \subset \mathbf{R}^3$ where $\mathbf{v}_1 = (1,0,0)$, $\mathbf{v}_2 = (0,1,0)$, and $\mathbf{v}_3 = (0,0,1)$. Then

$$\begin{aligned}\mathrm{Span}(S) &= \{s_1(1,0,0) + s_2(0,1,0) + s_3(0,0,1) \mid s_1, s_2, s_3 \in \mathbf{R}\}\\ &= \{(s_1, s_2, s_3) \mid s_1, s_2, s_3 \in \mathbf{R}\}\\ &= \mathbf{R}^3.\end{aligned}$$

Example 2.3.8. Let $S = \{\mathbf{v}_1, \mathbf{v}_2, \mathbf{v}_3, \mathbf{v}_4\} \subset \mathbf{R}^3$, where $\mathbf{v}_1 = (1,1,1)$, $\mathbf{v}_2 = (-1,1,0)$, $\mathbf{v}_3 = (1,3,2)$, and $\mathbf{v}_4 = (-3,1,-1)$. Then

$$\begin{aligned}\mathrm{Span}(S) &= \{s_1(1,1,1) + s_2(-1,1,0)\\ &\qquad + s_3(1,3,2) + s_4(-3,1,-1) \mid s_1, s_2, s_3, s_4 \in \mathbf{R}\}\\ &= \{(s_1 - s_2 + s_3 - 3s_4, s_1 + s_2 + 3s_3 + s_4,\\ &\qquad s_1 + 2s_3 - s_4) \mid s_1, s_2, s_3, s_4 \in \mathbf{R}\}.\end{aligned}$$

For example, consider $\mathbf{w} = (13,3,8) \in \mathbf{R}^3$. Then $\mathbf{w} \in \mathrm{Span}(S)$, since $\mathbf{w} = \mathbf{v}_1 - \mathbf{v}_2 + 2\mathbf{v}_3 - 3\mathbf{v}_4$.

Note that this set of four vectors S in $\mathbf{R}^3$ *does not* span $\mathbf{R}^3$. To see this, take an arbitrary $\mathbf{w} \in \mathbf{R}^3$, $\mathbf{w} = (w_1, w_2, w_3)$. If $\mathbf{w}$ is a linear combination of S, then there are scalars s_1, s_2, s_3, s_4 such that $\mathbf{w} = s_1\mathbf{v}_1 + s_2\mathbf{v}_2 + s_3\mathbf{v}_3 + s_4\mathbf{v}_4$. In other words, if $\mathbf{w} \in \mathrm{Span}(S)$, then the system

$$\begin{cases} s_1 - s_2 + s_3 - 3s_4 = w_1,\\ s_1 + s_2 + 3s_3 + s_4 = w_2,\\ s_1 \phantom{{}+s_2} + 2s_3 - s_4 = w_3,\end{cases}$$

is consistent: we can solve for s_1, s_2, s_3, s_4 in terms of w_1, w_2, w_3. Gaussian elimination of the corresponding augmented matrix

$$\left[\begin{array}{cccc|c} 1 & -1 & 1 & -3 & w_1\\ 1 & 1 & 3 & 1 & w_2\\ 1 & 0 & 2 & -1 & w_3\end{array}\right]$$

yields

$$\left[\begin{array}{cccc|c} 1 & 0 & 2 & -1 & w_3\\ 0 & 1 & 1 & 2 & -w_1 + w_3\\ 0 & 0 & 0 & 0 & w_1 + w_2 - 2w_3\end{array}\right].$$

Hence for every vector $\mathbf{w}$ such that $w_1 + w_2 - 2w_3 \neq 0$, the system is not consistent and $\mathbf{w} \notin \mathrm{Span}(S)$. For example, $(1,1,2) \notin \mathrm{Span}(S)$.

We return to this example below.

Note that a given subspace may have many different spanning sets. For example, consider $S = \{(1,0,0),(1,1,0),(1,1,1)\} \subset \mathbf{R}^3$. The reader may verify that S is a spanning set for $\mathbf{R}^3$. But in Example 2.3.7, we exhibited a different spanning set for $\mathbf{R}^3$.

2.4 Linear Independence, Basis, and Dimension

In the preceding section, we started with a finite set $S \subset V$ in order to generate a subspace $W = \text{Span}(S)$ in V. This procedure prompts the following question: For a subspace W, can we find a spanning set for W? If so, what is the "smallest" such set? These questions lead naturally to the notion of a basis. Before defining that notion, however, we introduce the concepts of linear dependence and independence.

For a vector space V, a finite set of vectors $S = \{\mathbf{v}_1, \ldots \mathbf{v}_n\}$, and a vector $\mathbf{w} \in V$, we have already considered the question whether $\mathbf{w} \in \text{Span}(S)$. Intuitively, we might say that $\mathbf{w}$ "depends linearly" on S if $\mathbf{w} \in \text{Span}(S)$, i.e., if $\mathbf{w}$ can be written as a linear combination of elements of S. In the simplest case, for example, that $S = \{\mathbf{v}\}$, then $\mathbf{w}$ "depends on" S if $\mathbf{w} = s\mathbf{v}$, or, what is the same, $\mathbf{w}$ is "independent" of S if $\mathbf{w}$ is not a scalar multiple of $\mathbf{v}$.

The following definition aims to make this sense of dependence precise.

Definition 2.4.1. A finite set of vectors $S = \{\mathbf{v}_1, \ldots \mathbf{v}_n\}$ is *linearly dependent* if there are scalars $s_1, \ldots, s_n$, *not all zero*, such that

$$s_1\mathbf{v}_1 + \cdots + s_n\mathbf{v}_n = \mathbf{0}.$$

If S is not linearly dependent, then it is *linearly independent*.

The positive way of defining linear independence, then, is that a finite set of vectors $S = \{\mathbf{v}_1, \ldots, \mathbf{v}_n\}$ is linearly independent if the condition that there are scalars $s_1, \ldots, s_n$ satisfying $s_1\mathbf{v}_1 + \cdots + s_n\mathbf{v}_n = \mathbf{0}$ implies that

$$s_1 = \cdots = s_n = 0.$$

Example 2.4.2. We refer back to the set $S = \{\mathbf{v}_1, \mathbf{v}_2, \mathbf{v}_3, \mathbf{v}_4\} \subset \mathbf{R}^3$, where $\mathbf{v}_1 = (1,1,1)$, $\mathbf{v}_2 = (-1,1,0)$, $\mathbf{v}_3 = (1,3,2)$, and $\mathbf{v}_4 = (-3,1,-1)$, in Example 2.3.8. We will show that the set S is linearly dependent. In other words, we will find scalars s_1, s_2, s_3, s_4, not all zero, such that $s_1\mathbf{v}_1 + s_2\mathbf{v}_2 + s_3\mathbf{v}_3 + s_4\mathbf{v}_4 = \mathbf{0}$.

This amounts to solving the homogeneous system

$$\begin{cases} s_1 - s_2 + s_3 - 3s_4 = 0, \\ s_1 + s_2 + 3s_3 + s_4 = 0, \\ s_1 \qquad + 2s_3 - s_4 = 0. \end{cases}$$

Gaussian elimination of the corresponding augmented matrix yields

$$\left[\begin{array}{cccc|c} 1 & 0 & 2 & -1 & 0 \\ 0 & 1 & 1 & 2 & 0 \\ 0 & 0 & 0 & 0 & 0 \end{array}\right].$$

This system has nontrivial solutions of the form $s_1 = -2t + u$, $s_2 = -t - 2u$, $s_3 = t$, $s_4 = u$. The reader can verify, for example, that

$$(-1)\mathbf{v}_1 + (-3)\mathbf{v}_2 + (1)\mathbf{v}_3 + (1)\mathbf{v}_4 = \mathbf{0}.$$

Hence S is linearly dependent.

Example 2.4.2 illustrates the fact that deciding whether a set is linearly dependent amounts to deciding whether a corresponding *homogeneous* system of linear equations has nontrivial solutions.

The following facts are consequences of Definition 2.4.1. The reader is invited to supply proofs.

Theorem 2.4.3. *Let S be a finite set of vectors in a vector space V. Then:*

1. *If $\mathbf{0} \in S$, then S is linearly dependent.*
2. *If $S = \{\mathbf{v}\}$ and $\mathbf{v} \neq \mathbf{0}$, then S is linearly independent.*
3. *Suppose S has at least two vectors. Then S is a linearly dependent set of nonzero vectors if and only if there exists a vector in S that can be written as a linear combination of the others.*

Linear dependence or independence has important consequences related to the notion of spanning sets. For example, the following theorem asserts that enlarging a set by adding linearly dependent vectors does not change the spanning set.

Theorem 2.4.4. *Let S be a finite set of vectors in a vector space V. Let $\mathbf{w} \in \mathrm{Span}(S)$, and let $S' = S \cup \{\mathbf{w}\}$. Then $\mathrm{Span}(S') = \mathrm{Span}(S)$.*

Proof. Exercise. □

Generating "larger" subspaces thus requires adding vectors that are *linearly independent* of the original spanning set.

We return to a version of the question at the outset of this section: If we are given a subspace, what is the "smallest" subset that can serve as a spanning set for this subspace? This motivates the definition of a basis.

Definition 2.4.5. Let V be a vector space. A *basis* for V is a set $B \subset V$ such that (1) $\mathrm{Span}(B) = V$ and (2) B is a linearly independent set.

Example 2.4.6. For the vector space $V = \mathbf{R}^n$, the set $B_0 = \{\mathbf{e}_1, \ldots, \mathbf{e}_n\}$, where $\mathbf{e}_1 = (1, 0, \ldots, 0)$, $\mathbf{e}_2 = (0, 1, 0, \ldots, 0)$, $\ldots$, $\mathbf{e}_n = (0, \ldots, 0, 1)$, is a basis for $\mathbf{R}^n$. The set B_0 is called the *standard basis* for $\mathbf{R}^n$.

Example 2.4.7. Let $V = \mathbf{R}^3$ and let $S = \{\mathbf{v}_1, \mathbf{v}_2, \mathbf{v}_3\}$, where $\mathbf{v}_1 = (1, 4, -1)$, $\mathbf{v}_2 = (1, 1, 1)$, and $\mathbf{v}_3 = (2, 0, -1)$. To show that S is a basis for $\mathbf{R}^3$, we need to show that S spans $\mathbf{R}^3$ and that S is linearly independent. To show that S spans $\mathbf{R}^3$ requires choosing an arbitrary vector $\mathbf{w} = (w_1, w_2, w_3) \in \mathbf{R}^3$ and finding scalars c_1, c_2, c_3 such that $\mathbf{w} = c_1\mathbf{v}_1 + c_2\mathbf{v}_2 + c_3\mathbf{v}_3$. To show that S is linearly independent requires showing that the equation $c_1\mathbf{v}_1 + c_2\mathbf{v}_2 + c_3\mathbf{v}_3 = \mathbf{0}$ has only the trivial solution $c_1 = c_2 = c_3 = 0$.

Both requirements involve analyzing systems of linear equations with coefficient matrix

$$A = \begin{bmatrix} \mathbf{v}_1 \ \mathbf{v}_2 \ \mathbf{v}_3 \end{bmatrix} = \begin{bmatrix} 1 & 1 & 2 \\ 4 & 1 & 0 \\ -1 & 1 & -1 \end{bmatrix},$$

in the first case the equation $A\mathbf{c} = \mathbf{w}$ (to determine whether it is consistent for all $\mathbf{w}$) and in the second case $A\mathbf{c} = \mathbf{0}$ (to determine whether it has only the trivial solution). Here $\mathbf{c} = (c_1, c_2, c_3)$ is the vector of coefficients. Both conditions are established by noting that $\det(A) \neq 0$. Hence S spans $\mathbf{R}^3$ and S is linearly independent, so S is a basis for $\mathbf{R}^3$.

The computations in Example 2.4.7 in fact point to a proof of a powerful technique for determining whether a set of vectors in $\mathbf{R}^n$ forms a basis for $\mathbf{R}^n$.

Theorem 2.4.8. *A set of n vectors $S = \{\mathbf{v}_1, \ldots, \mathbf{v}_n\} \subset \mathbf{R}^n$ forms a basis for $\mathbf{R}^n$ if and only if $\det(A) \neq 0$, where $A = [\mathbf{v}_1 \cdots \mathbf{v}_n]$ is the matrix formed by the column vectors $\mathbf{v}_i$.*

Just as we noted earlier that a vector space may have many spanning sets, the previous two examples illustrate that a vector space does not have a *unique* basis.

By definition, a basis B for a vector space V spans V, and so every element of V can be written as a linear combination of elements of B. However, the requirement that B be a linearly independent set has an important consequence.

Theorem 2.4.9. *Let B be a finite basis for a vector space V. Then each vector $\mathbf{v} \in V$ can be written* uniquely *as a linear combination of elements of B.*

Proof. Suppose that there are two different ways of expressing a vector $\mathbf{v}$ as a linear combination of elements of $B = \{\mathbf{b}_1, \ldots, \mathbf{b}_n\}$, so that there are scalars $c_1, \ldots, c_n$ and $d_1, \ldots, d_n$ such that

$$\mathbf{v} = c_1\mathbf{b}_1 + \cdots + c_n\mathbf{b}_n$$
$$\mathbf{v} = d_1\mathbf{b}_1 + \cdots + d_n\mathbf{b}_n.$$

Then

$$(c_1 - d_1)\mathbf{b}_1 + \cdots + (c_n - d_n)\mathbf{b}_n = \mathbf{0}.$$

By the linear independence of the set B, this implies that

$$c_1 = d_1, \ldots, c_n = d_n;$$

in other words, the two representations of $\mathbf{v}$ were in fact the same. □

As a consequence of Theorem 2.4.9, we introduce the following notation. Let $B = \{\mathbf{b}_1, \ldots, \mathbf{b}_n\}$ be a basis for a vector space V. Then for every $\mathbf{v} \in V$, let $[\mathbf{v}]_B \in \mathbf{R}^n$ be defined to be

$$[\mathbf{v}]_B = (v_1, \ldots, v_n),$$

where $\mathbf{v} = v_1\mathbf{b}_1 + \cdots + v_n\mathbf{b}_n$.

The following theorem is fundamental.

Theorem 2.4.10. *Let V be a vector space and let B be a basis for V that contains n vectors. Then no set with fewer than n vectors spans V, and no set with more than n vectors is linearly independent.*

Proof. Let $S = \{\mathbf{v}_1, \ldots, \mathbf{v}_m\}$ be a finite set of nonzero vectors in V. Since B is a basis, for each $i = 1, \ldots, m$ there are unique scalars $a_{i1}, \ldots, a_{in}$ such that

$$\mathbf{v}_i = a_{i1}\mathbf{b}_1 + \cdots + a_{in}\mathbf{b}_n.$$

Let A be the $m \times n$ matrix of components $A = [a_{ij}]$.

Suppose first that $m < n$. For $\mathbf{w} \in V$, suppose that there are scalars $c_1, \ldots, c_m$ such that

$$\begin{aligned}\mathbf{w} &= c_1\mathbf{v}_1 + \cdots + c_m\mathbf{v}_m \\ &= c_1(a_{11}\mathbf{b}_1 + \cdots + a_{1n}\mathbf{b}_n) + \cdots + c_m(a_{m1}\mathbf{b}_1 + \cdots + a_{mn}\mathbf{b}_n) \\ &= (c_1a_{11} + \cdots + c_ma_{m1})\mathbf{b}_1 + \cdots + (c_1a_{1n} + \cdots + c_ma_{mn})\mathbf{b}_n.\end{aligned}$$

Writing $[\mathbf{w}]_B = (w_1, \ldots, w_n)$ relative to the basis B, the above vector equation can be written in matrix form $A^T\mathbf{c} = [\mathbf{w}]_B$, where $\mathbf{c} = (c_1, \ldots, c_m)$. But since $m < n$, the row echelon form of the $(n \times m)$ matrix A^T must have a row of zeros, and so there exists a vector $\mathbf{w}_0 \in V$ such that $A^T\mathbf{c} = [\mathbf{w}_0]_B$ is not consistent. But this means that $\mathbf{w}_0 \notin \mathrm{Span}(S)$, and so S does not span V.

Likewise, if $m > n$, then the row echelon form of A^T has at most n leading ones. Then the vector equation $A^T\mathbf{c} = \mathbf{0}$ has nontrivial solutions, and S is not linearly independent. □

Corollary 2.4.11. *Let V be a vector space and let B be a basis of n vectors for V. Then every other basis B' of V must also have n elements.*

The corollary prompts the following definition.

Definition 2.4.12. Let V be a vector space. If there is no finite subset of V that spans V, then V is said to be *infinite-dimensional*. On the other hand, if V has a basis of n vectors (and hence, by Corollary 2.4.11, every basis has n vectors), then V is *finite-dimensional*, We call n the *dimension* of V and we write $\dim(V) = n$. By definition, $\dim(\{\mathbf{0}\}) = 0$.

Most of the examples we consider here will be finite-dimensional. However, of the vector spaces listed in Example 2.2.7, only P_n is finite-dimensional.

We conclude this section by considering the dimension of a subspace. Since a subspace is itself a vector space, Definition 2.4.12 makes sense in this context.

Theorem 2.4.13. *Let V be a finite-dimensional vector space, and let W be a subspace of V. Then $\dim(W) \leq \dim(V)$, with $\dim(W) = \dim(V)$ if and only if $W = V$. In particular, W is finite-dimensional.*

Proof. Exercise. □

Example 2.4.14. Recall $W_2 \subset \mathbf{R}^3$ from Example 2.2.4:

$$W_2 = \{(sa, sb, sc) \mid s \in \mathbf{R}\},$$

where $(a, b, c) \neq \mathbf{0}$. We have $W_2 = \mathrm{Span}(\{(a, b, c)\})$, and also the set $\{(a, b, c)\}$ is linearly independent by Theorem 2.4.3, so $\dim(W_2) = 1$.

Example 2.4.15. Recall $W_4 \subset \mathbf{R}^3$ from Example 2.2.6:

$$W_4 = \{(x, y, z) \mid ax + by + cz = 0\}$$

for some $(a, b, c) \neq \mathbf{0}$. Assume without loss of generality that $a \neq 0$. Then W_4 can be seen to be spanned by the set $S = \{(-b, a, 0), (-c, 0, a)\}$. Since S is a linearly independent set, $\dim(W_4) = 2$.

Example 2.4.16. We now justify the statement at the end of Sect. 2.3: Every proper, nontrivial subspace of $\mathbf{R}^3$ is of the form W_2 or W_4 above. Let W be a subspace of $\mathbf{R}^3$. If it is a proper subspace, then $\dim(W) = 1$ or $\dim(W) = 2$. If $\dim(W) = 1$, then W has a basis consisting of one element $\mathbf{a} = (a, b, c)$, and so W has the form of W_2.

If $\dim(W) = 2$, then W has a basis of two linearly independent vectors $\{\mathbf{a}, \mathbf{b}\}$, where $\mathbf{a} = (a_1, a_2, a_3)$ and $\mathbf{b} = (b_1, b_2, b_3)$. Let

$$\mathbf{c} = \mathbf{a} \times \mathbf{b} = (a_2 b_3 - a_3 b_2, a_3 b_1 - a_1 b_3, a_1 b_2 - a_2 b_1),$$

obtained using the vector cross product in $\mathbf{R}^3$. Note that $\mathbf{c} \neq \mathbf{0}$ by virtue of the linear independence of $\mathbf{a}$ and $\mathbf{b}$. The reader may verify that $\mathbf{w} = (x, y, z) \in W$ exactly when

$$\mathbf{c} \cdot \mathbf{w} = 0,$$

and so W has the form W_4 above.

Example 2.4.17. Recall the set $S = \{\mathbf{v}_1, \mathbf{v}_2, \mathbf{v}_3, \mathbf{v}_4\} \subset \mathbf{R}^3$, where $\mathbf{v}_1 = (1, 1, 1)$, $\mathbf{v}_2 = (-1, 1, 0)$, $\mathbf{v}_3 = (1, 3, 2)$, and $\mathbf{v}_4 = (-3, 1, -1)$, from Example 2.3.8. In that example we showed that S did not span $\mathbf{R}^3$, and so S cannot be a basis for $\mathbf{R}^3$. In fact, in Example 2.4.2, we showed that S is linearly dependent. A closer look at that example shows that the rank of the matrix $A = \begin{bmatrix} \mathbf{v}_1 & \mathbf{v}_2 & \mathbf{v}_3 & \mathbf{v}_4 \end{bmatrix}$ is two. A basis for $W = \text{Span}(S)$ can be obtained by choosing vectors in S whose corresponding column in the row-echelon form has a leading one. In this case, $S' = \{\mathbf{v}_1, \mathbf{v}_2\}$ is a basis for W, and so $\dim(W) = 2$.

2.5 Linear Transformations

For a set along with some extra structure, the next notion to consider is a function between the sets that in some suitable sense "preserves the structure." In the case of linear algebra, such functions are known as *linear transformations*. The structure they preserve should be the vector space operations of addition and scalar multiplication.

In what follows, we consider two vector spaces V and W. The reader might benefit at this point from reviewing Sect. 1.2 on functions in order to review the terminology and relevant definitions.

Definition 2.5.1. A function $T : V \to W$ is a *linear transformation* if (1) for all $\mathbf{u}, \mathbf{v} \in V$, $T(\mathbf{u} + \mathbf{v}) = T(\mathbf{u}) + T(\mathbf{v})$; and (2) for all $s \in \mathbf{R}$ and $\mathbf{v} \in V$, $T(s\mathbf{v}) = sT(\mathbf{v})$ (Fig. 2.2).

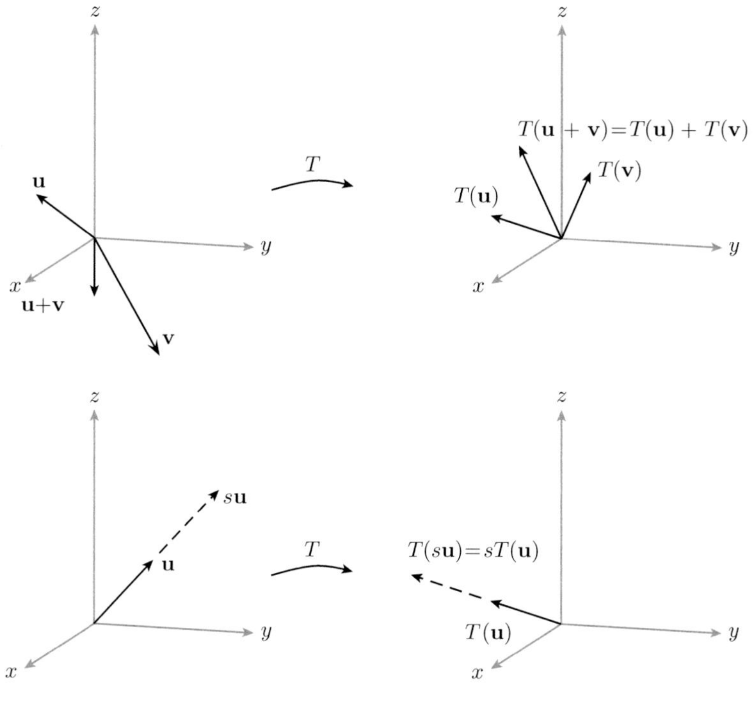

Fig. 2.2 The two conditions defining a linear transformation.

The two requirements for a function to be a linear transformation correspond exactly to the two vector space operations—the "structure"—on the sets V and W. The correct way of understanding these properties is to think of the function as "commuting" with the vector space operations: Performing the operation first (in V) and then applying the function yields the same result as applying the function first and then performing the operations (in W). It is in this sense that linear transformations "preserve the vector space structure."

We recall some elementary properties of linear transformations that are consequences of Definition 2.5.1.

Theorem 2.5.2. *Let V and W be vector spaces with corresponding zero vectors $\mathbf{0}_V$ and $\mathbf{0}_W$. Let $T : V \to W$ be a linear transformation. Then*

1. $T(\mathbf{0}_V) = \mathbf{0}_W$.
2. *For all* $\mathbf{u} \in V$, $T(-\mathbf{u}) = -T(\mathbf{u})$.

Proof. Keeping in mind Theorem 2.1.2, both of these statements are consequences of the second condition in Definition 2.5.1, using $s = 0$ and $s = -1$ respectively. □

The one-to-one, onto linear transformations play a special role in linear algebra. They allow one to say that two different vector spaces are "the same."

Definition 2.5.3. Suppose V and W are vector spaces. A linear transformation $T : V \to W$ is a *linear isomorphism* if it is one-to-one and onto. Two vector spaces V and W are said to be *isomorphic* if there is a linear isomorphism $T : V \to W$.

The most basic example of a linear isomorphism is the identity transformation $\mathrm{Id}_V : V \to V$ given by $\mathrm{Id}_V(\mathbf{v}) = \mathbf{v}$. We shall see other examples shortly.

The concept of linear isomorphism is an example of a recurring notion in this text. The fact that an isomorphism between vector spaces V and W is one-to-one and onto says that V and W are the "same" *as sets*; there is a pairing between vectors in V and W. The fact that a linear isomorphism is in fact a linear transformation further says that V and W have the same *structure*. Hence when V and W are isomorphic as vector spaces, they have the "same" sets and the "same" structure, making them mathematically the same (different only possibly in the names or characterizations of the vectors). This notion of isomorphism as sameness pervades mathematics. We shall see it again later in the context of geometric structures.

One important feature of one-to-one functions is that they admit an inverse function from the range of the original function to the domain of the original function. In the case of a one-to-one, onto function $T : V \to W$, the inverse $T^{-1} : W \to V$ is defined on all of W, where $T \circ T^{-1} = \mathrm{Id}_W$ and $T^{-1} \circ T = \mathrm{Id}_V$. We summarize this in the following theorem.

Theorem 2.5.4. *Let $T : V \to W$ be a linear isomorphism. Then there is a unique linear isomorphism $T^{-1} : W \to V$ such that $T \circ T^{-1} = \mathrm{Id}_W$ and $T^{-1} \circ T = \mathrm{Id}_V$.*

Proof. Exercise. The most important fact to be proved is that the inverse of a linear transformation, which exists purely on set-theoretic grounds, is in fact a linear transformation. □

We conclude with one sense in which isomorphic vector spaces have the same structure. We will see others throughout the chapter.

Theorem 2.5.5. *Suppose that V and W are finite-dimensional vector spaces, and suppose there is a linear isomorphism $T : V \to W$. Then* $\dim V = \dim W$.

Proof. If $\{\mathbf{v}_1, \ldots, \mathbf{v}_n\}$ is a basis for V, the reader can show that

$$\{T(\mathbf{v}_1), \ldots, T(\mathbf{v}_n)\}$$

is a basis for W and that $T(\mathbf{v}_1), \ldots, T(\mathbf{v}_n)$ are distinct. □

2.6 Constructing Linear Transformations

In this section we present two theorems that together generate a wealth of examples of linear transformations. In fact, for pairs of finite-dimensional vector spaces, these give a method that generates all possible linear transformations between them.

The first theorem should be familiar to readers who have been exposed to a first course in linear algebra. It establishes a basic correspondence between $m \times n$ matrices and linear transformations between Euclidean spaces.

Theorem 2.6.1. *Every linear transformation $T : \mathbf{R}^n \to \mathbf{R}^m$ can be expressed in terms of matrix multiplication in the following sense: There exists an $m \times n$ matrix $A_T = [T]$ such that $T(\mathbf{x}) = A_T\mathbf{x}$, where $\mathbf{x} \in \mathbf{R}^n$ is understood as a column vector. Conversely, every $m \times n$ matrix A gives rise to a linear transformation $T_A : \mathbf{R}^n \to \mathbf{R}^m$ by defining $T_A(\mathbf{x}) = A\mathbf{x}$.*

For example, the linear transformation $T : \mathbf{R}^3 \to \mathbf{R}^2$ given by

$$T(x, y, z) = (2x + y - z, x + 3z)$$

can be expressed as $T(\mathbf{x}) = A_T\mathbf{x}$, where

$$A_T = \begin{bmatrix} 2 & 1 & -1 \\ 1 & 0 & 3 \end{bmatrix}.$$

The proof of the first, main, statement of this theorem will emerge in the course of this section. The second statement is a consequence of the basic properties of matrix multiplication.

The most important of several basic features of the correspondence between matrices and linear transformations is that matrix multiplication corresponds to composition of linear transformations:

$$[S \circ T] = [S]\,[T],$$

and as a result, if $T : \mathbf{R}^n \to \mathbf{R}^n$ is a linear isomorphism, then

$$\left[T^{-1}\right] = [T]^{-1}.$$

We also note from the outset that the matrix representation of a linear transformation is not unique; it will be seen to depend on a choice of basis in both the domain and codomain. We return to this point later in the section.

The second theorem on its face gives a far more general method for constructing linear transformations, in the sense that it applies to the setting of linear transformations between arbitrary finite-dimensional vector spaces, not just between Euclidean spaces. It says that a linear transformation is uniquely defined by its action on a basis. The reader should compare this theorem to Theorem 2.5.5.

Theorem 2.6.2. *Let V be a finite-dimensional vector space with basis $B = \{\mathbf{e}_i, \dots, \mathbf{e}_n\}$. Let W be a vector space, and let $\mathbf{w}_1, \dots, \mathbf{w}_n$ be any n vectors in W, not necessarily distinct. Then there is a unique linear transformation $T : V \to W$ such that $T(\mathbf{e}_i) = \mathbf{w}_i$ for $i = 1, \dots, n$.*

If the set $\{\mathbf{w}_1, \dots, \mathbf{w}_n\}$ is in fact a basis for W, then T is a linear isomorphism.

Proof. By Theorem 2.4.9, every element $\mathbf{v} \in V$ can be uniquely written as a linear combination of elements of the basis B, which is to say there exist unique scalars $v_1, \dots, v_n$ such that $\mathbf{v} = v_1\mathbf{e}_1 + \cdots + v_n\mathbf{e}_n$. Then define $T(\mathbf{v}) = v_1\mathbf{w}_1 + \cdots + v_n\mathbf{w}_n$; the reader may check that T so defined is in fact a linear transformation.

If $B' = \{\mathbf{w}_1, \dots, \mathbf{w}_n\}$ is a basis, then T so defined is one-to-one and onto. Both statements follow from the fact that if $\mathbf{w} \in W$ is written according to B' as $\mathbf{w} = s_1\mathbf{w}_1 + \cdots + s_n\mathbf{w}_n$, then the vector $\mathbf{v} = s_1\mathbf{e}_1 + \cdots + s_n\mathbf{e}_n$ can be shown to be the unique vector such that $T(\mathbf{v}) = \mathbf{w}$. □

Example 2.6.3. Consider the basis $B = \{\mathbf{e}_1, \mathbf{e}_2\}$ for $\mathbf{R}^2$, where $\mathbf{e}_1 = (-1, 1)$ and $\mathbf{e}_2 = (2, 1)$. Define a linear transformation $T : \mathbf{R}^2 \to \mathbf{R}^4$ in the manner of Theorem 2.6.2 by setting $T(\mathbf{e}_1) = (1, 2, 3, 4)$ and $T(\mathbf{e}_2) = (-2, -4, -6, -8)$. More explicitly, let $\mathbf{v} = (v_1, v_2)$ be an arbitrary vector in $\mathbf{R}^2$. Writing $\mathbf{v} = c_1\mathbf{e}_1 + c_2\mathbf{e}_2$ uniquely as a linear combination of $\mathbf{e}_1, \mathbf{e}_2$ amounts to solving the system

$$\begin{cases} c_1(-1) + c_2(2) = v_1, \\ c_1(1) + c_2(1) = v_2, \end{cases}$$

to obtain $c_1 = \frac{1}{3}(-v_1 + 2v_2)$ and $c_2 = \frac{1}{3}(v_1 + v_2)$. Hence

$$\begin{aligned} T(\mathbf{v}) &= c_1 T(\mathbf{e}_1) + c_2 T(\mathbf{e}_2) \\ &= \frac{1}{3}\big(-v_1 + 2v_2\big)\Big(1, 2, 3, 4\Big) + \frac{1}{3}(v_1 + v_2)\Big(-2, -4, -6, -8\Big) \\ &= (-v_1, -2v_1, -3v_1, -4v_1). \end{aligned}$$

As a matrix, $T(\mathbf{v}) = A_T\mathbf{v}$, where

$$A_T = \begin{bmatrix} -1 & 0 \\ -2 & 0 \\ -3 & 0 \\ -4 & 0 \end{bmatrix}.$$

Note that the method of Theorem 2.6.2 illustrated in Example 2.6.3 gives rise to a general method of representing linear transformations between finite-dimensional vector spaces as matrices, as we did in the case of Euclidean spaces in Theorem 2.6.1.

Suppose we are given a linear transformation $T : V \to W$ as well as a basis $B = \{\mathbf{e}_1, \ldots, \mathbf{e}_n\}$ for V and a basis $B' = \{\mathbf{e}'_1, \ldots, \mathbf{e}'_m\}$ for W. Each of the vectors $T(\mathbf{e}_i)$ can be written uniquely as a linear combination of elements of B':

$$\begin{aligned} T(\mathbf{e}_1) &= a_{11}\mathbf{e}'_1 + \cdots + a_{1m}\mathbf{e}'_m, \\ &\vdots \\ T(\mathbf{e}_n) &= a_{n1}\mathbf{e}'_1 + \cdots + a_{nm}\mathbf{e}'_m. \end{aligned} \tag{2.1}$$

It is a straightforward exercise to verify that if $\mathbf{x} \in V$, where $\mathbf{x} = x_1\mathbf{e}_1 + \cdots + x_n\mathbf{e}_n$, and if $\mathbf{y} = T(\mathbf{x}) = y_1\mathbf{e}'_1 + \cdots + y_m\mathbf{e}'_m$, then $\mathbf{y} = A\mathbf{x}$, where $\mathbf{x} = (x_1, \ldots, x_n)$, $\mathbf{y} = (y_1, \ldots, y_m)$, and $A = [a_{ij}]$ with entries a_{ij} given in (2.1) above. Then A is called *the matrix of T relative to the bases B, B'* and will be denoted by $A = [T]_{B',B}$. The reader may verify that if $T : V \to W$ is a linear isomorphism, then $\left[T^{-1}\right]_{B,B'} = [T]^{-1}_{B',B}$.

Example 2.6.4. For the linear transformation $T : \mathbf{R}^3 \to \mathbf{R}^2$ defined by $T(x, y, z) = (2x + y - z, x + 3z)$, let $B = \{\mathbf{e}_1, \mathbf{e}_2, \mathbf{e}_3\}$, where $\mathbf{e}_1 = (1, 0, 0)$, $\mathbf{e}_2 = (1, 1, 0)$, and $\mathbf{e}_3 = (1, 1, 1)$, and let $B' = \{\mathbf{e}'_1, \mathbf{e}'_2\}$, where $\mathbf{e}'_1 = (-1, 1)$ and $\mathbf{e}'_2 = (2, 1)$. It is an exercise to check that B is a basis for $\mathbf{R}^3$ and B' is a basis for $\mathbf{R}^2$. We now compute $[T]_{B',B}$.

Note that for a general vector $\mathbf{w} = (w_1, w_2) \in \mathbf{R}^2$, writing $\mathbf{w}$ as a linear combination of B', $\mathbf{w} = c_1\mathbf{e}'_1 + c_2\mathbf{e}'_2$ amounts to solving the system

$$\begin{cases} c_1(-1) + c_2(2) &= w_1, \\ c_1(1) + c_2(1) &= w_2. \end{cases}$$

This is precisely the calculation we performed in Example 2.6.3. However, to illustrate an efficient general method for finding the matrix representation of a linear transformation, let us solve this system simultaneously for $T(\mathbf{e}_1) = (2, 1)$, $T(\mathbf{e}_2) = (3, 1)$, and $T(\mathbf{e}_3) = (2, 4)$ by Gaussian elimination of the matrix

$$\left[\begin{array}{cc:ccc} -1 & 2 & 2 & 3 & 2 \\ 1 & 1 & 1 & 1 & 4 \end{array}\right],$$

yielding

$$\left[\begin{array}{cc:ccc} 1 & 0 & 0 & -1/3 & 2 \\ 0 & 1 & 1 & 4/3 & 2 \end{array}\right].$$

In other words, $T(\mathbf{e}_1) = 0\mathbf{e}'_1 + 1\mathbf{e}'_2$, $T(\mathbf{e}_2) = (-1/3)\mathbf{e}'_1 + (4/3)\mathbf{e}'_2$, and $T(\mathbf{e}_3) = 2\mathbf{e}'_1 + 2\mathbf{e}'_2$. Hence the matrix for T relative to the bases B, B' is

$$[T]_{B',B} = \begin{bmatrix} 0 & -1/3 & 2 \\ 1 & 4/3 & 2 \end{bmatrix}.$$

A number of conclusions can be drawn from this example. First, comparing the matrix for T in Example 2.6.4 with the matrix for the same T given following Theorem 2.6.1 illustrates the dependence of the matrix for T on the bases involved. In particular, it illustrates the comment immediately following Theorem 2.6.1, that the matrix representation of a linear transformation is not unique.

Second, Theorem 2.6.2 in fact provides a proof for Theorem 2.6.1. The standard matrix representation of a linear transformation $T : \mathbf{R}^n \to \mathbf{R}^m$ is obtained by applying Theorem 2.6.2 using the standard bases for $\mathbf{R}^n$ and $\mathbf{R}^m$.

Recall that Theorem 2.5.5 shows that if two vector spaces are isomorphic, then they have the same dimension. Theorem 2.6.2 shows that the converse is also true, again only for finite-dimensional vector spaces.

Corollary 2.6.5. *Let V and W be vector spaces with the same finite dimension n. Then V and W are isomorphic.*

The above theorem justifies the statement following Example 2.1.3: Every n-dimensional vector space is isomorphic to the familiar example $\mathbf{R}^n$.

We remind the reader of the following basic result from matrix algebra, expressed in these new terms.

Theorem 2.6.6. *Let $T : V \to W$ be a linear transformation between vector spaces of the same finite dimension. Then T is a linear isomorphism if and only if $\det(A) \neq 0$, where $A = [T]_{B',B}$ is the matrix representation of T relative to any bases B of V and B' of W.*

Finally, we recall that for linear transformations $T : V \to V$, the determinant of T is independent of the basis in the following sense.

Theorem 2.6.7. *Let V be a finite-dimensional vector space, and let $T : V \to V$ be a linear transformation. Then for any two bases B_1, B_2 of V, we have*

$$\det\, [T]_{B_1,B_1} = \det\, [T]_{B_2,B_2}\,.$$

Proof. The result is a consequence of the fact that

$$[T]_{B_2,B_2} = [\mathrm{Id}]_{B_2,B_1} [T]_{B_1,B_1} [\mathrm{Id}]_{B_1,B_2},$$

and that $[\mathrm{Id}]_{B_2,B_1} = [\mathrm{Id}]_{B_1,B_2}^{-1}$, where $\mathrm{Id} : V \to V$ is the identity transformation. □

For this reason, we refer to the *determinant of the linear transformation* $T : V \to V$ and write $\det(T)$ to be the value of $\det(A)$, where $A = [T]_{B,B}$ for any basis B of V.

2.7 Constructing Subspaces II: Subspaces and Linear Transformations

There are several subspaces naturally associated with a linear transformation $T : V \to W$.

Definition 2.7.1. The *kernel* of a linear transformation of $T : V \to W$, denoted by $\ker(T)$, is defined to be the set

$$\ker(T) = \{\mathbf{v} \in V \mid T(\mathbf{v}) = \mathbf{0}\} \subset V.$$

Definition 2.7.2. The *range* of a linear transformation $T : V \to W$, denoted by $R(T)$, is defined to be the set

$$R(T) = \{\mathbf{w} \in W \mid \text{there is } \mathbf{v} \in V \text{ such that } T(\mathbf{v}) = \mathbf{w}\} \subset W.$$

Theorem 2.7.3. *Let $T : V \to W$ be a linear transformation. Then* $\ker(T)$ *and* $R(T)$ *are subspaces of V and W respectively.*

Proof. Exercise. □

It is a standard exercise in a first course in linear algebra to find a basis for the kernel of a given linear transformation.

Example 2.7.4. Let $T : \mathbf{R}^3 \to \mathbf{R}$ be given by $T(x, y, z) = ax + by + cz$, where a, b, c are not all zero. Then

$$\ker(T) = \{(x, y, z) \mid ax + by + cz = 0\},$$

the subspace we encountered in Example 2.2.6. Suppose that $a \neq 0$. Then we can also write

$$\ker(T) = \{(-bs - ct, as, at) \mid s, t \in \mathbf{R}\}.$$

The set $S = \{\mathbf{b}_1, \mathbf{b}_2\}$, where $\mathbf{b}_1 = (-b, a, 0)$ and $\mathbf{b}_2 = (-c, 0, a)$, is a basis for $\ker(T)$, and so $\dim(\ker(T)) = 2$.

For a linear transformation $T : V \to W$, the subspaces $\ker(T)$ and $R(T)$ are closely related to basic properties of T as a function. For example, by definition, T is onto if $R(T) = W$.

The following example highlights what might be thought of as the prototypical onto and one-to-one linear transformations.

Example 2.7.5. Consider Euclidean spaces $\mathbf{R}^n$, $\mathbf{R}^m$ with $m < n$.

The projection map $\mathrm{Pr} : \mathbf{R}^n \to \mathbf{R}^m$, given by

$$\mathrm{Pr}(x_1, \dots, x_n) = (x_1, \dots, x_m),$$

is a linear transformation that is onto but not one-to-one.

The inclusion map $\mathrm{In} : \mathbf{R}^m \to \mathbf{R}^n$ given by

$$\mathrm{In}(x_1, \dots, x_m) = (x_1, \dots, x_m, 0, \dots, 0)$$

is a linear transformation that is one-to-one but not onto.

We illustrate a powerful characterization of one-to-one linear transformations that has no parallel for general functions.

Theorem 2.7.6. *A linear transformation $T : V \to W$ is one-to-one if and only if $\ker(T) = \{\mathbf{0}\}$.*

Proof. Exercise. □

There is an important relationship between the dimensions of the kernel and range of a given linear transformation.

Theorem 2.7.7. *Let V be a finite-dimensional vector space, W another vector space, and $T : V \to W$ a linear transformation. Then*

$$\dim(R(T)) + \dim(\ker(T)) = \dim(V).$$

Proof. The proof involves a standard technique in linear algebra known as *completing a basis*. Let $\{\mathbf{e}_1, \dots \mathbf{e}_n\}$ be a basis for V. Then $\{T(\mathbf{e}_1), \dots, T(\mathbf{e}_n)\}$ spans $R(T)$, and so $\dim(R(T)) = r \leq n$. We will assume for the remainder of the proof that $1 \leq r < n$, and leave the special cases $r = 0, n$ to the reader.

Let $\{\mathbf{f}_1', \dots, \mathbf{f}_r'\}$ be a basis for $R(T)$. There is a corresponding set

$$E = \{\mathbf{e}_1', \dots \mathbf{e}_r'\} \subset V$$

such that $T(\mathbf{e}_i') = \mathbf{f}_i'$ for each $i = 1, \dots, r$. The reader may check that E must be linearly independent.

Since $r < n$, we now "complete" E by adding $(n - r)$ elements $\mathbf{e}_{r+1}', \dots, \mathbf{e}_n'$ to E in such a way that first, the new set

$$E' = \left\{\mathbf{e}_1', \dots, \mathbf{e}_r', \mathbf{e}_{r+1}', \dots, \mathbf{e}_n'\right\}$$

forms a basis for V, and second, that the set $\{\mathbf{e}'_{r+1}, \ldots, \mathbf{e}'_n\}$ forms a basis for $\ker(T)$. We illustrate the first step of this process. Choose $\mathbf{b}_{r+1} \notin \operatorname{Span}\{\mathbf{e}'_1, \ldots, \mathbf{e}'_r\}$. Since $\{\mathbf{f}'_1, \ldots, \mathbf{f}'_r\}$ is a basis for $R(T)$, write $T(\mathbf{b}_{r+1}) = \sum a_i \mathbf{f}'_i$ and define $\mathbf{e}'_{r+1} = \mathbf{b}_{r+1} - \sum a_i \mathbf{e}'_i$. Then the reader can verify that $\mathbf{e}'_{r+1}$ is still independent of E and that $T(\mathbf{e}'_{r+1}) = \mathbf{0}$, so $\mathbf{e}'_{r+1} \in \ker(T)$. Form the new set $E \cup \{\mathbf{e}'_{r+1}\}$. Repeated application of this process yields E'. We leave to the reader the verification that $\{\mathbf{e}'_{r+1}, \ldots, \mathbf{e}'_n\}$ forms a basis for $\ker(T)$. □

We will frequently refer to the dimension of the range of a linear transformation.

Definition 2.7.8. The *rank of a linear transformation* $T : V \to W$ is the dimension of $R(T)$.

The reader can verify that this definition of rank matches exactly that of the rank of any matrix representative of T relative to bases for V and W.

The following example illustrates both the statement of Theorem 2.7.7 and the notion of completing a basis used in the theorem's proof.

Example 2.7.9. Let V be a vector space with dimension n and let W be a subspace of V with dimension r, with $1 \leq r < n$. Let $B' = \{\mathbf{e}_1, \ldots \mathbf{e}_r\}$ be a basis for W. Complete this basis to a basis $B = \{\mathbf{e}_1, \ldots, \mathbf{e}_r, \mathbf{e}_{r+1}, \ldots, \mathbf{e}_n\}$ for V.

We define a linear transformation $\operatorname{Pr}_{B',B} : V \to V$ as follows: For every vector $\mathbf{v} \in V$, there are unique scalars $v_1, \ldots, v_n$ such that $\mathbf{v} = v_1\mathbf{e}_1 + \cdots + v_n\mathbf{e}_n$. Define

$$\operatorname{Pr}_{B',B}(\mathbf{v}) = v_1\mathbf{e}_1 + \cdots + v_r\mathbf{e}_r.$$

We leave it as an exercise to show that $\operatorname{Pr}_{B',B}$ is a linear transformation. Clearly $W = R(\operatorname{Pr}_{B',B})$, and so $\dim(R(\operatorname{Pr}_{B',B})) = r$. Theorem 2.7.7 then implies that $\dim(\ker(\operatorname{Pr}_{B',B})) = n - r$, a fact that is also seen by noting that $\{\mathbf{e}_{r+1}, \ldots, \mathbf{e}_n\}$ is a basis for $\ker(\operatorname{Pr}_{B',B})$.

As the notation implies, the map $\operatorname{Pr}_{B',B}$ depends on the choices of bases B' and B, not just on the subspace W.

Note that this example generalizes the projection defined in Example 2.7.5 above.

Theorem 2.7.7 has a number of important corollaries for finite-dimensional vector spaces. We leave the proofs to the reader.

Corollary 2.7.10. *Let $T : V \to W$ be a linear transformation between finite-dimensional vector spaces. If T is one-to-one, then $\dim(V) \leq \dim(W)$. If T is onto, then $\dim(V) \geq \dim(W)$.*

Note that this corollary gives another proof of Theorem 2.5.5.

As an application of the above results, we make note of the following corollary, which has no parallel in the nonlinear context.

Corollary 2.7.11. *Let $T : V \to W$ be a linear transformation between vector spaces of the same finite dimension. Then T is one-to-one if and only if T is onto.*

2.8 The Dual of a Vector Space, Forms, and Pullbacks

This section, while fundamental to linear algebra, is not generally presented in a first course on linear algebra. However, it is the algebraic foundation for the basic objects of differential geometry, differential forms, and tensors. For that reason, we will be more explicit with our proofs and explanations.

Starting with a vector space V, we will construct a new vector space V^*. Further, given vector spaces V and W along with a linear transformation $\Psi : V \to W$, we will construct a new linear transformation $\Psi^* : W^* \to V^*$ associated to Ψ.

Let V be a vector space. Define the set V^* to be the set of all linear transformations from V to $\mathbf{R}$:

$$V^* = \{T : V \to \mathbf{R} \mid T \text{ is a linear transformation}\}.$$

Note that an element $T \in V^*$ is a *function*. Define the operations of addition and scalar multiplication on V^* pointwise in the manner of Example 2.1.4. In other words, for $T_1, T_2 \in V^*$, define $T_1 + T_2 \in V^*$ by $(T_1 + T_2)(\mathbf{v}) = T_1(\mathbf{v}) + T_2(\mathbf{v})$ for all $\mathbf{v} \in V$, and for $s \in \mathbf{R}$ and $T \in V^*$, define $sT \in V^*$ by $(sT)(\mathbf{v}) = sT(\mathbf{v})$ for all $\mathbf{v} \in V$.

Theorem 2.8.1. *The set V^*, equipped with the operations of pointwise addition and scalar multiplication, is a vector space.*

Proof. The main item requiring proof is to demonstrate the closure axioms. Suppose $T_1, T_2 \in V^*$. Then for every $\mathbf{v}_1, \mathbf{v}_2 \in V$, we have

$$\begin{aligned}(T_1 + T_2)(\mathbf{v}_1 + \mathbf{v}_2) &= T_1(\mathbf{v}_1 + \mathbf{v}_2) + T_2(\mathbf{v}_1 + \mathbf{v}_2)\\ &= (T_1(\mathbf{v}_1) + T_1(\mathbf{v}_2)) + (T_2(\mathbf{v}_1) + T_2(\mathbf{v}_2))\\ &= (T_1 + T_2)(\mathbf{v}_1) + (T_1 + T_2)(\mathbf{v}_2).\end{aligned}$$

We have relied on the linearity of T_1 and T_2 in the second equality. The proof that $(T_1 + T_2)(c\mathbf{v}) = c(T_1 + T_2)(\mathbf{v})$ for every $c \in \mathbf{R}$ and $\mathbf{v} \in V$ is identical. Hence $T_1 + T_2 \in V^*$.

The fact that sT_1 is also linear for every $s \in \mathbf{R}$ is proved similarly. Note that the zero "vector" $O \in V^*$ is defined by $O(\mathbf{v}) = 0$ for all $\mathbf{v} \in V$. □

The space V^* is called the *dual vector space* to V. Elements of V^* are variously called *dual vectors*, *linear one-forms*, or *covectors*.

The proof of the following theorem, important in its own right, includes a construction that we will rely on often: the basis dual to a given basis.

Theorem 2.8.2. *Suppose that V is a finite-dimensional vector space. Then*

$$\dim(V) = \dim(V^*).$$

Proof. Let $B = \{\mathbf{e}_1, \ldots, \mathbf{e}_n\}$ be a basis for V. We will construct a basis of V^* having n covectors.

For $i = 1, \ldots, n$, define covectors $\varepsilon_i \in V^*$ by how they act on the basis B according to Theorem 2.6.2: $\varepsilon_i(\mathbf{e}_i) = 1$ and $\varepsilon_i(\mathbf{e}_j) = 0$ for $j \neq i$. In other words, for $\mathbf{v} = v_1\mathbf{e}_1 + \cdots + v_n\mathbf{e}_n$,

$$\varepsilon_i(\mathbf{v}) = v_i.$$

We show that $B^* = \{\varepsilon_1, \ldots, \varepsilon_n\}$ is a basis for V^*. To show that B^* is linearly independent, suppose that $c_1\varepsilon_1 + \cdots + c_n\varepsilon_n = O$ (an equality of linear transformations). This means that for all $\mathbf{v} \in V$,

$$c_1\varepsilon_1(\mathbf{v}) + \cdots + c_n\varepsilon_n(\mathbf{v}) = O(\mathbf{v}) = 0.$$

In particular, for each $i = 1, \ldots, n$, setting $\mathbf{v} = \mathbf{e}_i$ gives

$$\begin{aligned} 0 &= c_1\varepsilon_1(\mathbf{e}_i) \quad + \cdots + c_n\varepsilon_n(\mathbf{e}_i) \\ &= c_i. \end{aligned}$$

Hence B^* is a linearly independent set.

To show that B^* spans V^*, choose an arbitrary $T \in V^*$, i.e., $T : V \to \mathbf{R}$ is a linear transformation. We need to find scalars $c_1, \ldots, c_n$ such that $T = c_1\varepsilon_1 + \cdots + c_n\varepsilon_n$. Following the idea of the preceding argument for linear independence, define $c_i = T(\mathbf{e}_i)$.

We need to show that for all $\mathbf{v} \in V$,

$$T(\mathbf{v}) = (c_1\varepsilon_1 + \cdots + c_n\varepsilon_n)(\mathbf{v}).$$

Let $\mathbf{v} = v_1\mathbf{e}_1 + \cdots + v_n\mathbf{e}_n$. On the one hand,

$$\begin{aligned} T(\mathbf{v}) &= T(v_1\mathbf{e}_1 + \cdots + v_n\mathbf{e}_i) \\ &= v_1T(\mathbf{e}_1) + \cdots + v_nT(\mathbf{e}_n) \\ &= v_1c_1 + \cdots + v_nc_n. \end{aligned}$$

On the other hand,

$$\begin{aligned} (c_1\varepsilon_1 + \cdots + c_n\varepsilon_n)(\mathbf{v}) &= c_1\varepsilon_1(\mathbf{v}) + \cdots + c_n\varepsilon_n(\mathbf{v}) \\ &= c_1v_1 + \cdots + c_nv_n. \end{aligned}$$

Hence $T = c_1\varepsilon_1 + \cdots + c_n\varepsilon_n$, and B^* spans V^*. □

Definition 2.8.3. Let $B = \{\mathbf{e}_1, \ldots, \mathbf{e}_n\}$ be a basis for V. The basis $B^* = \{\varepsilon_1, \ldots, \varepsilon_n\}$ for V^*, where $\varepsilon_i : V \to \mathbf{R}$ are linear transformations defined by their action on the basis vectors as

$$\varepsilon_i(\mathbf{e}_j) = \begin{cases} 1 & \text{if } i = j, \\ 0 & \text{if } i \neq j, \end{cases}$$

is called the *basis of V^* dual to the basis B*.

Example 2.8.4. Let $B_0 = \{\mathbf{e}_1, \dots, \mathbf{e}_n\}$ be the standard basis for $\mathbf{R}^n$, i.e.,

$$\mathbf{e}_i = (0, \dots, 0, 1, 0, \dots, 0),$$

with 1 in the ith component (see Example 2.4.6). The basis $B_0^* = \{\varepsilon_1, \dots, \varepsilon_n\}$ dual to B_0 is known as the *standard basis for* $(\mathbf{R}^n)^*$. Note that if $\mathbf{v} = (v_1, \dots, v_n)$, then $\varepsilon_i(\mathbf{v}) = v_i$. In other words, in the language of Example 2.7.5, ε_i is the projection onto the ith component.

We note that Theorem 2.6.1 gives a standard method of writing a linear transformation $T : \mathbf{R}^n \to \mathbf{R}^m$ as an $m \times n$ matrix. Linear one-forms $T \in (\mathbf{R}^n)^*$, $T : \mathbf{R}^n \to \mathbf{R}$, are no exception. In this way, elements of $(\mathbf{R}^n)^*$ can be thought of as $1 \times n$ matrices, i.e., as row vectors. For example, the standard basis B_0^* in this notation would appear as

$$[\varepsilon_1] = \begin{bmatrix} 1 \; 0 \cdots 0 \end{bmatrix},$$

$$\vdots$$

$$[\varepsilon_n] = \begin{bmatrix} 0 \; 0 \cdots 1 \end{bmatrix}.$$

We now apply the "dual" construction to linear transformations between vector spaces V and W. For a linear transformation $\Psi : V \to W$, we will construct a new linear transformation

$$\Psi^* : W^* \to V^*.$$

(Note that this construction "reverses the arrow" of the transformation Ψ.)

Take an element of the domain $T \in W^*$, i.e., $T : W \to \mathbf{R}$ is a linear transformation. We wish to assign to T a linear transformation $S = \Psi^*(T) \in V^*$. In other words, given $T \in W^*$, we want to be able to describe a map $S : V \to \mathbf{R}$, $S(\mathbf{v}) = (\Psi^*(T))(\mathbf{v})$ for $\mathbf{v} \in V$, in such a way that S has the properties of a linear transformation.

Theorem 2.8.5. *Let $\Psi : V \to W$ be a linear transformation and let $\Psi^* : W^* \to V^*$ be given by*

$$(\Psi^*(T))(\mathbf{v}) = T(\Psi(\mathbf{v}))$$

for all $T \in W^*$ and $\mathbf{v} \in V$. Then Ψ^* is a linear transformation.

The transformation $\Psi^* : W^* \to V^*$ so defined is called the *pullback map* induced by Ψ, and $\Psi^*(T)$ is called the *pullback of T by Ψ*.

Proof. The first point to be verified is that for a fixed $T \in W^*$, we have in fact $\Psi^*(T) \in V^*$. In other words, we need to show that if $T : W \to \mathbf{R}$ is a linear transformation, then $\Psi^*(T) : V \to \mathbf{R}$ is a linear transformation. For $\mathbf{v}_1, \mathbf{v}_2 \in V$, we have

$$\begin{aligned}(\Psi^*(T))(\mathbf{v}_1 + \mathbf{v}_2) &= T(\Psi(\mathbf{v}_1 + \mathbf{v}_2)) \\ &= T(\Psi(\mathbf{v}_1) + \Psi(\mathbf{v}_2)) \quad \text{since } \Psi \text{ is linear} \\ &= T(\Psi(\mathbf{v}_1)) + T(\Psi(\mathbf{v}_2)) \quad \text{since } T \text{ is linear} \\ &= (\Psi^*(T))(\mathbf{v}_1) + (\Psi^*(T))(\mathbf{v}_2).\end{aligned}$$

The proof that $(\Psi^*(T))(s\mathbf{v}) = s(\Psi^*(T))(\mathbf{v})$ for a fixed T and for any vector $\mathbf{v} \in V$ and scalar $s \in \mathbf{R}$ is similar.

To prove linearity of Ψ^* itself, suppose that $s \in \mathbf{R}$ and $T \in W^*$. Then for all $\mathbf{v} \in V$,

$$\begin{aligned}(\Psi^*(sT))(\mathbf{v}) &= (sT)(\Psi(\mathbf{v})) \\ &= sT(\Psi(\mathbf{v})) \\ &= s((\Psi^*(T))(\mathbf{v})),\end{aligned}$$

and so $\Psi^*(sT) = s\Psi^*(T)$.

We leave as an exercise the verification that for all $T_1, T_2 \in W^*$, $\Psi^*(T_1 + T_2) = \Psi^*(T_1) + \Psi^*(T_2)$. □

Note that $\Psi^*(T) = T \circ \Psi$. It is worth mentioning that the definition of the pullback in Theorem 2.8.5 is the sort of "canonical" construction typical of abstract algebra. It can be expressed by the diagram

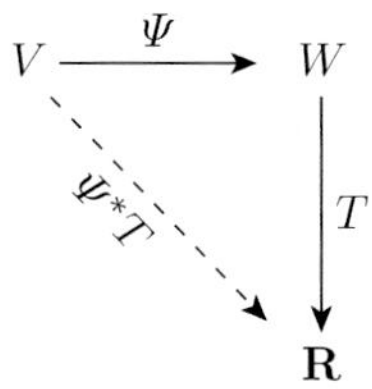

Example 2.8.6 (The matrix form of a pullback). Let $\Psi : \mathbf{R}^3 \to \mathbf{R}^2$ be given by $\Psi(x, y, z) = (2x + y - z, x + 3z)$ and let $T \in (\mathbf{R}^2)^*$ be given by $T(u, v) = u - 5v$. Then $\Psi^*(T) \in (\mathbf{R}^3)^*$ is given by

$$\begin{aligned}(\Psi^* T)(x, y, z) &= T(\Psi(x, y, z)) \\ &= T(2x + y - z, x + 3z) \\ &= (2x + y - z) - 5(x + 3z) \\ &= -3x + y - 16z.\end{aligned}$$

In the standard matrix representation of Theorem 2.6.1, we have

$$[\Psi] = \begin{bmatrix} 2 & 1 & -1 \\ 1 & 0 & 3 \end{bmatrix}, \; [T] = \begin{bmatrix} 1 & -5 \end{bmatrix} \text{ and } [\Psi^* T] = \begin{bmatrix} -3 & 1 & -16 \end{bmatrix} = [T]\,[\Psi].$$

Thus the pullback operation by Ψ on linear one-forms corresponds to matrix multiplication of a given row vector *on the right* by the matrix of Ψ.

This fact may seem strange to the reader who has become accustomed to linear transformations represented as matrices acting by multiplication on the left. It reflects the fact that all the calculations in the preceding paragraph were carried out by relying on the standard bases in $\mathbf{R}^n$ and $\mathbf{R}^m$ as opposed to the *dual bases* for $(\mathbf{R}^n)^*$ and $(\mathbf{R}^m)^*$.

Let us reconsider these calculations, this time using the dual basis from Example 2.8.4 and the more general matrix representation from the method following Theorem 2.6.2. Using the standard bases $B_0 = \{\varepsilon_1, \varepsilon_2\}$ for $(\mathbf{R}^2)^*$ and $B_0' = \{\varepsilon_1', \varepsilon_2', \varepsilon_3'\}$ for $(\mathbf{R}^3)^*$, where $\varepsilon_1 = \begin{bmatrix} 1 & 0 \end{bmatrix}$, $\varepsilon_2 = \begin{bmatrix} 0 & 1 \end{bmatrix}$, $\varepsilon_1' = \begin{bmatrix} 1 & 0 & 0 \end{bmatrix}$, $\varepsilon_2' = \begin{bmatrix} 0 & 1 & 0 \end{bmatrix}$, and $\varepsilon_3' = \begin{bmatrix} 0 & 0 & 1 \end{bmatrix}$, we note that $\Psi^*(\varepsilon_1) = 2\varepsilon_1' + \varepsilon_2' - \varepsilon_3'$ and $\Psi^*(\varepsilon_2) = \varepsilon_1' + 3\varepsilon_3'$. Hence

$$[\Psi^*]_{B_0', B_0} = \begin{bmatrix} 2 & 1 \\ 1 & 0 \\ -1 & 3 \end{bmatrix} = [\Psi]^T.$$

Now, when the calculations for the pullback of $T = \varepsilon_1 - 5\varepsilon_2$ by Ψ are written using the *column vector* $[T]_{B_0} = \begin{bmatrix} 1 \\ -5 \end{bmatrix}$, we see that

$$\begin{aligned} [\Psi^*(T)]_{B_0'} &= [\Psi^*]_{B_0', B_0} [T]_{B_0} \\ &= \begin{bmatrix} 2 & 1 \\ 1 & 0 \\ -1 & 3 \end{bmatrix} \begin{bmatrix} 1 \\ -5 \end{bmatrix} \\ &= \begin{bmatrix} -3 \\ 1 \\ -16 \end{bmatrix}. \end{aligned}$$

Since the pullback of a linear transformation is related to the matrix transpose, as the example illustrates, the following property is not surprising in light of the familiar property $(AB)^T = B^T A^T$.

Proposition 2.8.7. *Let $\Psi_1 : V_1 \to V_2$ and $\Psi_2 : V_2 \to V_3$ be linear transformations. Then*

$$(\Psi_2 \circ \Psi_1)^* = \Psi_1^* \circ \Psi_2^*.$$

Proof. Let $T \in V_3^*$ and choose $\mathbf{v} \in V_1$. Then on the one hand,

$$\begin{aligned}(\Psi_2 \circ \Psi_1)^*(T)(\mathbf{v}) &= T\left((\Psi_2 \circ \Psi_1)(\mathbf{v})\right) \\ &= T(\Psi_2(\Psi_1(\mathbf{v}))),\end{aligned}$$

while on the other hand,

$$\begin{aligned}(\Psi_1^* \circ \Psi_2^*)(T)(\mathbf{v}) &= (\Psi_1^*(\Psi_2^*(T)))(\mathbf{v}) \\ &= (\Psi_1^*(T \circ \Psi_2))(\mathbf{v}) \\ &= ((T \circ \Psi_2) \circ \Psi_1)(\mathbf{v}) \\ &= T(\Psi_2(\Psi_1(\mathbf{v}))). \qquad \square\end{aligned}$$

The construction of the dual space V^* is a special case of a more general construction. Suppose we are given several vector spaces $V_1, \dots, V_k$. Recall (see Sect. 1.1) that the *Cartesian product of* $V_1, \dots, V_k$ is the set of ordered k-tuples of vectors

$$V_1 \times \cdots \times V_k = \{(\mathbf{v}_1, \dots, \mathbf{v}_k) \mid \mathbf{v}_i \in V_i \text{ for all } i = 1, \dots, k\}.$$

The set $V = V_1 \times \cdots \times V_k$ can be given the structure of a vector space by defining vector addition and scalar multiplication componentwise.

Definition 2.8.8. Let $V_1, \dots, V_k$ and W be vector spaces. A function

$$T : V_1 \times \cdots \times V_k \to W$$

is *multilinear* if it is linear in each component:

$$T(\mathbf{x}_1 + \mathbf{y}, \mathbf{x}_2, \dots, \mathbf{x_k}) = T(\mathbf{x}_1, \mathbf{x}_2, \dots, \mathbf{x}_k) + T(\mathbf{y}, \mathbf{x}_2, \dots, \mathbf{x}_k),$$

$$\vdots$$

$$T(\mathbf{x}_1, \mathbf{x}_2, \dots, \mathbf{x}_{k-1}, \mathbf{x}_k + \mathbf{y}) = T(\mathbf{x}_1, \mathbf{x}_2, \dots, \mathbf{x}_{k-1}, \mathbf{x}_k) + T(\mathbf{x}_1, \mathbf{x}_2, \dots, \mathbf{x}_{k-1}, \mathbf{y}),$$

and

$$T(s\mathbf{x}_1, \mathbf{x}_2, \dots, \mathbf{x}_k) = sT(\mathbf{x}_1, \mathbf{x}_2, \dots, \mathbf{x}_k),$$

$$\vdots$$

$$T(\mathbf{x}_1, \mathbf{x}_2, \dots, s\mathbf{x}_k) = sT(\mathbf{x}_1, \mathbf{x}_2, \dots, \mathbf{x}_k).$$

In the special case that all the V_i are the same and $W = \mathbf{R}$, then a multilinear function $T : V \times \cdots \times V \to \mathbf{R}$ is called a *multilinear k-form on V*.

Example 2.8.9 (The zero k-form on V). The trivial example of a k-form on a vector space V is the zero form. Define $O(\mathbf{v}_1, \dots, \mathbf{v}_k) = 0$ for all $\mathbf{v}_1, \dots, \mathbf{v}_k \in V$. We leave it to the reader to show that O is multilinear.

Example 2.8.10 (The determinant as an n-form on $\mathbf{R}^n$). Define the map $\Omega : \mathbf{R}^n \times \cdots \times \mathbf{R}^n \to \mathbf{R}$ by

$$\Omega(\mathbf{a}_1, \ldots, \mathbf{a}_n) = \det A,$$

where A is the matrix whose columns are given by the vectors $\mathbf{a}_i \in \mathbf{R}^n$ relative to the standard basis: $A = [\mathbf{a}_1 \cdots \mathbf{a}_n]$. The fact that Ω is an n-form follows from properties of the determinant of matrices.

In the work that follows, we will see several important examples of *bilinear* forms (i.e., 2-forms) on $\mathbf{R}^n$.

Example 2.8.11. Let $G_0 : \mathbf{R}^n \times \mathbf{R}^n \to \mathbf{R}$ be the function defined by

$$G_0(\mathbf{x}, \mathbf{y}) = x_1 y_1 + \cdots + x_n y_n,$$

where $\mathbf{x} = (x_1, \ldots, x_n)$ and $\mathbf{y} = (y_1, \ldots, y_n)$. Then G_0 is a bilinear form. (Readers should recognize G_0 as the familiar "dot product" of vectors in $\mathbf{R}^n$.) We leave it as an exercise to verify the linearity of G_0 in each component. Note that $G_0(\mathbf{x}, \mathbf{y}) = G_0(\mathbf{y}, \mathbf{x})$ for all $\mathbf{x}, \mathbf{y} \in \mathbf{R}^n$.

Example 2.8.12. Let A be an $n \times n$ matrix and let G_0 be the bilinear form on $\mathbf{R}^n$ defined in the previous example. Then define $G_A : \mathbf{R}^n \times \mathbf{R}^n \to \mathbf{R}$ by $G_A(\mathbf{x}, \mathbf{y}) = G_0(A\mathbf{x}, A\mathbf{y})$. Bilinearity of G_A is a consequence of the bilinearity of G_0 and the linearity of matrix multiplication:

$$\begin{aligned} G_A(\mathbf{x}_1 + \mathbf{x}_2, \mathbf{y}) &= G_0(A(\mathbf{x}_1 + \mathbf{x}_2), A\mathbf{y}) \\ &= G_0(A\mathbf{x}_1 + A\mathbf{x}_2, A\mathbf{y}) \\ &= G_0(A\mathbf{x}_1, A\mathbf{y}) + G_0(A\mathbf{x}_2, A\mathbf{y}) \\ &= G_A(\mathbf{x}_1, \mathbf{y}) + G_A(\mathbf{x}_2, \mathbf{y}), \end{aligned}$$

and

$$\begin{aligned} G_A(s\mathbf{x}, \mathbf{y}) &= G_0(A(s\mathbf{x}), A\mathbf{y}) \\ &= G_0(sA\mathbf{x}, A\mathbf{y}) \\ &= sG_0(A\mathbf{x}, A\mathbf{y}) \\ &= sG_A(\mathbf{x}, \mathbf{y}). \end{aligned}$$

Linearity in the second component can be shown in the same way, or the reader may note that $G_A(\mathbf{x}, \mathbf{y}) = G_A(\mathbf{y}, \mathbf{x})$ for all $\mathbf{x}, \mathbf{y} \in \mathbf{R}^n$.

Example 2.8.13. Define $S : \mathbf{R}^2 \times \mathbf{R}^2 \to \mathbf{R}$ by $S(\mathbf{x}, \mathbf{y}) = x_1 y_2 - x_2 y_1$, where $\mathbf{x} = (x_1, x_2)$ and $\mathbf{y} = (y_1, y_2)$. For $\mathbf{z} = (z_1, z_2)$, we have

$$\begin{aligned} S(\mathbf{x} + \mathbf{z}, \mathbf{y}) &= (x_1 + z_1) y_2 - (x_2 + z_2) y_1 \\ &= (x_1 y_2 - x_2 y_1) + (z_1 y_2 - z_2 y_1) \\ &= S(\mathbf{x}, \mathbf{y}) + S(\mathbf{z}, \mathbf{y}). \end{aligned}$$

Similarly, for every $c \in \mathbf{R}$, $S(c\mathbf{x}, \mathbf{y}) = cS(\mathbf{x}, \mathbf{y})$. Hence S is linear in the first component. Linearity in the second component then follows from the fact that $S(\mathbf{y}, \mathbf{x}) = -S(\mathbf{x}, \mathbf{y})$ for all $\mathbf{x}, \mathbf{y} \in \mathbf{R}^2$. This shows that S is a bilinear form.

Let V be a vector space of dimension n, and let $b : V \times V \to \mathbf{R}$ be a bilinear form. There is a standard way to represent b by means of an $n \times n$ matrix B, assuming that a basis is specified.

Proposition 2.8.14. *Let V be a vector space with basis $\mathcal{E} = \{\mathbf{e}_1, \ldots, \mathbf{e}_n\}$ and let $b : V \times V \to \mathbf{R}$ be a bilinear form. Let $B = [b_{ij}]$, where $b_{ij} = b(\mathbf{e}_i, \mathbf{e}_j)$. Then for every $\mathbf{v}, \mathbf{w} \in V$, we have*

$$b(\mathbf{v}, \mathbf{w}) = \mathbf{w}^T B \mathbf{v},$$

where $\mathbf{v}$ and $\mathbf{w}$ are written as column vectors relative to the basis $\mathcal{E}$.

Proof. On each side, write $\mathbf{v}$ and $\mathbf{w}$ as linear combinations of the basis vectors $\mathbf{e}_1, \ldots, \mathbf{e}_n$. The result follows from the bilinearity of b and the linearity of matrix multiplication. □

This proposition allows us to study properties of the bilinear form b by means of properties of its matrix representation B, a fact that we will use in the future. Note that the matrix representation for G_A in Example 2.8.12 relative to the standard basis for $\mathbf{R}^n$ is $A^T A$.

Finally, the pullback operation can be extended to multilinear forms. We illustrate this in the case of bilinear forms, although we will return to this topic in more generality in Chap. 4.

Definition 2.8.15. Suppose $T : V \to W$ is a linear transformation between vector spaces V and W. Let $B : W \times W \to \mathbf{R}$ be a bilinear form on W. Then the *pullback of B by T* is the bilinear form $T^*B : V \times V \to \mathbf{R}$ defined by

$$(T^*B)(\mathbf{v}_1, \mathbf{v}_2) = B(T(\mathbf{v}_1), T(\mathbf{v}_2))$$

for all $\mathbf{v}_1, \mathbf{v}_2 \in V$.

The reader may check that T^*B so defined is in fact a bilinear form.

Proposition 2.8.16. *Let U, V, and W be vector spaces and let $T_1 : U \to V$ and $T_2 : V \to W$ be linear transformations. Let $B : W \times W \to \mathbf{R}$ be a bilinear form on W. Then*

$$(T_2 \circ T_1)^* B = T_1^*(T_2^* B).$$

Proof. The proof is a minor adaptation of the proof of Proposition 2.8.7. □

2.9 Geometric Structures I: Inner Products

There are relatively few traditional geometric concepts that can be defined strictly within the axiomatic structure of vector spaces and linear transformations as presented above. One that we might define, for example, is the notion of two vectors being parallel: For two vectors $\mathbf{v}, \mathbf{w}$ in a vector space V, we could say that $\mathbf{v}$ is *parallel* to $\mathbf{w}$ if there is a scalar $s \in \mathbf{R}$ such that $\mathbf{w} = s\mathbf{v}$.

The notion of vectors being *perpendicular* might also be defined in a crude way by means of the projection map of Example 2.7.9. Namely, a vector $\mathbf{v} \in V$ could be defined to be perpendicular to a nonzero vector $\mathbf{w} \in V$ if $\mathbf{v} \in \ker \mathrm{Pr}_{B',B}$, where $B' = \{\mathbf{w}\}$ and B is chosen to be a basis for V whose first vector is $\mathbf{w}$. This has the distinct disadvantage of being completely dependent on the choice of basis B, in the sense that a vector might be perpendicular to another vector with respect to one basis but not with respect to another.

The reader should note that both these attempts at definitions of basic geometric notions are somewhat stilted, since we are defining the terms parallel and perpendicular without reference to the notion of angle. In fact, we have already noted that in the entire presentation of linear algebra up to this point, two notions traditionally associated with vectors—magnitude and direction—have not been defined at all. These notions do not have a natural description using the vector space axioms alone.

The notions of magnitude and direction can be described easily, however, by means of an additional mathematical structure that generalizes the familiar "dot product" (Example 2.8.11).

Definition 2.9.1. An *inner product* on a vector space V is a function $G : V \times V \to \mathbf{R}$ with the following properties:

(I1) G is a bilinear form;
(I2) G is symmetric: For all $\mathbf{v}, \mathbf{w} \in V$, $G(\mathbf{v}, \mathbf{w}) = G(\mathbf{w}, \mathbf{v})$;
(I3) G is positive definite: For all $\mathbf{v} \in V$, $G(\mathbf{v}, \mathbf{v}) \geq 0$, with $G(\mathbf{v}, \mathbf{v}) = 0$ if and only if $\mathbf{v} = \mathbf{0}$.

The pair (V, G) is called an *inner product space*.

We mention that the conditions (I1)–(I3) imply that the matrix A corresponding to the bilinear form G according to Proposition 2.8.14 must be symmetric ($A^T = A$) and positive definite ($\mathbf{x}^T A\mathbf{x} \geq 0$ for all $\mathbf{x} \in \mathbf{R}^n$, with equality only when $\mathbf{x} = \mathbf{0}$).

Example 2.9.2 (The dot product). On the vector space $\mathbf{R}^n$, define $G_0(\mathbf{v}, \mathbf{w}) = v_1w_1 + \cdots + v_nw_n$, where $\mathbf{v} = (v_1, \ldots, v_n)$ and $\mathbf{w} = (w_1, \ldots, w_n)$. We saw in Example 2.8.11 that G_0 is a bilinear form on $\mathbf{R}^n$. The reader may verify property (I2). To see property (I3), note that $G_0(\mathbf{v}, \mathbf{v}) = v_1^2 + \cdots + v_n^2$ is a quantity that is always nonnegative and is zero exactly when $v_1 = \cdots = v_n = 0$, i.e., when $\mathbf{v} = \mathbf{0}$.

Note that Example 2.9.2 can be generalized to any finite-dimensional vector space V. Starting with any basis $B = \{\mathbf{e}_1, \ldots, \mathbf{e}_n\}$ for V, define $G_B(\mathbf{v}, \mathbf{w}) = v_1w_1 + \cdots + v_nw_n$, where $\mathbf{v} = v_1\mathbf{e}_1 + \cdots + v_n\mathbf{e}_n$ and $\mathbf{w} = w_1\mathbf{e}_1 + \cdots + w_n\mathbf{e}_n$. This function G_B is well defined because of the unique representation of $\mathbf{v}$ and $\mathbf{w}$ in the basis B. This observation proves the following:

Theorem 2.9.3. *Every finite-dimensional vector space carries an inner product structure.*

Of course, there is no *unique* inner product structure on a given vector space. The geometry of an inner product space will be determined by the choice of inner product.

It is easy to construct new inner products from a given inner product structure. We illustrate one method in $\mathbf{R}^n$, starting from the standard inner product in Example 2.9.2 above.

Example 2.9.4. Let A be any *invertible* $n \times n$ matrix. Define a bilinear form G_A on $\mathbf{R}^n$ as in Example 2.8.12: $G_A(\mathbf{v}, \mathbf{w}) = G_0(A\mathbf{v}, A\mathbf{w})$, where G_0 is the standard inner product from Example 2.9.2. Then G_A is symmetric, since G_0 is symmetric. Similarly, $G_A(\mathbf{v}, \mathbf{v}) \geq 0$ for all $\mathbf{v} \in V$ because of the corresponding property of G_0. Now suppose that $G_A(\mathbf{v}, \mathbf{v}) = 0$. Since $0 = G_A(\mathbf{v}, \mathbf{v}) = G_0(A\mathbf{v}, A\mathbf{v})$, we have $A\mathbf{v} = \mathbf{0}$ by property (I3) for G_0. Since A is invertible, $\mathbf{v} = \mathbf{0}$. This completes the verification that G_A is an inner product on $\mathbf{R}^n$.

To illustrate this construction with a simple example in $\mathbf{R}^2$, consider the matrix $A = \begin{bmatrix} 2 & -1 \\ 1 & 0 \end{bmatrix}$. Then if $\mathbf{v} = (v_1, v_2)$ and $\mathbf{w} = (w_1, w_2)$, we have

$$\begin{aligned} G_A(\mathbf{v}, \mathbf{w}) &= G_0(A\mathbf{v}, A\mathbf{w}) \\ &= G_0\big((2v_1 - v_2, v_1), (2w_1 - w_2, w_1)\big) \\ &= (2v_1 - v_2)(2w_1 - w_2) + v_1 w_1 \\ &= 5v_1 w_1 - 2v_1 w_2 - 2v_2 w_1 + v_2 w_2. \end{aligned}$$

Note that the matrix representation for G_A as described in Proposition 2.8.14 is given by $A^T A$.

An inner product allows us to define geometric notions such as length, distance, magnitude, angle, and direction.

Definition 2.9.5. Let (V, G) be an inner product space. The *magnitude* (also called the *length* or the *norm*) of a vector $\mathbf{v} \in V$ is given by

$$||\mathbf{v}|| = G(\mathbf{v}, \mathbf{v})^{1/2}.$$

The *distance* between vectors $\mathbf{v}$ and $\mathbf{w}$ is given by

$$d(\mathbf{v}, \mathbf{w}) = ||\mathbf{v} - \mathbf{w}|| = G(\mathbf{v} - \mathbf{w}, \mathbf{v} - \mathbf{w})^{1/2}.$$

To define the notion of direction, or angle between vectors, we first state a fundamental property of inner products.

Theorem 2.9.6 (Cauchy–Schwarz). *Let (V, G) be an inner product space. Then for all $\mathbf{v}, \mathbf{w} \in V$,*

$$|G(\mathbf{v}, \mathbf{w})| \leq ||\mathbf{v}|| \cdot ||\mathbf{w}||.$$

The standard proof of the Cauchy–Schwarz inequality relies on the non-intuitive observation that the discriminant of the quadratic expression (in t) $G(t\mathbf{v}+\mathbf{w}, t\mathbf{v}+\mathbf{w})$ must be nonpositive by property (I3).

Definition 2.9.7. Let (V, G) be an inner product space. For every two nonzero vectors $\mathbf{v}, \mathbf{w} \in V$, the *angle* $\angle_G(\mathbf{v}, \mathbf{w})$ between $\mathbf{v}$ and $\mathbf{w}$ is defined to be

$$\angle_G(\mathbf{v}, \mathbf{w}) = \cos^{-1}\left(\frac{G(\mathbf{v}, \mathbf{w})}{||\mathbf{v}|| \cdot ||\mathbf{w}||}\right).$$

Note that this definition of angle is well defined as a result of Theorem 2.9.6.

As a consequence of this definition of angle, it is possible to define a notion of orthogonality: Two vectors $\mathbf{v}, \mathbf{w} \in V$ are *orthogonal* if $G(\mathbf{v}, \mathbf{w}) = 0$, since then $\angle_G(\mathbf{v}, \mathbf{w}) = \pi/2$. The notion of orthogonality, in turn, distinguishes "special" bases for V and provides a further method for producing new subspaces of V from a given set of vectors in V.

Theorem 2.9.8. *Let (V, G) be an inner product space with $\dim(V) = n > 0$. There exists a basis $B = \{\mathbf{u}_1, \dots, \mathbf{u}_n\}$ satisfying the following two properties:*

(O1) For each $i = 1, \dots, n$, $G(\mathbf{u}_i, \mathbf{u}_i) = 1$;
(O2) For each $i \neq j$, $G(\mathbf{u}_i, \mathbf{u}_j) = 0$.

Such a basis is known as an orthonormal *basis.*

The reader is encouraged to review the proof of this theorem, which can be found in any elementary linear algebra text. It relies on an important procedure, similar in spirit to the proof of Theorem 2.7.7, known as *Gram–Schmidt orthonormalization*. Beginning with any given basis, the procedure gives a way of constructing a new basis satisfying (O1) and (O2). In the next section, we will carry out the details of an analogous procedure in the symplectic setting.

As with bases in general, there is no unique orthonormal basis for a given inner product space (V, G).

We state without proof a kind of converse to Theorem 2.9.8. This theorem is actually a restatement of the comment following Example 2.9.2.

Theorem 2.9.9. *Let V be a vector space and let $B = \{\mathbf{e}_1, \dots, \mathbf{e}_n\}$ be a finite basis for V. Define a function G_B by requiring that*

$$G_B(\mathbf{e}_i, \mathbf{e}_j) = \begin{cases} 1 & \textit{if } i = j, \\ 0 & \textit{if } i \neq j, \end{cases}$$

and extending linearly in both components in the manner of Theorem 2.6.2. Then G_B is an inner product.

For any vector $\mathbf{v}$ in an inner product space (V, G), the set W of all vectors orthogonal to $\mathbf{v}$ can be seen to be a subspace of V. One could appeal directly to Theorem 2.2.2 (since $\mathbf{0} \in W$), or one could note that W is the kernel of the linear transformation $i_{\mathbf{v}} : V \to \mathbf{R}$ given by $i_{\mathbf{v}}(\mathbf{w}) = G(\mathbf{v}, \mathbf{w})$. More generally, we have the following:

Theorem 2.9.10. *Let S be any nonempty set of vectors in an inner product space (V, G). The set $S^\perp$ defined by*

$$S^\perp = \{\mathbf{w} \in V \mid \textit{For all } \mathbf{v} \in S,\ G(\mathbf{v}, \mathbf{w}) = 0\}$$

is a subspace of V.

Proof. Exercise. □

The set $S^\perp$ is called the *orthogonal complement to S*.

Example 2.9.11. Let $S = \{\mathbf{v}\} \subset \mathbf{R}^3$, where $\mathbf{v} = (a, b, c) \neq \mathbf{0}$. Let G_0 be the standard inner product on R^3 (see Example 2.9.2). Then

$$S^\perp = \{(x, y, z) \mid ax + by + cz = 0\}.$$

See Example 2.2.4.

Example 2.9.12. Let $A = \begin{bmatrix} 2 & -1 \\ 1 & 0 \end{bmatrix}$ and let G_A be the inner product defined on $\mathbf{R}^2$ according to Example 2.9.4. Let $\mathbf{v} = (1, 0)$, and let $S = \{\mathbf{v}\}$. Then the reader may verify that

$$S^\perp = \{(w_1, w_2) \mid 5w_1 - 2w_2 = 0\},$$

which is spanned by the set $\{(2, 5)\}$. See Fig. 2.3.

Theorem 2.9.13. *Let (V, G) be an inner product space. Let S be a finite subset of V and let $W = \operatorname{Span}(S)$. Then $W^\perp = S^\perp$.*

Proof. Let $S = \{\mathbf{w}_1, \ldots, \mathbf{w}_k\}$. Take a vector $\mathbf{v} \in W^\perp$, so that $G(\mathbf{w}, \mathbf{v}) = 0$ for all $\mathbf{w} \in W$. In particular, since $S \subset W$, for each $\mathbf{w}_i \in S$, $G(\mathbf{w}_i, \mathbf{v}) = 0$. So $\mathbf{v} \in S^\perp$ and $W^\perp \subset S^\perp$.

Now take a vector $\mathbf{v} \in S^\perp$. Let $\mathbf{w} \in W$, so that there are scalars $c_1, \ldots, c_k$ such that $\mathbf{w} = c_1\mathbf{w}_1 + \cdots + c_k\mathbf{w}_k$. Relying on the linearity of G in the first component, we obtain

$$\begin{aligned} G(\mathbf{w}, \mathbf{v}) &= G(c_1\mathbf{w}_1 + \cdots + c_k\mathbf{w}_k, \mathbf{v}) \\ &= G(c_1\mathbf{w}_1, \mathbf{v}) + \cdots + G(c_k\mathbf{w}_k, \mathbf{v}) \\ &= c_1G(\mathbf{w}_1, \mathbf{v}) + \cdots + c_kG(\mathbf{w}_k, \mathbf{v}) \\ &= 0, \quad \text{since } \mathbf{v} \in S^\perp. \end{aligned}$$

Hence $\mathbf{v} \in W^\perp$, and so $S^\perp \subset W^\perp$.

Together, these two statements show that $W^\perp = S^\perp$. □

Corollary 2.9.14. *Let B be a basis for a subspace $W \subset V$. Then $W^\perp = B^\perp$.*

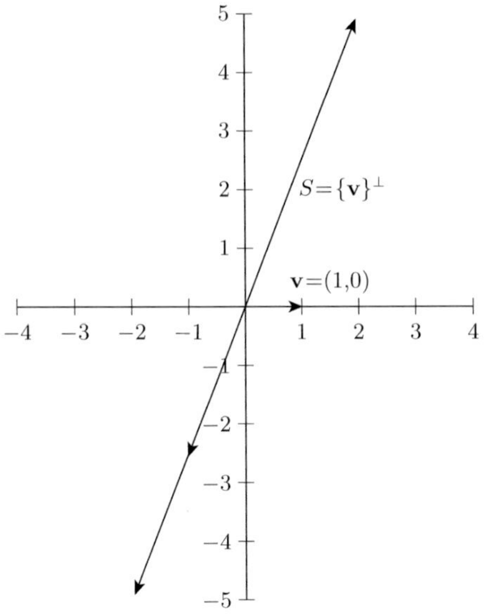

Fig. 2.3 The orthogonal complement to $\{(1, 0)\}$ relative to G in Example 2.9.12.

The following theorems discuss the relationship between a vector subspace W and its orthogonal complement $W^{\perp}$.

Theorem 2.9.15. *Let W be a subspace of an inner product space (V, G). Then* $W \cap W^{\perp} = \{\mathbf{0}\}$.

Proof. Exercise. □

Theorem 2.9.16. *Let W be a subspace of a finite-dimensional inner product space (V, G). Then*

$$\dim(W) + \dim(W^{\perp}) = \dim(V).$$

Proof. Consider the map $S : V \to W^*$ given by $S(\mathbf{v}) = i_{\mathbf{v}}$, where for every $\mathbf{w} \in W$, $i_{\mathbf{v}}(\mathbf{w}) = G(\mathbf{v}, \mathbf{w})$. The reader can check that S is a linear transformation, that $\ker S = W^{\perp}$, and that S is onto. The result then follows from Theorem 2.7.7. □

We now turn our attention to linear transformations of an inner product space that preserve the additional inner product structure.

Definition 2.9.17. Let (V_1, G_1) be an inner product space. A linear transformation $T : V \to V$ is a *linear isometry* if for all $\mathbf{v}, \mathbf{w} \in V$, $G(T(\mathbf{v}), T(\mathbf{w})) = G(\mathbf{v}, \mathbf{w})$. Stated in the language of Sect. 2.8, T is a linear isometry if $T^*G = G$.

Note that a linear isometry preserves all quantities defined in terms of the inner product: distance, magnitude, the angle between vectors, and orthogonality.

The reader may verify the following property of linear isometries.

Proposition 2.9.18. *Let (V, G) be an inner product space. If T_1, T_2 are linear isometries of V, then $T_2 \circ T_1$ is also a linear isometry.*

The following theorem, which we state without proof, gives a matrix characterization of linear isometries.

Theorem 2.9.19. *Let (V, G) be a finite-dimensional inner product space with $\dim(V) = n > 0$, and let $T : V \to V$ be a linear isometry. Then the matrix representation $A = [T]$ of T relative to any orthonormal basis of V satisfies $A^T A = I_n$, where I_n is the $n \times n$ identity matrix.*

A matrix with the property that $A^T A = I_n$ is called an *orthogonal* matrix.

Corollary 2.9.20. *Let $T : V \to V$ be a linear isometry of a finite-dimensional inner product space (V, G). Then $\det(T) = \pm 1$. In particular, T is invertible.*

Proposition 2.9.21. *Let (V, G) be a finite-dimensional inner product space, and let $T : V \to V$ be a linear isometry. Then its inverse T^{-1} is also a linear isometry.*

Proof. Assuming T is a linear isometry, apply Proposition 2.8.16 to $G = (\mathrm{Id})^* G = (T \circ T^{-1})^* G$ and use the assumption that $T^* G = G$. □

We conclude this section with an important technical theorem, a consequence of the positive definite property of inner products. Recall that V and V^* have the same dimension by Theorem 2.8.2, and so by Corollary 2.6.5, the two vector spaces are isomorphic. A choice of an inner product on V, however, induces a distinguished isomorphism between them.

Theorem 2.9.22. *Let G be an inner product on a finite-dimensional vector space V. For every $\mathbf{v} \in V$, define $i_{\mathbf{v}} \in V^*$ by $i_{\mathbf{v}}(\mathbf{w}) = G(\mathbf{v}, \mathbf{w})$ for $\mathbf{w} \in V$. Then the function*

$$\Phi : V \to V^*$$

defined by $\Phi(\mathbf{v}) = i_{\mathbf{v}}$ is a linear isomorphism.

Proof. The fact that Φ is linear is a consequence of the fact that G is bilinear. For example, for $\mathbf{v} \in V$ and $s \in \mathbf{R}$, $\Phi(s\mathbf{v}) = i_{s\mathbf{v}}$, and so for all $\mathbf{w} \in V$, $i_{s\mathbf{v}}(\mathbf{w}) = G(s\mathbf{v}, \mathbf{w}) = sG(\mathbf{v}, \mathbf{w}) = s i_{\mathbf{v}}(\mathbf{w})$. Hence $i_{s\mathbf{v}} = s i_{\mathbf{v}}$, and so $\Phi(s\mathbf{v}) = s\Phi(\mathbf{v})$. Likewise, $\Phi(\mathbf{v} + \mathbf{w}) = \Phi(\mathbf{v}) + \Phi(\mathbf{w})$ for all $\mathbf{v}, \mathbf{w} \in V$.

To show that Φ is one-to-one, we show that $\ker(\Phi) = \{\mathbf{0}\}$. Let $\mathbf{v} \in \ker(\Phi)$. Then $\Phi(\mathbf{v}) = O$, i.e., $G(\mathbf{v}, \mathbf{w}) = 0$ for all $\mathbf{w} \in V$. In particular, $G(\mathbf{v}, \mathbf{v}) = 0$, and so $\mathbf{v} = \mathbf{0}$ by positive definiteness. Hence $\ker(\Phi) = \{\mathbf{0}\}$, and so by Theorem 2.7.6, Φ is one-to-one.

The fact that Φ is onto now follows from the fact that a one-to-one linear map between vector spaces of the same dimension must be onto (Corollary 2.7.11). However, we will show directly that Φ is onto in order to exhibit the inverse transformation $\Phi^{-1} : V^* \to V$.

Let $T \in V^*$. We need to find $\mathbf{v}_T \in V$ such that $\Phi(\mathbf{v}_T) = T$. Let $\{\mathbf{u}_1, \dots, \mathbf{u}_n\}$ be an orthonormal basis for (V, G), as guaranteed by Theorem 2.9.8. Define c_i by

$c_i = T(\mathbf{u}_i)$, and define $\mathbf{v}_T = c_1\mathbf{u}_1 + \cdots + c_n\mathbf{u}_n$. By the linearity of G in the first component, we have $\Phi(\mathbf{v}_T) = T$, or, what is the same, $\mathbf{v}_T = \Phi^{-1}(T)$. Hence Φ is onto. □

The reader should notice the similarity between the construction of Φ^{-1} and the procedure outlined in the proof of Theorem 2.8.2.

The fact that the map Φ in Theorem 2.9.22 is one-to-one can be rephrased by saying that the inner product G is *nondegenerate*: If $G(\mathbf{v}, \mathbf{w}) = 0$ for all $\mathbf{w} \in V$, then $\mathbf{v} = \mathbf{0}$. We will encounter this condition again shortly in the symplectic setting.

2.10 Geometric Structures II: Linear Symplectic Forms

In this section, we outline the essentials of linear symplectic geometry, which will be the starting point for one of the main differential-geometric structures that we will present later in the text. The presentation here will parallel the development of inner product structures in Sect. 2.9 in order to emphasize the similarities and differences between the two structures, both of which are defined by bilinear forms. We will discuss more about the background of symplectic geometry in Chap. 7.

Unlike most of the material in this chapter so far, what follows is not generally presented in a first course in linear algebra. As in Sect. 2.8, we will be more detailed in the presentation and proof of the statements in this section.

Definition 2.10.1. A *linear symplectic form* on a vector space V is a function $\omega : V \times V \to \mathbf{R}$ satisfying thc following properties:

(S1) ω is a bilinear form on V;
(S2) ω is skew-symmetric: For all $\mathbf{v}, \mathbf{w} \in V$, $\omega(\mathbf{w}, \mathbf{v}) = -\omega(\mathbf{v}, \mathbf{w})$;
(S3) ω is nondegenerate: If $\mathbf{v} \in V$ has the property that $\omega(\mathbf{v}, \mathbf{w}) = 0$ for all $\mathbf{w} \in V$, then $\mathbf{v} = \mathbf{0}$.

The pair (V, ω) is called a *symplectic vector space*.

Note that the main difference between (S1)–(S3) and (I1)–(I3) in Definition 2.9.1 is that a linear symplectic form is skew-symmetric, in contrast to the symmetric inner product. We can summarize properties (S1) and (S2) by saying that ω is an *alternating bilinear form* on V. We will discuss the nondegeneracy condition (S3) in more detail below. Note that in sharp contrast to inner products, $\omega(\mathbf{v}, \mathbf{v}) = 0$ for all $\mathbf{v} \in V$ as a consequence of (S2).

Example 2.10.2. On the vector space $\mathbf{R}^2$, define $\omega_0(\mathbf{v}, \mathbf{w}) = v_1w_2 - v_2w_1$, where $\mathbf{v} = (v_1, v_2)$ and $\mathbf{w} = (w_1, w_2)$. The reader may recognize this as the determinant of the matrix whose column vectors are $\mathbf{v}, \mathbf{w}$. That observation, or direct verification, will confirm properties (S1) and (S2). To verify (S3), suppose $\mathbf{v} = (v_1, v_2)$ is such that $\omega_0(\mathbf{v}, \mathbf{w}) = 0$ for all $\mathbf{w} \in \mathbf{R}^2$. In particular, $0 = \omega_0(\mathbf{v}, (1, 0)) = (v_1)(0) - (1)(v_2) = -v_2$, and so $v_2 = 0$. Likewise,

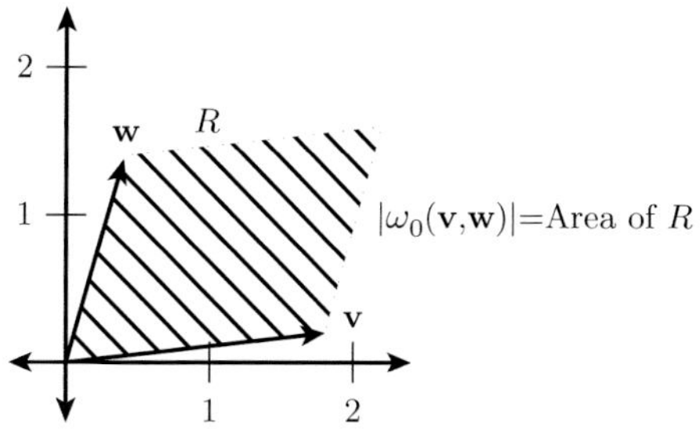

Fig. 2.4 The standard symplectic form on $\mathbf{R}^2$.

$0 = \omega_0(\mathbf{v}, (0,1)) = v_1$. Together, these show that $\mathbf{v} = \mathbf{0}$, and so (S3) is satisfied. In this case, ω_0 measures the *oriented area* of the parallelogram defined by two vectors. See Fig. 2.4.

Example 2.10.3. Generalizing Example 2.10.2, consider the Euclidean vector space $\mathbf{R}^{2n}$. Define the function

$$\omega_0(\mathbf{v},\mathbf{w}) = (a_1t_1 - b_1s_1) + \cdots + (a_nt_n - b_ns_n),$$

where $\mathbf{v} = (a_1, b_1, \ldots, a_n, b_n)$ and $\mathbf{w} = (s_1, t_1, \ldots, s_n, t_n)$. The verification that ω_0 is a symplectic form proceeds exactly as in Example 2.10.2; it will be called the *standard symplectic form* on $\mathbf{R}^{2n}$. Similarly, the pair $(\mathbf{R}^{2n}, \omega_0)$ will be called the *standard symplectic vector space*.

Before proceeding to more examples, we immediately prove in detail a consequence of the existence of a linear symplectic form on a finite-dimensional vector space: *The dimension of a symplectic vector space must be even.* In other words, there can be no linear symplectic form on an odd-dimensional vector space. This is the first significant difference between symmetric and skew-symmetric bilinear forms.

Theorem 2.10.4 below should be read alongside Theorems 2.9.3 and 2.9.8, which show that every finite-dimensional vector space carries an inner product, to which correspond distinguished orthonormal bases.

Theorem 2.10.4. *Let (V, ω) be a finite-dimensional symplectic vector space. Then V has a basis $\{\mathbf{e}_1, \mathbf{f}_1, \ldots, \mathbf{e}_n, \mathbf{f}_n\}$ with the following properties:*

(SO1) $\omega(\mathbf{e}_i, \mathbf{f}_i) = 1$ *for all* $i = 1, \ldots, n$*;*

(SO2) $\omega(\mathbf{e}_i, \mathbf{e}_j) = 0$ *for all* $i, j = 1, \ldots, n$*;*

(SO3) $\omega(\mathbf{f}_i, \mathbf{f}_j) = 0$ *for all* $i, j = 1, \ldots, n$*;*

(SO4) $\omega(\mathbf{e}_i, \mathbf{f}_j) = 0$ *for* $i \neq j$*.*

In particular, the dimension of V is even.

Proof. The inductive process of constructing a basis with properties (SO1)–(SO4) is a modified version of the Gram–Schmidt orthonormalization process that is at the heart of the proof of Theorem 2.9.8. The reader should also compare the technique here with the technique in the proof of Theorem 2.7.7.

Choose any nonzero $\mathbf{v} \in V$ and define $\mathbf{e}_1 = \mathbf{v}$. Since $\mathbf{e}_1 \neq \mathbf{0}$, property (S3) of the symplectic form guarantees the existence of a vector $\mathbf{w}_1$ such that $\omega(\mathbf{e}_1, \mathbf{w}_1) = c_1 \neq 0$. Define $\mathbf{f}_1 = (1/c_1)\mathbf{w}_1$. By the bilinearity condition (S1), we have $\omega(\mathbf{e}_1, \mathbf{f}_1) = 1$. Note also that $\{\mathbf{e}_1, \mathbf{f}_1\}$ is a linearly independent set, since otherwise, by Theorem 2.4.3 we would have a scalar s such that $\mathbf{f}_1 = s\mathbf{e}_1$, and so by (S2), $\omega(\mathbf{e}_1, \mathbf{f}_1) = 0$, contradicting the definition of $\mathbf{f}_1$. Further, by (S2), we have $\omega(\mathbf{e}_1, \mathbf{e}_1) = \omega(\mathbf{f}_1, \mathbf{f}_1) = 0$, and so the set $\{\mathbf{e}_1, \mathbf{f}_1\}$ satisfies properties (SO1)–(SO4).

Now suppose we have constructed k pairs of linearly independent vectors

$$B_k = \{\mathbf{e}_1, \mathbf{f}_1, \dots, \mathbf{e}_k, \mathbf{f}_k\}$$

satisfying (SO1)–(SO4). If $\mathrm{Span}(B_k) = V$, then B_k is the desired basis and we are done.

If $\mathrm{Span}(B_k) \subsetneq V$, then there is a nonzero vector $\mathbf{v}_{k+1} \notin \mathrm{Span}(B_k)$. Define

$$\mathbf{e}_{k+1} = \mathbf{v}_{k+1} - \sum_{j=1}^{k} \omega(\mathbf{v}_{k+1}, \mathbf{f}_j)\mathbf{e}_j - \sum_{j=1}^{k} \omega(\mathbf{e}_j, \mathbf{v}_{k+1})\mathbf{f}_j.$$

Note that $\mathbf{e}_{k+1} \neq \mathbf{0}$, since otherwise $\mathbf{v}_{k+1}$ would be a linear combination of vectors in B_k, contradicting the definition of $\mathbf{v}_{k+1}$. Since ω is bilinear, we have for each $i = 1, \dots, k$ that

$$\begin{aligned}
\omega(\mathbf{e}_{k+1}, \mathbf{e}_i) &= \omega(\mathbf{v}_{k+1}, \mathbf{e}_i) - \sum_{j=1}^{k} \omega(\mathbf{v}_{k+1}, \mathbf{f}_j)\omega(\mathbf{e}_j, \mathbf{e}_i) - \sum_{j=1}^{k} \omega(\mathbf{e}_j, \mathbf{v}_{k+1})\omega(\mathbf{f}_j, \mathbf{e}_i) \\
&= \omega(\mathbf{v}_{k+1}, \mathbf{e}_i) - \omega(\mathbf{e}_i, \mathbf{v}_{k+1})\omega(\mathbf{f}_i, \mathbf{e}_i) \quad \text{by the inductive hypothesis} \\
&= \omega(\mathbf{v}_{k+1}, \mathbf{e}_i) - \omega(\mathbf{v}_{k+1}, \mathbf{e}_i) \quad \text{by the inductive hypothesis, (S2)} \\
&= 0,
\end{aligned}$$

and similarly,

$$\begin{aligned}
\omega(\mathbf{e}_{k+1}, \mathbf{f}_i) &= \omega(\mathbf{v}_{k+1}, \mathbf{f}_i) - \sum_{j=1}^{k} \omega(\mathbf{v}_{k+1}, \mathbf{f}_j)\omega(\mathbf{e}_j, \mathbf{f}_i) - \sum_{j=1}^{k} \omega(\mathbf{e}_j, \mathbf{v}_{k+1})\omega(\mathbf{f}_j, \mathbf{f}_i) \\
&= \omega(\mathbf{v}_{k+1}, \mathbf{f}_i) - \omega(\mathbf{v}_{k+1}, \mathbf{f}_i) \\
&= 0.
\end{aligned}$$

Now, by property (S3), there is a vector $\mathbf{w}_{k+1}$ such that $\omega(\mathbf{e}_{k+1}, \mathbf{w}_{k+1}) = c_{k+1} \neq 0$. Note that $\mathbf{w}_{k+1}$ cannot be in $\mathrm{Span}(B_k)$, since otherwise, the previous calculations would imply that $\omega(\mathbf{e}_{k+1}, \mathbf{w}_{k+1}) = 0$. Define further

$$\mathbf{u}_{k+1} = \mathbf{w}_{k+1} - \sum_{j=1}^{k} \omega(\mathbf{w}_{k+1}, \mathbf{f}_j)\mathbf{e}_j - \sum_{j=1}^{k} \omega(\mathbf{e}_j, \mathbf{w}_{k+1})\mathbf{f}_j.$$

Again we have $\mathbf{u}_{k+1} \neq \mathbf{0}$, since if $\mathbf{u}_{k+1} = \mathbf{0}$, then $\mathbf{w}_{k+1} \in \mathrm{Span}(B_k)$.

Let $\mathbf{f}_{k+1} = (1/c_{k+1})\mathbf{u}_{k+1}$, which makes $\omega(\mathbf{e}_{k+1}, \mathbf{f}_{k+1}) = 1$. As in the preceding argument, $\mathbf{f}_{k+1} \notin \mathrm{Span}(B_k \cup \{\mathbf{e}_{\mathbf{k+1}}\})$. Properties (SO3) and (SO4) about $\mathbf{f}_{k+1}$ are proved in the same way as the analagous properties were derived above for $\mathbf{e}_{k+1}$. Hence, defining $B_{k+1} = B_k \cup \{\mathbf{e}_{k+1}, \mathbf{f}_{k+1}\}$, we have a linearly independent set satisfying properties (SO1)–(SO4). But since V is finite-dimensional, there must be an n such that $V = \mathrm{Span}(B_n)$, which completes the proof. □

Note that this proof relies in an essential way on the nondegeneracy condition (S3) of ω. We will see another proof of this result below in Theorem 2.10.24.

A basis of a symplectic vector space (V, ω) satisfying (SO1)–(SO4) is called a *symplectic basis* for V.

For example, for the standard symplectic space $(\mathbf{R}^{2n}, \omega_0)$, the set

$$B_0 = \{\mathbf{e}_1, \mathbf{f}_1, \ldots, \mathbf{e}_n, \mathbf{f}_n\} \tag{2.2}$$

given by

$$\begin{aligned}
\mathbf{e}_1 &= (1, 0, 0, 0, \ldots, 0, 0), \\
\mathbf{f}_1 &= (0, 1, 0, 0, \ldots, 0, 0), \\
&\vdots \\
\mathbf{e}_n &= (0, 0, \ldots, 0, 0, 1, 0), \\
\mathbf{f}_n &= (0, 0, \ldots, 0, 0, 0, 1),
\end{aligned}$$

is a symplectic basis. We will call B_0 the *standard symplectic basis for* $(\mathbf{R}^{2n}, \omega_0)$.

As in the case of orthonormal bases for an inner product space, there is no unique symplectic basis for a given symplectic vector space (V, ω).

The following is another indication of the importance of the nondegeneracy condition (S3) of Definition 2.10.1. It gives further evidence of why (S3) is the correct symplectic analogue to the inner product positive definite condition (I3) in Definition 2.9.1.

Theorem 2.10.5. *Let (V, ω) be a finite-dimensional symplectic vector space. The map $\Psi : V \to V^*$ defined by $\Psi(\mathbf{v}) = i_{\mathbf{v}} \in V^*$ for all $\mathbf{v} \in V$, where $i_{\mathbf{v}}(\mathbf{w}) = \omega(\mathbf{v}, \mathbf{w})$ for $\mathbf{w} \in V$, is a linear isomorphism.*

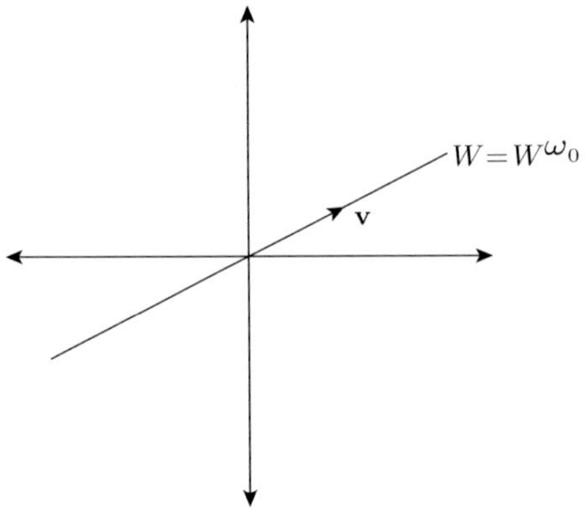

Fig. 2.5 Symplectic orthogonality in $(\mathbf{R}^2, \omega_0)$; see Example 2.10.8.

The proof exactly parallels the proof of Theorem 2.9.22.

The geometric consequences of the existence of a symplectic structure are quite different from those of an inner product structure. There is no sense, for example, of the length of a vector or of the angle between two vectors; it is enough to recall again that $\omega(\mathbf{v}, \mathbf{v}) = 0$ for all $\mathbf{v} \in V$. There is, however, a notion corresponding to the inner product notion of orthogonality.

Definition 2.10.6. Let (V, ω) be a symplectic vector space. Two vectors $\mathbf{v}, \mathbf{w} \in V$ are called *ω-orthogonal* (or *skew-orthogonal*) if $\omega(\mathbf{v}, \mathbf{w}) = 0$.

In contrast to inner product orthogonality, as noted earlier, every vector is ω-orthogonal to itself. This consequence of the skew-symmetry of ω is in marked contrast to the symmetric case; compare, for example, to Theorem 2.9.10.

Following the idea of Sect. 2.9, we have the following:

Theorem 2.10.7. *Let S be a set of vectors in a symplectic vector space (V, ω). The set*

$$S^\omega = \{\mathbf{w} \in V \mid \textit{for all } \mathbf{v} \in S,\ \omega(\mathbf{v}, \mathbf{w}) = 0\}$$

is a subspace of V.

Proof. Exercise. □

The set S^ω is called the *ω-orthogonal complement to S*.

Example 2.10.8. Let $S = \{\mathbf{v}\} \subset \mathbf{R}^2$, where $\mathbf{v} = (a, b) \neq \mathbf{0}$. Let ω_0 be the standard linear symplectic form on $\mathbf{R}^2$. Then $S^{\omega_0} = \{(sa, sb) \mid s \in \mathbf{R}\}$. Indeed, if $(x, y) \in S^{\omega_0}$, then $\omega_0\big((a, b), (x, y)\big) = ay - bx = 0$. Hence $ay = bx$, and since either a or b is not 0, we can write (for example if $a \neq 0$) $y = (b/a)x$, and so $(x, y) = \big(x, (b/a)x\big) = (x/a)\big(a, b\big)$. In fact, if $W = \mathrm{Span}(S)$, then $W = W^{\omega_0}$. See Fig. 2.5.

Example 2.10.9. Let $S = \{\mathbf{v}\} \subset (\mathbf{R}^4, \omega_0)$, where $\mathbf{v} = (1, 2, 3, 4)$. Then

$$\begin{aligned} S^{\omega_0} &= \{(x_1, y_1, x_2, y_2) \mid \omega_0((1,2,3,4),(x_1,y_1,x_2,y_2)) = 0\} \\ &= \{(x_1, y_1, x_2, y_2) \mid -2x_1 + y_1 - 4x_2 + 3y_2 = 0\}. \end{aligned}$$

We can also write

$$S^{\omega_0} = \{(s, 2s + 4t - 3u, t, u) \mid s, t, u \in \mathbf{R}\},$$

so the set $\{(1, 2, 0, 0), (0, 4, 1, 0), (0, -3, 0, 1)\}$ is a basis for S^{ω_0}.

The following theorems parallel the corresponding results for inner product spaces.

Theorem 2.10.10. *Let S be a finite set of vectors in a symplectic vector space (V, ω) and let $W = \text{Span}(S)$. Then $W^{\omega} = S^{\omega}$.*

Proof. The proof follows mutatis mutandis that of Theorem 2.9.13. □

Corollary 2.10.11. *Let B be a basis for a subspace $W \subset V$. Then $W^{\omega} = B^{\omega}$.*

Despite the significant differences between the notions of orthogonality and ω-orthogonality, Theorem 2.9.16 concerning the dimension of the orthogonal complement has a direct parallel in the symplectic setting.

Theorem 2.10.12. *Let W be a subspace of a finite-dimensional symplectic vector space (V, ω). Then*

$$\dim(W) + \dim(W^{\omega}) = \dim(V).$$

Proof. Let $k = \dim(W)$. Note that if $k = 0$, then $W^{\omega} = V$, and the conclusion of the theorem trivially holds.

If $k \geq 1$, we rely on the isomorphism $\Psi : V \to V^*$ given by $\Psi(\mathbf{v}) = i_{\mathbf{v}}$ of Theorem 2.10.5, where $i_{\mathbf{v}} : V \to \mathbf{R}$ is defined to be $T_{\mathbf{v}}(\mathbf{w}) = \omega(\mathbf{v}, \mathbf{w})$. Consider the map $S : V \to W^*$ given by $S(\mathbf{v}) = \Psi(\mathbf{v})|_W$. On the one hand, S is onto. To see this, let $\mathcal{B} = \{\mathbf{w}_1, \ldots, \mathbf{w}_{2n}\}$ be a basis for V such that $\{\mathbf{w}_1, \ldots, \mathbf{w}_k\}$ is a basis for W. For any $\alpha \in W^*$, define $\tilde{\alpha} \in V^*$ to be $\tilde{\alpha}(\mathbf{v}) = c_1\alpha(\mathbf{w_1}) + \cdots + c_k\alpha(\mathbf{w}_k)$, where we are writing $\mathbf{v} = c_1\mathbf{w}_1 + \cdots + c_{2n}\mathbf{w}_{2n}$ according to the basis $\mathcal{B}$. The reader can check that $\tilde{\alpha} \in V^*$; in fact, $\tilde{\alpha} = \alpha \circ \text{Pr}$, where $\text{Pr} : V \to W$ is the projection onto W relative to the bases chosen above, defined in Example 2.7.9. Let $\mathbf{v}_\alpha \in V$ be such that $\Psi(\mathbf{v}_\alpha) = \tilde{\alpha}$. Then $S(\mathbf{v}_\alpha) = \alpha$, and so S is onto.

In addition, practically by definition,

$$\begin{aligned} \ker S &= \{\mathbf{v} \in V \mid (\Psi(\mathbf{v}))(\mathbf{w}) = 0 \text{ for all } \mathbf{w} \in W\} \\ &= \{\mathbf{v} \in V \mid \omega(\mathbf{v}, \mathbf{w}) = 0 \text{ for all } \mathbf{w} \in W\} \\ &= W^{\omega}. \end{aligned}$$

We thus rely on Theorem 2.7.7:

$$\dim(V) = \dim\ker S + \dim R(S) = \dim W^{\omega} + \dim W^{*} = \dim W^{\omega} + \dim W,$$

the last equality due to Theorem 2.8.2. □

The analogue of Theorem 2.9.15 about the trivial intersection of W with $W^{\perp}$ does not always hold in a symplectic vector space. In fact, we can identify a number of possible relationships between a subspace W and its ω-orthogonal complement W^{ω}.

Definition 2.10.13. Let W be a subspace of a symplectic vector space (V, ω). Then W is called

- *isotropic* if $W \subset W^{\omega}$;
- *coisotropic* if $W^{\omega} \subset W$;
- *Lagrangian* if $W = W^{\omega}$;
- *symplectic* if $W \cap W^{\omega} = \{\mathbf{0}\}$.

Proposition 2.10.14. *Let W be a subspace of the symplectic vector space (V, ω) with $\dim V = 2n$. Then*

- *If W is isotropic, then $\dim W \leq n$.*
- *If W is coisotropic, then $\dim W \geq n$.*
- *If W is Lagrangian, then $\dim W = n$.*
- *If W is symplectic, then $\dim W = 2m$ for some $m \leq n$.*

Proof. The first three statements are corollaries of Theorem 2.10.12. For example, if W is isotropic, then $\dim W \leq \dim W^{\omega}$, and so if $\dim W > n$, then $\dim V > 2n$, a contradiction.

To prove the last statement, note that the symplectic condition amounts to saying that ω is nondegenerate on W: If not, then there is a $\mathbf{w}_0 \in W$ having the property that $\mathbf{w}_0 \neq \mathbf{0}$ and $\omega(\mathbf{w}_0, \mathbf{w}) = 0$ for all $\mathbf{w} \in W$. But this means that $\mathbf{w}_0 \in W^{\omega}$, contradicting the assumption that W is symplectic.

Because ω is nondegenerate on W, we can apply the argument of the proof of Theorem 2.10.4 to construct a symplectic basis for W that necessarily has an even number of elements, as claimed. □

Example 2.10.15 (Examples of ω-orthogonal complements). The subspace $W_1 = \mathrm{Span}(S) \subset \mathbf{R}^2$, where S is the set in Example 2.10.8, is Lagrangian; note that this means it is both isotropic and coisotropic.

The subspace $W_2 = \mathrm{Span}(S) \subset \mathbf{R}^4$, where S is the set in Example 2.10.9, is isotropic.

If (V, ω) is a symplectic vector space and $\{\mathbf{e}_1, \mathbf{f}_1, \ldots, \mathbf{e}_n, \mathbf{f}_n\}$ is a symplectic basis for V, then every subspace $W_3 = \mathrm{Span}(\{\mathbf{e}_1, \mathbf{f}_1, \ldots, \mathbf{e}_k, \mathbf{f}_k\})$ where $1 \leq k \leq n$ is a symplectic subspace of V.

As with inner product spaces, a linear symplectic form on a vector space (V, ω) distinguishes special linear transformations on V, namely those that preserve the symplectic structure.

Definition 2.10.16. Let (V, ω) be a symplectic vector space. A linear transformation $T : V \to V$ is a *linear symplectomorphism* (or a *linear symplectic transformation*) if for all $\mathbf{v}, \mathbf{w} \in V$ we have $\omega(T(\mathbf{v}), T(\mathbf{w})) = \omega(\mathbf{v}, \mathbf{w})$. In the language of Sect. 2.8, T is a linear symplectomorphism if $T^*\omega = \omega$.

We list here some basic properties of linear symplectomorphisms.

Proposition 2.10.17. *Let (V, ω) be a finite-dimensional symplectic vector space. Then:*

- *If $T_1, T_2 : V \to V$ are linear symplectomorphisms, then $T_2 \circ T_1$ is also a linear symplectomorphism.*
- *If $T : V \to V$ is a linear symplectomorphism, then T has an inverse T^{-1}. Moreover, T^{-1} is a linear symplectomorphism.*

Proof. The reader should refer to Proposition 2.8.16 on the pullback of the composition of linear maps.

The first statement follows immediately from the fact that

$$(T_2 \circ T_1)^*\omega = T_1^*(T_2^*\omega).$$

To prove the second statement, we will show that $\ker(T) = \{\mathbf{0}\}$; by Theorem 2.7.6, this means that T is one-to-one, and so by Corollary 2.7.11, T is also onto.

To do this, suppose $\mathbf{v} \in \ker(T)$ and assume $\mathbf{v} \neq \mathbf{0}$. By the nondegeneracy of ω, there exists a vector $\mathbf{w}$ such that $\omega(\mathbf{v}, \mathbf{w}) \neq 0$. But since T is a linear symplectomorphism,

$$\begin{aligned} \omega(\mathbf{v}, \mathbf{w}) &= (T^*\omega)(\mathbf{v}, \mathbf{w}) \\ &= \omega(T(\mathbf{v}), T(\mathbf{w})) \\ &= 0 \quad \text{since } \mathbf{v} \in \ker T, \end{aligned}$$

a contradiction. Hence $\ker(T) = \{\mathbf{0}\}$, proving the claim. So T is invertible with inverse T^{-1}. That T^{-1} is itself a linear symplectomorphism follows from the fact that

$$\omega = (\mathrm{Id})^*\omega = (T \circ T^{-1})^*\omega = (T^{-1})^*(T^*\omega) = (T^{-1})^*\omega,$$

assuming that T is a linear symplectomorphism. □

Linear symplectomorphisms can be characterized in terms of the concept of a symplectic basis.

Theorem 2.10.18. *Let (V, ω) be a symplectic vector space with dimension $\dim V = 2n$ and symplectic basis $\{\mathbf{e}_1, \mathbf{f}_1, \dots, \mathbf{e}_n, \mathbf{f}_n\}$. If $T : V \to V$ is a linear symplectomorphism, then*

$$\{T(\mathbf{e}_1), T(\mathbf{f}_1), \dots, T(\mathbf{e}_n), T(\mathbf{f}_n)\}$$

is also a symplectic basis for V.

Conversely, suppose that

$$\mathcal{B} = \{\mathbf{e}_1, \mathbf{f}_1, \ldots, \mathbf{e}_n, \mathbf{f}_n\}$$

and

$$\mathcal{B}' = \{\mathbf{u}_1, \mathbf{v}_1, \ldots, \mathbf{u}_n, \mathbf{v}_n\}$$

are two symplectic bases for (V, ω) *and* $T : V \to V$ *is the linear isomorphism defined (according to Theorem 2.6.2) by* $T(\mathbf{e}_i) = \mathbf{u}_i$ *and* $T(\mathbf{f}_i) = \mathbf{v}_i$*. Then* T *is a linear symplectomorphism.*

Proof. The first statement follows from the fact that

$$\omega(T(\mathbf{v}), T(\mathbf{w})) = (T^*\omega)(\mathbf{v}, \mathbf{w}) = \omega(\mathbf{v}, \mathbf{w}),$$

assuming that T is a linear symplectomorphism.

To prove the converse, note that if vectors $\mathbf{v}$ and $\mathbf{w}$ are written according to the symplectic basis $\mathcal{B}$, i.e.,

$$\mathbf{v} = \sum (s_i \mathbf{e}_i + t_i \mathbf{f}_i), \quad \mathbf{w} = \sum (a_i \mathbf{e}_i + b_i \mathbf{f}_i),$$

then a calculation shows that

$$\omega(\mathbf{v}, \mathbf{w}) = \sum (s_i b_i - t_i a_i).$$

This calculation holds for any symplectic basis. In particular, we have

$$(T^*\omega)(\mathbf{v}, \mathbf{w}) = \omega(T(\mathbf{v}), T(\mathbf{w})) = \sum (s_i b_i - t_i a_i) = \omega(\mathbf{v}, \mathbf{w}). \quad \square$$

Proposition 2.10.19. *If* T *is a linear symplectomorphism of* (V, ω) *and* $\mathbf{v}, \mathbf{w}$ *are* ω*-orthogonal, then so are* $T(\mathbf{v}), T(\mathbf{w})$.

Proof. Exercise. □

We turn now to the matrix representation of the standard symplectic form ω_0 on $\mathbf{R}^{2n}$. This case in fact covers the matrix representation for any linear symplectic form on a symplectic vector space (V, ω), as long as vectors are represented in components relative to a symplectic basis.

Recall that the standard symplectic basis B_0 for $(\mathbf{R}^{2n}, \omega_0)$ was described above, following Theorem 2.10.4. Writing vectors $\mathbf{v}, \mathbf{w}$ as column vectors relative to B_0, the reader can verify using Proposition 2.8.14 that

$$\omega_0(\mathbf{v}, \mathbf{w}) = \mathbf{w}^T J \mathbf{v},$$

where, using block matrix notation,

$$J = \begin{bmatrix} J_0 & O & \cdots & O \\ O & J_0 & \cdots & O \\ O & O & \ddots & O \\ O & \cdots & O & J_0 \end{bmatrix}, \qquad J_0 = \begin{bmatrix} 0 & -1 \\ 1 & 0 \end{bmatrix}.$$

The matrix J, representing the standard symplectic form, also allows a matrix characterization of a linear symplectomorphism.

Theorem 2.10.20. *Let $T : \mathbf{R}^{2n} \to \mathbf{R}^{2n}$ be a linear transformation. Then T is a linear symplectomorphism of $(\mathbf{R}^{2n}, \omega_0)$ if and only if its matrix representation $A = [T]$ relative to the standard symplectic basis satisfies*

$$A^T J A = J.$$

Proof. The condition that $T^*\omega_0 = \omega_0$ means that for all $\mathbf{v}, \mathbf{w} \in \mathbf{R}^{2n}$, $\omega_0(T(\mathbf{v}), T(\mathbf{w})) = \omega_0(\mathbf{v}, \mathbf{w})$. But, in matrix notation,

$$\omega_0(T(\mathbf{v}), T(\mathbf{w})) = (A\mathbf{w})^T J(A\mathbf{v}) = \mathbf{w}^T (A^T J A)\mathbf{v},$$

while $\omega_0(\mathbf{v}, \mathbf{w}) = \mathbf{w}^T J\mathbf{v}$. Hence $T^*\omega_0 = \omega_0$ is equivalent to the matrix equation $A^T J A = J$. □

A $2n \times 2n$ matrix satisfying the condition that $A^T J A = J$ will be called a *symplectic matrix*. We write $\mathrm{Sp}(2n)$ to denote the set of all $2n \times 2n$ symplectic matrices. A number of properties of symplectic matrices will be explored in the exercises. The following theorem indicates only the most important properties.

Theorem 2.10.21. *Let $A \in \mathrm{Sp}(2n)$. Then:*

1. *A is invertible;*
2. *$A^T \in \mathrm{Sp}(2n)$;*
3. *$A^{-1} \in \mathrm{Sp}(2n)$.*

Proof. In light of Theorem 2.10.20, statements (1) and (3) are in fact corollaries of Proposition 2.10.17. However, we prove the statements here using matrix techniques.

Suppose $A \in \mathrm{Sp}(2n)$, i.e., $A^T J A = J$. Then since $\det J = 1$, we have

$$\begin{aligned} 1 &= \det J \\ &= \det(A^T J A) \\ &= (\det A)^2, \end{aligned}$$

and so $\det A = \pm 1 \neq 0$. Hence A is invertible.

Since $J^{-1} = -J$ and $J^2 = -I$, and using the fact that $A^T J A = J$, we have

$$J A^T J A = J^2 = -I,$$

which shows that

$$-JA^T = (JA)^{-1} = A^{-1}J^{-1} = -A^{-1}J,$$

and hence $AJA^T = (A^T)^T J(A^T) = J$. So $A^T \in \mathrm{Sp}(2n)$.

We leave the proof of (3) as an exercise. □

We saw in the context of the preceding proof that the determinant of a symplectic matrix is ± 1. In fact, a stronger results holds.

Theorem 2.10.22. *If* $A \in \mathrm{Sp}(2n)$, *then* $\det(A) = 1$.

We will defer the proof, however, to Chap. 7. We will ultimately rely on the tools of exterior algebra that we present in Chap. 3.

The following statement concerns the eigenvalues of a symplectic matrix.

Theorem 2.10.23. *Suppose* λ *is an eigenvalue of the symplectic matrix* $A \in \mathrm{Sp}(2n)$ *with multiplicity* k. *Then* $1/\lambda$, $\overline{\lambda}$, *and* $1/\overline{\lambda}$ *are also eigenvalues of* A *with multiplicity* k. *(Here* $\overline{\lambda}$ *is the complex conjugate of* λ.*)*

Proof. Consider the characteristic polynomial $p(x) = \det(A - xI)$; note that 0 cannot be a root, since then A would not be invertible. It is always the case that $\overline{\lambda}$ is a root of p if λ is, since p is a real polynomial, and that the multiplicities of λ and $\overline{\lambda}$ are the same.

For every nonzero x, we have the following:

$$\begin{aligned}
p(x) &= \det(A - xI) \\
&= \det(J(A - xI)J^{-1}) \\
&= \det(JAJ^{-1} - xI) \\
&= \det((A^{-1})^T - xI) \quad \text{since } A^T JA = J \\
&= \det((A^{-1} - xI)^T) \\
&= \det(A^{-1} - xI) \\
&= \det(A^{-1}(I - xA)) \\
&= \det(A^{-1})\det(I - xA) \\
&= \det(I - xA) \quad \text{by Theorem 2.10.22} \\
&= x^{2n}\det\left(\frac{1}{x}I - A\right) \\
&= x^{2n}p\left(\frac{1}{x}\right).
\end{aligned}$$

This shows that if λ is a root of p, then so is $1/\lambda$ (and hence $1/\overline{\lambda}$ also).

Now assume that λ is a root of the characteristic polynomial p with multiplicity k, so that

$$p(x) = (x - \lambda)^k q(x)$$

for some polynomial q satisfying $q(\lambda) \neq 0$. But then for $x \neq 0$ we have

$$\begin{aligned} p(x) &= x^{2n} p\left(\frac{1}{x}\right) \quad \text{by the above calculation} \\ &= x^{2n}\left(\frac{1}{x} - \lambda\right)^k q\left(\frac{1}{x}\right) \\ &= \lambda^k x^{2n-k}\left(\frac{1}{\lambda} - x\right)^k q\left(\frac{1}{x}\right). \end{aligned}$$

Hence, since $q(\lambda) \neq 0$, we have that $1/\lambda$ is a root of p with multiplicity k. □

We will have occasion to consider the case of a vector space V that has both a symplectic linear form and an inner product. Unfortunately, the Gram–Schmidt methods of Theorems 2.9.8 and 2.10.4 are not compatible, in the sense that they cannot produce a basis that is simultaneously symplectic and orthogonal. Nevertheless, it is possible to construct such a basis by resorting to techniques particular to complex vector spaces—vector spaces whose scalars are complex numbers.

For basic results about complex vector spaces, the reader may consult any textbook in linear algebra, for example [2]. In the proof of the following theorem, *Hermitian* matrices will play a prominent role. A Hermitian matrix A is a square $(n \times n)$ matrix with complex entries having the property that

$$A = (\overline{A})^T,$$

where the bar represents componenentwise complex conjugation. The most important property of Hermitian matrices for our purposes is that they have n linearly independent (over **C**) eigenvectors that are orthonormal with respect to the standard Hermitian product $\langle \mathbf{x}, \mathbf{y} \rangle = \overline{\mathbf{x}}^T \mathbf{y}$ and whose corresponding eigenvalues are real.

Theorem 2.10.24. *Let (V, ω) be a symplectic vector space with $\dim V = 2n$. Suppose that G is an inner product on V. Then there is a symplectic basis*

$$\mathcal{B} = \{\tilde{\mathbf{u}}_1, \tilde{\mathbf{v}}_1, \ldots, \tilde{\mathbf{u}}_n, \tilde{\mathbf{v}}_n\}$$

that is also G-orthogonal, i.e.,

$$G(\tilde{\mathbf{u}}_j, \tilde{\mathbf{v}}_k) = 0 \quad \text{for all } j, k = 1, \ldots, n; \quad G(\tilde{\mathbf{u}}_j, \tilde{\mathbf{u}}_k) = G(\tilde{\mathbf{v}}_j, \tilde{\mathbf{v}}_k) = 0 \quad \text{for } j \neq k.$$

Moreover, the basis can be chosen so that $G(\tilde{\mathbf{u}}_j, \tilde{\mathbf{u}}_j) = G(\tilde{\mathbf{v}}_j, \tilde{\mathbf{v}}_j)$ for all $j = 1, \ldots, n$.

Proof. We begin with an orthonormal basis

$$\mathcal{B}' = \{\mathbf{e}_1, \ldots, \mathbf{e}_{2n}\}$$

of V relative to G, which exists according to Theorem 2.9.8. Let A be the $2n \times 2n$ matrix defined by the symplectic form ω relative to $\mathcal{B}'$ as follows:

$$\omega(\mathbf{v}, \mathbf{w}) = G(\mathbf{v}, A\mathbf{w})$$

for all $\mathbf{v}, \mathbf{w} \in V$. We write $A = [a_{jk}]$, where $a_{jk} = \omega(\mathbf{e}_j, \mathbf{e}_k)$. Due to the skew-symmetry of ω, the matrix A is skew-symmetric: $A^T = -A$.

Throughout this proof, we will consider vectors $\mathbf{v}, \mathbf{w}$ to be column vectors written using components relative to the basis $\mathcal{B}'$. In particular,

$$G(\mathbf{v}, \mathbf{w}) = \mathbf{v}^T \mathbf{w},$$

and $\omega(\mathbf{v}, \mathbf{w}) = \mathbf{v}^T A\mathbf{w}$.

The skew-symmetry of A implies that the $2n \times 2n$ matrix iA with purely imaginary entries ia_{jk} is Hermitian:

$$\begin{aligned}
\left(\overline{iA}\right)^T &= (-iA)^T && \text{since the entries of } iA \text{ are purely imaginary} \\
&= -iA^T \\
&= -i(-A) && \text{since } A \text{ is skew-symmetric} \\
&= iA.
\end{aligned}$$

Since ω is nondegenerate, A must be invertible, and hence the eigenvalues of A are nonzero. By the property of Hermitian matrices mentioned above, there are in fact $2n$ linearly independent eigenvectors of iA that are orthonormal with respect to the Hermitian product and with real corresponding eigenvalues. In fact, the reader can verify that the eigenvectors of iA occur in pairs

$$\mathbf{y}_1, \overline{\mathbf{y}}_1, \ldots, \mathbf{y}_n, \overline{\mathbf{y}}_n$$

(these are vectors with complex components, and the bar represents componenent-wise complex conjugation). The corresponding eigenvalues will be denoted by

$$\mu_1, -\mu_1, \ldots, \mu_n, -\mu_n.$$

The orthonormality is expressed in matrix notation as

$$\overline{\mathbf{y}}_j^T \mathbf{y}_k = \delta_k^j = \begin{cases} 1, & j = k, \\ 0, & j \neq k. \end{cases}$$

Note that since $(iA)\mathbf{y}_j = \mu_j \mathbf{y}_j$, we have $A\mathbf{y}_j = (-i\mu_j)\mathbf{y}_j$; in other words, the eigenvalues of A are $\pm i\mu_j$. For each $j = 1, \dots, n$, we choose pairs λ_j and $\mathbf{x}_j$ as follows: From each pair of eigenvectors $\mathbf{y}_j, \overline{\mathbf{y}}_j$ with corresponding nonzero eigenvalues $\mu_j, -\mu_j$, choose $\lambda_j = \pm\mu_j$ so that $\lambda_j > 0$, and then if $\lambda_j = \mu_j$, choose $\mathbf{x}_j = \overline{\mathbf{y}}_j$, while if $\lambda_j = -\mu_j$, choose $\mathbf{x}_j = \mathbf{y}_j$. In this way, we have $A\mathbf{x}_j = i\lambda_j \mathbf{x}_j$ with $\lambda_j > 0$.

Write

$$\mathbf{x}_j = \mathbf{u}_j + i\mathbf{v}_j$$

with vectors $\mathbf{u}_j$ and $\mathbf{v}_j$ having have real components. We claim that the set $\mathcal{B}'' = \{\mathbf{u}_1, \mathbf{v}_1, \dots, \mathbf{u}_n, \mathbf{v}_n\}$ is a G-orthogonal basis for V. The fact that $\mathcal{B}''$ is a basis for V is a consequence of the above-mentioned property of the eigenvectors of a Hermitian matrix. To show that $\mathcal{B}''$ is G-orthogonal, we note that the Hermitian orthonormality condition $\overline{\mathbf{x}}_j^T \mathbf{x}_k = \delta_k^j$ can be expressed as

$$\mathbf{u}_j^T \mathbf{u}_k + \mathbf{v}_j^T \mathbf{v}_k = \delta_k^j, \quad \mathbf{u}_j^T \mathbf{v}_k - \mathbf{v}_j^T \mathbf{u}_k = 0.$$

Also, the fact that $A\mathbf{x}_j = i\lambda_j \mathbf{x}_j$ means that

$$A\mathbf{u}_j = -\lambda_j \mathbf{v}_j, \quad A\mathbf{v}_j = \lambda_j \mathbf{u}_j.$$

Hence, for $j \neq k$, we have

$$\begin{aligned}
\mathbf{u}_j^T \mathbf{u}_k &= \mathbf{u}_j^T \left(\frac{1}{\lambda_k} A\mathbf{v}_k \right) \\
&= \frac{1}{\lambda_k} \left(\mathbf{u}_j^T A \mathbf{v}_k \right)^T \quad \text{since the quantity in parentheses is a scalar} \\
&= \frac{1}{\lambda_k} \left(\mathbf{v}_k^T A^T \mathbf{u}_j \right) \\
&= -\frac{1}{\lambda_k} \left(\mathbf{v}_k^T A \mathbf{u}_j \right) \quad \text{since } A \text{ is skew-symmetric} \\
&= \frac{\lambda_j}{\lambda_k} \left(\mathbf{v}_k^T \mathbf{v}_j \right) \quad \text{since } A\mathbf{u}_j = -\lambda_j \mathbf{v}_j \\
&= -\frac{\lambda_j}{\lambda_k} \mathbf{u}_k^T \mathbf{u}_j \quad \text{since } \mathbf{u}_j^T \mathbf{u}_k + \mathbf{v}_j^T \mathbf{v}_k = 0 \text{ for } j \neq k \\
&= -\frac{\lambda_j}{\lambda_k} (\mathbf{u}_k^T \mathbf{u}_j)^T \quad \text{since the quantity in parentheses is a scalar} \\
&= -\frac{\lambda_j}{\lambda_k} \mathbf{u}_j^T \mathbf{u}_k,
\end{aligned}$$

which implies, since $\lambda_j, \lambda_k > 0$, that $\mathbf{u}_j^T \mathbf{u}_k = 0$.

In the same way,

$$\mathbf{v}_j^T \mathbf{v}_k = 0 \quad \text{for } j \neq k.$$

We leave it to the reader to show, in a similar way, that for all j and k,

$$\mathbf{u}_j^T \mathbf{v}_k = \mathbf{v}_j^T \mathbf{u}_k = 0.$$

All this shows that $\mathcal{B}''$ is G-orthogonal, since $G(\mathbf{v}, \mathbf{w}) = \mathbf{v}^T \mathbf{w}$.

Note that

$$\begin{aligned}\omega(\mathbf{u}_j, \mathbf{v}_j) &= \mathbf{u}_j^T A \mathbf{v}_j \\ &= \lambda_j \mathbf{u}_j^T \mathbf{u}_j \\ &= \lambda_j |\mathbf{u}_j|^2 \\ &> 0.\end{aligned}$$

We leave it as an exercise for the reader to find scalars c_j and d_j such that for $\tilde{\mathbf{u}}_j = c_j \mathbf{u}_j$ and $\tilde{\mathbf{v}}_j = d_j \mathbf{v}_j$,

$$\omega(\tilde{\mathbf{u}}_j, \tilde{\mathbf{v}}_j) = 1 \quad \text{and} \quad G(\tilde{\mathbf{u}}_j, \tilde{\mathbf{u}}_j) = G(\tilde{\mathbf{v}}_j, \tilde{\mathbf{v}}_j)$$

for all $j = 1, \ldots, n$. The set

$$\mathcal{B} = \{\tilde{\mathbf{u}}_1, \tilde{\mathbf{v}}_1, \ldots, \tilde{\mathbf{u}}_n, \tilde{\mathbf{v}}_n\}$$

is the desired basis. □

We will see that for the standard symplectic vector space $(\mathbf{R}^{2n}, \omega_0)$, ellipsoids play an important role in measuring linear symplectomorphisms. By an *ellipsoid*, we mean a set $E \subset \mathbf{R}^{2n}$ defined by a positive definite symmetric matrix A in the following way:

$$E = \left\{ \mathbf{x} \in \mathbf{R}^{2n} \mid \mathbf{x}^T A \mathbf{x} \leq 1 \right\}.$$

An important fact about ellipsoids is that they can be brought into a "normal form" by means of linear symplectomorphisms.

Theorem 2.10.25. *Let $E \subset \mathbf{R}^{2n}$ be an ellipsoid defined by the positive definite symmetric matrix A. Then there are positive constants $r_1, \ldots, r_n$ and a linear symplectomorphism $\Phi : (\mathbf{R}^{2n}, \omega_0) \to (\mathbf{R}^{2n}, \omega_0)$ such that $\Phi(E(r_1, \ldots, r_n)) = E$, where*

$$E(r_1, \ldots, r_n) = \left\{ (x_1, y_1, \ldots, x_n, y_n) \,\middle|\, \sum \left(\frac{x_i^2 + y_i^2}{r_i^2} \right) \leq 1 \right\}.$$

The constants are uniquely determined when ordered $0 < r_1 \leq \cdots \leq r_n$.

Proof. Since A is a positive definite symmetric matrix, it defines an inner product G by $G(\mathbf{x},\mathbf{y}) = \mathbf{x}^T A\mathbf{y}$. The ellipsoid E is then characterized as

$$E = \left\{\mathbf{b} \in \mathbf{R}^{2n} \mid G(\mathbf{b},\mathbf{b}) \leq 1\right\}.$$

According to Theorem 2.10.24, there is a basis

$$\{\mathbf{u}_1, \mathbf{v}_1, \ldots, \mathbf{u}_n, \mathbf{v}_n\}$$

that is both symplectic relative to ω_0 and G-orthogonal, with $G(\mathbf{u}_i,\mathbf{u}_i) = G(\mathbf{v}_i,\mathbf{v}_i)$ for all $i = 1,\ldots,n$. So define the positive constants r_i by

$$\frac{1}{r_i^2} = G(\mathbf{u}_i,\mathbf{u}_i).$$

Let $\Phi : \mathbf{R}^{2n} \to \mathbf{R}^{2n}$ be the linear symplectomorphism defined by its action on the standard symplectic basis $\{\mathbf{e}_1, \mathbf{f}_1, \ldots, \mathbf{e}_n, \mathbf{f}_n\}$ for $(\mathbf{R}^{2n}, \omega_0)$:

$$\Phi(\mathbf{e}_i) = \mathbf{u}_i, \quad \Phi(\mathbf{f}_i) = \mathbf{v}_i.$$

More explicitly, since

$$(x_1, y_1, \ldots, x_n, y_n) = x_1\mathbf{e}_1 + y_1\mathbf{f}_1 + \cdots + x_n\mathbf{e}_n + y_n\mathbf{f}_n,$$

we have

$$\Phi(x_1, y_1, \ldots, x_n, y_n) = x_1\mathbf{u}_n + y_1\mathbf{v}_1 + \cdots + x_n\mathbf{u}_n + y_n\mathbf{v}_n.$$

We will show that $\Phi(E(r_1,\ldots,r_n)) = E$. On the one hand, suppose $\mathbf{b} \in \Phi(E(r_1,\ldots,r_n))$. In other words, there is $\mathbf{a} \in E(r_1,\ldots,r_n)$ such that $\Phi(\mathbf{a}) = \mathbf{b}$. Writing

$$\mathbf{a} = (x_1, y_1, \ldots, x_n, y_n) = x_1\mathbf{e}_1 + y_1\mathbf{f}_1 + \cdots + x_n\mathbf{e}_n + y_n\mathbf{f}_n,$$

we then have

$$\mathbf{b} = \Phi(\mathbf{a}) = x_1\mathbf{u}_1 + y_1\mathbf{v}_1 + \cdots + x_n\mathbf{u}_n + y_n\mathbf{v}_n,$$

and so

$$\begin{aligned} G(\mathbf{b},\mathbf{b}) &= \sum \left(x_i^2 G(\mathbf{u}_i,\mathbf{u}_i) + y_i^2 G(\mathbf{v}_i,\mathbf{v}_i)\right) \\ &= \sum \left(x_i^2\left(\frac{1}{r_i^2}\right) + y_i^2\left(\frac{1}{r_i^2}\right)\right) \\ &\leq 1 \quad \text{since } \mathbf{a} \in E(r_1,\ldots,r_n). \end{aligned}$$

Hence $\mathbf{b} \in E$, and so $\Phi(E(r_1,\ldots,r_n)) \subset E$.

On the other hand, suppose that $\mathbf{b} \in E$, so that $G(\mathbf{b}, \mathbf{b}) \leq 1$. There is $\mathbf{a} \in \mathbf{R}^{2n}$ such that $\Phi(\mathbf{a}) = \mathbf{b}$, since Φ is a linear isomorphism. Writing $\mathbf{b}$ according to the basis above, we obtain

$$\mathbf{b} = \tilde{x}_1 \mathbf{u}_1 + \tilde{y}_1 \mathbf{v}_1 + \cdots + \tilde{x}_n \mathbf{u}_n + \tilde{y}_n \mathbf{v}_n,$$

so

$$\mathbf{a} = (\tilde{x}_1, \tilde{y}_1, \ldots, \tilde{x}_n, \tilde{y}_n).$$

But

$$\sum \left(\frac{\tilde{x}_i^2 + \tilde{y}_i^2}{r_i^2} \right) = G(\mathbf{b}, \mathbf{b}) \leq 1,$$

and so $\mathbf{a} \in E(r_1, \ldots, r_n)$ and $E \subset \Phi(E(r_1, \ldots, r_n))$.

All this shows that $\Phi(E(r_1, \ldots, r_n)) = E$.

To show that the constants r_i are uniquely determined up to ordering, suppose that there are linear symplectomorphisms $\Phi_1, \Phi_2 : \mathbf{R}^{2n} \to \mathbf{R}^{2n}$ and n-tuples $(r_1, \ldots, r_n)$, $(r_1', \ldots, r_n')$ with $0 < r_1 \leq \cdots \leq r_n$ and $0 < r_1' \leq \cdots \leq r_n'$ such that

$$\Phi_1(E(r_1, \ldots, r_n)) = E, \quad \Phi_2(E(r_1', \ldots, r_n')) = E.$$

Then, writing $\Phi = \Phi_1^{-1} \circ \Phi_2$, we have

$$\Phi(E(r_1', \ldots, r_n')) = E(r_1, \ldots, r_n).$$

In matrix notation, this says that $\mathbf{x}^T D' \mathbf{x} \leq 1$ if and only if $(\Phi \mathbf{x})^T D (\Phi \mathbf{x}) = \mathbf{x}^T (\Phi^T D \Phi) \mathbf{x} \leq 1$, where $\mathbf{x} = (x_1, y_1, \ldots, x_n, y_n)$ is a column vector, D is the diagonal matrix $D = \operatorname{diag}\left[1/(r_1)^2, 1/(r_1)^2, \ldots, 1/(r_n)^2, 1/(r_n)^2\right]$, and D' is the diagonal matrix $D' = \operatorname{diag}\left[1/(r_1')^2, 1/(r_1')^2, \ldots, 1/(r_n')^2, 1/(r_n')^2\right]$. This implies that

$$\Phi^T D \Phi = D'.$$

The facts that Φ satisfies $\Phi^T J \Phi = J$ and $J^{-1} = -J$ together imply $\Phi^T = -J\Phi^{-1}J$, and so

$$\Phi^{-1} J D \Phi = J D'.$$

This shows that JD is similar to JD', and so the two matrices have the same eigenvalues. The reader may verify that the eigenvalues of JD are $\pm i r_j$ and those of JD' are $\pm i r_j'$. Since the r_i and r_i' are ordered from least to greatest, we must have $r_j = r_j'$ for all $j = 1, \ldots, n$. □

Theorem 2.10.25 prompts the following definition.

Definition 2.10.26. Let $E \subset \mathbf{R}^{2n}$ be an ellipsoid in the standard symplectic space $(\mathbf{R}^{2n}, \omega_0)$. The *symplectic spectrum of* E is the unique n-tuple $\sigma(E) = (r_1, \ldots, r_n)$, $0 < r_1 \leq \cdots \leq r_n$, such that there is a linear symplectomorphism Φ with $\Phi(E(r_1, \ldots, r_n)) = E$ (Fig. 2.6).

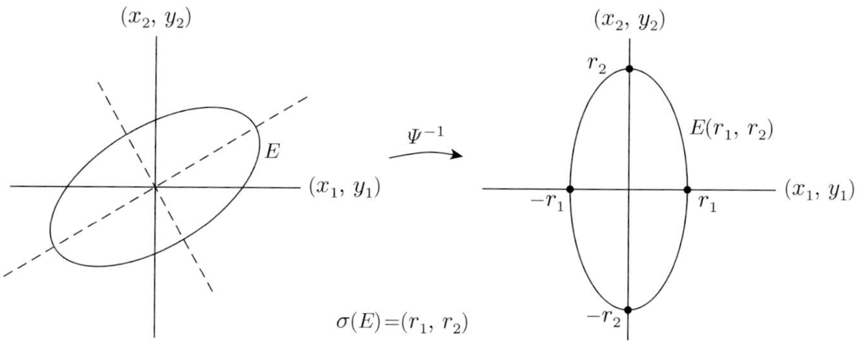

Fig. 2.6 Linear symplectomorphisms and the symplectic spectrum.

We will continue to develop some topics in linear symplectic geometry in Sect. 7.7 as motivation for a key concept in (nonlinear) symplectic geometry, the symplectic capacity.

2.11 For Further Reading

With the exception of Sects. 2.8 and 2.10, much of the material in this chapter can be found in any textbook on linear algebra. The notation here generally follows that in [2].

While many linear algebra textbooks have detailed presentations of inner product spaces, symplectic vector spaces are usually presented only as introductory matter in the context of specialized texts. We refer to A. Banyaga's summary in [4, Chap. 1] or to [31].

2.12 Exercises

The exercises in this chapter emphasize topics not usually presented in a first elementary linear algebra course.

2.1. Prove Theorem 2.4.3.

2.2. Prove Theorem 2.4.10.

2.3. Prove Theorem 2.4.13.

2.4. Let $T : V \to W$ be a linear isomorphism between vector spaces V and W, and let $T^{-1} : W \to V$ be the inverse of T. Show that T^{-1} is a *linear* transformation.

2.5. Consider the basis $\mathcal{B} = \{\mathbf{b}_1, \mathbf{b}_2, \mathbf{b}_3\}$ of $\mathbf{R}^3$, where

$$\mathbf{b}_1 = (1,0,1), \quad \mathbf{b}_2 = (1,1,0), \quad \mathbf{b}_3 = (0,2,1).$$

(a) Write the components of $\mathbf{w} = (2,3,5)$ relative to the basis $\mathcal{B}$.
(b) Let $\{\beta_1, \beta_2, \beta_3\}$ be the basis of $(\mathbf{R}^3)^*$ dual to $\mathcal{B}$. Compute $\beta_i(\mathbf{w})$ for each $i = 1,2,3$, where $\mathbf{w}$ is the vector given in part (a).
(c) For each $i = 1,2,3$, compute $\beta_i(\mathbf{v})$, where $\mathbf{v} = (v_1, v_2, v_3)$ is an arbitrary vector in $\mathbf{R}^3$.

2.6. For each of the linear transformations Ψ and linear one-forms T below, compute $\Psi^* T$:

(a) $\Psi : \mathbf{R}^3 \to \mathbf{R}^3$, $\Psi(u,v,w) = (2u, 3u - v - w, u + 2w)$,

$$T(x,y,z) = 3x + y - z.$$

(b) $\Psi : \mathbf{R}^3 \to \mathbf{R}^2$, $\Psi(u,v,w) = (v, 2u - w)$,

$$T(x,y) = x + 3y.$$

(c) $\Psi : \mathbf{R}^4 \to \mathbf{R}^3$, $\Psi(x,y,z,w) = (x + y - z - 2w, w - 4x - z, y + 3z)$,

$$T(x,y,z) = x - 2y + 3z.$$

2.7. Let $\alpha \in (\mathbf{R}^3)^*$ be given by $\alpha(x,y,z) = 4y + z$.

(a) Find $\ker \alpha$.
(b) Find all linear transformations $\Psi : \mathbf{R}^3 \to \mathbf{R}^3$ with the property that $\Psi^* \alpha = \alpha$.

2.8. Consider the linear transformation $T : \mathbf{R}^2 \to \mathbf{R}^2$ given by $T(x_1, x_2) = (2x_1 - x_2, x_1 + 3x_2)$.

(a) Compute $T^* G_0$, where G_0 is the standard inner product defined in Example 2.8.11.
(b) Compute $T^* S$, where S is the bilinear form in Example 2.8.13.

2.9. Let $T : \mathbf{R}^n \to \mathbf{R}^n$ be a linear transformation described in matrix form $[T]$ relative to the standard basis for $\mathbf{R}^n$. Show that for every $n \times n$ matrix A, one has

$$T^* G_A = G_{A[T]},$$

where G_A and $G_{A[T]}$ are defined according to Example 2.8.12.

2.10. Prove the following converse to Proposition 2.8.14: Let B be an $n \times n$ matrix and let $\mathcal{E}$ be a basis for the n-dimensional vector space V. Then the function $b : V \times V \to \mathbf{R}$ defined by

$$b(\mathbf{v}, \mathbf{w}) = \mathbf{w}^T B \mathbf{v},$$

where $\mathbf{v}$ and $\mathbf{w}$ are written as column vectors relative to the basis $\mathcal{E}$, is a bilinear form.

2.11. Use Exercise 2.10 to give five examples of bilinear forms on $\mathbf{R}^3$ and five examples of bilinear forms on $\mathbf{R}^4$.

2.12. Let b, B, and $\mathcal{E}$ be as given in Exercise 2.10, and let $T : V \to V$ be a linear transformation. Show that $T^*b = \tilde{b}$, where $\tilde{b}$ is the bilinear form corresponding to the matrix $A^T BA$ for $A = [T]_{\mathcal{E},\mathcal{E}}$, the matrix representation of T relative to the basis $\mathcal{E}$.

2.13. For each of the following 2×2 matrices, write the coordinate expression for the inner product G_A relative to the standard basis as in Example 2.9.4. For each, compute $G_A(\mathbf{e}_1, \mathbf{e}_1)$, $G_A(\mathbf{e}_1, \mathbf{e}_2)$, and $G_A(\mathbf{e}_2, \mathbf{e}_2)$ along with $\angle(\mathbf{e}_1, \mathbf{e}_2)$, where $\mathbf{e}_1 = (1, 0)$ and $\mathbf{e}_2 = (0, 1)$:

(a) $\begin{bmatrix} 2 & 1 \\ 1 & -1 \end{bmatrix}$;

(b) $\begin{bmatrix} 2 & -1 \\ 1 & 3 \end{bmatrix}$;

(c) $\begin{bmatrix} 1 & 2 \\ 1 & 3 \end{bmatrix}$.

2.14. Show that the function $G(\mathbf{v}, \mathbf{w}) = v_1 w_1 + 2v_1 w_2 + 2v_2 w_1 + 5v_2 w_2$ is an inner product on $\mathbf{R}^2$, where $\mathbf{v} = (v_1, v_2)$ and $\mathbf{w} = (w_1, w_2)$. Find an orthonormal basis $\{\mathbf{u}_1, \mathbf{u}_2\}$ for $\mathbf{R}^2$ relative to G.

2.15. Let $\{\mathbf{u}_1, \mathbf{u}_2\}$ be the basis for $\mathbf{R}^2$ given by $\mathbf{u}_1 = (3, 2)$ and $\mathbf{u}_2 = (1, 1)$. Let G be the inner product on $\mathbf{R}^2$ such that $\{\mathbf{u}_1, \mathbf{u}_2\}$ is orthonormal (see Theorem 2.9.9). Find $G(\mathbf{v}, \mathbf{w})$, where $\mathbf{v} = (v_1, v_2)$ and $\mathbf{w} = (w_1, w_2)$. Find $\angle((1, 0), (0, 1))$.

2.16. Prove Theorem 2.9.10.

2.17. For the following subspaces W of $\mathbf{R}^n$, find a basis for $W^\perp$, the orthogonal complement of W relative to the standard inner product on $\mathbf{R}^n$

(a) $W = \operatorname{Span}\{(1, 2)\} \subset \mathbf{R}^2$;
(b) $W = \operatorname{Span}\{(1, 2, 3)\} \subset \mathbf{R}^3$;
(c) $W = \operatorname{Span}\{(1, 0, 1), (-1, 1, 0)\} \subset \mathbf{R}^3$;
(d) $W = \operatorname{Span}\{(1, -2, 2, 1), (0, 1, 1, -3)\} \subset \mathbf{R}^4$.

2.18. Provide the details for the proof of Theorem 2.9.16.

2.19. Let (V, G) be an inner product space and let W be a subset of V.

(a) Show that $W \subset (W^\perp)^\perp$.
(b) Show that if V is finite-dimensional, then there is an orthonormal basis $\{\mathbf{u}_1, \ldots, \mathbf{u}_n\}$ of V such that $\{\mathbf{u}_1, \ldots, \mathbf{u}_k\}$ is a basis for W and $\{\mathbf{u}_{k+1}, \ldots, \mathbf{u}_n\}$ is a basis for $W^\perp$. (See Theorem 2.9.16.)
(c) Show that if V is finite-dimensional, then $(W^\perp)^\perp \subset W$, and so by (a), $W = (W^\perp)^\perp$.

2.20. Prove Proposition 2.9.18.

2.21. Prove Proposition 2.9.21.

2.22. Let (V, G) be a finite-dimensional inner product space. Show that a linear transformation $T : V \to V$ is a linear isometry if and only if for every orthonormal basis $\{\mathbf{e}_1, \ldots, \mathbf{e}_n\}$ of V, the set $\{T(\mathbf{e}_1), \ldots, T(\mathbf{e}_n)\}$ is also an orthonormal basis for V.

2.23. Give three examples of linear symplectic forms on $\mathbf{R}^4$.

2.24. Suppose $\mathcal{B} = \{\mathbf{a}_1, \mathbf{b}_1, \ldots, \mathbf{a}_n, \mathbf{b}_n\}$ is a basis for $\mathbf{R}^{2n}$. Define the alternating bilinear form $\omega_{\mathcal{B}}$ by specifying its action on the basis vectors,

$$\begin{aligned} \omega_{\mathcal{B}}(\mathbf{a}_i, \mathbf{a}_j) = \omega_{\mathcal{B}}(\mathbf{b}_i, \mathbf{b}_j) = 0 \quad &\text{for all } i, j, \\ \omega_{\mathcal{B}}(\mathbf{a}_i, \mathbf{b}_j) = 0 \quad &\text{for } i \neq j, \\ \omega_{\mathcal{B}}(\mathbf{a}_i, \mathbf{b}_i) = 1, \end{aligned}$$

and extending bilinearly. Show that $\omega_{\mathcal{B}}$ is a linear symplectic form.

2.25. Define a bilinear form S on $\mathbf{R}^4$ by $S(\mathbf{v}, \mathbf{w}) = \mathbf{w}^T A \mathbf{v}$, where

$$A = \begin{bmatrix} 0 & 1 & 1 & 1 \\ -1 & 0 & 2 & 0 \\ -1 & -2 & 0 & 3 \\ -1 & 0 & -3 & 0 \end{bmatrix}.$$

(a) Show that S is a linear symplectic form.
(b) Use the process outlined in Theorem 2.10.4 to find a symplectic basis $\{\mathbf{e}_1, \mathbf{f}_1, \mathbf{e}_2, \mathbf{f}_2\}$ for $\mathbf{R}^4$ relative to S.

2.26. Use the procedure in Theorem 2.10.4 to construct three different symplectic bases for $\mathbf{R}^4$ by making appropriate choices at different stages of the process.

2.27. Consider $(\mathbf{R}^4, \omega_0)$, where ω_0 is the standard linear symplectic form on $\mathbf{R}^4$. Decide whether the following subspaces of $\mathbf{R}^4$ are isotropic, coisotropic, Lagrangian, or symplectic:

(a) $W_1 = \operatorname{Span}\{(1, 0, -1, 3)\}$;
(b) $W_2 = \operatorname{Span}\{(3, 1, 0, -1), (2, 1, 2, 1)\}$;
(c) $W_3 = \operatorname{Span}\{(1, 0, 2, -1), (0, 1, 1, -1)\}$;
(d) $W_4 = \operatorname{Span}\{(1, 1, 1, 0), (2, -1, 0, 1), (0, 2, 0, -1)\}$;
(e) $W_5 = \ker T$, where $T : \mathbf{R}^4 \to \mathbf{R}^2$ is given by

$$T(x_1, y_1, x_2, y_2) = (2x_2 - y_1, x_1 + x_2 + y_1 + y_2).$$

2.28. Prove Theorem 2.10.7.

2.29. Prove Theorem 2.10.12.

2.30. Let W_1 and W_2 be subspaces of a symplectic vector space (V, ω). Show that if $W_1 \subset W_2$, then $(W_2)^\omega \subset (W_1)^\omega$.

2.31. Show that if W is a subspace of a finite-dimensional symplectic vector space (V, ω), then $(W^\omega)^\omega = W$. (See Exercise 2.19.)

2.32. Is it possible for a 2-dimensional subspace of a 4-dimensional symplectic vector space to be neither symplectic nor Lagrangian? If so, find necessary conditions for this to occur. If not, state and prove the corresponding result. To what extent can this question be generalized to higher dimensions?

2.33. Prove Proposition 2.10.19.

2.34. Prove Theorem 2.10.5.

2.35. For each of the examples in Exercise 2.23, write the isomorphism Ψ described in Theorem 2.10.5 explicitly in terms of the standard bases of $\mathbf{R}^4$ and $(\mathbf{R}^4)^*$.

2.36. Let W be a subspace of a finite-dimensional symplectic vector space (V, ω). Let $\Psi : V \to V^*$ be the isomorphism described in Theorem 2.10.5.

(a) Let
$$W^0 = \{\alpha \in V^* \mid \alpha(\mathbf{w}) = 0 \text{ for all } \mathbf{w} \in W\}.$$
Show that W^0 is a subspace of V^*.

(b) Show that $\Psi(W^\omega) = W^0$.

(c) Show that $\Psi(W) = (W^\omega)^0$.

2.37. Provide the details for the proof of Theorem 2.10.24. In particular:

(a) Show that the set $\mathcal{B}''$ is a basis for V.

(b) Show that $\mathbf{u}_j^T \mathbf{v}_k = \mathbf{v}_j^T \mathbf{u}_k = 0$.

(c) Find scalars c_j and d_j such that for $\tilde{\mathbf{u}}_j = c_j \mathbf{u}_j$ and $\tilde{\mathbf{v}}_j = d_j \mathbf{v}_j$,
$$\omega(\tilde{\mathbf{u}}_j, \tilde{\mathbf{v}}_j) = 1 \quad \text{and} \quad G(\tilde{\mathbf{u}}_j, \tilde{\mathbf{u}}_j) = G(\tilde{\mathbf{v}}_j, \tilde{\mathbf{v}}_j)$$
for all $j = 1, \ldots, n$.

2.38. Verify directly that the matrix
$$A = \begin{bmatrix} 1 & -1 & 1 & 0 \\ 1 & -1 & 0 & 1 \\ 1 & -1 & 0 & 0 \\ 2 & -1 & 1 & -1 \end{bmatrix}$$
is a symplectic matrix, i.e., that $A^T J A = J$.

2.39. Let $\mathcal{B} = \{\mathbf{a}_1, \mathbf{b}_1, \ldots, \mathbf{a}_n, \mathbf{b}_n\}$ be a symplectic basis for the standard symplectic space $(\mathbf{R}^{2n}, \omega_0)$. Show that the matrix

$$A = \left[\, \mathbf{a}_1 \,\vdots\, \mathbf{b}_1 \,\vdots\, \cdots \,\vdots\, \mathbf{a}_n \,\vdots\, \mathbf{b}_n \,\right]$$

is a symplectic matrix.

2.40. Show that if $A \in \mathrm{Sp}(2n)$, then $A^{-1} \in \mathrm{Sp}(2n)$.

2.41. Show that if $A \in \mathrm{Sp}(2n)$, then $A^{-1} = -JA^TJ$.

Chapter 3
Advanced Calculus

This chapter primarily concerns nonlinear functions between Euclidean spaces. Even basic properties of such functions, such as whether they are one-to-one or onto, can be much more difficult to verify than the corresponding verification for linear functions. For one thing, the powerful machinery of matrix representation of linear transformations is not available.

The tradeoff is that the breadth of geometric objects that can be modeled by nonlinear functions increases dramatically. The world of lines and planes, for example, gives way to the more general world of curves and surfaces.

We consider differential calculus to be the study of nonlinear objects by means of "linearization." This is, in essence, the approach of differential geometry.

From this point of view, we begin this chapter with a definition of the derivative as a linear transformation $Df : \mathbf{R}^n \to \mathbf{R}^m$ that arises from a nonlinear function $f : \mathbf{R}^n \to \mathbf{R}^m$. This definition will at first blur the distinct notions in Euclidean space of geometric "points" and algebraic "vectors."

One of the main purposes of this chapter, though, is to establish the distinction between points and vectors by associating to each (geometric) point in $\mathbf{R}^n$ a vector space of "tangent vectors" at that point. In fact, we present this vector space from two alternative but equivalent points of view, one geometric and the other analytic.

Once this has been accomplished, we revisit the definition of the derivative in its proper setting, as a linear transformation between tangent spaces.

Finally, we conclude the chapter with some of the geometric questions that arise immediately from this presentation of calculus, especially the problem of "integrating" vector fields.

As will become clear, this outline encompasses two somewhat distinct goals of the chapter. On the one hand, the first differential-geometric problems that come from calculus are presented with an eye toward technique. On the other hand, the core of this chapter—especially the first few sections—has a "point of view" character to it, more foundational than practical. The reader who is able to grasp this point of view will have made the conceptual transition necessary toward the more abstract setting of manifolds.

A. McInerney, *First Steps in Differential Geometry: Riemannian, Contact, Symplectic*, Undergraduate Texts in Mathematics, DOI 10.1007/978-1-4614-7732-7_3,

This having been said, there are two areas in which we deliberately gloss over important foundational questions as "technical details." First, we deliberately obscure topological concepts such as open neighborhoods and questions of continuity—clearly a limitation in view of the local nature of calculus. For example, we will talk about properties that hold "near p" as opposed to the more precise "in an open neighborhood containing p."

Second, we avoid the question of "how differentiable" a function is, for example, whether its derivative is continuous, whether the derivative itself is differentiable, etc. The technical definition of a C^r function, which is r times differentiable and whose rth derivative is continuous, is an important one in a formal presentation of advanced calculus or differential geometry, but would detract from the main line of presentation here. We adopt the convention of assuming that all functions have "enough derivatives" for all questions under consideration, or are "infinitely differentiable" with continuous derivatives of all orders. We will adopt the terminology of loosely referring to such functions as *smooth*. We caution the reader that this convention blurs precise hypotheses of several key theorems. For example, our "smooth" hypothesis of the implicit function theorem (p. 88) actually means "continuously differentiable." We have made an effort to insert cautions, along with references to precise statements at appropriate places.

To conclude this introduction, we note that the term "advanced calculus" has many senses not encompassed by this chapter. Traditionally, a course in advanced calculus is aimed in one of two directions. In one, it may serve as a prelude to a course in analysis, emphasizing foundational theorems concerning limits and continuity in Euclidean space. We will not follow that approach here, although we will assume such results throughout. The reader may refer to [30] or [37] for a collection of the relevant theorems and proofs.

In the other, more in keeping with the goals of this text, advanced calculus is seen as an introduction to differential geometry. Most of the texts with this approach, however, are structured around motivating and presenting the definition of a manifold. We will not make this leap, despite the fact that it is only a short step away once we have developed in detail the concept of a tangent space, which is the main goal of this chapter.

3.1 The Derivative and Linear Approximation

In a typical first-year calculus course, the derivative is presented in its historical setting as the slope of the tangent line to a curve or as the instantaneous velocity of a moving particle. This gives rise to the definition of the derivative of a real-valued function $f : \mathbf{R} \to \mathbf{R}$ at a point $a \in \mathbf{R}$ in the domain of the function:

$$Df(a) = f'(a) = \lim_{x \to a} \frac{f(x) - f(a)}{x - a} = \lim_{h \to 0} \frac{f(a+h) - f(a)}{h}.$$

A closely related concept, the *linear approximation to f at a*,

$$L_{f,a}(x) = f(a) + f'(a)(x - a),$$

is typically presented only in passing as an application of the derivative. It is this latter notion, the derivative as the linear function that "best approximates" f near a, that offers the path to generalizing the definition of the derivative to a multivariable setting.

In this section, we consider the Euclidean spaces $\mathbf{R}^n$ equipped with their standard vector space structure along with the standard inner product. In particular, we will be most concerned with the magnitude (or norm) of $\mathbf{x} = (x_1, \dots, x_n) \in \mathbf{R}^n$ given by

$$||\mathbf{x}|| = \sqrt{x_1^2 + \cdots + x_n^2};$$

see Sect. 2.9.

For a function $f : \mathbf{R}^n \to \mathbf{R}^m$ and $\mathbf{x} \in \mathbf{R}^n$, we write

$$f(\mathbf{x}) = (f^1(\mathbf{x}), \dots, f^m(\mathbf{x})),$$

where the real-valued functions $f^i : \mathbf{R}^n \to \mathbf{R}$ are called the *component functions* of f.

Definition 3.1.1. A function $f : \mathbf{R}^n \to \mathbf{R}^m$ is *differentiable at* $\mathbf{a} \in \mathbf{R}^n$ if there is a linear transformation $T_{\mathbf{a}} : \mathbf{R}^n \to \mathbf{R}^m$ such that

$$\lim_{||\mathbf{h}|| \to 0} \frac{||f(\mathbf{a} + \mathbf{h}) - f(\mathbf{a}) - T_{\mathbf{a}}(\mathbf{h})||}{||\mathbf{h}||} = 0.$$

The linear transformation $T_{\mathbf{a}}$ is called the *derivative of f at* $\mathbf{a}$, and we adopt the customary notation $Df(\mathbf{a}) = T_{\mathbf{a}}$.

Note that the norms that appear in the numerator and the denominator are computed in different vector spaces, in $\mathbf{R}^m$ and $\mathbf{R}^n$ respectively.

Also note that in the case of a single-variable function $f : \mathbf{R} \to \mathbf{R}$, Definition 3.1.1 is essentially the standard definition of the derivative, keeping in mind that $\lim_{x \to a} f(x) = 0$ exactly when $\lim_{x \to a} |f(x)| = 0$. In that case, the linear transformation $T_a : \mathbf{R} \to \mathbf{R}$ is given by $T_a(h) = f'(a) \cdot h$.

Definition 3.1.1 prompts the following definition.

Definition 3.1.2. Let $f : \mathbf{R}^n \to \mathbf{R}^m$ be a function that is differentiable at $\mathbf{a} \in \mathbf{R}^n$. The *linearization of f at* $\mathbf{a}$ is the function $L_{f,\mathbf{a}} : \mathbf{R}^n \to \mathbf{R}^m$ defined by

$$L_{f,\mathbf{a}}(\mathbf{x}) = f(\mathbf{a}) + Df(\mathbf{a})(\mathbf{x} - \mathbf{a}).$$

It is a consequence of the basic properties of limits that if a function is differentiable at $\mathbf{a}$, then the derivative at $\mathbf{a}$ is unique.

We now illustrate Definition 3.1.1 with several basic but fundamental examples.

Example 3.1.3. Let $f : \mathbf{R}^n \to \mathbf{R}^m$ be a constant function: $f(\mathbf{x}) = \mathbf{b}$ for all $\mathbf{x} \in \mathbf{R}^n$. Then f is differentiable for all $\mathbf{a} \in \mathbf{R}^n$ with

$$Df(\mathbf{a}) = \mathbf{O},$$

where $\mathbf{O} : \mathbf{R}^n \to \mathbf{R}^m$ is the zero transformation. We have $f(\mathbf{a}+\mathbf{h}) - f(\mathbf{a}) = \mathbf{b} - \mathbf{b} = \mathbf{0}$ for all $\mathbf{h}$, and so

$$\lim_{||\mathbf{h}||\to 0} \frac{||f(\mathbf{a}+\mathbf{h}) - f(\mathbf{a}) - \mathbf{O}(\mathbf{h})||}{||\mathbf{h}||} = 0.$$

Example 3.1.4. Let $f : \mathbf{R}^n \to \mathbf{R}^m$ be a linear function, $f(\mathbf{x}) = A\mathbf{x}$, where A is an $m \times n$ matrix and $\mathbf{x}$ is considered an $n \times 1$ column vector. Then f is differentiable for all $\mathbf{a} \in \mathbf{R}^n$, and $Df(\mathbf{a})(\mathbf{h}) = A(\mathbf{h})$. This follows from the fact that

$$f(\mathbf{a}+\mathbf{h}) - f(\mathbf{a}) = A(\mathbf{a}+\mathbf{h}) - A(\mathbf{a}) = A(\mathbf{a}) + A(\mathbf{h}) - A(\mathbf{a}) = A(\mathbf{h})$$

by the linearity of A, and so again,

$$\lim_{||\mathbf{h}||\to 0} \frac{||f(\mathbf{a}+\mathbf{h}) - f(\mathbf{a}) - A(\mathbf{h})||}{||\mathbf{h}||} = 0.$$

Example 3.1.5. Let $\mu : \mathbf{R}^2 \to \mathbf{R}$ be defined by $\mu(x,y) = xy$. Then μ is differentiable for all $\mathbf{a} = (a_1, a_2) \in \mathbf{R}^2$, and $D\mu(a_1,a_2)(h_1,h_2) = a_1h_2 + a_2h_1$: For every $\mathbf{h} = (h_1, h_2)$,

$$\begin{aligned}
&\lim_{||\mathbf{h}||\to 0} \frac{||\mu(\mathbf{a}+\mathbf{h}) - \mu(\mathbf{a}) - D\mu(\mathbf{a})(\mathbf{h})||}{||\mathbf{h}||} \\
&\qquad = \lim_{||\mathbf{h}||\to 0} \frac{|(a_1+h_1)(a_2+h_2) - a_1a_2 - (a_1h_2 + a_2h_1)|}{\sqrt{h_1^2+h_2^2}} \\
&\qquad\qquad = \lim_{||\mathbf{h}||\to 0} \frac{|h_1h_2|}{\sqrt{h_1^2+h_2^2}} = 0.
\end{aligned}$$

In Exercise 3.4, the reader is asked to prove the last equality.

The derivative defined as a linear transformation enjoys a number of familiar properties, which we list here. Detailed proofs of these theorems, most of which involve techniques of analysis relying on the definition of the derivative as a limit, can be found in any text in advanced calculus emphasizing elementary real analysis; see, for example, [37]. As with many such arguments, they rely on properties of the norm, and in particular on the triangle inequality:

$$||\mathbf{a}+\mathbf{b}|| \le ||\mathbf{a}|| + ||\mathbf{b}||. \tag{3.1}$$

It is worth mentioning that many of these proofs rely in an essential way on the following:

Proposition 3.1.6. *Let $T : \mathbf{R}^n \to \mathbf{R}^m$ be a linear transformation. Then T is bounded, i.e., there is a constant $M > 0$ such that for all $\mathbf{h} \in \mathbf{R}^n$, $||T(\mathbf{h})|| \leq M||\mathbf{h}||$.*

In fact, both Eq. (3.1) and Proposition 3.1.6 can be seen to be consequences of Theorem 2.9.6, the Cauchy–Schwarz inequality.

Theorem 3.1.7 (Chain rule). *Let $f : \mathbf{R}^n \to \mathbf{R}^m$ be differentiable at $\mathbf{a} \in \mathbf{R}^n$ and let $g : \mathbf{R}^m \to \mathbf{R}^k$ be differentiable at $\mathbf{b} = f(\mathbf{a}) \in \mathbf{R}^m$. Then $g \circ f : \mathbf{R}^n \to \mathbf{R}^k$ is differentiable at $\mathbf{a}$, with*

$$D(g \circ f)(\mathbf{a}) = Dg(\mathbf{b}) \circ Df(\mathbf{a}).$$

Note that the composition on the right side of the equation in Theorem 3.1.7 is a composition of *linear* maps. If these linear maps are written as matrices, the composition corresponds to matrix multiplication.

Proof. The goal is to show that under the hypotheses of differentiability of f and g at $\mathbf{a}$ and $\mathbf{b}$ respectively,

$$\lim_{||\mathbf{h}|| \to 0} \frac{||(g \circ f)(\mathbf{a}+\mathbf{h}) - (g \circ f)(\mathbf{a}) - (Dg(\mathbf{b}) \circ Df(\mathbf{a}))(\mathbf{h})||}{||\mathbf{h}||} = 0.$$

Introducing extra terms and applying the triangle inequality implies that the numerator satisfies

$$\begin{aligned} &||(g \circ f)(\mathbf{a}+\mathbf{h}) - (g \circ f)(\mathbf{a}) - Dg(\mathbf{b})\,[f(\mathbf{a}+\mathbf{h}) - f(\mathbf{a})] \\ &\qquad + Dg(\mathbf{b})\,[f(\mathbf{a}+\mathbf{h}) - f(\mathbf{a}) - Df(\mathbf{a})(\mathbf{h})]\,|| \\ &\quad \leq ||(g \circ f)(\mathbf{a}+\mathbf{h}) - (g \circ f)(\mathbf{a}) - Dg(\mathbf{b})\,[f(\mathbf{a}+\mathbf{h}) - f(\mathbf{a})]\,|| \\ &\qquad + ||Dg(\mathbf{b})\,[f(\mathbf{a}+\mathbf{h}) - f(\mathbf{a}) - Df(\mathbf{a})(\mathbf{h})]\,||. \end{aligned}$$

The differentiability assumptions and Proposition 3.1.6 then show that we have both

$$\lim_{||\mathbf{h}|| \to 0} \frac{||(g \circ f)(\mathbf{a}+\mathbf{h}) - (g \circ f)(\mathbf{a}) - Dg(\mathbf{b})\,[f(\mathbf{a}+\mathbf{h}) - f(\mathbf{a})]\,||}{||\mathbf{h}||} = 0$$

and

$$\lim_{||\mathbf{h}|| \to 0} \frac{||Dg(\mathbf{b})\big[f(\mathbf{a}+\mathbf{h}) - f(\mathbf{a}) - Df(\mathbf{a})(\mathbf{h})\big]\,||}{||\mathbf{h}||} = 0,$$

which together imply the result. See [30] or [37] for details. □

Theorem 3.1.8. *Let $f : \mathbf{R}^n \to \mathbf{R}^m$ have component functions $f^i : \mathbf{R}^n \to \mathbf{R}$, i.e., $f(\mathbf{x}) = \big(f^1(\mathbf{x}), f^2(\mathbf{x}), \dots, f^m(\mathbf{x})\big)$. Then f is differentiable at $\mathbf{a} \in \mathbf{R}^n$ if and only if the component functions f^i are differentiable at $\mathbf{a}$. Moreover, using the standard bases for $\mathbf{R}^n$ and $\mathbf{R}^m$, the matrix representation for $Df(\mathbf{a})$ is given by*

$$\big[Df(\mathbf{a})\big] = \begin{bmatrix} Df^1(\mathbf{a}) \\ \vdots \\ Df^m(\mathbf{a}) \end{bmatrix},$$

considering the linear maps $Df^i(\mathbf{a}) : \mathbf{R}^n \to \mathbf{R}$ as row vectors.

Proof. Assuming that the component functions f^i are differentiable at $\mathbf{a}$ with derivatives $Df^i(\mathbf{a})$, the differentiability of f follows from applying the inequality $\sqrt{x_1^2 + \dots + x_n^2} \le \sum |x_i|$ to the numerator in the definition of the derivative.

The converse follows from applying the chain rule to the functions $f^i = \pi^i \circ f$, where, for $i = 1, \dots, n$, $\pi^i : \mathbf{R}^n \to \mathbf{R}$ given by $\pi^i(x_1, \dots, x_n) = x_i$ is the linear transformation describing projection onto the ith coordinate (whose derivative exists and is given by Example 3.1.4). □

We will return to this "componentwise" differentiation after establishing several familiar properties of the derivative.

Theorem 3.1.9. *Let $f, g : \mathbf{R}^n \to \mathbf{R}^m$ be differentiable functions at $\mathbf{a} \in \mathbf{R}^n$. Let $c \in \mathbf{R}$. Then:*

1. *$f + g$ is differentiable at $\mathbf{a}$, with $D(f+g)(\mathbf{a}) = Df(\mathbf{a}) + Dg(\mathbf{a})$.*
2. *cf is differentiable at $\mathbf{a}$, with $D(cf)(\mathbf{a}) = cDf(\mathbf{a})$.*

Proof. Both statements follow immediately from the corresponding properties of limits. □

Theorem 3.1.10. *Let $f, g : \mathbf{R}^n \to \mathbf{R}$ be real-valued functions defined on $\mathbf{R}^n$ both of which are differentiable at $\mathbf{a} \in \mathbf{R}^n$. Then*

1. *$f \cdot g$ is differentiable at $\mathbf{a}$, with*

$$D(f \cdot g)(\mathbf{a}) = g(\mathbf{a}) \cdot Df(\mathbf{a}) + f(\mathbf{a}) \cdot Dg(\mathbf{a}).$$

2. *If $g(\mathbf{a}) \neq 0$, then f/g is differentiable at $\mathbf{a}$, with*

$$D(f/g)(\mathbf{a}) = \frac{g(\mathbf{a}) \cdot Df(\mathbf{a}) - f(\mathbf{a}) \cdot Dg(\mathbf{a})}{[g(\mathbf{a})]^2}.$$

Proof. Both statements can be proved directly from the definition. More elegantly, however, the product $p(\mathbf{a}) = f(\mathbf{a}) \cdot g(\mathbf{a})$ can be expressed as the composition $p = \mu \circ F$, where $F : \mathbf{R}^n \to \mathbf{R}^2$ is given by $F(\mathbf{x}) = (f(\mathbf{x}), g(\mathbf{x}))$ and $\mu : \mathbf{R}^2 \to \mathbf{R}$ is given by $\mu(x, y) = xy$. After computing the derivatives of each of those functions separately (see Example 3.1.5), the statement follows immediately from the chain rule.

Similarly, the quotient $q(\mathbf{a}) = f(\mathbf{a})/g(\mathbf{a})$ can be written as the product $q = f \cdot (1/g)$, and $D(1/g)(\mathbf{a}) = \left(-1/(g(\mathbf{a}))^2\right) Dg(\mathbf{a})$. □

The property known as the product rule is sometimes also called *Leibniz's rule*.

We conclude this section on a practical note. Recall the familiar notion of the partial derivative: For a real-valued function $f : \mathbf{R}^n \to \mathbf{R}$ and for $\mathbf{a} = (a_1, \ldots, a_n) \in \mathbf{R}^n$, the *partial derivative* of f at $\mathbf{a}$ with respect to the ith component is given by

$$D_i f(\mathbf{a}) = \frac{\partial f}{\partial x_i}(\mathbf{a}) = \lim_{h \to 0} \frac{f(a_1, \ldots, a_i + h, \ldots, a_n) - f(\mathbf{a})}{h}.$$

Note that in terms of the development of the derivative as a linear map, we have

$$D_i f(\mathbf{a}) = \big(Df(\mathbf{a})\big)(\mathbf{e}_i),$$

where $\mathbf{e}_i = (0, \ldots, 0, 1, 0, \ldots, 0)$, with the 1 in the ith place. This follows from Definition 3.1.1 by replacing $\mathbf{h}$ with $h\mathbf{e}_i$. In this way, we can write the linear map $Df(\mathbf{a}) : \mathbf{R}^n \to \mathbf{R}$ as a $1 \times n$ row vector using the standard bases, according to the method of Sect. 2.6:

$$\begin{aligned} [Df(\mathbf{a})] &= \big[Df(\mathbf{a})(\mathbf{e}_1) \cdots Df(\mathbf{a})(\mathbf{e}_n)\big] \\ &= \big[D_1 f(\mathbf{a}) \cdots D_n f(\mathbf{a})\big]. \end{aligned}$$

Combining this observation with Theorem 3.1.8 yields the following result:

Theorem 3.1.11. *Let $f : \mathbf{R}^n \to \mathbf{R}^m$ be a function with components $f^j : \mathbf{R}^n \to \mathbf{R}$, i.e., $f(\mathbf{x}) = (f^1(\mathbf{x}), \ldots, f^m(\mathbf{x}))$. If f is differentiable at $\mathbf{a} \in \mathbf{R}^n$, then each of the partial derivatives $D_i f^j(\mathbf{a})$ exists, and the matrix form of the linear transformation $Df(\mathbf{a})$ is given by*

$$\big[Df(\mathbf{a})\big] = \big[D_i f^j(\mathbf{a})\big],$$

for $i = 1, \ldots, n$ and $j = 1, \ldots, m$.

The matrix described in the conclusion of Theorem 3.1.11 is known as the *Jacobian matrix* of the function f.

Example 3.1.12. Let $f : \mathbf{R}^3 \to \mathbf{R}^2$ be given by $f(x, y, z) = (x^2 + y^2, xyz)$. Each component function, as a polynomial, is differentiable for all $\mathbf{a} = (a, b, c) \in \mathbf{R}^3$. Hence by Theorem 3.1.8, f is differentiable for all $\mathbf{a}$, and so by Theorem 3.1.11,

$$\big[Df(\mathbf{a})\big] = \begin{bmatrix} \partial f^1/\partial x(\mathbf{a}) & \partial f^1/\partial y(\mathbf{a}) & \partial f^1/\partial z(\mathbf{a}) \\ \partial f^2/\partial x(\mathbf{a}) & \partial f^2/\partial y(\mathbf{a}) & \partial f^2/\partial z(\mathbf{a}) \end{bmatrix}$$

$$= \begin{bmatrix} 2a & 2b & 0 \\ bc & ac & ab \end{bmatrix}.$$

For example,

$$[Df(1,-2,3)] = \begin{bmatrix} 2 & -4 & 0 \\ -6 & 3 & -2 \end{bmatrix},$$

or, what is the same,

$$(Df(1,-2,3))(h_1,h_2,h_3) = (2h_1 - 4h_2, -6h_1 + 3h_2 - 2h_3).$$

The linearization of f near $\mathbf{a} = (1,-2,3)$ is given, according to Definition 3.1.2, by

$$\begin{aligned} L_{f,\mathbf{a}}(x,y,z) &= f(1,-2,3) + Df(1,-2,3)(x-1, y+2, z-3) \\ &= (5,-6) \\ &\quad + (2(x-1) - 4(y+2), -6(x-1) + 3(y+2) - 2(z-3)) \\ &= (2x - 4y - 5, -6x + 3y - 2z + 12). \end{aligned}$$

Choosing for example $\mathbf{x} = (0.9, -2.1, 3.2)$, we have $f(\mathbf{x}) = (5.22, -6.048)$, while $L_{f,\mathbf{a}}(\mathbf{x}) = (5.2, -6.1)$. The linear approximation here gives an accurate estimate of the function value.

It is a standard observation in multivariable calculus that the converse of Theorem 3.1.11 is not true without an additional hypothesis. The relevant theorem is this:

Theorem 3.1.13. *Let $f : \mathbf{R}^n \to \mathbf{R}^m$ be a function with component functions $f^j : \mathbf{R}^n \to \mathbf{R}$. Suppose that for all $i = 1,\dots,n$ and $j = 1,\dots,m$, the partial derivatives $D_i f^j$ are defined at all points near $\mathbf{a} \in \mathbf{R}^n$, and in addition are continuous at $\mathbf{a}$. Then f is differentiable at $\mathbf{a}$, and*

$$\left[Df(\mathbf{a})\right] = \left[D_i f^j(\mathbf{a})\right].$$

A function f that satisfies the hypotheses of Theorem 3.1.13 is said to be *continuously differentiable at* $\mathbf{a}$.

3.2 The Tangent Space I: A Geometric Definition

The goal of the previous section was to show how the tools of calculus can be used to study nonlinear functions by means of linear approximation. Indeed, a nonlinear function and its linearization agree "to the first order" at every point where the derivative is defined.

In emphasizing the construction of the derivative $Df(\mathbf{a})$ of a function $f : \mathbf{R}^n \to \mathbf{R}^m$ at a point $\mathbf{a} \in \mathbf{R}^n$, however, we have glossed over a more foundational question: How do we envision the domain and the range of the linear transformation $Df(\mathbf{a})$? Answering this question is one of the main conceptual chasms in passing from calculus to differential geometry. The next sections offer different but equivalent bridges to cross this chasm.

The starting point for these discussions is a divide with deep roots in the history of mathematics: the divide between geometry and algebra. This manifests itself in two different "views" of sets. In geometry, we consider elements of sets to be "points," along with various subsets that represent curves, surfaces, or their higher-dimensional analogues. The sets themselves are referred to as "spaces." For us, this can be considered the setting for nonlinear objects. We can also consider functions defined on these geometric spaces.

In algebra, however, we consider elements of sets to be objects with which we can perform operations. In our context, the algebraic objects will be vectors, and so the operations will be vector addition and scalar multiplication. This will allow us the use of the techniques and theorems of Chap. 2.

The process of "linearization" or linear approximation that we began in the first section will now be extended to the sets in question. In differential geometry, the key concept for carrying out that process is the tangent space. If the derivative is the key to linearizing a nonlinear function, the tangent space can be considered a "linear approximation" to the geometric sets under consideration.

More precisely, we will associate an (algebraic) vector space to each point in the (geometric) space. Roughly speaking, at each point p in space, the tangent space will consist of the set of possible "infinitesimal displacements" that a particle "moving through" the set at p might travel. It is, to use a somewhat archaic concept from physics, the set of "virtual displacements."

We will be working with the set $\mathbf{R}^n$, which we consider now as geometric "space." Elements $p \in \mathbf{R}^n$ will be viewed as "points" in space. As mentioned above, we abandon any algebraic characterizations of $\mathbf{R}^n$ here, and in particular we set aside all of the vector space characterization of $\mathbf{R}^n$ that we developed in the first chapter. We retain only notions such as "adding points" and "multiplying a point by a scalar" to the extent that such operations are required to discuss limits and continuity—most of which will be in the background of the presentation here.

In short, our first step is to reintroduce the conceptual wall between geometry and algebra that Descartes and the school of analytic geometry have so thoroughly eroded over centuries. A disadvantage to beginning our presentation in $\mathbf{R}^n$ is that this is where the distinction between geometry and algebra is least visible. We ourselves blurred the distinction in the last section. In this most basic case, some of the constructions of this section may seem artificial at first. However, they will be better understood in the next section, on geometric sets.

Let us outline in brief the objectives we have in mind to define the tangent space.

At each point $p \in \mathbf{R}^n$, we will associate a vector space, which we will denote by $T_p(\mathbf{R}^n)$. Elements of $T_p(\mathbf{R}^n)$ will be called *tangent vectors at* p and will be denoted temporarily using vector notation with a subscript denoting the point to which they are associated, $\mathbf{v}_p \in T_p(\mathbf{R}^n)$.

Once we have defined the tangent vectors $\mathbf{v}_p$ themselves, we will define vector addition $\mathbf{v}_p + \mathbf{w}_p \in T_p(\mathbf{R}^n)$ and scalar multiplication $s\mathbf{v}_p \in T_p(\mathbf{R}^n)$. We note from the outset that we will not perform operations on tangent vectors associated to different points: The expression $\mathbf{v}_p + \mathbf{w}_q$ is meaningless when $p \neq q$.

Our first effort in this direction will begin with a familiar object from vector calculus. Let $I \subset \mathbf{R}$ be an open interval containing 0. A *parameterized curve* is a smooth function $c : I \to \mathbf{R}^n$. We will write $c(t) = \big(x_1(t), \dots, x_n(t)\big)$, with smooth component functions $x_1, \dots, x_n : I \to \mathbf{R}$. Note in particular that a parameterized curve refers to the function c, and not to its image $c(I)$, which is a set of points in $\mathbf{R}^n$.

For a parameterized curve $c : I \to \mathbf{R}^n$, let $p = c(0)$. We temporarily adopt the notation

$$\mathbf{v}_p = c'(0) = \big\langle x_1'(0), \dots, x_n'(0)\big\rangle_p,$$

where the angle brackets are meant to emphasize the conceptual difference between the *point* $p = c(0)$ with coordinates $c(0) = \big(x_1(0), \dots, x_n(0)\big)$ and the *vector* $c'(0) = \big\langle x_1'(0), \dots, x_n'(0)\big\rangle_p$.

The notation is motivated by the following:

Definition 3.2.1. Let $p \in \mathbf{R}^n$. A *tangent vector at* p is an ordered n-tuple of real numbers $\mathbf{v}_p = \langle a_1, \dots, a_n\rangle_p$ such that there exists a smooth parameterized curve $c : I \to \mathbf{R}^n$ having the properties that $c(0) = p$ and that

$$c'(0) = \mathbf{v}_p = \langle a_1, \dots, a_n\rangle_p .$$

There are several notable features to this definition. First, motivated by the geometric picture from vector calculus of the vector $c'(0)$ as tangent to the image curve $c(I)$ at $c(0)$ (see Fig. 3.1), we have defined a tangent vector as tangent *to something*. Second, this definition generalizes well to more general settings than the Euclidean spaces $\mathbf{R}^n$, provided of course that there is sufficient structure to talk about the derivative. We will begin to see this in the next section, on geometric sets.

Definition 3.2.2. For every $p \in \mathbf{R}^n$, the set of all tangent vectors at p, denoted by $T_p(\mathbf{R}^n)$, is called the *tangent space* to $\mathbf{R}^n$ at p.

The following theorem shows that in the case of $\mathbf{R}^n$, the property that defines a tangent vector actually imposes no restriction on n-tuples. In other words, for every point $p \in \mathbf{R}^n$, every n-tuple can be viewed as a tangent vector in $\mathbf{R}^n$ at p. This will not be true in more general settings.

Theorem 3.2.3. *For all* $p \in \mathbf{R}^n$, *every* n*-tuple* $\mathbf{v} = \langle a_1, \dots, a_n\rangle$ *can be regarded as a tangent vector at* p.

Proof. For $p = (p_1, \dots, p_n)$, define $c : \mathbf{R} \to \mathbf{R}^n$ by

$$c(t) = \big(p_1 + a_1 t, \dots, p_n + a_n t\big).$$

Then $c(0) = p$ and $c'(0) = \mathbf{v}$, so $\mathbf{v} = \mathbf{v}_p \in T_p(\mathbf{R}^n)$. □

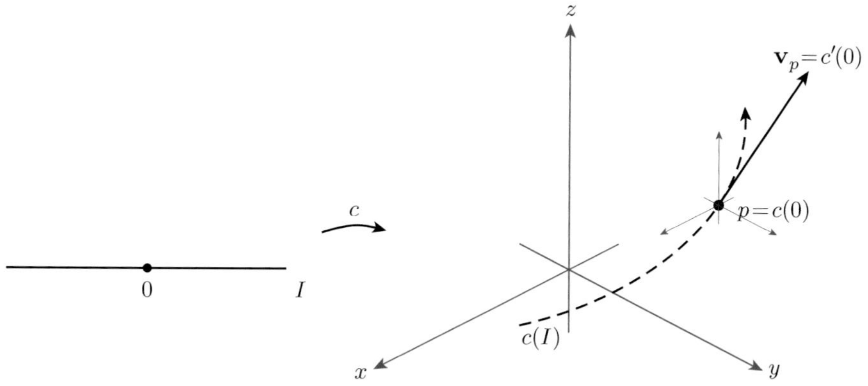

Fig. 3.1 The geometric definition of a tangent vector in $\mathbf{R}^3$.

One can think of $T_p(\mathbf{R}^n)$ as a copy of the vector space $\mathbf{R}^n$ "attached to" the point p. Said differently, each point p in $\mathbf{R}^n$ has a set of tangent vectors associated to it, where each tangent vector can be visualized as an arrow or directed line segment in $\mathbf{R}^n$ "based at" p.

Theorem 3.2.4. *Let $p \in \mathbf{R}^n$. The tangent space $T_p(\mathbf{R}^n)$, equipped with the operations of componentwise addition and scalar multiplication, is a vector space.*

Proof. Keeping in mind Theorem 3.2.3, the statement follows immediately from the fact that $\mathbf{R}^n$ with its usual operations is a vector space. However, in order to prepare for the general setting of the next section, we present a more axiomatic verification of the closure and existence axioms.

For every $\mathbf{v}_p, \mathbf{w}_p \in T_p(\mathbf{R}^n)$, we first aim to show that $\mathbf{v}_p + \mathbf{w}_p \in T_p(\mathbf{R}^n)$. To do so, we will produce a parameterized curve $c : I \to \mathbf{R}^n$ with the property that $c(0) = p$ and $c'(0) = \mathbf{v}_p + \mathbf{w}_p$.

Assuming $\mathbf{v}_p \in T_p(\mathbf{R}^n)$, there is a parameterized curve $c_1 : I_1 \to \mathbf{R}^n$ satisfying $c_1(0) = p$ and $c_1'(0) = \mathbf{v}_p$. Likewise, we assume that there is $c_2 : I_2 \to \mathbf{R}^n$ with $c_2(0) = p$ and $c_2'(0) = \mathbf{w}_p$. We write $c_1(t) = (x_1(t), \ldots, x_n(t))$, $c_2(t) = (y_1(t), \ldots, y_n(t))$, and $p = (p_1, \ldots, p_n) = c_1(0) = c_2(0)$.

Now define $c : I \to \mathbf{R}^n$, where $I = I_1 \cap I_2$, by

$$c(t) = (x_1(t) + y_1(t) - p_1, \ldots, x_n(t) + y_n(t) - p_n).$$

The reader may verify that $c(0) = p$ and that

$$c'(0) = \langle x_1'(0) + y_1'(0), \ldots, x_n'(0) + y_n'(0) \rangle_p = \mathbf{v}_p + \mathbf{w}_p,$$

so $\mathbf{v}_p + \mathbf{w}_p \in T_p(\mathbf{R}^n)$.

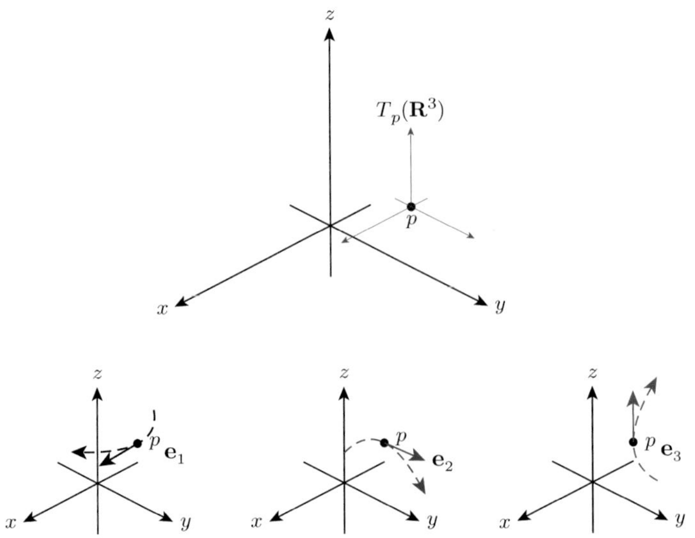

Fig. 3.2 The standard basis for $T_p(\mathbf{R}^3)$ (geometric definition).

In order to show that $T_p(\mathbf{R}^n)$ is closed under scalar multiplication, we adopt the notation that if $I = (a, b) \subset \mathbf{R}$ is an interval, then for $s > 0$, $sI = (sa, sb)$. On the other hand, for $s < 0$, we write $sI = (sb, sa)$. In particular, $-I = (-b, -a)$.

Suppose $\mathbf{v}_p \in T_p(\mathbf{R}^n)$, so that there is $c_1 : I_1 \to \mathbf{R}^n$ with $c_1(0) = p$ and $c_1'(0) = \mathbf{v}_p$. For $s \in \mathbf{R}$, define $c : I \to \mathbf{R}^n$, where $I = (1/s)I_1$ ($I = \mathbf{R}$ if $s = 0$) and $c(t) = c_1(st)$. Then $c(0) = c_1(0) = p$ and $c'(0) = sc_1'(0) = s\mathbf{v}_p$ by the chain rule. So $s\mathbf{v}_p \in T_p(\mathbf{R}^n)$.

To verify the existence of the zero vector in $T_p(\mathbf{R}^n)$, we observe that $\mathbf{0}_p = \langle 0, \ldots, 0 \rangle_p \in T_p(\mathbf{R}^n)$, since $c_0 : \mathbf{R} \to \mathbf{R}^n$ defined by $c_0(t) = p$ for all t satisfies $c_0(0) = p$ and $c_0'(0) = \mathbf{0}_p$.

Finally, we show the existence of additive inverses. For $\mathbf{v}_p \in T_p(\mathbf{R}^n)$ with $c_1 : I_1 \to \mathbf{R}^n$, $c_1(0) = p$, $c_1'(0) = \mathbf{v}_p$, define $c : I \to \mathbf{R}^n$, where $I = -I_1$ and $c(t) = c_1(-t)$. Then $c(0) = c_1(0) = p$ and $c'(0) = -c_1'(0) = -\mathbf{v}_p$. Hence $-\mathbf{v}_p \in T_p(\mathbf{R}^n)$. □

Interpreted in the light of Theorem 3.2.4, Theorem 3.2.3 can be rephrased by saying that $T_p(\mathbf{R}^n)$ is *isomorphic to* the vector space $\mathbf{R}^n$. In particular, we have the following corollary:

Corollary 3.2.5. $\dim(T_p(\mathbf{R}^n)) = n$.

Proof. We take advantage of this corollary to call the reader's attention, among all the possible bases for $T_p(\mathbf{R}^n)$, to what we will call the *standard basis* for $T_p(\mathbf{R}^n)$:

$$(B_0)_p = \{(\mathbf{e}_1)_p, \ldots, (\mathbf{e}_n)_p\},$$

where $(\mathbf{e}_1)_p = \langle 1, 0, \ldots, 0 \rangle_p$, $(\mathbf{e}_2)_p = \langle 0, 1, 0, \ldots, 0 \rangle_p$, etc. See Fig. 3.2. □

3.3 Geometric Sets and Subspaces of $T_p(\mathbf{R}^n)$

The tangent vector has been defined in an essentially *local* way, depending on the geometric point p to which it is "attached." From the outset, we have no way of comparing—much less performing operations with—tangent vectors in different tangent spaces. The passage from the locally defined tangent space to more global geometric questions concerning relationships among different points and subsets of $\mathbf{R}^n$ requires substantially more effort.

As a first step in that direction, we illustrate the interplay of local and global by showing how certain "geometric sets" S of $\mathbf{R}^n$—roughly, those subsets for which we can define a sense of linear approximation—give rise to vector subspaces of the tangent spaces $T_p(\mathbf{R}^n)$ at each point $p \in S$. The relationship between these (algebraic) subspaces and the geometric set S is a major theme in differential geometry, to which we will return in later chapters.

We adopt the following loose definition:

> A *geometric set* $S \subset \mathbf{R}^n$ is a set having the property that for each point $p \in S$, there is a vector subspace $T_pS \subset T_p\mathbf{R}^n$. Moreover, these subspaces should vary smoothly with p and should all have the same dimension.

The goal of this section is to make this loose definition more precise.

The first example of a geometric set in $\mathbf{R}^n$ can be thought of as coming from generalizing aspects of the familiar parameterized curve. In particular, instead of considering differentiable functions $c : I \to \mathbf{R}^n$, where $I \subset \mathbf{R}^1$, we consider functions $\phi : U \to \mathbf{R}^n$, where $U \subset \mathbf{R}^k$ for $k \leq n$. However, we need to place restrictions on both U and ϕ.

Definition 3.3.1. Let $U \subset \mathbf{R}^k$ be a subset of $\mathbf{R}^k$. We will call U a *domain* if it has the topological properties of being a connected open set.

As mentioned in the introduction, we will not dwell on these topological restrictions. The reader who is not familiar with the terms in the definition can think of a domain as the interior of a small ball in $\mathbf{R}^k$.

Definition 3.3.2. Let $U \subset \mathbf{R}^k$ be a domain and let $\phi : U \to \mathbf{R}^n$ $(k \leq n)$ be a smooth, one-to-one function that is *regular*: $D\phi(p)$ must be a one-to-one linear map (and so must have constant, maximal rank k) for all $p \in U$. A *parameterized set* $S = \phi(U)$ is defined to be the image of U in $\mathbf{R}^n$ by ϕ.

Note that the existence of a regular parameterization implies the existence of n smooth, real-valued component functions $\phi^i : \mathbf{R}^k \to \mathbf{R}$,

$$\phi = (\phi^1, \dots, \phi^n).$$

The geometric features of the parameterized set S come from "encoding" features of the parameter space U through the function ϕ. We will discuss the condition of ϕ being regular below.

Example 3.3.3. Let $U = \mathbf{R}^2$ and let $\phi : U \to \mathbf{R}^3$ be given by

$$\phi(u, v) = (u, v, 2u - 3v).$$

Note that

$$D\phi(a, b) = \begin{bmatrix} 1 & 0 \\ 0 & 1 \\ 2 & -3 \end{bmatrix},$$

which has maximal rank 2 for each $(a, b) \in U$ and is hence, by Theorem 2.7.7, one-to-one. The parameterized set $S = \phi(U)$ is the plane through the origin described by the equation $2x - 3y - z = 0$.

Example 3.3.4. Let

$$U = \{(r, \theta) \mid 0 < r < 2, -\pi < \theta < \pi\} \subset \mathbf{R}^2$$

and let $\phi : U \to \mathbf{R}^3$ be given by

$$\phi(r, \theta) = (r \cos\theta, r \sin\theta, r^2).$$

We have

$$D\phi(r, \theta) = \begin{bmatrix} \cos\theta & -r\sin\theta \\ \sin\theta & r\cos\theta \\ 2r & 0 \end{bmatrix},$$

which has rank 2 (since $r > 0$) and so is one-to-one on U again by Theorem 2.7.7. The parameterized set $S = \phi(U)$ is the part of the paraboloid described by $z = x^2 + y^2$ between the planes $z = 0$ and $z = 4$, not including the segment intersecting the half-plane $y = 0, x < 0$ (corresponding to $\theta = \pi$). See Fig. 3.3.

As mentioned above, we would like a way to formalize the notion of "linearizing" the parameterized set $S = \phi(U)$. That is the essence of the following definition.

Definition 3.3.5. Let $S = \phi(U)$ be a parameterized set and let $p \in S$. The *tangent space to S at p*, denoted by $T_p(S)$, is the subset of $T_p(\mathbf{R}^n)$ defined by

$$T_p(S) = \{\mathbf{v}_p \in T_p(\mathbf{R}^n) \mid \exists c : I \to S \text{ smooth with } c(0) = p, \;\; c'(0) = \mathbf{v}_p\}.$$

Here, when we say that a parameterized curve $c : I \to S$ is smooth, we mean that there is a smooth function $\tilde{c} : I \to U$ such that $c = \phi \circ \tilde{c}$.

The key to this definition is that the codomain of the representative parameterized curve c is restricted to the set $S \subset \mathbf{R}^n$.

Theorem 3.3.6. *The set $T_p(S)$ as defined in Definition 3.3.5 is a vector subspace of $T_p(\mathbf{R}^n)$.*

Proof. Exercise. □

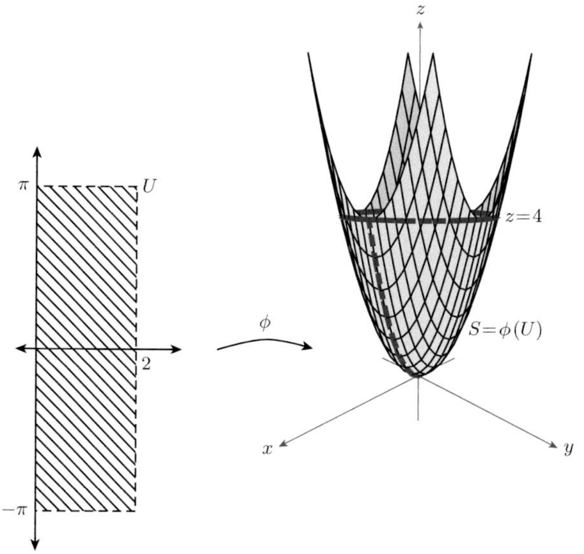

Fig. 3.3 A parameterized paraboloid in $\mathbf{R}^3$.

We illustrate Definition 3.3.5 in detail with a simple example.

Example 3.3.7. Let $S = \phi(U)$, where $U = \mathbf{R}^2$ and $\phi : U \to \mathbf{R}^3$ is given by $\phi(u,v) = (u, v, 2u - 3v)$; see Example 3.3.3. Suppose $c : I \to S$ is a smooth parameterized curve in S. Then there is a smooth function $\tilde{c} : I \to U$ such that $c = \phi \circ \tilde{c}$. In other words, for each $t \in I$, there are smooth functions $u, v : I \to \mathbf{R}$ such that

$$\tilde{c}(t) = (u(t), v(t)) \in \mathbf{R}^2$$

and such that $c(t) = \phi(u(t), v(t)) = (u(t), v(t), 2u(t) - 3v(t))$. That is, the component functions of a curve $c : I \to S$ given by $c(t) = (x(t), y(t), z(t))$ must satisfy the relation

$$2x(t) - 3y(t) - z(t) = 0 \text{ for all } t \in I \tag{3.2}$$

or, what is the same,

$$z(t) = 2x(t) - 3y(t) \text{ for all } t \in I. \tag{3.3}$$

Let $p = (x_0, y_0, z_0) \in S$, i.e., $2x_0 - 3y_0 - z_0 = 0$. To say that $\mathbf{v}_p = \langle a, b, c \rangle_p \in T_p(\mathbf{R}^3)$ means that there is $c : I \to \mathbf{R}^3$ satisfying $c(0) = p$ and $c'(0) = \langle a, b, c \rangle_p$.

To say further that $\mathbf{v}_p \in T_p(S)$ means we have the restriction that $c : I \to S \subset \mathbf{R}^3$, which, as shown above, means that Eq. (3.2) holds. Differentiation of Eq. (3.2) with respect to t shows that $2x'(t) - 3y'(t) - z'(t) = 0$ for all $t \in I$. In particular,

$$\begin{aligned}0 &= 2x'(0) - 3y'(0) - z'(0) \\ &= 2a - 3b - c.\end{aligned}$$

Moreover, if $\langle a, b, c\rangle_p \in T_p(\mathbf{R}^3)$ satisfies $2a - 3b - c = 0$, then the curve defined by $c(t) = (x_0 + at, y_0 + bt, z_0 + ct)$ is a curve $c : \mathbf{R} \to S$:

$$\begin{aligned}2(x_0 + at) - 3(y_0 + bt) - (z_0 + ct) &= (2x_0 - 3y_0 - z_0) + t(2a - 3b - c) \\ &= 0.\end{aligned}$$

This discussion shows that for all $p \in S$,

$$T_p(S) = \left\{\langle a, b, c\rangle_p \mid 2a - 3b - c = 0\right\} \subset T_p(\mathbf{R}^3).$$

We now illustrate the definition with a nonlinear example.

Example 3.3.8. Let S be the paraboloid of Example 3.3.4, i.e., $S = \phi(U)$, where $U = \{(r, \theta) \mid 0 < r < 2, \ -\pi < \theta < \pi\}$ and $\phi : U \to \mathbf{R}^3$ is given by

$$\phi(r, \theta) = (r \cos\theta, r \sin\theta, r^2).$$

We investigate the tangent space to S at the point $p = \phi(\sqrt{2}, \pi/4) = (1, 1, 2)$. We first consider two curves $\tilde{c}_i : I_i \to U$, $i = 1, 2$, where $I_1 = (-1/2, 1/2)$, $I_2 = (-\pi/2, \pi/2)$ given by

$$\tilde{c}_1(t) = \left(t + \sqrt{2}, \pi/4\right),$$

and

$$\tilde{c}_2(t) = \left(\sqrt{2}, t + \pi/4\right).$$

Geometrically, the image curves $\tilde{c}_1(I_1)$ and $\tilde{c}_2(I_2)$ are parallel to the coordinate axes in U and intersect at $(\sqrt{2}, \pi/4) = \tilde{c}_1(0) = \tilde{c}_2(0)$.

Now define $c_1 : I_1 \to S$ and $c_2 : I_2 \to S$ by $c_1 = \phi \circ \tilde{c}_1$ and $c_2 = \phi \circ \tilde{c}_2$. Explicitly,

$$\begin{aligned}c_1(t) &= \left((t + \sqrt{2}) \cos(\pi/4), (t + \sqrt{2}) \sin(\pi/4), (t + \sqrt{2})^2\right) \\ &= \left(\frac{t}{\sqrt{2}} + 1, \frac{t}{\sqrt{2}} + 1, (t + \sqrt{2})^2\right), \\ c_2(t) &= \left(\sqrt{2} \cos(t + \pi/4), \sqrt{2} \sin(t + \pi/4), 2\right) \\ &= (\cos t - \sin t, \cos t + \sin t, 2).\end{aligned}$$

The image curves $c_1(I_1)$ and $c_2(I_2)$ are, respectively, the curve of intersection of the plane $x = y$ with the paraboloid $z = x^2 + y^2$ and the curve of intersection of the plane $z = 2$ with the paraboloid $z = x^2 + y^2$.

Note that both $c_1(0) = p$ and $c_2(0) = p$, so that the tangent vectors $\mathbf{v}_p = c_1'(0)$ and $\mathbf{w}_p = c_2'(0)$ are by definition elements of $T_p(S)$. Explicitly,

$$\begin{aligned}\mathbf{v}_p &= \left\langle 1/\sqrt{2}, 1/\sqrt{2}, 2\sqrt{2} \right\rangle_p \\ &= \frac{1}{\sqrt{2}} \left\langle 1, 1, 4 \right\rangle_p\end{aligned}$$

and

$$\mathbf{w}_p = \langle -1, 1, 0 \rangle_p .$$

In fact, we show that every linear combination of $\mathbf{v}_p$ and $\mathbf{w}_p$ is a tangent vector to S at p. Let $\mathbf{x}_p = s_1 \mathbf{v}_p + s_2 \mathbf{w}_p$, i.e.,

$$\mathbf{x}_p = \left\langle \frac{s_1}{\sqrt{2}} - s_2, \frac{s_1}{\sqrt{2}} + s_2, \frac{4s_1}{\sqrt{2}} \right\rangle_p .$$

Adopting the interval notation from the proof of Theorem 3.2.4 above, we let $\tilde{I}_1 = \dfrac{1}{s_1}(-1/2, 1/2)$, $\tilde{I}_2 = \dfrac{1}{s_2}(-\pi/2, \pi/2)$, and $\tilde{I} = \tilde{I}_1 \cap \tilde{I}_2$. Define $\tilde{c} : \tilde{I} \to U$ and $c : I \to S$ by

$$\tilde{c}(t) = \left(s_1 t + \sqrt{2}, s_2 t + \pi/4 \right)$$

and

$$\begin{aligned}c(t) &= (\phi \circ \tilde{c}(t))\,(t) \\ &= \left((s_1 t + \sqrt{2}) \cos (s_2 t + \pi/4) , (s_1 t + \sqrt{2}) \sin (s_2 t + \pi/4) , (s_1 t + \sqrt{2})^2 \right) .\end{aligned}$$

The reader can verify that $c'(0) = \mathbf{x}_p$, and so $\mathbf{x}_p \in T_p(S)$.

We note further that if $\mathbf{y}_p = \langle a, b, c \rangle_p \in T_p(S)$, then

$$-2a - 2b + c = 0. \tag{3.4}$$

This is a consequence of the fact that if $\mathbf{y}_p = c'(0)$, where $c : I \to S$ is given by $c(t) = (x(t), y(t), z(t))$ with $c(0) = p = (1, 1, 2)$, then for all $t \in I$, the component functions must satisfy the equation

$$z(t) = [x(t)]^2 + [y(t)]^2 , \tag{3.5}$$

and so differentiating with respect to t yields

$$z'(t) = 2x(t)x'(t) + 2y(t)y'(t). \tag{3.6}$$

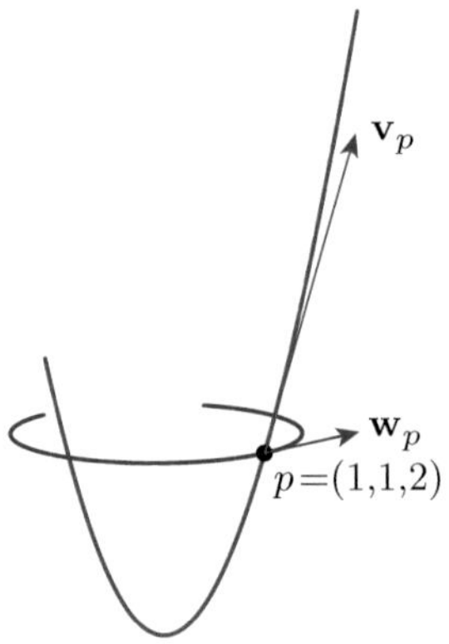

Fig. 3.4 In Example 3.3.8, $T_p(S) = \text{Span}\{\mathbf{v}_p, \mathbf{w}_p\}$.

Evaluating at $t = 0$ gives Eq. (3.4).

Comparing the sets $T_p(S)$ and $W = \left\{\langle a,b,c\rangle_p \mid -2a - 2b + c = 0\right\} \subset T_p(\mathbf{R}^3)$, we note that the vectors $\mathbf{v}_p = \frac{1}{\sqrt{2}}\langle 1,1,4\rangle_p$ and $\mathbf{w}_p = \langle -1,1,0\rangle$ studied above are two linearly independent vectors in the 2-dimensional subspace W, and so

$$T_p(S) = W = \text{Span}\{\mathbf{v}_p, \mathbf{w}_p\}.$$

See Fig. 3.4.

We have developed these two previous examples in some detail in order to illustrate the general situation, which we summarize in the following theorem and corollaries.

Theorem 3.3.9. *Let $U \subset \mathbf{R}^k$, $\phi : U \to \mathbf{R}^n$, and $S = \phi(U)$ be as above. Let $p \in S$, i.e., $p = \phi(\tilde{p})$ for $\tilde{p} = (\tilde{p}_1, \ldots, \tilde{p}_k) \in U$. For each $i = 1, \ldots, k$, define $\tilde{c}_i : I_i \to U$ by*

$$\tilde{c}_i(t) = (\tilde{p}_1, \ldots, \tilde{p}_i + t, \ldots, \tilde{p}_k),$$

where I_i is an appropriate interval containing 0 satisfying the condition that $\tilde{c}_i(I_i) \subset U$. Define $c_i : I_i \to S$ by $c_i = \phi \circ \tilde{c}_i$, and let $(\mathbf{e}_i)_p \in T_p(S)$ be defined by $(\mathbf{e}_i)_p = c_i'(0)$.

Then $\{(\mathbf{e}_1)_p, \ldots, (\mathbf{e}_k)_p\}$ is a basis for $T_p(S)$.

The set in the conclusion of Theorem 3.3.9 will be called the *standard basis for the parameterized set S.*

Proof. To prove that $\{(\mathbf{e}_1)_p, \ldots, (\mathbf{e}_k)_p\}$ is a spanning set for $T_p(S)$ amounts to adapting the methods used in Example 3.3.8 to the general case. Care should be taken in appropriately modifying the intervals on which the c_i are defined in order to construct a new curve.

The fact that the set is linearly independent derives essentially from the fact that the vectors $\tilde{c}_i'(0)$ are the standard basis vectors, viewed as vectors in $\mathbf{R}^k$, along with the regularity of ϕ. □

This theorem has some simple but notable corollaries. Both depend on the fact that the parameterization ϕ is regular.

Corollary 3.3.10. *Suppose $U \subset \mathbf{R}^k$ is a domain and $\phi : U \to \mathbf{R}^n$ is a regular parameterization, and let $S = \phi(U)$. Then for all $p \in S$, $\dim T_p(S) = k$.*

Corollary 3.3.11. *If $S = \phi(U)$ is a parameterized set, then*

$$T_p(S) = \text{Image}\,(D\phi(\tilde{p}))\,,$$

where $p = \phi(\tilde{p})$.

As can be seen from the definitions and examples above, parameterization plays a key role in studying geometric objects using the techniques of differential calculus. However, there are some disadvantages to overemphasizing the role played by parameterized sets. It has been a hallmark of differential geometry to isolate "intrinsic" properties of geometric objects—those that are properties of the objects themselves and are not reliant on the "extrinsic" coordinate systems or parameterizations used to study them. For example, supposing that a set S can be parameterized by two different parameterizations $\phi_1 : U \to S$ and $\phi_2 : V \to S$, where $S = \phi_1(U) = \phi_2(V)$. Then $T_p(S)$ defined by either ϕ_1 or ϕ_2 should yield the same subspace of $T_p(\mathbf{R}^n)$.

The practice of isolating concepts that are intrinsically geometric is a fundamental goal of modern differential geometry, and leads to the abstract formulations of manifolds, submanifolds, and related concepts. We do not emphasize that perspective here, however.

From a practical perspective, many objects in geometry—even analytic geometry, with its use of coordinate systems—are not described as parameterized sets. We now discuss some other common ways that geometric sets are encountered.

Let $U \subset \mathbf{R}^{n-1}$ be a domain and let $f : U \to \mathbf{R}$ be a smooth function. The *graph of* f is the set $S_f \subset \mathbf{R}^n$ defined by

$$S_f = \{(x_1, \ldots, x_{n-1}, f(x_1, \ldots, x_{n-1})) \mid (x_1, \ldots, x_{n-1}) \in U\}\,.$$

We will show that graphs so defined are geometric sets in the sense defined at the beginning of the section. See Fig. 3.5. We begin by mimicking Definition 3.3.5 of the tangent space above:

Definition 3.3.12. Let $U \subset \mathbf{R}^{n-1}$ be a domain and let $f : U \to \mathbf{R}$ be a smooth function. Let $S_f \subset \mathbf{R}^n$ be the graph of f and let $p \in S_f$. The *tangent space to S_f at p*, denoted by $T_p(S_f)$, is the subset of $T_p(\mathbf{R}^n)$ defined by

$$T_p(S) = \{\mathbf{v}_p \in T_p(\mathbf{R}^n) \mid \exists c : I \to S_f \text{ smooth with } c(0) = p,\ \ c'(0) = \mathbf{v}_p\}\,.$$

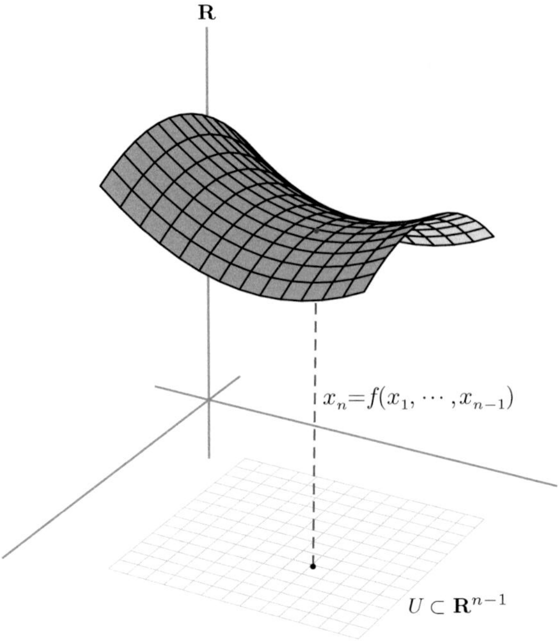

Fig. 3.5 The graph of a function $f : U \to \mathbf{R}$ in $\mathbf{R}^n$.

As in Definition 3.3.5, the essential aspect of this definition is the restriction of the codomain of the parameterized curves.

Theorem 3.3.13. *Let S_f be the graph of a smooth function $f : \mathrm{U} \to \mathbf{R}$, where $U \subset \mathbf{R}^{n-1}$ is a domain. Then the tangent space $T_p(S_f)$ is a vector subspace of $T_p(\mathbf{R}^n)$ with $\dim T_p(S_f) = n - 1$.*

Proof. The statement can be reduced to Theorem 3.3.9 and its corollaries by exhibiting a parameterization of S_f, i.e., a one-to-one, smooth, and regular function $\phi : U \to S_f$. This can be done by defining $\phi : U \to S_f$ by

$$\phi(x_1, \ldots, x_{n-1}) = (x_1, \ldots, x_{n-1}, f(x_1, \ldots, x_{n-1})) .$$

The fact that ϕ has rank $n-1$ follows by computing the Jacobian matrix at any point in U; hence ϕ is regular. □

Theorem 3.3.14. *For the graph of a function $f : \mathbf{R}^{n-1} \to \mathbf{R}$, we have*

$$T_p(S_f) = \left\{ \langle a_1, \ldots, a_{n-1}, a_n \rangle_p \mid a_n = \sum_{i=1}^{n-1} a_i \frac{\partial f}{\partial x_i}(\tilde{p}) \right\},$$

where $p = (\tilde{p}, f(\tilde{p}))$, $\tilde{p} \in U$.

Proof. Let $\tilde{p} \in U$ and let $p = (\tilde{p}, f(\tilde{p}))$. Note that the standard basis vectors (cf. Theorem 3.3.9) for $T_p(S_f)$ considered with the parameterization mentioned in the proof of Theorem 3.3.13 are

$$(\mathbf{e}_1)_p = \left\langle 1, 0, \ldots, 0, \frac{\partial f}{\partial x_1}(\tilde{p}) \right\rangle_p,$$

$$(\mathbf{e}_2)_p = \left\langle 0, 1, 0, \ldots, 0, \frac{\partial f}{\partial x_2}(\tilde{p}) \right\rangle_p,$$

$$\vdots$$

$$(\mathbf{e}_{n-1})_p = \left\langle 0, \ldots, 0, 1, \frac{\partial f}{\partial x_{n-1}}(\tilde{p}) \right\rangle_p.$$

The assertion of the theorem is then just a restatement of saying that $\mathbf{v}_p$ is a linear combination of the basis vectors $\mathbf{e}_1, \ldots, \mathbf{e}_{n-1}$. □

Example 3.3.15. Returning to the paraboloid of Example 3.3.8, we note that the surface S can be expressed as the graph $z = f(x, y)$ of the function $f : \mathbf{R}^2 \to \mathbf{R}$ given by $f(x, y) = x^2 + y^2$. By Theorem 3.3.14,

$$T_p(S) = \left\{ \langle a, b, c \rangle_p \mid c = a(2p_1) + b(2p_2) \right\},$$

where $p = (p_1, p_2, p_3) \in S$ and so $p_3 = p_1^2 + p_2^2$. For example, with $p = (1, 1, 2)$ as in Example 3.3.8,

$$T_p(S) = \left\{ \langle a, b, c \rangle_p \mid c = 2a + 2b \right\},$$

corresponding to the result we obtained previously.

Our definition of a graph can easily be generalized. The ease of parameterization for the graph in $\mathbf{R}^n$ is really a consequence of the fact that the sum of the dimensions of the domain of f (which is $n - 1$) and the range of f (which is 1) is n. More generally, for $0 < k < n$, we could consider a smooth function $F : \mathbf{R}^{n-k} \to \mathbf{R}^k$. All of the corresponding statements are easily reformulated in these terms, with $\dim T_p(S_F) = n - k$.

Yet another standard way of describing a geometric set in $\mathbf{R}^n$ is as the level set of a smooth function $f : \mathbf{R}^n \to \mathbf{R}$. Given such a function f, let $a \in \mathbf{R}$ be in the range of f. Define the *level set* of f at a to be the set

$$S_{f,a} = \{(x_1, \ldots, x_n) \in \mathbf{R}^n \mid f(x_1, \ldots, x_n) = a\}.$$

We say that a is a *regular value* of f if for all $p \in S_{f,a}$, $Df(p)$ is an onto linear map or, what is the same, $Df(p)$ is not the zero transformation for any $p \in S_{f,a}$.

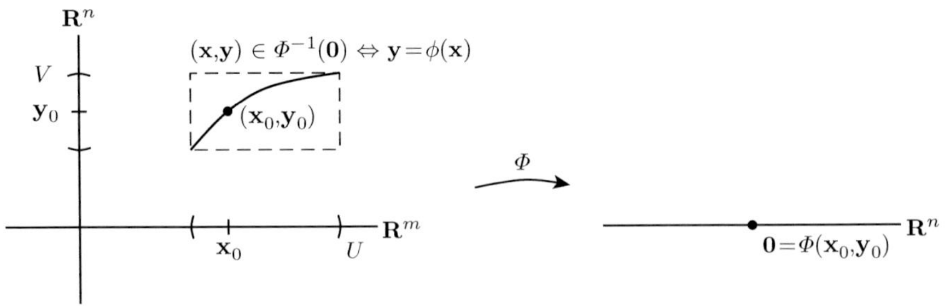

Fig. 3.6 The implicit function theorem.

For example, the unit circle in $\mathbf{R}^2$, $\{(x,y) \mid x^2+y^2=1\}$, can be seen as the level set for the function $f(x,y)=x^2+y^2$ with $a=1$.

Definition 3.3.16. Let $S_{f,a}$ be the level set of a smooth function $f:\mathbf{R}^n \to \mathbf{R}$ at $a \in \mathbf{R}$. The *tangent space to $S_{f,a}$ at $p \in S_{f,a}$* is defined to be the set

$$T_p(S_{f,a}) = \{\mathbf{v}_p \in T_p(\mathbf{R}^n) \mid \exists\, c : I \to S_{f,a} \text{ smooth with } c(0)=p,\ c'(0)=\mathbf{v}_p\}\,.$$

Theorem 3.3.17. *Let $f:\mathbf{R}^n \to \mathbf{R}$ be a smooth function and let $a \in \mathbf{R}$ be a regular value of f. Then $T_p(S_{f,a})$ is a vector subspace of $T_p(\mathbf{R}^n)$ and $\dim T_p(S_{f,a}) = n-1$.*

To prove this theorem, we will rely on a fundamental tool of differential geometry, the implicit function theorem. We state a version of the theorem here for reference. A more precise statement and proof can be found in [37].

Theorem 3.3.18 (Implicit function theorem). *Suppose that $\Phi:\mathbf{R}^m \times \mathbf{R}^n \to \mathbf{R}^n$ is a function that is smooth near the point $(\mathbf{x}_0,\mathbf{y}_0) \in \mathbf{R}^m \times \mathbf{R}^n$. Writing Φ by means of the component functions*

$$\Phi(\mathbf{x},\mathbf{y}) = (\Phi^1(\mathbf{x},\mathbf{y}),\ldots,\Phi^n(\mathbf{x},\mathbf{y})),$$

where $\mathbf{x}=(x_1,\ldots,x_m) \in \mathbf{R}^m$ and $\mathbf{y}=(y_1,\ldots,y_n) \in \mathbf{R}^n$, suppose further that we have

- *$\Phi(\mathbf{x}_0,\mathbf{y}_0)=\mathbf{0}$, and*
- *the $n \times n$ matrix $\left[\dfrac{\partial \Phi^i}{\partial y_j}(\mathbf{x}_0,\mathbf{y}_0)\right]$ $(i,j=1,\ldots,n)$ is invertible.*

Then there exist a domain $U \subset \mathbf{R}^m$ containing $\mathbf{x}_0$, a domain $V \subset \mathbf{R}^n$ containing $\mathbf{y}_0$, and a unique smooth function $\phi: U \to V$ such that $\phi(\mathbf{x}_0)=\mathbf{y}_0$ and

$$\{(\mathbf{x},\mathbf{y}) \mid (\mathbf{x},\mathbf{y}) \in U \times V \text{ and } \Phi(\mathbf{x},\mathbf{y})=\mathbf{0}\} = \{(\mathbf{x},\phi(\mathbf{x})) \mid \mathbf{x} \in U\}\,.$$

See Fig. 3.6 for an illustration of the implicit function theorem.

Proof (of Theorem 3.3.17). Let $p = (p_1, \ldots, p_n) \in S_{f,a}$. Since a is a regular value of f, we have $Df(p) = \left[\frac{\partial f}{\partial x_1}(p) \cdots \frac{\partial f}{\partial x_n}(p)\right] \neq \mathbf{0}$. Assume without loss of generality that $\frac{\partial f}{\partial x_n}(p) \neq 0$. We can then apply the implicit function theorem to

$$f : \mathbf{R}^{n-1} \times \mathbf{R} \to \mathbf{R}$$

near p to obtain a domain $U \subset \mathbf{R}^{n-1}$ containing $(p_1, \ldots, p_{n-1})$ and a smooth function $g : U \to \mathbf{R}$ such that the graph S_g coincides with the level set $S_{f,a}$ near p. In particular, tangent vectors to $S_{f,a}$ at p in the sense of Definition 3.3.12 coincide with those defined in the sense of Definition 3.3.16. The theorem then follows by Theorem 3.3.13. □

In the course of Example 3.3.7, we showed that the tangent plane to the level set S given by $f(x, y, z) = 2x - 3y - z = 0$ at a point p is

$$T_p(S) = \left\{ \langle a, b, c \rangle_p \mid 2a - 3b - c = 0 \right\}.$$

The following theorem shows that this is a symptom of a general phenomenon.

Theorem 3.3.19. *Let $f : \mathbf{R}^n \to \mathbf{R}$ be a smooth function, $a \in \mathbf{R}$ a regular value of f, and $p \in S_{f,a} \subset \mathbf{R}^n$. Then*

$$\begin{aligned} T_p(S_{f,a}) &= \ker Df(p) \\ &= \left\{ \langle v_1, \ldots, v_n \rangle_p \;\middle|\; \frac{\partial f}{\partial x_1}(p) \cdot v_1 + \cdots + \frac{\partial f}{\partial x_n}(p) \cdot v_n = 0 \right\}. \end{aligned}$$

Proof. Assume first that $\mathbf{v}_p \in T_p(S_{f,a})$. Then there is a parameterized curve $c : I \to S_{f,a}$ such that $c(0) = p$ and $c'(0) = \mathbf{v}_p$. Since $f(c(t)) = a$ for all $t \in I$, we have by the chain rule that $\mathbf{0} = Df(c(0))(c'(0)) = Df(p)(\mathbf{v}_p)$, showing that $\mathbf{v}_p \in \ker Df(p)$ and so $T_p(S_{f,a}) \subset \ker Df(p)$.

Conversely, assume that $\mathbf{v}_p = \langle v_1, \ldots, v_n \rangle_p \in \ker Df(p)$, i.e.,

$$\frac{\partial f}{\partial x_1}(p) \cdot v_1 + \cdots \frac{\partial f}{\partial x_n}(p) \cdot v_n = 0. \tag{3.7}$$

To show that $\mathbf{v}_p \in T_p(S_{f,a})$, we need to construct a parameterized curve $c : I \to S_{f,a}$ such that $c(0) = p = (p_1, \ldots, p_n)$ and $c'(0) = \mathbf{v}_p$. Since a is a regular value, the implicit function theorem above guarantees that there exist a domain $U \subset \mathbf{R}^{n-1}$ with $(p_1, \ldots, p_{n-1}) \in U$ and a smooth function $g : U \to \mathbf{R}$ with the property that $p_n = g(p_1, \ldots, p_{n-1})$ and

$$f\left(x_1, \ldots, x_{n-1}, g(x_1, \ldots, x_{n-1})\right) = a \tag{3.8}$$

for all $(x_1, \ldots, x_{n-1}) \in U$. We have made the assumption, without loss of generality, that $\frac{\partial f}{\partial x_n}(p) \neq 0$. Define an interval $I \subset \mathbf{R}$ in such a way that $0 \in I$

and that for all $t \in I$, $(p_1 + v_1 t, \dots, p_{n-1} + v_{n-1} t) \in U$. Define the parameterized curve $c : I \to S_{f,a}$ by

$$c(t) = (p_1 + v_1 t, \dots, p_{n-1} + v_{n-1} t, g(p_1 + v_1 t, \dots, p_{n-1} + v_{n-1} t)) .$$

Using the chain rule in the last component, we have

$$c'(0) = \left\langle v_1, \dots, v_{n-1}, \frac{\partial g}{\partial x_1}(\tilde{p}) \cdot v_1 + \cdots + \frac{\partial g}{\partial x_{n-1}}(\tilde{p}) \cdot v_{n-1} \right\rangle_p , \tag{3.9}$$

where $\tilde{p} = (p_1, \dots, p_{n-1})$. However, we note that by applying the chain rule to Eq. (3.8), we have

$$\frac{\partial f}{\partial x_i} + \frac{\partial f}{\partial x_n} \cdot \frac{\partial g}{\partial x_i} = 0, \quad \text{for } i = 1, \dots, n-1. \tag{3.10}$$

Using Eq. (3.10) and recalling the assumption that $\frac{\partial f}{\partial x_n}(p) \neq 0$, we obtain

$$\frac{\partial g}{\partial x_i} = -\frac{\frac{\partial f}{\partial x_i}}{\frac{\partial f}{\partial x_n}}. \tag{3.11}$$

Substituting Eq. (3.11) into Eq. (3.9) yields

$$\begin{aligned} c'(0) &= \left\langle v_1, \dots, v_{n-1}, -\frac{1}{\frac{\partial f}{\partial x_n}} \left[\frac{\partial f}{\partial x_1} \cdot v_1 + \cdots + \frac{\partial f}{\partial x_{n-1}} \cdot v_{n-1} \right] \right\rangle_p \\ &= \langle v_1, \dots, v_{n-1}, v_n \rangle_p , \end{aligned}$$

the last equality due to Eq. (3.7). This shows that $\mathbf{v}_p = c'(0) \in T_p(S_{f,a})$. □

As was the case for graphs of functions, the presentation of level sets here emphasized *hypersurfaces*, namely geometric sets whose tangent space at each point has dimension one less than the dimension of the ambient space. This, of course, is a consequence of considering functions $f : \mathbf{R}^n \to \mathbf{R} = \mathbf{R}^1$. However, all the definitions and theorems here can be generalized to functions $F : \mathbf{R}^n \to \mathbf{R}^k$ where $0 < k < n$. In this case, the level set $S_{F,q}$ of a regular value $q \in \mathbf{R}^k$ will be a geometric set $S_{F,q}$ with $\dim T_p(S_{F,q}) = n - k$ for each $p \in S_{F,q}$.

We close this section with a summary of what we have accomplished thus far. In Sect. 3.2, we defined tangent vectors at a point $p \in \mathbf{R}^n$ to be vectors that are tangent *to something*, namely, tangent to parameterized curves. This is a pointwise construction, leading to the definition of the tangent space at p as the vector space of all tangent vectors at p. In this section, we illustrated how certain "geometric sets" $S \subset \mathbf{R}^n$ define subsets of $T_p(\mathbf{R}^n)$ at each point $p \in S$. For practical purposes, the important theorems for calculating the tangent space to a geometric set at a point p are Theorems 3.3.9, 3.3.14, and 3.3.19 along with Corollary 3.3.11.

The reader may have noticed, however, that as appealing to geometric intuition as this definition of a tangent vector is, it is somewhat clumsy from a mathematical perspective. To show that a vector is really a tangent vector, one must exhibit a parameterized curve that gives rise to the vector in question. We have not even touched the subject of the question of different parameterized curves giving rise to the same tangent vector, a technicality that can be addressed by appealing to more abstract mathematical constructions such as equivalence classes. We will ask the reader to investigate some of these questions in the exercises at the end of the chapter.

In the next section, we give an alternative definition of tangent vectors, one that is in some sense dual to the definition given in this section. The fact that the two definitions are equivalent will permit us two ways of envisioning tangent vectors, and so also the tangent space.

3.4 The Tangent Space II: An Analytic Definition

In the geometric definition of tangent vectors in Sect. 3.2, a tangent vector is the algebraic object that results when the operation of differentiation is performed on a geometric object, namely on a parameterized curve. In this section, we shift the perspective completely. The tangent vector will be an analytic object that in some sense "performs an operation of differentiation" on other objects, which will now be functions defined on some appropriate domain. The motivating idea in this perspective will be the notion of a *directional derivative*. The approach here will have the advantage of offering a more coordinate-free approach to vector fields in later sections, as well as a lesser reliance on the seemingly extraneous (although geometrically appealing) parameterized curves of Sect. 3.2.

Let $f : U \to \mathbf{R}$ be a smooth function, where $U \subset \mathbf{R}^n$ is a domain containing the point $p \in U$. Let $c : I \to \mathbf{R}^n$ be a smooth parameterized curve such that $c(0) = p$. Then, by the chain rule,

$$\begin{aligned} \left.\frac{d}{dt} f\left(c(t)\right)\right|_{t=0} &= Df\left(c(0)\right) c'(0) \\ &= \nabla f(p) \cdot \mathbf{v}_p, \end{aligned}$$

where $\mathbf{v}_p = c'(0)$. We are here using the common notation for the gradient and "dot product" from a first course in vector calculus. This corresponds, up to the usual convention of $\mathbf{v}_p$ being chosen as a unit vector, to the standard definition of the *directional derivative of f in the direction* $\mathbf{v}_p$. See, for example, [30, p. 164].

Even without the suggestive notation, the reader will have noticed the presence of the tangent vector $\mathbf{v}_p \in T_p(\mathbf{R}^n)$ in the preceding paragraph, defined in the sense of Sect. 3.2—the tangent vector to the parameterized curve $c(t)$ at $c(0) = p$.

However, we now view the tangent vector as an operator—a "function" whose domain consists of smooth functions defined near p and whose range consists of real numbers:

$$\mathbf{v}_p : f \mapsto \nabla f(p) \cdot \mathbf{v}_p \in \mathbf{R}.$$

We now formalize this discussion by defining tangent vectors in such a way as to capture the essential properties of the directional derivative. Let $C_p^\infty(\mathbf{R}^n)$ be the set of smooth real-valued functions defined on some domain containing p. (We will be vague about the domains of different elements of $C_p^\infty(\mathbf{R}^n)$; a precise definition requires considering equivalence classes of functions known as *germs*.)

Definition 3.4.1. Let $p \in \mathbf{R}^n$. A *tangent vector at* p is an operator

$$\mathbf{v}_p : C_p^\infty(\mathbf{R}^n) \to \mathbf{R}$$

satisfying the following three properties:

1. For all $f, g \in C_p^\infty(\mathbf{R}^n)$, $\mathbf{v}_p[f+g] = \mathbf{v}_p[f] + \mathbf{v}_p[g]$.

2. For all $f \in C_p^\infty(\mathbf{R}^n)$ and $c \in \mathbf{R}$, $\mathbf{v}_p[cf] = c\mathbf{v}_p[f]$.

3. For all $f, g \in C_p^\infty(\mathbf{R}^n)$, $\mathbf{v}_p[f \cdot g] = \mathbf{v}_p[f] \cdot g(p) + f(p) \cdot \mathbf{v}_p[g]$.

The three properties defining a tangent vector should be familiar to any calculus student. The first two assert that the tangent vector, as an operator, is *linear*. The third asserts that tangent vectors obey the "product rule" (or Leibniz rule). An operator that satisfies (1)–(3) is known as a *linear derivation*.

While this definition may appear abstract, we present an important example immediately to show that linear derivations appear routinely in multivariable calculus. The prototypical tangent vectors (as linear derivations) are the partial differentiation operators.

Example 3.4.2. For $i = 1, \ldots, n$, let $(\partial_i)_p : C_p^\infty(\mathbf{R}^n) \to \mathbf{R}$ be defined by $(\partial_i)_p[f] = \left.\dfrac{\partial f}{\partial x_i}\right|_p$. Then $(\partial_i)_p$ is a linear derivation. The verification amounts to restating basic properties of the partial derivative.

The following theorem shows that a familiar property of derivatives is in fact a formal consequence of the properties defining a linear derivation.

Theorem 3.4.3. *Let $\mathbf{v}_p$ be a tangent vector at p and let f be a function that is constant near p, i.e., there exist a domain $U \subset \mathbf{R}^n$ containing p and a real number $c \in \mathbf{R}$ such that $f(x) = c$ for all $x \in U$. Then $\mathbf{v}_p[f] = 0$.*

Proof. We prove the theorem when $f(x) = 1$ for all $x \in U$, since the result then follows using property (2) of Definition 3.4.1.

We can write

$$\begin{aligned}\mathbf{v}_p[f] &= \mathbf{v}_p[f \cdot f] \quad \text{since } f \equiv 1 \\ &= \mathbf{v}_p[f] \cdot f(p) + f(p) \cdot \mathbf{v}_p[f] \quad \text{by property (3) of Definition 3.4.1} \\ &= 2\mathbf{v}_p[f] \quad \text{since } f(p) = 1,\end{aligned}$$

and so $\mathbf{v}_p[f] = 0$. □

Starting from this new analytic definition of tangent vectors, we now begin the process of reconstructing the key results of Sects. 3.2 and 3.3. Since our goal is ultimately to view tangent vectors interchangeably as linear derivations or as tangents to parameterized curves, we purposely blur the distinctions in notation between the two (as yet distinct) objects. For example, we denote $T_p(\mathbf{R}^n)$ in this section to be the set of all linear derivations at p, which is the same notation we used in the last section to denote the set of all vectors that arise from differentiating parameterized curves.

Theorem 3.4.4. *The set $T_p(\mathbf{R}^n)$ of all linear derivations at $p \in \mathbf{R}^n$, with the usual operations of addition and scalar multiplication of functions, forms a vector space. The space $T_p(\mathbf{R}^n)$ will be called the* (analytic) tangent space at p.

Proof. Exercise. □

In general, sets of real-valued functions like $\mathcal{F}(\mathbf{R})$ from Example 2.2.7 or any of its subsets listed there are very large. One would expect that sets of operators whose domains consist of sets of real-valued functions would be that much larger. However, the linearity and derivation properties described above ensure that the tangent space to $\mathbf{R}^n$ at a point p is finite-dimensional.

Theorem 3.4.5. *The set $\mathcal{B}_0 = \left\{(\partial_1)_p, \ldots, (\partial_n)_p\right\}$ of partial derivative operators is a basis for $T_p(\mathbf{R}^n)$.*

We call $\mathcal{B}_0$ the *standard basis for $T_p(\mathbf{R}^n)$* (Fig. 3.7).

This result is a consequence of the following lemma of advanced calculus, a version of Taylor's theorem for functions of many variables.

Lemma 3.4.6. *Let $f : U \to \mathbf{R}$ be a smooth function defined on a convex domain $U \subset \mathbf{R}^n$ containing $p = (p_1, \ldots, p_n) \in \mathbf{R}^n$. Then there are n smooth functions $a_i : U \to \mathbf{R}$ ($i = 1, \ldots, n$) satisfying $a_i(p) = 0$ and*

$$f(x) = f(p) + \sum_{i=1}^{n} (x_i - p_i) \left[\frac{\partial f}{\partial x_i}(p) + a_i(x)\right],$$

for all $x = (x_1, \ldots, x_n) \in U$.

Proof (of Lemma 3.4.6). Convexity implies that for all $x \in U$, all points on the segment

$$\{(1-t)p + tx \mid 0 \leq t \leq 1\}$$

are also contained in U.

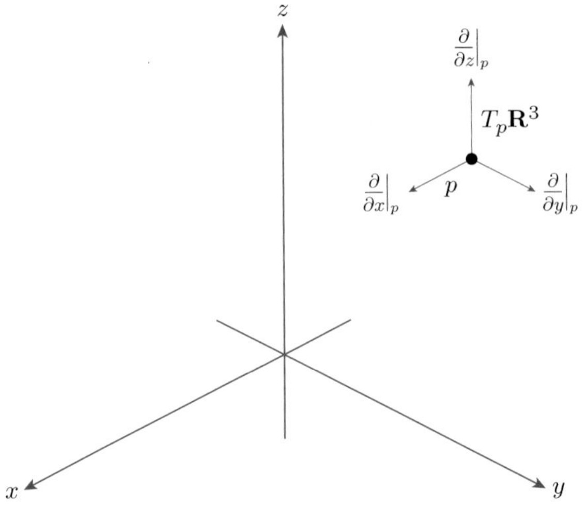

Fig. 3.7 The standard basis for $T_p(\mathbf{R}^3)$ (analytic version).

We appeal to the fundamental theorem of calculus and the chain rule:

$$\begin{aligned} f(x) - f(p) &= \int_0^1 \frac{d}{dt} f\left(p + t(x-p)\right) dt \\ &= \int_0^1 \left[\frac{\partial f}{\partial x_1}\Big(p + t(x-p)\Big) \cdot (x_1 - p_1) + \cdots \right. \\ &\qquad \left. \cdots + \frac{\partial f}{\partial x_n}\Big(p + t(x-p)\Big) \cdot (x_n - p_n) \right] dt \\ &= \sum_{i=1}^n (x_i - p_i) \int_0^1 \frac{\partial f}{\partial x_i}(p + t(x-p))\, dt \\ &= \sum_{i=1}^n (x_i - p_i) \left[\frac{\partial f}{\partial x_i}(p) + \int_0^1 \left(\frac{\partial f}{\partial x_i}(p + t(x-p)) - \frac{\partial f}{\partial x_i}(p) \right) dt \right]. \end{aligned}$$

The reader may show that the functions $a_i : \mathbf{R}^n \to \mathbf{R}$ defined by $a_i(x) = \int_0^1 \left(\frac{\partial f}{\partial x_i}(p + t(x-p)) - \frac{\partial f}{\partial x_i}(p) \right) dt$ are the desired functions. □

Proof (of Theorem 3.4.5). The set $B = \{(\partial_1)_p, \ldots, (\partial_n)_p\}$ can be seen to be linearly independent by applying an arbitrary linear combination

$$c_1(\partial_1)_p + \cdots + c_n(\partial_n)_p$$

to each of the n functions $\pi_i : \mathbf{R}^n \to \mathbf{R}$ $(i = 1, \ldots, n)$ defined by $\pi_i(x_1, \ldots, x_n) = x_i$.

To show that B spans $T_p(\mathbf{R}^n)$, we must show that for an arbitrary linear derivation $\mathbf{v}_p \in T_p(\mathbf{R}^n)$, there are scalars c_i such that $\mathbf{v}_p = c_1(\partial_1)_p + \cdots + c_n(\partial_n)_p$. To do so, let f be a smooth function defined in a convex domain containing p (for example, choosing a small ball centered at p). By Lemma 3.4.6, there are smooth functions a_i satisfying $a_i(p) = 0$ and also satisfying

$$f(x) = f(p) + \sum_{i=1}^{n} (\pi_i(x) - p_i)\Big[\frac{\partial f}{\partial x_i}\Big|_p + a_i(x)\Big].$$

Apply $\mathbf{v}_p$ to f:

$$\begin{aligned}\mathbf{v}_p[f] &= \mathbf{v}_p[f(p)] + \sum_{i=1}^{n} \mathbf{v}_p\Big[\frac{\partial f}{\partial x_i}\Big|_p (\pi_i - p_i) + a_i(x)(\pi_i - p_i)\Big] \\ &= 0 + \sum_{i=1}^{n} \left[\frac{\partial f}{\partial x_i}\Big|_p \mathbf{v}_p[\pi_i - p_i] + (\pi_i - p_i)\big|_p \mathbf{v}_p[a_i(x)] + a_i(p)\mathbf{v}_p[\pi_i - p_i]\right] \\ &= \sum_{i=1}^{n} \mathbf{v}_p[\pi_i]\frac{\partial f}{\partial x_i}\Big|_p .\end{aligned}$$

We have relied throughout these calculations on the properties of $\mathbf{v}_p$ as a derivation and Theorem 3.4.3. Setting $c_i = \mathbf{v}_p[\pi_i]$, we thus have

$$\mathbf{v}_p = \sum c_i(\partial_i)_p,$$

and so $\mathbf{v}_p \in \mathrm{Span}(B)$. Hence $T_p(\mathbf{R}^n) \subset \mathrm{Span}(B)$, and together with the rest, B is a basis for $T_p(\mathbf{R}^n)$. □

Corollary 3.4.7. *For all Euclidean spaces* $\mathbf{R}^n$, $\dim T_p(\mathbf{R}^n) = n$.

Recall that we motivated the definition of tangent vectors as linear derivations by observing some key properties of the directional derivative. We note now that using the standard basis for $T_p(\mathbf{R}^n)$, a tangent vector $\mathbf{v}_p$ can be written as $\mathbf{v}_p = \sum c_i(\partial_i)_p$, and so

$$\mathbf{v}_p[f] = \sum c_i(\partial_i)_p[f] = \sum c_i \frac{\partial f}{\partial x_i}\Big|_p,$$

which agrees with the standard definition of the directional derivative in the direction $\mathbf{v} = \langle c_1, \ldots, c_n \rangle$ in the case that $\mathbf{v}$ is a *unit* vector.

From this definition, we now turn to the implications for defining the tangent space for geometric sets such as the parameterized sets, graphs, and level sets described in the previous section. Using the geometric definition of tangent vectors as tangent vectors to curves, we defined the tangent space to a geometric subset

$S \subset \mathbf{R}^n$ by restricting the *codomain* of the curves under consideration. In other words, the restriction on tangent vectors in $T_p(\mathbf{R}^n)$ to tangent vectors in $T_p(S)$ amounts to restricting from the more general curves $c : I \to \mathbf{R}^n$ to the more specific $c : I \to S \subset \mathbf{R}^n$.

In the new analytic definition, the restriction will be on the *domain* of functions on which the tangent vectors as derivations operate. Namely, we will consider a tangent vector to the set S at a point $p \in S$ to be a linear derivation acting on real-valued functions whose domains lie in $S \subset \mathbf{R}^n$.

There are some technical and nontrivial considerations here. In particular, we want to consider the set $C_p^\infty(S)$ as a subset of the set $C_p^\infty(\mathbf{R}^n)$. That is, we want to consider a function defined only at points of S (near a point p) to be a function defined at all points of $\mathbf{R}^n$ near p. In general, this is a kind of "extension" problem. We will address this problem on a case-by-case basis for geometric sets described as parameterized sets, graphs, or level sets, although we will always assume that S satisfies the requirements for considering $C_p^\infty(S)$ a subset of $C_p^\infty(\mathbf{R}^n)$.

Definition 3.4.8. Let S be a geometric subset of $\mathbf{R}^n$ with $p \in S$ and consider $C_p^\infty(S) \subset C_p^\infty(\mathbf{R}^n)$. A *tangent vector to S at p* is a linear derivation $\mathbf{v}_p : C_p^\infty(S) \to \mathbf{R}$. The set of all tangent vectors to S at p will be called the *tangent space to S at p* and is denoted by $T_p(S)$.

As we have mentioned, we have purposely adopted the same notation for tangent vectors and the tangent space as in the prior two sections. Note that considering $C_p^\infty(S) \subset C_p^\infty(\mathbf{R}^n)$ is essential to ensuring that $T_p(S)$ can be considered a subset of $T_p(\mathbf{R}^n)$.

Theorem 3.4.9. *Let S be a geometric subset of $\mathbf{R}^n$. Then $T_p(S)$ is a vector subspace of $T_p(\mathbf{R}^n)$.*

Proof. Exercise. □

We now illustrate this analytic definition in the cases of the geometric sets we have seen so far. The development will give a preview of the more precise definition of the derivative that we are aiming for in the next section.

The first type of geometric set we encountered in the last section was a parametrized set $S = \phi(U) \subset \mathbf{R}^n$, where $U \subset \mathbf{R}^k$ and $\phi : U \to \mathbf{R}^n$ is a smooth, regular parameterization. In this case, for every $p = \phi(a) \in S$ we will say that $f \in C_p^\infty(S) \subset C_p^\infty(\mathbf{R}^n)$ if and only if there is a smooth function $\tilde{f} \in C_a(\mathbf{R}^k)$ such that $\tilde{f} = f \circ \phi$. This leads to the following characterization of the tangent space.

Theorem 3.4.10. *Let $S = \phi(U)$ be a parameterized set described by a domain $U \subset \mathbf{R}^k$ and a smooth, regular parameterization $\phi : U \to \mathbf{R}^n$ ($1 \leq k \leq n$). For every $p = \phi(a) \in S$, define the linear transformation $\Phi_a : T_a(\mathbf{R}^k) \to T_p(\mathbf{R}^n)$ as that whose matrix relative to the standard bases*

$$\left\{ (\tilde{\partial}_1)_a = \left.\frac{\partial}{\partial u_1}\right|_a, \dots, (\tilde{\partial}_k)_a = \left.\frac{\partial}{\partial u_k}\right|_a \right\} \subset T_a(\mathbf{R}^k),$$

$$\left\{(\partial_1)_p = \left.\frac{\partial}{\partial x_1}\right|_p, \ldots, (\partial_n)_p = \left.\frac{\partial}{\partial x_n}\right|_p\right\} \subset T_p(\mathbf{R}^n)$$

is given by $[D\phi(a)]$, *the Jacobian matrix for* ϕ *at* $a \in \mathbf{R}^k$. *Then*

$$T_p(S) = \operatorname{Image}(\Phi_a).$$

Proof. We will show that the set $B_p = \{(\mathbf{e}_1)_p, \ldots, (\mathbf{e}_k)_p\}$, where $(\mathbf{e}_i)_p = \Phi_a((\tilde{\partial}_i)_a)$, is a basis for $T_p(S)$. The set B_p is linearly independent by virtue of the corresponding property of the set $\left\{(\tilde{\partial}_1)_a, \ldots, (\tilde{\partial}_k)_a\right\}$ and the fact that Φ_a is one-to-one (since ϕ is regular).

To show that B_p spans $T_p(S)$, we note that linear derivations on $C_p^\infty(S)$ are in one-to-one correspondence with linear derivations on $C_a^\infty(\mathbf{R}^k)$, a consequence of the fact that $f \in C_p^\infty(S)$ if and only if there is a smooth function $\tilde{f} \in C_a(\mathbf{R}^k)$ such that $\tilde{f} = f \circ \phi$. For any such f and corresponding $\tilde{f}$, the chain rule gives

$$\begin{aligned}
(\tilde{\partial}_i)_a \left[\tilde{f}\right] &= \left.\frac{\partial}{\partial u_i}\right|_a [f \circ \phi] \\
&= \sum_{k=1}^{n} \left.\frac{\partial f}{\partial x_k}\right|_p \left.\frac{\partial \phi^k}{\partial u_i}\right|_a \\
&= (\mathbf{e}_i)_p \, [f] \, .
\end{aligned}$$

Hence to say that B_p spans $T_p(S)$ is equivalent to saying that $\left\{(\tilde{\partial}_1)_a, \ldots, (\tilde{\partial}_k)_a\right\}$ spans $T_a(\mathbf{R}^k)$. □

Corollary 3.4.11. *For a parameterized set* $S = \phi(U)$, *we have* $\dim T_p(S) = k$, *where* $U \subset \mathbf{R}^k$ *is a domain and* $\phi : U \to \mathbf{R}^n$ *is a smooth, regular parameterization.*

We now turn our attention to a geometric set defined as the graph of a real-valued function.

Theorem 3.4.12. *Let* $U \subset \mathbf{R}^{n-1}$ *be a domain and let* $g : U \to \mathbf{R}$ *be a smooth, real-valued function on* U. *Let* $S_g \subset \mathbf{R}^n$ *be the graph of* g:

$$S_g = \{(x_1, \ldots, x_n) \mid (x_1, \ldots, x_{n-1}) \in U, \; x_n = g(x_1, \ldots, x_{n-1})\} \, .$$

Then for $p = (\tilde{p}, g(\tilde{p}))$, $\tilde{p} \in U$, *we have*

$$T_p(S_g) = \left\{ a_1(\partial_1)_p + \cdots + a_n(\partial_n)_p \;\middle|\; a_n = \sum_{i=1}^{n-1} a_i \left.\frac{\partial g}{\partial x_i}\right|_{\tilde{p}} \right\} \subset T_p(\mathbf{R}^n).$$

Proof. The proof is essentially identical to the proof of Theorem 3.3.14. In particular, using the parameterization of S by the function $\phi : U \to S$ given by

$$\phi(x_1,\ldots,x_{n-1}) = (x_1,\ldots,x_{n-1},g(x_1,\ldots,x_{n-1}))\,,$$

Theorem 3.4.10 gives

$$(\mathbf{e}_i)_p = (\partial_i)_p + \left.\frac{\partial g}{\partial x_i}\right|_{\tilde{p}} (\partial_n)_p,$$

for $i = 1,\ldots,n-1$. □

Our last result of this section is the analytic analogue of Theorem 3.3.17 describing the tangent space of a geometric set described as a level set.

Theorem 3.4.13. *Let $\Phi : \mathbf{R}^n \to \mathbf{R}$ be a smooth, real-valued function with a regular value c, and let $S = \{(x_1,\ldots,x_n) \mid \Phi(x_1,\ldots,x_n) = c\}$. Then for $p \in S$,*

$$T_p(S) = \left\{ c_1(\partial_1)_p + \cdots + c_n(\partial_n)_p \;\middle|\; c_1 \left.\frac{\partial \Phi}{\partial x_1}\right|_p + \cdots + c_n \left.\frac{\partial \Phi}{\partial x_n}\right|_p = 0 \right\}.$$

Proof. On the one hand, for any $p \in S$, we can consider Φ to be an element of $C_p^\infty(S)$ that is constant. Theorem 3.4.3 implies that for all $\mathbf{v}_p \in T_p(S)$, $\mathbf{v}_p[\Phi] = 0$. Writing $\mathbf{v}_p$ using the standard basis vectors $\mathbf{v}_p = \sum c_i(\delta_i)_p$ shows that $c_1 \frac{\partial \Phi}{\partial x_1}\big|_p + \cdots + c_n \frac{\partial \Phi}{\partial x_n}\big|_p = 0$. So

$$T_p(S) \subset \left\{ c_1(\partial_1)_p + \cdots + c_n(\partial_n)_p \;\middle|\; c_1 \left.\frac{\partial \Phi}{\partial x_1}\right|_p + \cdots + c_n \left.\frac{\partial \Phi}{\partial x_n}\right|_p = 0 \right\}.$$

Now assume that $\mathbf{v}_p = \sum c_i(\delta_i)_p \in T_p(\mathbf{R}^n)$ satisfies $\mathbf{v}_p[\Phi] = 0$ for $p = (p_1,\ldots,p_n) \in S$. We may assume that $\dfrac{\partial \Phi}{\partial x_n} \neq 0$, since c is a regular value for Φ. Hence

$$c_n = -\frac{1}{\partial \Phi / \partial x_n} \left[\sum_{j=1}^{n-1} c_j \frac{\partial \Phi}{\partial x_j} \right].$$

Following the strategy of Theorem 3.3.19, by the implicit function theorem, there exist a domain $U \subset \mathbf{R}^{n-1}$ containing $(p_1,\ldots,p_{n-1})$ and a smooth function $g : U \to \mathbf{R}$ such that $\Phi(\tilde{p}, g(\tilde{p})) = 0$ for all $\tilde{p} \in U$. By the chain rule, we have for all $j = 1,\ldots,n-1$ that

$$\frac{\partial \Phi}{\partial x_j} + \frac{\partial \Phi}{\partial x_n}\frac{\partial g}{\partial x_j} = 0.$$

This combined with the assumption shows that

$$c_n = \sum_{j=1}^{n-1} c_j \frac{\partial g}{\partial x_j},$$

and so $\mathbf{v}_p \in T_p(S)$ by Theorem 3.4.12. □

We conclude by noting that this section has essentially replicated the line of reasoning in the previous two sections. Despite the differences between the geometric Definition 3.2.1 and the analytic Definition 3.4.1, there is an essential unity that can be seen in the pairs

Theorem 3.2.4 $\longleftrightarrow$ Theorem 3.4.4,
Theorem 3.3.9 $\longleftrightarrow$ Theorem 3.4.10,
Theorem 3.3.14 $\longleftrightarrow$ Theorem 3.4.12,
Theorem 3.3.19 $\longleftrightarrow$ Theorem 3.4.13.

In his classic text, Spivak presents the more general theorem that "all reasonable candidates" for the tangent space "turn out to be essentially the same" [38, p. 122] by working in the abstract framework of vector bundles and vector bundle equivalences.

Let us summarize what we have achieved to this point. To each point $p \in \mathbf{R}^n$, we have associated a vector space $T_p(\mathbf{R}^n)$. Following either route—the geometric or the analytic—leads to a vector space that is easily identified with $\mathbf{R}^n$. For certain "geometric" subsets $S \subset \mathbf{R}^n$, differentiability gives rise to subspaces $T_p(S)$ of $T_p(\mathbf{R}^n)$.

One of the main aims of differential geometry is to use information about the linear spaces $T_p(S)$, along with additional "structures," to study geometric properties of the nonlinear spaces S.

3.5 The Derivative as a Linear Map Between Tangent Spaces

Having spent some effort in defining the tangent space at a point in a Euclidean space $\mathbf{R}^n$, we now return to the definition of the derivative of a function $f : \mathbf{R}^n \to \mathbf{R}^m$ that has been at the heart of many of our constructions thus far. The goal in this section is to understand the derivative in its proper setting: not just as a linear map, which is what we emphasized in Sect. 3.1, but as a linear map *between tangent spaces*.

Definition 3.5.1. Let $f : \mathbf{R}^n \to \mathbf{R}^m$ be a function that is differentiable at a point $p \in \mathbf{R}^n$. The *tangent map to f at p* is a function, denoted by $(f_*)_p : T_p(\mathbf{R}^n) \to T_{f(p)}(\mathbf{R}^m)$, defined in either of the following equivalent ways (according to the definition of the tangent space):

1. For $\mathbf{v}_p \in T_p(\mathbf{R}^n)$, where $\mathbf{v}_p = c'(0)$ for a differentiable curve $c : I \to \mathbf{R}^n$ with $c(0) = p$, define
$$(f_*)_p(\mathbf{v}_p) = (f \circ c)'(0).$$
— or —
2. For $\mathbf{v}_p \in T_p(\mathbf{R}^n)$ given as a linear derivation, define
$$\big((f_*)_p(\mathbf{v}_p)\big)[\phi] = \mathbf{v}_p[\phi \circ f],$$
where $\phi \in C_p^\infty(\mathbf{R}^m)$.

The fact that the two definitions are equivalent is due the equivalence of the two definitions of the tangent space discussed in the previous section. Despite the fact that we are aiming to work interchangeably with two different definitions, we will not prove their equivalence. Nor will we choose one or the other definition as "the" definition.

The most subtle point is to verify that this definition does not depend on the representative curve or germ of the function that represents the tangent vector $\mathbf{v}_p$. We will not discuss this here, since we are not emphasizing the nature of a tangent vector as an equivalence class. See, however, Exercise 3.18.

The following result about the role of the Jacobian matrix should provide more evidence of the essential unity of the two approaches.

Theorem 3.5.2. *Let $f : \mathbf{R}^n \to \mathbf{R}^m$ be a function that is differentiable at $p \in \mathbf{R}^n$. Then the tangent map*

$$(f_*)_p : T_p(\mathbf{R}^n) \to T_{f(p)}(\mathbf{R}^m)$$

is a linear transformation of vector spaces. Moreover, using either definition, the linear transformation $(f_*)_p$ can be represented in matrix form (relative to the standard bases) by the Jacobian matrix $[Df(p)]$. See Fig. 3.8.

Proof. There are several considerations to prove. First, we need to show that by either definition, $(f_*)_p(\mathbf{v}_p) \in T_{f(p)}(\mathbf{R}^m)$. We leave the details to the reader, although we will illustrate the product rule property for the operator $(f_*)_p(\mathbf{v}_p)$ at $f(p)$ in the case of the analytic definition. Note first that for $\phi_1, \phi_2 \in C^\infty_{f(p)}(\mathbf{R}^m)$, we have $\phi_1 \circ f \in C^\infty_p(\mathbf{R}^n)$ and $\phi_2 \circ f \in C^\infty_p(\mathbf{R}^n)$. This is due to the differentiability of f. Also, $(\phi_1 \cdot \phi_2) \circ f = (\phi_1 \circ f) \cdot (\phi_2 \circ f)$. So

$$\begin{aligned}
\big((f_*)_p(\mathbf{v}_p)\big)[\phi_1 \cdot \phi_2] &= \mathbf{v}_p\big[(\phi_1 \cdot \phi_2) \circ f\big] \\
&= \mathbf{v}_p\big[(\phi_1 \circ f) \cdot (\phi_2 \circ f)\big] \\
&= (\phi_1 \circ f)(p)\mathbf{v}_p\left[\phi_2 \circ f\right] + (\phi_2 \circ f)(p)\mathbf{v}_p\left[\phi_1 \circ f\right] \\
&= \phi_1(f(p))\big((f_*)_p(\mathbf{v}_p)\big)[\phi_2] + \phi_2(f(p))\big((f_*)_p(\mathbf{v}_p)\big)[\phi_1].
\end{aligned}$$

Here we have relied on the fact that $\mathbf{v}_p$ is a linear derivation at p. This, together with the linearity properties to be verified by the reader, shows that $(f_*)_p(\mathbf{v}_p)$ is a linear derivation at $f(p)$.

The second statement to verify is that $(f_*)_p$ is a linear transformation. We will illustrate part of this verification in the case of the geometric definition. Let $\mathbf{v}_p, \mathbf{w}_p \in T_p(\mathbf{R}^n)$. Hence there are curves $c_1 : I_1 \to \mathbf{R}^n$ and $c_2 : I_2 \to \mathbf{R}^n$ such that $c_1(0) = c_2(0) = p$, $c_1'(0) = \mathbf{v}_p$, and $c_2'(0) = \mathbf{w}_p$. Let $c : I \to \mathbf{R}^n$ be such that $c(0) = p$ and $c'(0) = c_1'(0) + c_2'(0)$. Then, using the chain rule,

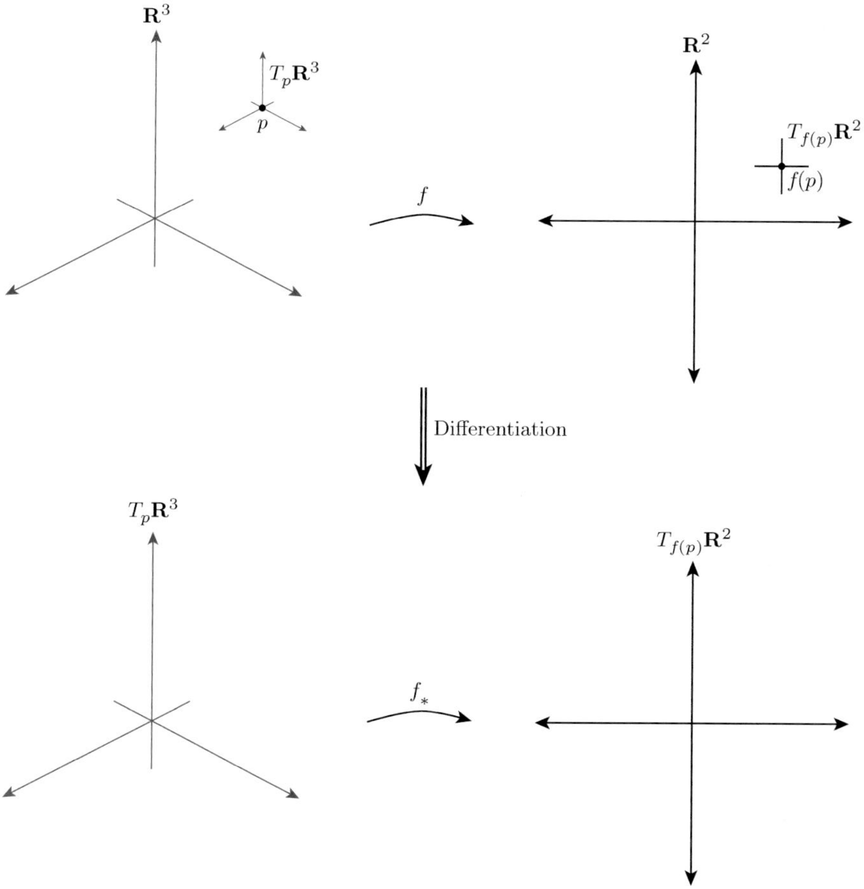

Fig. 3.8 The tangent map of a smooth function.

$$\begin{aligned}(f_*)_p(\mathbf{v}_p + \mathbf{w}_p) &= (f \circ c)'(0) \\ &= \big(Df(c(0))\big)\,(c'(0)) \\ &= \big(Df(p)\big)\,(c_1'(0) + c_2'(0)) \\ &= \big(Df(p)\big)\,(c_1'(0)) + \big(Df(p)\big)\,(c_2'(0)) \quad (Df(p) \text{ is linear}) \\ &= (f \circ c_1)'(0) + (f \circ c_2)'(0) \\ &= (f_*)_p(\mathbf{v}_p) + (f_*)_p(\mathbf{w}_p).\end{aligned}$$

The fact that $(f_*)_p$ preserves scalar multiplication follows similarly.

The final statement to address, which was implicit in the previous verification, is that the matrix representation of $(f_*)_p$ using the standard basis vectors is the

Jacobian matrix. In the analytic definition, the matrix representation of $(f_*)_p$ is essentially given by Theorem 3.4.10 which, read in this context, implies that the columns of the matrix representation of $(f_*)_p$ are given by

$$\begin{bmatrix} \frac{\partial f_i}{\partial x_1} \\ \vdots \\ \frac{\partial f_i}{\partial x_m} \end{bmatrix}.$$

□

We note that the chain rule, Theorem 3.1.7, has a particularly nice form in this framework.

Theorem 3.5.3. *Let $f : \mathbf{R}^n \to \mathbf{R}^m$ and $g : \mathbf{R}^m \to \mathbf{R}^k$ be smooth functions. Then*

$$\left((g \circ f)_*\right)_p = (g_*)_{f(p)} \circ (f_*)_p.$$

The presentation here adapts nicely to the situation of geometric sets. We need only agree on what is meant for a function $f : S_n \to S_m$ to be differentiable.

Definition 3.5.4. Let $S_n \subset \mathbf{R}^n$ and $S_m \subset \mathbf{R}^m$ be geometric sets of the respective Euclidean spaces. Then a function $f : S_n \to S_m$ is said to be *differentiable at* $p \in S_n$ if there is a smooth function $g : U \to \mathbf{R}^m$ defined on a domain $U \subset \mathbf{R}^n$ containing p such that $g(x) = f(x)$ for all $x \in S_n$. In other words, f is the restriction of a smooth function $g : \mathbf{R}^n \to \mathbf{R}^m$.

Compare this definition with the remark prior to Definition 3.4.8. The point is to be able to rely on the definition of differentiation in the "surrounding" set $\mathbf{R}^n$ (by referring to the function g defined on a "larger" set) despite the fact that f is defined (or considered) strictly as a function of the "smaller" set S_n.

Definition 3.5.5. Let $f : S_n \to S_m$ be a smooth function at $p \in S_n$, where $S_n \subset \mathbf{R}^n$ and $S_m \subset \mathbf{R}^m$ are geometric sets. The *tangent map to f at p* is the function $(f_*)_p : T_p(S_n) \to T_{f(p)}(S_m)$ defined exactly as in Definition 3.5.1.

So defined, the tangent map between geometric sets is again a linear transformation, in the manner of Theorem 3.5.2.

Finally, this definition allows us to recast Theorem 3.4.10, which characterizes the tangent space of a parameterized set, in its proper context. We begin by proving an obvious but important property of domains in $\mathbf{R}^n$.

Theorem 3.5.6. *Let $U \subset \mathbf{R}^n$ be a domain. Then for all $p \in U$, we have*

$$T_p(U) = T_p(\mathbf{R}^n).$$

Proof. The theorem is a consequence of considering U as a parameterized set in $\mathbf{R}^n$ with the parameterization $\phi : U \to \mathbf{R}^n$ by

$$\phi(x) = x,$$

for all $x \in U$. Then $D\phi(p) = \text{Id}$ for all $p \in U$, and so the result follows from Definition 3.5.5 and Theorem 3.5.2. □

More generally, the above theorem is true when U is any open set.

Corollary 3.5.7. *Let $S = \phi(U)$ be a parameterized set in $\mathbf{R}^n$, where $U \subset \mathbf{R}^k$ is a domain ($1 \leq k < n$), $a \in U$, and $\phi : U \to \mathbf{R}^n$ is a smooth, one-to-one regular function with $\phi(a) = p$. Then*

$$T_p(S) = (\phi_*)_a \left(T_a(\mathbf{R}^k)\right) = \text{Image}((\phi_*)_a).$$

Proof. This is essentially Theorem 3.4.10 in the language of Definition 3.5.5 and Theorem 3.5.6. □

3.6 Diffeomorphisms

Having placed the derivative of a function in its proper context—as a linear map between tangent spaces—we immediately turn to a special class of differentiable functions. These will be the "isomorphisms" in the differentiable category. They appear naturally whenever we discuss "change of variables." We shall encounter examples of these later in our treatment of geometry as the "symmetries" of a given geometric structure.

In this section, we consider functions $f : \mathbf{R}^n \to \mathbf{R}^n$ between Euclidean spaces of the same dimension. In Chap. 2, linear functions (transformations) $T : \mathbf{R}^n \to \mathbf{R}^n$ were studied by means of square matrices, with bases being chosen for the domain and range. In that case, properties of T such as whether it is one-to-one or (equivalently, by Corollary 2.7.11) onto could be ascertained, for example by considering the determinant of the matrix representation of T.

The case we are considering in this chapter, in which $f : \mathbf{R}^n \to \mathbf{R}^n$ is nonlinear, is more intricate. For example, there is no general method available even to determine whether such a function is one-to-one or onto, short of falling back on the definitions directly. We list here a few examples and encourage the reader to verify the stated properties.

Example 3.6.1. Let $f : \mathbf{R}^2 \to \mathbf{R}^2$ be given by

$$f(u, v) = (u^3 - v^3, u).$$

Then f is one-to-one and onto.

Example 3.6.2. Let $f : \mathbf{R}^2 \to \mathbf{R}^2$ be given by

$$f(s, t) = (s \cos t, s \sin t).$$

Then f is onto but not one-to-one.

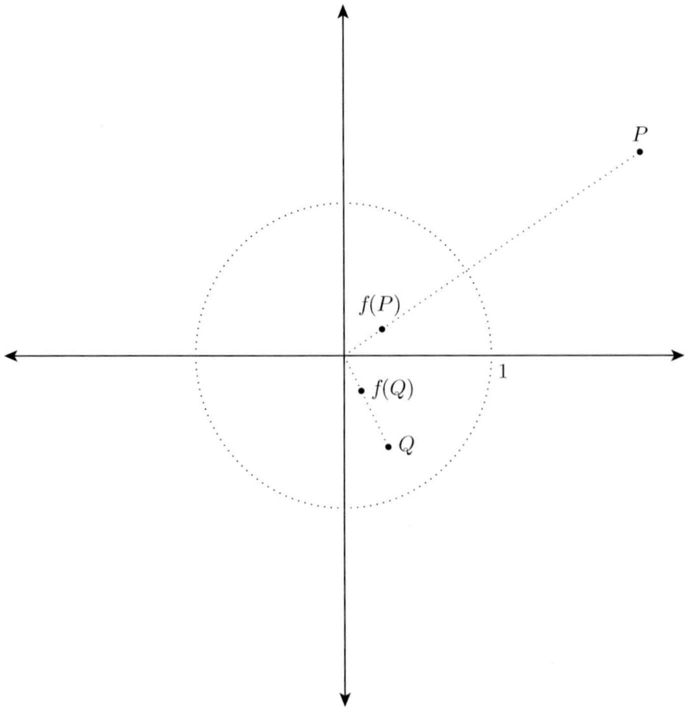

Fig. 3.9 The function $f : \mathbf{R}^2 \to \mathbf{R}^2$ of Example 3.6.3.

Example 3.6.3. Let $f : \mathbf{R}^2 \to \mathbf{R}^2$ be given by

$$f(x,y) = \begin{cases} \left(\frac{x(-1+\sqrt{1+4x^2+4y^2})}{2(x^2+y^2)}, \frac{y(-1+\sqrt{1+4x^2+4y^2})}{2(x^2+y^2)} \right), & \text{if } (x,y) \neq (0,0), \\ (0,0), & \text{if } (x,y) = (0,0). \end{cases}$$

Geometrically, this function maps a point P in the plane to a point $f(P)$ inside the unit circle by scaling the vector $\overrightarrow{OP}$ by a factor depending on the distance from the origin. See Fig. 3.9. In fact, in polar coordinates, the function f is given by

$$(r,\theta) \mapsto \left(\frac{-1+\sqrt{1+4r^2}}{2r}, \theta \right).$$

Then f is one-to-one but not onto. However, if we let $U \subset \mathbf{R}^2$ be the open unit disk

$$U = \left\{ (u,v) \mid u^2 + v^2 < 1 \right\},$$

then the function $f : \mathbf{R}^2 \to U$ is one-to-one and onto. In fact, the inverse function $f^{-1} : U \to \mathbf{R}^2$ is given by

$$f^{-1}(u,v) = \left(\frac{u}{1-u^2-v^2}, \frac{v}{1-u^2-v^2}\right),$$

which maps the open disk onto the plane.

One-to-one, onto functions play a special role in mathematics. For example, they generalize the notion of *permutations* or "rearrangements" of finite sets, which can be thought of as functions that change the "order" of elements but not the number of elements.

When a one-to-one, onto function in addition preserves some additional structure on the domain and range, we are led to the notion of *isomorphism*. Two sets with some structure are isomorphic if there is a one-to-one, structure-preserving function between them. In Sect. 2.7, we saw the example of a *linear isomorphism* as a one-to-one, onto function between two vector spaces that preserves the vector space operations. Linear isometries (Sect. 2.9) and linear symplectomorphisms (Sect. 2.10) are special cases of this.

In the context of differential geometry, the sets we consider are given enough structure so that we are able to define derivatives, along with the concepts that can be defined in terms of differentiability, such as tangent vectors and tangent space. That is what motivates, at a more abstract level, the definition of a manifold.

We now consider isomorphisms that preserve the differential structure.

Definition 3.6.4. Let $U, V \subset \mathbf{R}^n$ be domains. A function $f : U \to V$ is a *diffeomorphism* if:

1. f is one-to-one and onto, and so the inverse $f^{-1} : V \to U$ exists;
2. f is smooth on U;
3. f^{-1} is smooth on V.

The second and third conditions of Definition 3.6.4 are independent of each other, as the following example shows.

Example 3.6.5 (A one-to-one, onto function that is smooth but whose inverse is not smooth). Consider the function $f : \mathbf{R}^1 \to \mathbf{R}^1$ given by $f(x) = x^3$, which the reader can confirm is one-to-one and onto. Moreover, f is differentiable for all $x \in \mathbf{R}^1$. However, the inverse $f^{-1}(y) = y^{1/3}$ is not differentiable at 0, so f is not a diffeomorphism of $\mathbf{R}^1$.

We will see that this example reflects a more general fact about the relationship between the tangent map of a function and that of its inverse.

Theorem 3.6.6. *Let $U, V \subset \mathbf{R}^n$ be domains and let $f : U \to V$ be a diffeomorphism. Then for all $p \in U$, $(f_*)_p : T_p(\mathbf{R}^n) \to T_{f(p)}(\mathbf{R}^n)$ is a linear* isomorphism *of vector spaces with inverse*

$$(f_*)_p^{-1} = ((f^{-1})_*)_{f(p)}.$$

Proof. The statement is a consequence of applying the chain rule (Theorem 3.5.3) and the fact that $(\mathrm{Id}_{\mathbf{R}^n})_*(p) = \mathrm{Id}_{T_p(\mathbf{R}^n)}$ to the identities

$$g \circ f = \mathrm{Id}, \quad f \circ g = \mathrm{Id},$$

where $g = f^{-1}$. □

Corollary 3.6.7. *Suppose $f : U \to V$ is a diffeomorphism. Then for all $p \in U$, $\det(f_*)_p \neq 0$.*

Example 3.6.8. Let

$$U = \left\{(p_1, p_2) \mid p_1^2 + p_2^2 > 0\right\} = \mathbf{R}^2 \setminus \{(0,0)\},$$

and define the function $f : U \to U$ by

$$f(p_1, p_2) = \left(\frac{p_1}{p_1^2 + p_2^2}, \frac{p_2}{p_1^2 + p_2^2} \right).$$

Geometrically, f assigns to each point p in U that is $r > 0$ units away from the origin another point $f(p)$ that is on the same ray as $\overrightarrow{0p}$ but $1/r$ units away from the origin. In particular, all points on the unit circle $S^1 = \left\{(p_1, p_2) \mid p_1^2 + p_2^2 = 1\right\}$ are *fixed points*, i.e., for $p \in S^1$, $f(p) = p$. See Fig. 3.10.

It can be proved directly that f is one-to-one and onto. It is more convenient to observe, either geometrically or by the appropriate computation, that

$$(f \circ f)(p) = p,$$

and so $f^{-1} = f$.

Computing the Jacobian matrix yields

$$[f_*](p_1, p_2) = \begin{bmatrix} \frac{p_2^2 - p_1^2}{(p_1^2+p_2^2)^2} & \frac{-2p_1p_2}{(p_1^2+p_2^2)^2} \\ \frac{-2p_1p_2}{(p_1^2+p_2^2)^2} & \frac{p_1^2 - p_2^2}{(p_1^2+p_2^2)^2} \end{bmatrix}.$$

This can be seen to be smooth on U. Since $f^{-1} = f$, f^{-1} is differentiable also. Hence f is a diffeomorphism.

Corollary 3.6.7 implies that if f has a *critical point*, then it is not a diffeomorphism. This gives another way of viewing Example 3.6.5, since $f(x) = x^3$ has a critical point at $x = 0$.

It is not hard to describe the same type of obstacle in higher dimensions.

Example 3.6.9. Let $f : \mathbf{R}^2 \to \mathbf{R}^2$ be the function given by $f(x, y) = (x^3 - y^3, x)$. We mentioned in Example 3.6.1 that f is one-to-one and onto. However, the Jacobian matrix is given by

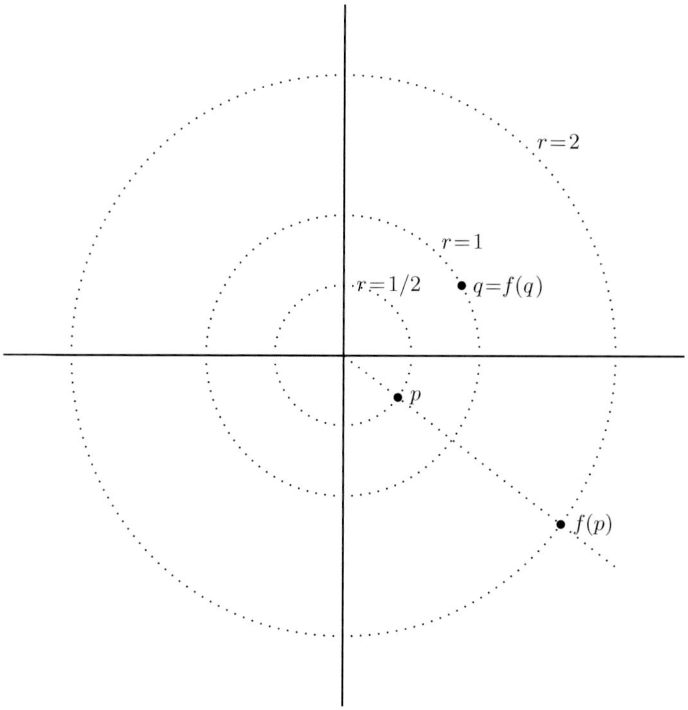

Fig. 3.10 The diffeomorphism of Example 3.6.8.

$$Df(x,y) = \begin{bmatrix} 3x^2 & 1 \\ -3y^2 & 0 \end{bmatrix},$$

which is not invertible whenever $y = 0$, i.e., along the x-axis. Hence f is not a diffeomorphism of $\mathbf{R}^2$.

Example 3.6.10. The following functions are diffeomorphisms:

1. $f : \mathbf{R}^2 \to \mathbf{R}^2$ given by $f(x,y) = (xe^y, y)$;
2. $g : \mathbf{R}^3 \to \mathbf{R}^3$ given by $g(x,y,z) = (2x, y, z - xy)$.

The verifications are left as an exercise.

We mention that there is an "algebra" of diffeomorphisms in the sense of group theory. Namely, we have the following result, the proof of which is a routine application of the definitions and the chain rule.

Theorem 3.6.11. *The set of all diffeomorphisms of* $\mathbf{R}^n$ *forms a* group *with the operation of function composition. That is,*

1. *If* $f, g : \mathbf{R}^n \to \mathbf{R}^n$ *are diffeomorphisms, then so is* $f \circ g$;
2. *The identity function* Id $: \mathbf{R}^n \to \mathbf{R}^n$ *given by* $\mathrm{Id}(x) = x$ *is a diffeomorphism;*
3. *If* $f : \mathbf{R}^n \to \mathbf{R}^n$ *is a diffeomorphism, then so is* f^{-1}.

We will devote significant attention in this and later chapters to developing methods of constructing diffeomorphisms.

We close this section by stating a partial converse to Theorem 3.6.6. This is a key result in advanced calculus that implies (in fact, it is equivalent to) the implicit function theorem (Theorem 3.3.18) above.

Theorem 3.6.12 (Inverse function theorem). *Let $f : \mathbf{R}^n \to \mathbf{R}^n$ be a function and $p \in \mathbf{R}^n$. Suppose that f is smooth in a domain containing p and that $(f_*)_p$ is a linear isomorphism. Then there exist a domain U containing p, a domain V containing $f(p)$, and a function $g : V \to U$ such that:*

1. *g is smooth on V;*
2. *$g \circ f = \mathrm{Id}_U$ and $f \circ g = \mathrm{Id}_V$.*

In other words, a function that is differentiable near p and with a one-to-one and onto tangent map at p has a differentiable inverse *near* p. For this reason, a function $f : \mathbf{R}^n \to \mathbf{R}^n$ satisfying the conditions of the inverse function theorem is sometimes called a *local diffeomorphism*.

The proof of the inverse function theorem is quite technical. It can be found, for example, in Spivak's *Calculus on Manifolds* [37]. In this case, as in the case of the implicit function theorem, we have stated the theorem for smooth functions, even though the result is true for functions that are continuously differentiable.

We will illustrate the power of the inverse function theorem by showing that a geometric set S described as a parameterized set may also be described locally as the level set of some function.

Proposition 3.6.13. *Let $S = \phi(U)$ be a parameterized geometric set, where $U \subset \mathbf{R}^k$ ($1 \leq k < n$) is a domain and $\phi : U \to \mathbf{R}^n$ is a regular parameterization. Then for all $p \in U$, there exist a domain $V \subset U$ containing p, a domain $W \subset \mathbf{R}^n$ containing $\phi(p)$ such that $\phi(V) \subset W$, and a diffeomorphism $\Phi : W \to \Phi(W) \subset \mathbf{R}^n$ such that*

$$\phi(V) = \left\{ y \in W \mid \Phi^i(y) = 0 \textit{ for } i = k+1, \dots, n \right\},$$

where the maps $\Phi^j : W \to \mathbf{R}$ ($j = 1, \dots, n$) are the component functions of $\Phi = (\Phi^1, \dots, \Phi^n)$.

Proof. Let $\phi^i : U \to \mathbf{R}$ be the component functions of ϕ, so that for $(u_1, \dots, u_k) \in U$,

$$\phi(u_1, \dots, u_k) = (\phi^1(u_1, \dots, u_k), \dots, \phi^n(u_1, \dots, u_k)).$$

Since ϕ is a regular parameterization, the derivative $(\phi_*)(p)$ can be represented as an $n \times k$ matrix with rank k at every point p. We may assume, by permuting the coordinates if necessary, that

$$\left[\frac{\partial \phi^i}{\partial u_j} \right] (p), \quad i, j = 1, \dots, k,$$

is an invertible matrix for $p \in U$. Define $\tilde{\phi} : U \to \mathbf{R}^k$ as

$$\tilde{\phi}(u_1, \ldots, u_k) = (\phi^1(u_1, \ldots, u_k), \ldots, \phi^k(u_1, \ldots, u_k)).$$

Then by the inverse function theorem (Theorem 3.6.12), there exist a domain $V' \subset U$ containing p and a domain $V'' \subset \mathbf{R}^k$ containing $\tilde{\phi}(p)$ such that $\tilde{\phi} : V' \to V''$ is a diffeomorphism with inverse $\tau = \tilde{\phi}^{-1} : V'' \to V'$.

Let $W' \subset \mathbf{R}^n$ be a domain with the property that $\pi(W') \subset V''$, where $\pi : \mathbf{R}^n \to \mathbf{R}^k$ is the projection $\pi(x_1, \ldots, x_n) = (x_1, \ldots, x_k)$. Note that $\tilde{\phi} = \pi \circ \phi$.

Define the map $\Phi : W' \to \mathbf{R}^n$ by means of the component functions $\Phi = (\Phi^1, \ldots, \Phi^n)$ as follows:

$$\Phi^i(x) = \begin{cases} x_i & \text{for } i = 1, \ldots, k, \\ x_i - \phi^i(\tau(\pi(x))) & \text{for } i = k+1, \ldots, n. \end{cases}$$

Since $\frac{\partial \pi}{\partial x_j} = 0$ for $j = k+1, \ldots, n$, the reader can check that $\det(\Phi_*) = 1$. Hence, again by the inverse function theorem, there is a domain $W \subset W'$ containing $\phi(p)$ such that $\Phi : W \to \Phi(W)$ is a diffeomorphism.

Now set $V = \tau(\pi(W)) \cap \phi^{-1}(W)$. A topological argument shows that V is a domain. Further, since $\phi(p) \in W$, we have $p \in V$. By construction, $\phi(V) \subset W$.

We will show that Φ, along with the domains V and W, has the desired properties, namely that

$$\phi(V) = \left\{ x \in \mathbf{R}^n \mid \Phi^i(x) = 0 \text{ when } i = k+1, \ldots, n \right\}.$$

Suppose first that $y = (y_1, \ldots, y_n) \in \phi(V)$, i.e., there is $x \in V$ such that $y = \phi(x)$. Then for every $i = k+1, \ldots, n$,

$$\begin{aligned} \Phi^i(y) &= y_i - \phi^i(\tau(\pi(y))) \\ &= \phi^i(x) - \phi^i(\tau(\pi(\phi(x)))) \\ &= \phi^i(x) - \phi^i(\tau(\tilde{\phi}(x))) \\ &= \phi^i(x) - \phi^i(x) \\ &= 0. \end{aligned}$$

Hence

$$\phi(V) \subset \left\{ y \in W \mid \Phi^i(y) = 0 \text{ when } i = k+1, \ldots, n \right\}.$$

Now suppose that $y = (y_1, \ldots, y_n) \in W$ is such that $\Phi^i(y) = 0$ for all $i = k+1, \ldots, n$. In other words, for $i = k+1, \ldots, n$, we have

$$y_i = \phi^i(\tau(\pi(y))).$$

Let $x = \tau(\pi(y))$. For $i = 1, \ldots, k$,

$$\begin{aligned}\phi^i(x) &= \tilde{\phi}^i(x) \\ &= \tilde{\phi}^i(\tau(\pi(y))) \\ &= \left(\tilde{\phi}(\tau(\pi(y)))\right)_i \\ &= (\pi(y))_i \\ &= y_i.\end{aligned}$$

Together with the supposition, we have $y_i = \phi^i(x)$ for all $i = 1, \ldots, n$, i.e., $y = \phi(x)$. Since $x \in \tau(\pi(W))$ by construction and since $\phi(x) = y \in W$, we have $x \in V$, and so $y \in \phi(V)$. This shows that

$$\left\{y \in W \mid \Phi^i(y) = 0 \text{ when } i = k+1, \ldots, n\right\} \subset \phi(V),$$

which together with the preceding argument yields

$$\left\{y \in W \mid \Phi^i(y) = 0 \text{ when } i = k+1, \ldots, n\right\} = \phi(V),$$

as desired. □

It is in the sense of the following corollary that a parameterized set can be realized locally as a level set.

Corollary 3.6.14. *Let S be a geometric set parameterized as $S = \phi(U)$, where $U \subset \mathbf{R}^k$ $(1 \leq k < n)$ is a domain and $\phi : U \to \mathbf{R}^n$ is a regular parameterization. Then for each point $p \in S$, there exist a domain $W \subset \mathbf{R}^n$ containing p and a smooth map $F : \mathbf{R}^n \to \mathbf{R}^{n-k}$ such that $S \cap W = F^{-1}(\mathbf{0})$.*

Proof. For $p \in S$, by Proposition 3.6.13, there exist a domain W containing p and a map $\Phi : W \to \mathbf{R}^n$ such that for all $x \in S \cap W$, $\Phi^i(x) = 0$ for $i = k+1, \ldots, n$. Then the map

$$F(x) = (\Phi^{k+1}(x), \ldots, \Phi^n(x))$$

is the desired function. □

3.7 Vector Fields: From Local to Global

Up to this point, the core idea of this chapter has been associating to each point of some (possibly) nonlinear object—be it a function or a geometric set—a linear object that "approximates" the nonlinear one. In this section, we begin to consider how constructions at the level of linear approximations can give rise to nonlinear objects. It is passage from "local" to "global." In the course of these considerations, we will encounter many of the objects fundamental to differential geometry.

The first step in the transition from local considerations, or constructions *at a point* p, to global ones involving whole regions or sets of points will be constructing the proper context for such considerations—the *tangent bundle*. Up to now, for each point $p \in \mathbf{R}^n$ we have constructed a vector space $T_p(\mathbf{R}^n)$. We now consider the collection of *all* such tangent spaces.

Definition 3.7.1. The *tangent bundle to* $\mathbf{R}^n$, denoted by $T\mathbf{R}^n$, is the set of all ordered pairs of the form $(p, \mathbf{v}_p)$, where $p \in \mathbf{R}^n$ and $\mathbf{v}_p \in T_p(\mathbf{R}^n)$.

Using the standard bases for $\mathbf{R}^n$ and the tangent spaces $T_p(\mathbf{R}^n)$, elements of the tangent bundle "look like" elements of $\mathbf{R}^{2n}$:

$$(p, \mathbf{v}_p) = \big((p_1, \ldots, p_n), \langle v_1, \ldots, v_n \rangle_p\big),$$

where for $i = 1, \ldots, n$, p_i is a coordinate of the point p and v_i is a component of the tangent vector $\mathbf{v}_p$, considered in either of the two senses we have defined: either $\langle v_1, \ldots, v_n \rangle_p = c'(0)$, where c is a smooth curve passing through $c(0) = p$, or

$$\mathbf{v}_p = \sum_{i=1}^{n} v_i \frac{\partial}{\partial x_i}\bigg|_p.$$

It is conceptually important, however, to note that $T\mathbf{R}^n$ is "not" $\mathbf{R}^{2n}$, at least not as vector spaces. In particular, there is no sense of addition of elements $(p, \mathbf{v}_p), (q, \mathbf{v}_q) \in T\mathbf{R}^n$ unless $p = q$. The tangent bundle $T\mathbf{R}^n$ has the structure of what is known as a *vector bundle*, where there is a vector space structure on the "fiber" $T_p(\mathbf{R}^n)$ over each point p in the "base space" $\mathbf{R}^n$. We refer the interested reader to [38] for an in-depth presentation of this mathematical structure in the context of differential geometry.

For our purposes, however, the main importance of the tangent bundle is its role as the setting in which to describe vector fields.

Definition 3.7.2. For any domain $U \subset \mathbf{R}^n$, a (smooth) *vector field on* U is a function $V : U \to T\mathbf{R}^n$ such that:

1. For each $p \in U$, $V(p) = (p, \mathbf{v}_p)$, where $\mathbf{v}_p \in T_p(\mathbf{R}^n)$;
2. V is differentiable in the following sense: for all $p \in U$ and for all smooth functions f defined near p, the function $F : U \to \mathbf{R}$ given by

$$F(p) = \mathbf{v}_p[f] \text{ where } V(p) = (p, \mathbf{v}_p)$$

is a smooth function of p.

The first condition essentially assigns to each point p a tangent vector at p. The second ensures that this assignment is done smoothly. In terms of components, it amounts to saying that V can be written in the form

$$V(p) = \left(p, \sum_{i=1}^{n} v_i(p) \frac{\partial}{\partial x_i}\bigg|_p\right),$$

where the functions $v_i : U \to \mathbf{R}$ are smooth functions of p. See Fig. 3.11.

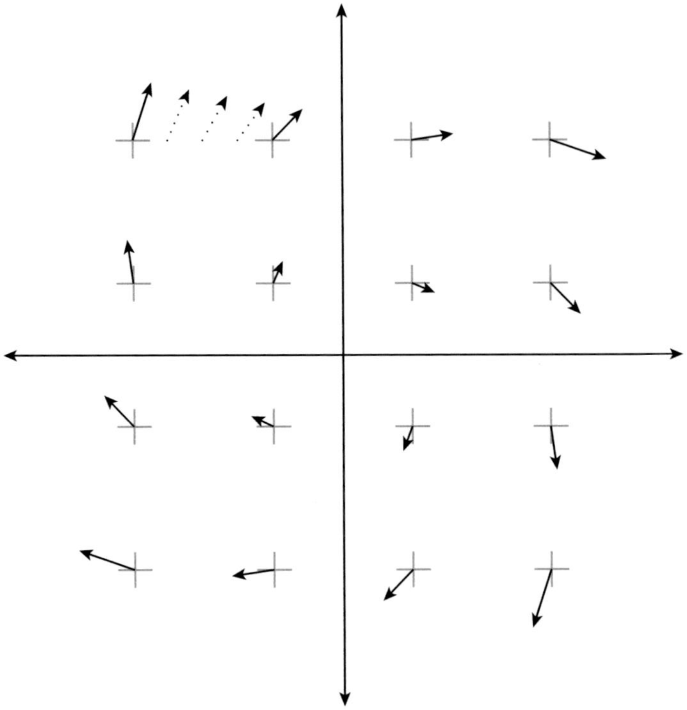

Fig. 3.11 A vector field in $\mathbf{R}^2$.

Definition 3.7.2 relies on the analytic presentation of tangent vectors as linear derivations on $C_p^\infty(\mathbf{R}^n)$. There is also a natural geometric interpretation of vector fields, however. A vector field can be pictured as a smooth assignment of a vector (an "arrow") to each point $p \in \mathbf{R}^n$. The component functions then describe how the arrows change with position. In physical terms, the vector field might represent a *force field*, where the arrow at each point represents a force acting on a particle located at that point, with the force depending on position. The vector field could also represent a *velocity field* of a fluid, where the vector at each point represents the velocity that a particle "dropped into" the fluid at that point would experience.

We now illustrate the definition with some typical vector fields that arise in various applications.

Example 3.7.3 (Constant vector fields). Let $V : \mathbf{R}^n \to T\mathbf{R}^n$ be given by

$$\begin{aligned} V(p) &= \left(p, \sum_{i=1}^n k_i \frac{\partial}{\partial x_i}\bigg|_p\right) \\ &= \Big((p_1, \ldots, p_n), \langle k_1, \ldots, k_n\rangle_p\Big), \end{aligned}$$

where the k_i are constants. In terms of the component functions $v_i : \mathbf{R}^n \to \mathbf{R}$, these vector fields correspond to $v_i(p) = k_i$ for all $p \in \mathbf{R}^n$ and each $i = 1, \ldots, n$. Geometrically, these vector fields can be envisioned by drawing the same vector at each point $p \in \mathbf{R}^n$.

In particular, we can consider the constant vector fields

$$E_i(p) = \left(p, \left.\frac{\partial}{\partial x_i}\right|_p\right),$$

for $i = 1, \ldots, n$, where $v_j(p) = 1$ if $j = i$ and $v_j(p) = 0$ if $j \neq i$. At each point p, the set of tangent vectors $\{E_1(p), \ldots, E_n(p)\}$ is the standard basis for $T_p(\mathbf{R}^n)$.

We have explicitly written the dependence of vector fields on the "base point" p by writing them as an ordered pair. This notation emphasizes the role of the tangent bundle as the proper codomain of a vector field:

$$V : U \to T\mathbf{R}^n.$$

However, it is cumbersome and ultimately somewhat redundant to emphasize the dependence on the point p with a separate component as we have been doing when we write $V(p) = (p, \mathbf{v}_p)$. Hence, in the following, we will write $V(p) = \mathbf{v}_p \in T_p(\mathbf{R}^n)$, with the understanding that the dependence on p is seen in the components of $\mathbf{v}_p$ (as functions of p) and with the agreement that it is not possible to perform vector space operations on vectors $V(p), V(q)$ when $p \neq q$.

Example 3.7.4 (Radial vector fields). Let $W_0 : \mathbf{R}^3 \to T\mathbf{R}^3$ be given by

$$\begin{aligned} W_0(p_1, p_2, p_3) &= \sum_{i=1}^{n} p_i \left.\frac{\partial}{\partial x_i}\right|_p \\ &= \langle p_1, p_2, p_3 \rangle_p, \end{aligned}$$

where $p = (p_1, p_2, p_3)$. In the notation above, $v_i(p) = p_i$. Geometrically, $W_0(p)$ can be represented as a tangent vector at p that points away from the origin with a magnitude equal to the distance from p to the origin. In fact, $W_0(p)$ is precisely the position vector of p (the vector beginning at the origin $O = (0, 0, 0)$ and ending at p) but affixed to the point p instead of the origin. Note that $W_0(O) = \mathbf{0}_O$.

Similarly, we can define $W_1 : \mathbf{R}^3 \setminus \{\mathbf{0}\} \to TR^3$ by

$$\begin{aligned} W_1(p_1, p_2, p_3) &= \sum_{i=1}^{n} v_i(p) \left.\frac{\partial}{\partial x_i}\right|_p \\ &= \langle v_1(p), v_2(p), v_3(p) \rangle_p, \end{aligned}$$

where

$$v_i(p) = \frac{p_i}{(p_1^2 + p_2^2 + p_3^2)^{1/2}}.$$

In this case, $W_1(p)$ can be visualized as a unit vector pointing away from the origin at p. However, $W_1(O)$ is not defined.

Example 3.7.5 (A rotational vector field). Let $X_\theta : \mathbf{R}^2 \to T\mathbf{R}^2$ be the vector field defined by

$$\begin{aligned} X_\theta(p_1, p_2) &= -p_2 \frac{\partial}{\partial x_1}\bigg|_p + p_1 \frac{\partial}{\partial x_2}\bigg|_p \\ &= \langle -p_2, p_1 \rangle_p \, . \end{aligned}$$

Note that $X_\theta(0,0) = \mathbf{0}_{(0,0)}$. Further, at any point p a distance r from the origin, $X_\theta(p)$ will be a vector tangent at p to the circle of radius r centered at the origin, oriented in a counterclockwise direction. This can be seen by noting that at each point p, $X_\theta(p)$ is perpendicular to the position vector. Notice that for $c(t) = (r\cos t, r\sin t)$, the tangent vector to $c(t)$ at $p = (p_1, p_2) = c(t_0) = (r\cos t_0, r\sin t_0)$, is given by

$$\begin{aligned} c'(t_0) &= \langle -r\sin t_0, r\cos t_0 \rangle_p \\ &= \langle -p_2, p_1 \rangle_p \\ &= X_\theta(c(t_0)). \end{aligned}$$

This vector field is closely tied to polar coordinates. Let $f \in C^\infty(\mathbf{R}^2)$ be a real-valued smooth function defined on $\mathbf{R}^2$ and consider polar coordinates

$$\begin{aligned} p_1 &= r\cos\theta, \\ p_2 &= r\sin\theta. \end{aligned}$$

The chain rule gives

$$\begin{aligned} \frac{\partial f}{\partial \theta} &= \frac{\partial f}{\partial p_1}\frac{\partial p_1}{\partial \theta} + \frac{\partial f}{\partial p_2}\frac{\partial p_2}{\partial \theta} \\ &= \frac{\partial f}{\partial p_1}\Big(-r\sin\theta\Big) + \frac{\partial f}{\partial p_2}\Big(r\cos\theta\Big) \\ &= \frac{\partial f}{\partial p_1}\left(-p_2\right) + \frac{\partial f}{\partial p_2}\left(p_1\right), \end{aligned}$$

or in other words,

$$\frac{\partial}{\partial \theta}[f] = \left(-p_2 \frac{\partial}{\partial p_1} + p_1 \frac{\partial}{\partial p_2}\right)[f],$$

which shows that in polar coordinates, $X_\theta = \frac{\partial}{\partial \theta}$.

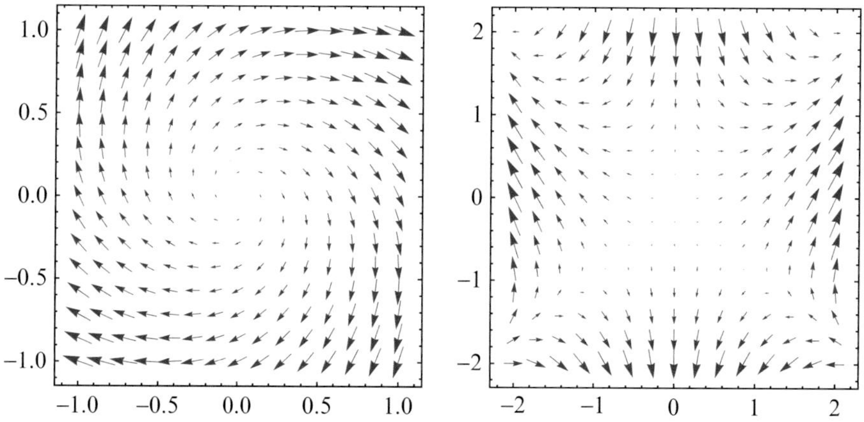

Fig. 3.12 Some vector fields plotted with Mathematica.

Since the advent of computer algebra systems like Mathematica, Maple, and MathCAD, picturing vector fields has become a routine task, at least in two dimensions. See Fig. 3.12. Readers with access to such software would benefit by spending 15 minutes plotting various vector fields of their choosing to develop an intuition for how the component functions affect the behavior of the vector field in question.

We will repeatedly confront the question of how a map $\phi : \mathbf{R}^n \to \mathbf{R}^m$ acts on vector fields.

Definition 3.7.6. Let $\phi : \mathbf{R}^n \to \mathbf{R}^m$ be a smooth map, and let X be a smooth vector field on $\mathbf{R}^n$. Then the *pushforward* of the vector field X, denoted by $\phi_* X$, is the vector field defined on the range of ϕ such that for every $g \in C_q^\infty(\mathbf{R}^m)$,

$$(\phi_* X)[g](q) = X[g \circ \phi](p),$$

where $q = \phi(p)$ is any point in the range of ϕ.

The following proposition is a smooth version of Theorem 3.5.2.

Proposition 3.7.7. *Let $\phi : \mathbf{R}^n \to \mathbf{R}^m$ be a smooth map with component functions ϕ^j, i.e., $\phi(p) = (\phi^1(p), \ldots, \phi^m(p))$, and let $\left\{ \frac{\partial}{\partial x_i} \mid i = 1, \ldots, n \right\}$ and $\left\{ \frac{\partial}{\partial y_j} \mid j = 1, \ldots, m \right\}$ be the standard bases for $T(\mathbf{R}^n)$ and $T(\mathbf{R}^m)$ respectively. Let $X = \sum X^i \frac{\partial}{\partial x_i}$ be a smooth vector field on $\mathbf{R}^n$. Then the components of $Y = \phi_* X = \sum Y^j \frac{\partial}{\partial y_j}$ are given by the smooth functions*

$$Y^j = \sum_{i=1}^n \frac{\partial \phi^j}{\partial x_i} \cdot X^i.$$

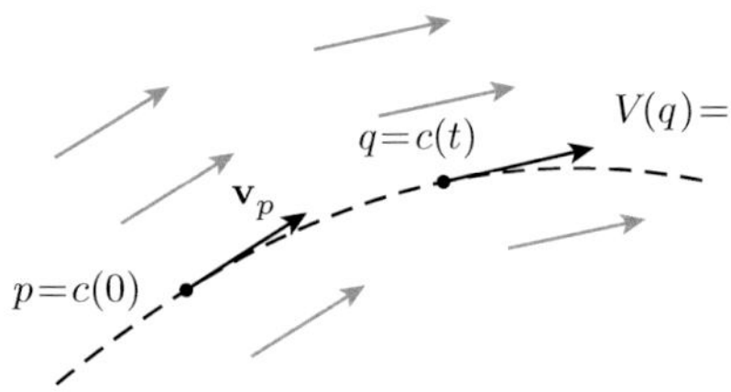

Fig. 3.13 The integral curve of a vector field.

Proof. Since the definition of the pushforward is essentially a pointwise definition, the only additional point to prove beyond Theorem 3.5.2 is the smoothness of the functions Y^j. This follows from the smoothness of ϕ and the component functions X^i. □

3.8 Integral Curves

The notion of a vector field gives rise to the following "integration" problem: Starting with a vector field $V : \mathbf{R}^n \to T\mathbf{R}^n$, can we construct a curve $c : I \to \mathbf{R}^n$ such that at every point along the curve, the tangent vector to c agrees with the vector field V? Such a curve will be called variously an "integral curve" or a "flow line" for V.

This problem is the first instance we have encountered of starting with a "local" object (defined at the level of the tangent space $T_p(\mathbf{R}^n)$) and producing a "global" object (defined at the level of the geometric space $\mathbf{R}^n$). It is the smoothness condition that allows this passage from local to global.

In physical terms, if we imagine V as the velocity vector field for a fluid at each point, the integral curve will represent the position of a particle "dropped into" the fluid at a given point as it is carried along by the fluid's currents. For example, in the examples of vector fields given in the previous section, an integral curve of a constant vector field should be a line in the direction of the field, while those of the radial fields should be a line pointing away from the origin and those of the rotational fields should be circles centered at the origin.

In order to frame the problem, we begin with a precise definition.

Definition 3.8.1. Let V be a smooth vector field on $\mathbf{R}^n$, so that $V(p) \in T_p(\mathbf{R}^n)$ for all $p \in \mathbf{R}^n$. An *integral curve of* V *through* p is a smooth parameterized curve $c : I \to \mathbf{R}^n$, where I is an interval containing 0, such that $c(0) = p$ and such that for all $t \in I$, $c'(t) = V(c(t))$.

See Fig. 3.13. It is customary to refer to the parameter t of an integral curve as "time."

Before discussing some of the important technical and theoretical aspects of this definition, we will attempt to construct some integral curves for vector fields in $\mathbf{R}^2$.

Example 3.8.2. Let V be the constant vector field on $\mathbf{R}^2$ given by

$$V(p) = \langle k_1, k_2 \rangle_p \in T_p(\mathbf{R}^2);$$

see Example 3.7.3. To find an integral curve through $p = (p_1, p_2)$, we consider a curve $c(t) = (x(t), y(t))$ and require that

$$c(0) = (x(0), y(0)) = (p_1, p_2)$$

and

$$c'(t) = \langle x'(t), y'(t) \rangle_{c(t)} = \langle k_1, k_2 \rangle_{c(t)} = V(c(t))$$

for all t where c is defined. In fact, the two conditions amount to solving the system of first-order differential equations

$$\begin{cases} \frac{dx}{dt} = k_1, \\ \frac{dy}{dt} = k_2, \end{cases}$$

with the initial conditions $x(0) = p_1$ and $y(0) = p_2$. The curve

$$c(t) = (p_1 + k_1 t, p_2 + k_2 t)$$

is a solution to this system and hence defines an integral curve for V. As predicted, the integral curves are all lines in the direction of $\langle k_1, k_2 \rangle$. Note that these curves are defined for all t, i.e., we can take $I = (-\infty, \infty)$.

Example 3.8.3. Let V be the radial vector field on $\mathbf{R}^2$ given by

$$V(p) = \langle p_1, p_2 \rangle_p \in T_p(\mathbf{R}^2),$$

where $p = (p_1, p_2)$; see Example 3.7.4. An integral curve $c : I \to \mathbf{R}^2$ of V through p given by $c(t) = (x(t), y(t))$ must satisfy $(x(0), y(0)) = (p_1, p_2)$ and

$$\langle x'(t), y'(t) \rangle_{c(t)} = \langle x(t), y(t) \rangle_{c(t)} .$$

This translates into the linear system of first-order differential equations

$$\begin{cases} \frac{dx}{dt} = x, \\ \frac{dy}{dt} = y, \end{cases}$$

with the initial conditions $x(0) = p_1$ and $y(0) = p_2$. This system is separable and has solution $(x(t), y(t)) = (p_1 e^t, p_2 e^t)$, i.e., $c(t) = (p_1 e^t, p_2 e^t)$ is an integral curve for V through (p_1, p_2). Note that the solution is the same even when p_1 or p_2 is zero. As in the previous example, $c(t)$ is defined for all t, and so the interval I can be chosen to be $(-\infty, \infty)$.

The curves themselves are in general curves of the form $y = mx$, i.e., lines through the origin with slope $m = p_2/p_1$ (or vertical when $p_1 = 0$). However, when $(p_1, p_2) \neq (0,0)$, the integral curve is in fact a ray, with $c(t) \to (0,0)$ as $t \to -\infty$ and $c(t)$ moving away from the origin as t increases. When $(p_1, p_2) = (0,0)$, the integral curve is the constant $c(t) = (0,0)$, representing a *stationary point* (or *equilibrium solution*).

Example 3.8.4. Let X_θ be the rotational vector field on $\mathbf{R}^2$ given by

$$X_\theta(p) = \langle -p_2, p_1 \rangle_p \in T_p(\mathbf{R}^2),$$

where again $p = (p_1, p_2)$; see Example 3.7.5. Without using the terminology, we already saw in that example that $c(t) = (r\cos t, r\sin t)$ was an integral curve through $(r\cos t_0, r\sin t_0)$. More generally, an integral curve $c(t) = (x(t), y(t))$ of X_θ through a point $p = (p_1, p_2)$ is a solution to the linear system of first-order differential equations

$$\begin{cases} \frac{dx}{dt} = -y, \\ \frac{dy}{dt} = \ x, \end{cases}$$

with the initial conditions $x(0) = p_1$ and $y(0) = p_2$. Such a system has solution

$$c(t) = (p_1 \cos t - p_2 \sin t, p_1 \sin t + p_2 \cos t).$$

Note again that c is defined for all t, so we can choose the interval $I = (-\infty, \infty)$.

Geometrically, the integral curves are all circles centered at the origin with radius $\sqrt{p_1^2 + p_2^2}$, at least when $p \neq (0,0)$. In the case $p = (0,0)$, the integral curve is the constant map defined by $c(t) = (0,0)$. Physically, one could picture that a particle dropped into the velocity field X_θ would rotate counterclockwise about the origin at a constant distance from the origin.

The examples above correspond to *linear* systems of differential equations. As a first course in differential equations illustrates, the solutions of such systems are particularly well behaved. For example, they are defined for all t. Equilibrium solutions are in general isolated (when 0 is not an eigenvalue of the associated matrix). A review of these topics can be found, for example, in [9].

To illustrate the subtleties involved in the case of a general vector field, we present one more example. Although the corresponding system of differential equations still admits analytic methods of producing solutions explicitly, as a partially decoupled system, it already illustrates the complicated nature of integral curves.

Example 3.8.5. Let V be the vector field on $\mathbf{R}^2$ given by

$$V(p) = \langle p_1^2, p_1 p_2 \rangle_p \in T_p(\mathbf{R}^2),$$

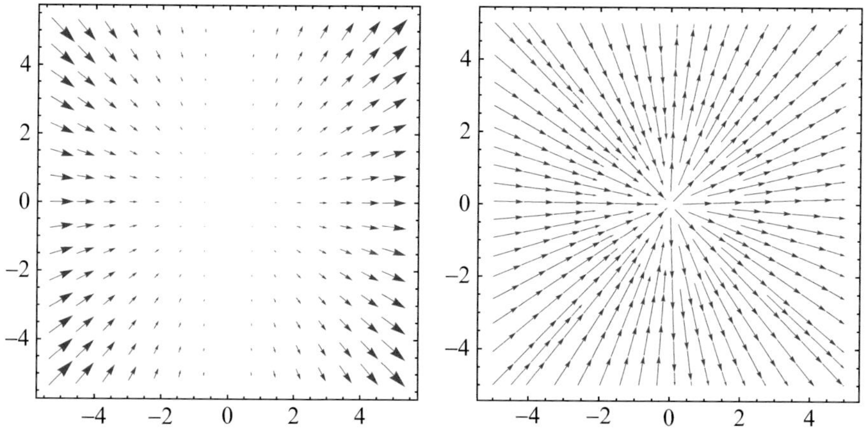

Fig. 3.14 A Mathematica plot of the vector field and integral curves from Example 3.8.5.

where $p = (p_1, p_2)$. Finding the integral curve $c(t) = (x(t), y(t))$ through p amounts to solving the partially decoupled system

$$\begin{cases} \frac{dx}{dt} = x^2, \\ \frac{dy}{dt} = xy, \end{cases}$$

with the initial conditions $x(0) = p_1$ and $y(0) = p_2$. In the case that $p_1 = 0$, the solution is simply $(x(t), y(t)) = (0, p_2)$ for all t. When $p_1 \neq 0$, the system admits as a solution $(x(t), y(t)) = \left(\frac{p_1}{1 - p_1 t}, \frac{p_2}{1 - p_1 t}\right)$. Note, however, that this solution is *not* defined for all t; in particular, it is not defined for $t = 1/p_1$. Since the interval I must contain 0, we choose I to be either $\left(-\infty, \frac{1}{p_1}\right)$ or $\left(\frac{1}{p_1}, +\infty\right)$ according to whether p_1 is positive or negative.

The integral curves $c : I \to \mathbf{R}^2$ of V given by

$$c(t) = \left(\frac{p_1}{1 - p_1 t}, \frac{p_2}{1 - p_1 t}\right)$$

represent rays with vertex at the origin (which is not part of the image of the integral curve). In the case that $p_1 > 0$, c is defined on the interval $I = \left(-\infty, \frac{1}{p_1}\right)$, and so a particle moving in the velocity field V moves away from the origin as t approaches $1/p_1$. In the case that $p_1 < 0$, c is defined on the interval $\left(\frac{1}{p_1}, \infty\right)$, so a particle would move toward the origin as t approaches ∞. Points on the p_2-axis, where $p_1 = 0$, are stationary points. See Fig. 3.14.

We now state the fundamental theorem of existence and uniqueness for first-order systems of differential equations, presented in the language of vector fields and integral curves. A detailed proof can be found in [38]; it is just a routine modification of the standard argument to prove existence and uniqueness for ordinary differential equations.

Theorem 3.8.6. *Let V be a smooth vector field on $\mathbf{R}^n$. For each point $p \in \mathbf{R}^n$, there exist an interval $I \subset \mathbf{R}$ containing 0 and a smooth curve $c : I \to \mathbf{R}^n$ such that $c(0) = p$ and $c'(t) = V(c(t))$ for all $t \in I$.*

Moreover, the integral curve for V is unique in the following sense: If $c_1 : I_1 \to \mathbf{R}^n$ and $c_2 : I_2 :\to \mathbf{R}^n$ are two integral curves for V through $p \in \mathbf{R}^n$, i.e., $c_1(0) = c_2(0) = p$, $c_1'(t) = V(c_1(t))$ for all $t \in I_1$ and $c_2'(t) = V(c_2(t))$ for all $t \in I_2$, then $c_1(t) = c_2(t)$ for all $t \in I_1 \cap I_2$.

We generally refer to the integral curve solely in terms of the parameterization c. In this case, we assume that the corresponding interval I is the *maximum* interval on which c is defined and has the required properties.

As usual with differential equations, explicitly producing an integral curve for a given vector field may be difficult or impossible. Nevertheless, the existence statement of Theorem 3.8.6 allows us to talk about integral curves for any smooth vector field. The uniqueness statement, in addition to allowing us to talk about *the* integral curve for a given vector field, has the following geometric consequence: Distinct integral curves do not cross.

Corollary 3.8.7. *Let V be a smooth vector field on $\mathbf{R}^n$. Let $c_1 : I_1 \to \mathbf{R}^n$ and $c_2 : I_2 \to \mathbf{R}^n$ be two integral curves for V. Then the images of c_1 and c_2 either do not intersect or they intersect for all t in an interval.*

Proof. Suppose that the images of two integral curves intersect at a point p. We may assume by reparameterizing by translation if necessary that $0 \in I_1 \cap I_2$ and that $c_1(0) = c_2(0) = p$. Then by the uniqueness of integral curves, $c_1(t) = c_2(t)$ for all $t \in I_1 \cap I_2$. □

We also note that diffeomorphisms "exchange" integral curves.

Proposition 3.8.8. *Let $\Phi : \mathbf{R}^n \to \mathbf{R}^n$ be a diffeomorphism and let X be a smooth vector field on $\mathbf{R}^n$. Let $c_1 : I_1 \to \mathbf{R}^n$ be the integral curve of X through $p \in \mathbf{R}^n$ and let $c_2 : I_2 \to \mathbf{R}^n$ be the integral curve of $Y = \Phi_* X$ through $\Phi(p) \in \mathbf{R}^n$. Then $\Phi(c_1(t)) = c_2(t)$ for all $t \in I_1 \cap I_2$.*

Proof. The proof is an application of the chain rule along with the uniqueness of integral curves. We have

$$\begin{aligned}\frac{d}{dt}(\Phi \circ c_1) &= \Phi_* \left(\frac{dc_1}{dt}\right) \\ &= \Phi_*(X(c_1(t))) \\ &= Y(\Phi(c_1(t))),\end{aligned}$$

so $(\Phi \circ c_1)(t)$ is the integral curve of Y through the point $\Phi(c_1(0)) = \Phi(p)$. By uniqueness of integral curves, then, $\Phi \circ c_1 = c_2$. □

From Example 3.8.5, it should be clear that the interval I on which an integral curve is defined cannot in general be chosen to be all of $\mathbf{R}$. We single out those vector fields whose integral curves are defined for all t.

Definition 3.8.9. A smooth vector field V on $\mathbf{R}^n$ is called *complete* if for all $p \in \mathbf{R}^n$, the interval I on which the integral curve through p is defined can be taken to be $I = \mathbf{R}$.

In this language, the vector fields in Examples 3.8.2–3.8.4 are complete, while the vector field in Example 3.8.5 is not complete.

Whether a vector field is complete depends on a number of factors and is in general not easy to determine. Topological conditions on the domain play a role. The growth of the component functions of the vector field also comes into play. The interested reader may consult [1, pp. 248–252] for a detailed discussion of this question to get a sense of some of the issues at stake.

3.9 Diffeomorphisms Generated by Vector Fields

In the previous section, we saw how a vector field naturally gives rise to an integral curve. Taking a more global perspective, we now show how a vector field gives rise to a family of diffeomorphisms, at least near a given point. We begin with an example.

Example 3.9.1. Let V be the vector field on $\mathbf{R}^2$ defined by

$$V(p) = V(p_1, p_2) = \langle k_1, k_2 \rangle_p \in T_p(\mathbf{R}^2),$$

where k_1 and k_2 are constants. We saw in Example 3.8.2 that through any given point p, the integral curve through p is given by $c_p : \mathbf{R} \to \mathbf{R}^2$,

$$c_p(t) = (p_1 + k_1 t, p_2 + k_2 t).$$

Now, for fixed t, define the function $\phi_t : \mathbf{R}^2 \to \mathbf{R}^2$ by

$$\phi_t(x, y) = (x + k_1 t, y + k_2 t).$$

We might consider, for example, the "time-1 flow" $\phi_1 : \mathbf{R}^2 \to \mathbf{R}^2$ given by $\phi_1(x, y) = (x + k_1, y + k_2)$ or the "time-(-3) flow" $\phi_{-3} : \mathbf{R}^2 \to \mathbf{R}^2$ given by $\phi_{-3}(x, y) = (x - 3k_1, y - 3k_2)$.

This function is closely related to the integral curves of V. The image of a point (x, y) under ϕ_t is the point obtained by starting at (x, y) and "following" the integral curve through (x, y) for time t. Using the notation of the previous section,

$$\phi_t(x,y) = c_{(x,y)}(t).$$

Note that the notation $c_{(x,y)}(t)$ for the integral curve emphasizes dependence on t, whereas the notation $\phi_t(x,y)$ emphasizes dependence on p. We will discuss this dependence in more detail below.

Now we choose a fixed value of t, say for definiteness $t = 1$. Then $\phi = \phi_1$ is defined for all $p \in \mathbf{R}^2$, with

$$\phi(x,y) = (x + k_1, y + k_2)$$

representing the function given by translation by the constant vector $\langle k_1, k_2 \rangle$. This function is a diffeomorphism whose Jacobian matrix is given by

$$[\phi_*](x,y) = \begin{bmatrix} 1 & 0 \\ 0 & 1 \end{bmatrix},$$

and the inverse is given by

$$\phi^{-1}(x,y) = (x - k_1, y - k_2).$$

The point of this example is to show how the integral curves of a vector field define a diffeomorphism, which is obtained by "following" the curve through some given increment of time. The next theorem generalizes this construction. It is really just a restatement of Theorem 3.8.6, emphasizing the smooth dependence on both p and t.

Theorem 3.9.2. *Let V be a smooth vector field on $\mathbf{R}^n$. Then for each point $p \in \mathbf{R}^n$, there are:*

- *a domain $U \subset \mathbf{R}^n$ containing p;*
- *an open interval $I_p \subset \mathbf{R}$ containing 0; and*
- *a smooth map $\phi : U \times I_p \to \mathbf{R}^n$, with elements in the range written as $\phi(x,t)$ or $\phi_t(x)$,*

with the following properties:

1. *For all $x \in U$, $\phi(x,0) = x$;.*
2. *For all $x \in U$ and all $t \in I_p$, the parameterized curve $c : I_p \to \mathbf{R}^n$ given by*

$$c(t) = \phi(x,t)$$

is the integral curve of V through x.

3. *For all $t \in I_p$, the function $\phi_t : U \to \phi_t(U) \subset \mathbf{R}^n$ is a diffeomorphism onto its image.*
4. *For all $s, t \in I_p$ such that $s + t \in I_p$, we have*

$$\phi_s(\phi_t(x)) = \phi_{s+t}(x).$$

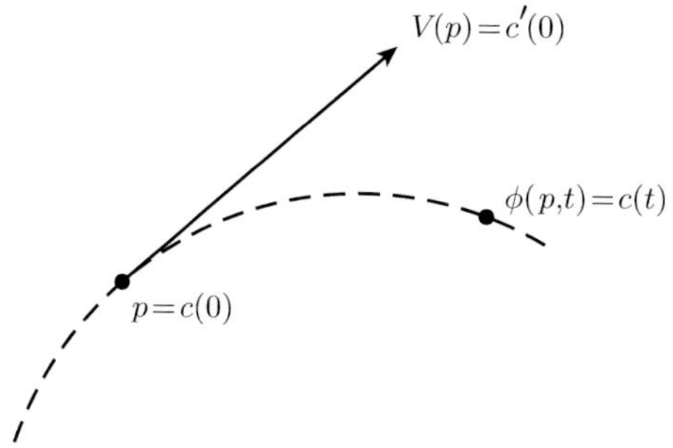

Fig. 3.15 A schematic of the flow of a vector field.

The function $\phi : U_p \times I_p \to \mathbf{R}^n$ is called the *flow* of the vector field V. See Fig. 3.15.

For a detailed discussion of the proof, we refer to [38, Chap. 5]. A more "modern" presentation, written in the language of calculus on Banach spaces, is given in [1].

Before proceeding with more examples of this construction, it is worth clarifying the content of this fundamental theorem. From the outset, the theorem shows that the differentiability of the vector field V guarantees the existence of *three* distinct data, all of which depend on p. The dependence on p, in the language of ordinary differential equations, is the dependence of solutions on the initial conditions. Condition (1) states that the diffeomorphism ϕ_0 is the identity map. Condition (2) specifies the way in which the family of diffeomorphisms depends on the vector field V. We have stated the theorem to emphasize condition (3): Vector fields give rise to a family of diffeomorphisms, which are local in the sense that their domain is not necessarily all of $\mathbf{R}^n$. Condition (3) also expresses the smooth dependence of solutions of differential equations on the initial conditions.

Condition (4) has a number of important consequences on the relationship between different diffeomorphisms in the "family" that can be considered parameterized by t. For example, every ϕ_t has an inverse given by $(\phi_t)^{-1} = \phi_{-t}$, as long as t and $-t$ are in I_p. Moreover, the diffeomorphisms in each family (for a given p) commute:

$$(\phi_s \circ \phi_t)(p) = (\phi_t \circ \phi_s)(p).$$

In fact, when the vector field V is *complete*, the interval can be chosen to be $\mathbf{R}$ for all p. In this case, the set

$$\{\phi_t \mid t \in \mathbf{R}\}$$

is a *group* under the operation of composition of diffeomorphisms. For this reason, the set of diffeomorphisms $\{\phi_t\}$ is often called the *one-parameter group of diffeomorphisms* of the vector field V.

Especially in this case that V is complete, we often choose a *particular* diffeomorphism from each family, i.e., we fix a value of t independently of the point p. For example, as in Example 3.9.1, we chose $t = 1$ for each p, yielding what is called the *time-one flow* ϕ_1 of a vector field V.

Example 3.9.3. Following Example 3.8.3, let

$$V(p) = V(p_1, p_2) = \langle p_1, p_2 \rangle_p .$$

As shown earlier, the integral curve through p is given by $c(t) = (p_1 e^t, p_2 e^t)$. Translated into the language of flows and diffeomorphisms, this means that, writing $\phi : \mathbf{R}^2 \times \mathbf{R} \to \mathbf{R}^2$ or $\phi_t : \mathbf{R}^2 \to \mathbf{R}^2$,

$$\phi(x, y, t) = \phi_t(x, y) = \left(xe^t, ye^t\right),$$

defined on all of $\mathbf{R}^2 \times \mathbf{R}$. Since V is complete, we can consider the time-one flow given by

$$\phi_1(x, y) = (x \cdot e, y \cdot e).$$

In fact, for any fixed t_0, ϕ_{t_0} is *linear*; written in terms of linear transformations on the vector space $\mathbf{R}^2$, we have $\phi_{t_0}(\mathbf{x}) = A\mathbf{x}$, where $\mathbf{x} = (x, y)$ is a column vector and $A = \begin{bmatrix} \lambda & 0 \\ 0 & \lambda \end{bmatrix}$ for $\lambda = e^{t_0}$.

The time-one flow is a *dilation* by the constant factor e. Note that all dilations by a factor $\lambda > 0$ are obtained in this way by choosing different (constant) values for t, e.g. the "time-2 flow" or the "time-$(\ln 5)$ flow."

Example 3.9.4. Let X_θ be the rotational vector field on $\mathbf{R}^2$ given by

$$X_\theta(p) = X_\theta(p_1, p_2) = \langle -p_2, p_1 \rangle_p \in T_p(\mathbf{R}^2);$$

see Example 3.8.4. An integral curve $c(t) = (x(t), y(t))$ of X_θ through a point $p = (p_1, p_2)$ is given by

$$c(t) = (p_1 \cos t - p_2 \sin t, p_1 \sin t + p_2 \cos t).$$

This corresponds to the flow

$$\phi(x, y, t) = \phi_t(x, y) = (x \cos t - y \sin t, x \sin t + y \cos t),$$

defined on all of $\mathbf{R}^2 \times \mathbf{R}$. Hence the time-one flow is given by

$$\phi_1(x, y) = (ax - by, bx + ay), \tag{3.12}$$

where $a = \cos 1$ and $b = \sin 1$. In fact, different (but fixed) values of t yield all possible functions having the form of Eq. (3.12) satisfying $a^2 + b^2 = 1$.

Geometrically, flows of X_θ yield rotations of $\mathbf{R}^2$.

The final example we will present here will be the flow of a vector field that is not complete, based on Example 3.8.5. The integral curves of that vector field arose from a nonlinear system of differential equations.

Example 3.9.5. Let V be the vector field on $\mathbf{R}^2$ described in Example 3.8.5 by

$$V(p) = V(p_1, p_2) = \langle p_1^2, p_1 p_2 \rangle_p \in T_p(\mathbf{R}^2).$$

Consider a given point $p = (p_1, p_2) \in \mathbf{R}^2$. For the sake of definiteness, we assume that $p_1 > 0$, with the understanding that the argument below can be adjusted to the case $p_1 < 0$. Let $U_p \subset \mathbf{R}^2$ be the set

$$U_p = \left\{ (x, y) \mid \frac{p_1}{2} < x < 2p_1 \right\}.$$

Let I_p be the interval

$$I_p = \left(-\frac{1}{2p_1}, \frac{1}{2p_1} \right).$$

In Example 3.8.5, we showed that the integral curve of V through p is given by

$$c(t) = \left(\frac{p_1}{1 - p_1 t}, \frac{p_2}{1 - p_1 t} \right),$$

defined for $t \in I_p$ (actually on a larger interval, but in particular on this interval for reasons noted below).

From this, we aim to find the (local) flow of V, a diffeomorphism defined not on all of $\mathbf{R}^2$, but only on the set U_p. In fact, the function

$$\phi : U_p \times I_p \to \mathbf{R}^2$$

given by

$$\phi(x, y, t) = \phi_t(x, y) = \left(\frac{x}{1 - xt}, \frac{y}{1 - xt} \right)$$

is defined for all $t \in I_p$ and $(x, y) \in U_p$, since the condition $1 - xt > 0$ is satisfied when $t < 1/x$, or, since we have $x > p_1/2$ for $(x, y) \in U_p$, when $t < 2/p_1$. This condition is satisfied in particular for all $t \in I_p$.

Depending on the initial choice of p, if $1 \notin I_p$, it will not make sense to talk about the time-one flow. However, for any $p = (p_1, p_2)$ with $p_1 > 0$, we can talk about the time-$(1/4p_1)$ flow $\phi : U_p \to \mathbf{R}^2$,

$$\phi(x, y) = \phi_{1/4p_1}(x, y) = \left(\frac{4p_1 x}{4p_1 - x}, \frac{4p_1 y}{4p_1 - x} \right).$$

We leave it as an exercise to verify that ϕ is a diffeomorphism onto its image

$$\phi(U_p) = \left\{ (x, y) \;\middle|\; \frac{4p_1}{7} < x < 4p_1 \right\}.$$

We have carried out these constructions of diffeomorphisms as flows of vector fields in order both to illustrate the practical difficulties of explicitly producing such flows (as the solutions of systems of differential equations) even in simple cases, and to exhibit the local nature of these diffeomorphisms, defined only on some subset that depends on the initial point chosen. Many times, however, the hope for an explicit expression for the local flow of a vector field is abandoned in light of Theorem 3.9.2. We just refer to "the flow ϕ of a vector field V near p."

One of the central questions in later chapters will be establishing properties of the flow ϕ from properties of the vector field V. This is again the passage from "local" (or "infinitesimal") to "global."

3.10 For Further Reading

Presenting the derivative as a linear transformation is a standard feature of most treatments of advanced calculus. Marsden and Tromba [30] do so nicely in the context of a first course in multivariable calculus. This is also the starting point for Spivak's *Calculus on Manifolds*, [37]. These treatments all exploit the vector space structure of $\mathbf{R}^n$, which allows the student to easily identify points with tangent vectors.

Putting the derivative in its geometric context as a transformation between tangent spaces is usually the first order of business in a course on differential geometry. There, the basic object is the more abstract notion of a manifold, which makes the formal treatment of the tangent space essential, in contrast to the somewhat more concrete setting in $\mathbf{R}^n$. The presentation closest to the one adopted here might be found in O'Neill's *Elementary Differential Geometry* [33], which is accessible to an undergraduate reader. By far the most thorough treatment, in terms of presenting the theoretical and technical details in the definition of the tangent space, can be found in Volume I of Spivak's *Comprehensive Introduction to Differential Geometry* [38]. The details there are carried out in the setting of manifolds.

Vector fields and flows are sometimes downplayed in a first course in differential geometry, where the emphasis is often on metric concepts such as length and curvature. The presentation here is influenced again by Spivak [38] as well as by Warner's *Foundations of Differentiable Manifolds and Lie Groups* [40].

After reading Sect. 3.9 above on constructing flows from vector fields, the reader might be inspired to review the basics of solving systems of differential equations. Blanchard, Devaney, and Hall's text [9] gives an introductory approach to the subject that emphasizes both techniques of explicitly solving systems and the limitations of such methods, along with qualitative techniques in studying solutions.

3.11 Exercises

3.1. Use Definition 3.1.1 to verify the following: If $f : \mathbf{R}^2 \to \mathbf{R}^2$ is given by $f(x, y) = (x^2 - y^2, xy)$, then

$$(Df(1,2))(v_1, v_2) = (2v_1 - 4v_2, 2v_1 + v_2).$$

Hint: Use the inequality

$$h_1^4 - h_1^2 h_2^2 + h_2^4 \leq h_1^4 + 2h_1^2 h_2^2 + h_2^4.$$

3.2. For each of the functions f below:

1. $f : \mathbf{R}^2 \to \mathbf{R}^2$ given by

$$f(x_1, x_2) = (x_1^2 - x_2^2, 2x_1 x_2);$$

2. $f : \mathbf{R}^3 \to \mathbf{R}^3$ given by

$$f(x_1, x_2, x_3) = (x_1, x_2, x_1^2 + x_2^2 + x_3^2);$$

3. $f : \mathbf{R}^2 \to \mathbf{R}^4$ given by

$$f(x_1, x_2) = (x_1 + x_2, x_1 - x_2, x_1^2 + x_2^2, x_1^2 - x_2^2);$$

(a) Compute $Df(\mathbf{x})$, where $\mathbf{x}$ is an arbitrary point in the domain;
(b) Find all vectors $\mathbf{b}$ such that $Df(\mathbf{b})$ does not have maximum rank;
(c) For each such vector $\mathbf{b}$ in part (b), find a basis for $\ker Df(\mathbf{b})$.

3.3. For each of the following functions, compute the Jacobian matrix at the specified point p and compute the rank of $(f_*)_p$.

(a) $f : \mathbf{R}^2 \to \mathbf{R}^2$ given by

$$f(x, y) = (x^2 + y^2, x - 3y),$$

$p = (1, -3)$.
(b) $f : \mathbf{R}^3 \to \mathbf{R}^2$ given by

$$f(x, y, z) = (x^2 + y^2 - z^2, \sin(xz)),$$

$p = (1, 2, 0)$.

(c) $f : \mathbf{R}^2 \to \mathbf{R}^3$ given by

$$f(x, y) = (x^2 - y^2, x - y, xy),$$

$p = (0, 0)$.

3.4. This exercise fills in some of the analysis gaps in Example 3.1.5.

(a) Show that $|h_1 \cdot h_2| \leq \frac{1}{2}(h_1^2 + h_2^2)$ for all real numbers h_1, h_2. Hint: Begin with $(|h_1| - |h_2|)^2 \geq 0$.

(b) Show that $\displaystyle\lim_{(h_1,h_2)\to(0,0)} \frac{|h_1 \cdot h_2|}{||(h_1, h_2)||} = 0.$

3.5. Suppose $f : \mathbf{R}^2 \to \mathbf{R}$ is *bilinear* in the sense of Sect. 2.8.

(a) Show that

$$\lim_{(h_1,h_2)\to(0,0)} \frac{|f(h_1, h_2)|}{||(h_1, h_2)||} = 0.$$

Hint: For each fixed vector $\mathbf{h} = (h_1, h_2) \neq \mathbf{0}$, consider

$$\lim_{s\to 0} \frac{|f(s\mathbf{h})|}{||s\mathbf{h}||}.$$

(b) Use Definition 3.1.1 to show that $(Df(a, b))\,(v_1, v_2) = f(a, v_2) + f(v_1, b)$.

3.6. Let $f : \mathbf{R}^n \to \mathbf{R}$. For each vector $\mathbf{v} \in \mathbf{R}^n$, define

$$D_{\mathbf{v}} f(\mathbf{a}) = \lim_{t\to 0} \frac{f(\mathbf{a} + t\mathbf{v}) - f(\mathbf{a})}{t},$$

if the limit exists. $D_{\mathbf{v}} f(\mathbf{a})$ is the *directional derivative of f with respect to* $\mathbf{v}$ *at* $\mathbf{a}$.

(a) Show that if $\{\mathbf{e}_1, \ldots, \mathbf{e}_n\}$ is the standard basis for $\mathbf{R}^n$, then

$$D_{\mathbf{e}_i} f(\mathbf{a}) = D_i f(\mathbf{a}),$$

where D_i is the appropriate partial derivative of f.

(b) For every scalar $s \in \mathbf{R}$, show that

$$D_{s\mathbf{v}} f(\mathbf{a}) = s D_{\mathbf{v}} f(\mathbf{a}).$$

(c) Show that for vectors $\mathbf{v}, \mathbf{w} \in \mathbf{R}^n$, one has

$$D_{\mathbf{v}+\mathbf{w}} f(\mathbf{a}) = D_{\mathbf{v}} f(\mathbf{a}) + D_{\mathbf{w}} f(\mathbf{a}).$$

3.7. In this exercise we make a more precise definition of the tangent space to $\mathbf{R}^n$ at a point p.

An *equivalence relation* is a relation $\sim$ on a set S satisfying the following properties:

- For all $s \in S$, $s \sim s$.
- For any $s_1, s_2 \in S$, if $s_1 \sim s_2$, then $s_2 \sim s_1$.
- For any $s_1, s_2, s_3 \in S$, if $s_1 \sim s_2$ and $s_2 \sim s_3$, then $s_1 \sim s_3$.

For every $p \in \mathbf{R}^n$, let $\mathcal{C}_p$ be the set of all parameterized curves $c : I \to \mathbf{R}^n$ through p, where I is an open interval satisfying $0 \in I$ and $c(0) = p$. Define a relation on $\mathcal{C}_p$ by saying that for $c_1, c_2 \in \mathcal{C}_p$, $c_1 \sim c_2$ if and only if $c_1'(0) = c_2'(0)$.

(a) Show that $\sim$ so defined is an equivalence relation.

(b) Recall that an *equivalence class* of a set S relative to an equivalence relation $\sim$ is a subset

$$[a] = \{s \in S \mid s \sim a\},$$

where $a \in S$. (Note that $[a] = [b]$ if and only if $a \sim b$.)

Now let V_p be the set of equivalence classes of the set $\mathcal{C}_p$ of parametrized curves through p under the equivalence relation $\sim$. Show that if $a, b \in \mathcal{C}_p$, then the operations

$$[a] + [b] = [c],$$

where $c \in \mathcal{C}_p$ is given by $c(t) = p + t(a'(0) + b'(0))$ for all t on which a and b are defined, and

$$s\,[a] = [d],$$

where $d \in \mathcal{C}_p$ is given by $d(t) = p + tsa'(0)$ for all t on which a is defined, are *well defined* is the sense that if $a \sim \tilde{a}$ and $b \sim \tilde{b}$, then $[a] + [b] = [\tilde{a}] + [\tilde{b}]$ and $s\,[a] = s\,[\tilde{a}]$.

(c) Show that V_p with the operations of addition and scalar multiplication defined in part (b) forms a vector space. The space V_p constructed in this way is a precise definition of the tangent space $T_p(\mathbf{R}^n)$.

3.8. Consider the surface S in $\mathbf{R}^3$ described by the equation $x + y = z^2$, a parabolic cylinder.

(a) Show that $\phi_1 : \mathbf{R}^2 \to \mathbf{R}^3$ given by $\phi_1(u, v) = (v^2 - u, u, v)$ is a regular parameterization for S.

(b) Show that $\phi_2 : \mathbf{R}^2 \to \mathbf{R}^3$ given by $\phi_2(u, v) = (u^2 + v^2, 2uv, u + v)$ is a parameterization of part of S that is not regular along the curve of intersection of S and the plane $y = x$.

3.9. The unit sphere S^2 in $\mathbf{R}^3$ is the set

$$S^2 = \left\{(x, y, z) \mid x^2 + y^2 + z^2 = 1\right\}.$$

Show that $\phi : U \to \mathbf{R}^3$, where $U = \{(u,v) \mid 0 < u < 2\pi,\ 0 < v < \pi\}$ and $\phi(u,v) = (\cos u \sin v, \sin u \sin v, \cos v)$, is a regular parameterization of a subset in S^2. Identify the subset of S^2 that is not parameterized by ϕ.

3.10. Find regular parameterizations for (parts of) each of the following surfaces in $\mathbf{R}^3$. Identify the subsets of each that are not parameterized, if any, according to the parameterization that you specify:

(a) the cylinder $x^2 + y^2 = 1$;
(b) the hyperbolic cylinder $x^2 - y^2 = 1$;
(c) the hyperboloid of one sheet $x^2 + y^2 - z^2 = 1$;
(d) the hyperboloid of two sheets $z^2 = 1 + x^2 + y^2$;
(e) the ellipsoid $9x^2 + 4y^2 + z^2 = 1$.

3.11. Prove Theorems 3.3.6 and 3.4.9. In other words, using the two definitions of tangent vectors given, show that $T_p(S)$ is a vector subspace of $T_p(\mathbf{R}^n)$ for every geometric set $S \subset \mathbf{R}^n$.

3.12. For the unit sphere S^2 as in Exercise 3.9, consider $p=(1/2, 1/2, \sqrt{2}/2) \in S^2$. Find a basis for $T_p(S^2)$ in the following ways:

(a) Using the parameterization in Exercise 3.9 by the method of Theorem 3.3.9.
(b) As the graph $z = f(x,y)$ of the function $f(x,y) = \sqrt{1 - x^2 - y^2}$, using the method of Theorem 3.3.14.
(c) As the level surface of the function $f(x,y,z) = x^2 + y^2 + z^2$, using the method of Theorem 3.3.19. Be sure to verify that 1 is a regular value for f.
(d) Confirm that $T_p(S^2)$ obtained in each of the three manners (a)–(c) always gives the same vector subspace of $T_p(\mathbf{R}^3)$.

3.13. It is a fact from topology that it is impossible to parameterize all of the unit sphere $S^2 \subset \mathbf{R}^3$ using a single domain $U \subset \mathbf{R}^2$. *Stereographic projection* represents a way to parameterize "as much of S^2 as possible" using one domain. Geometrically, $\phi : \mathbf{R}^2 \to \mathbf{R}^3$ is described as follows:

> $\phi(u,v) = p$, where p is the point of intersection of S^2 with the line passing through the points $(0,0,1)$ and $(u,v,0)$.

(a) Verify that ϕ is given by

$$\phi(u,v) = \left(\frac{2u}{1+u^2+v^2}, \frac{2v}{1+u^2+v^2}, \frac{u^2+v^2-1}{1+u^2+v^2} \right).$$

(b) Show that ϕ is regular by showing that all three of the possible 2×2 minors of $(\phi)_*$ cannot be simultaneously zero.
(c) Find $\phi^{-1}(x,y,z)$ assuming $(x,y,x) \in S^2$.
(d) If $p = (1/2, 1/2, \sqrt{2}/2)$, use the parameterization ϕ to compute $T_p(S^2)$.

3.14. For the parameterized surfaces S below, find a basis for the tangent space T_pS at an arbitrary point $p \in S$:

(a) the helicoid, parameterized by $\phi_1 : U \to \mathbf{R}^3$, where

$$U = (0, 2\pi) \times \mathbf{R} \subset \mathbf{R}^2$$

and

$$\phi_1(u, v) = (v \cos u, v \sin u, u);$$

(b) the catenoid, parameterized by $\phi_2 : U \to \mathbf{R}^3$, where

$$U = (0, 2\pi) \times \mathbf{R} \subset \mathbf{R}^2$$

and

$$\phi_2(u, v) = (\cosh v \cos u, \cosh v \sin u, v);$$

(c) the torus, parameterized by $\phi_3 : U \to \mathbf{R}^3$, where

$$U = (0, 2\pi) \times (0, 2\pi) \subset \mathbf{R}^2$$

and

$$\phi_3(u, v) = ((4 + \cos u) \cos v, (4 + \cos u) \sin v, \sin u).$$

3.15. For each of the surfaces S in Exercise 3.10, find a basis for the tangent space T_pS at an arbitrary point $p = (x, y, z) \in S$.

3.16. Fill in the details of the proof of Theorem 3.3.9.

3.17. This exercise introduces a smooth "localizing" function that is 1 near a given point p and 0 farther away from p.

(a) Show that the function $f : \mathbf{R} \to \mathbf{R}$ given by

$$f(t) = \begin{cases} e^{-1/t}, & t > 0, \\ 0, & t \leq 0, \end{cases}$$

is differentiable at $t = 0$ (and hence for all t).

(b) For a fixed $r > 0$, let $g_r : \mathbf{R} \to \mathbf{R}$ be the function defined as

$$g_r(s) = \frac{f(s)}{f(s) + f(r^2 - s)},$$

where f is the function described in part (a). Show that g_r is differentiable and satisfies the properties that g_r is nonnegative for all s, $g_r(s) = 0$ when $s \leq 0$ and $g_r(s) = 1$ when $s \geq r^2$.

(c) For given $r > 0$ and $p \in \mathbf{R}^n$, define $H_{p,r} : \mathbf{R}^n \to \mathbf{R}$ by

$$H_{p,r}(x) = g_r(2r^2 - ||x - p||^2).$$

Show that $H_{p,r}$ is differentiable, that $H_{p,r}(x) = 1$ when $||x - p|| \leq r$, and that $H_{p,r}(x) = 0$ when $||x - p|| \geq (\sqrt{2})r$.

3.18. Let $\mathbf{v}_p$ be a tangent vector at $p \in \mathbf{R}^n$, in the sense of Definition 3.4.1. Suppose two functions $f, g : \mathbf{R}^n \to \mathbf{R}$ agree near p, i.e., there is some distance r such that if $||x - p|| < r$, then $f(x) = g(x)$. Show that $\mathbf{v}_p[f] = \mathbf{v}_p[g]$. Hint: First use the properties of a linear derivation to show that if $h(x) = 0$ whenever $||x - p|| < r$, then $\mathbf{v}_p[h] = 0$ by computing

$$\mathbf{v}_p[H_{p,r/\sqrt{2}} \cdot h],$$

where $H_{p,r/\sqrt{2}}$ is the function described in the previous exercise. Note that $H_{p,r/\sqrt{2}}(x) \cdot h(x) = 0$ for all $x \in \mathbf{R}^n$. Then use that result along with the linearity of $\mathbf{v}_p$ applied to $h = f - g$.

3.19. Complete the proof of Theorem 3.4.4.

3.20. Let S^2 be the unit sphere in $\mathbf{R}^3$ and let P be the paraboloid in $\mathbf{R}^3$ given by

$$P = \left\{(x, y, z) \mid z = x^2 + y^2\right\}.$$

Let $f : \mathbf{R}^3 \to \mathbf{R}^3$ be given by

$$f(x, y, z) = (y, x, 1 - z^2).$$

(a) Show that f restricts to a map $f : S^2 \to P$. In other words, show that if $p \in S^2$, then $f(p) \in P$. Is $f : S^2 \to P$ one-to-one? onto?
(b) Let $p = (1/3, 1/2, \sqrt{23}/6)$. Describe both $T_p(S^2)$ and $T_{f(p)}(P)$. Find a basis for each. In addition, list five tangent vectors in each tangent space.
(c) Compute $(f_*)_p$ for p as in part (b). For each basis vector $\mathbf{v}_p$ you produced for $T_p(S^2)$, show that $(f_*)_p(\mathbf{v}_p) \in T_{f(p)}(P)$.

3.21. Show that the functions described in Example 3.6.10 are in fact diffeomorphisms.

3.22. For a constant $a \in \mathbf{R}$, let $f : \mathbf{R}^3 \to \mathbf{R}^3$ be given by

$$f(x, y, z) = \left(x - \frac{az}{\sqrt{1 + z^2}}, y + \frac{a}{\sqrt{1 + z^2}}, z\right).$$

Show that f is a diffeomorphism.

3.23. Suppose that $\phi_1 : U_1 \to S$ and $\phi_2 : U_2 \to S$ are two regular parameterizations of the same set $S \subset \mathbf{R}^n$, where $U_1, U_2 \subset \mathbf{R}^k$. Show that $\phi_2^{-1} \circ \phi_1 : U_1 \to U_2$ is a diffeomorphism.

3.24. Provide the details of the proof of Theorem 3.6.6.

3.25. Prove Theorem 3.6.11.

3.26. For $\mathbf{R}^2$ with coordinates $p = (p_1, p_2)$, find and graph the integral curve through the point $(1, 0)$ for each of the following vector fields:

(a) $X(p) = (4p_1 + 2p_2, p_1 + 3p_2)$.
(b) $X(p) = (-p_1 + 2p_2, -p_1 - p_2)$.
(c) $X(p) = (p_1 - 5p_2, p_1 - p_2)$.
(d) $X(p) = (-2p_1 - p_2, p_1 - 4p_2)$.

3.27. Let $f : \mathbf{R}^2 \to \mathbf{R}^2$ be defined by

$$f(x, y) = \left(x^2 - y^2, 2xy\right).$$

Compute $f_* X$ for each of the vector fields in Exercise 3.26.

3.28. A vector field $X(x, y) = (X^1(x, y), X^2(x, y))$ on $\mathbf{R}^2$ is called *Hamiltonian* if there is a function $H : \mathbf{R}^2 \to \mathbf{R}$ (called the *Hamiltonian function for* X) such that $X^1 = \frac{\partial H}{\partial y}$ and $X^2 = -\frac{\partial H}{\partial x}$.

(a) Give three examples of Hamiltonian vector fields.
(b) Show that $X = (2xy + x^2, -(x + y)^2)$ is a Hamiltonian vector field by finding the Hamiltonian function for X.

3.29. Suppose X is a Hamiltonian vector field with Hamiltonian function H, defined in the previous exercise. Show that all integral curves of X lie on level sets of H, i.e., if $c(t) = (x(t), y(t))$ is an integral curve of X, then for all t, $H(x(t), y(t)) = a$ for some constant $a \in \mathbf{R}$.

3.30. For each of the vector fields X on $\mathbf{R}^2$ in Exercise 3.26, find the flow and time-one flow for X.

3.31. Suppose that $U, V \subset \mathbf{R}^n$ are domains and let $f : U \to V$ be a diffeomorphism. Let X be a vector field on U with flow ϕ_t. Show that the flow ψ_t of the vector field $Y = f_* X$ on V is given by

$$\psi_t = f^{-1} \circ \phi_t \circ f.$$

Chapter 4
Differential Forms and Tensors

In the previous chapter, we emphasized the central role of the tangent space in differential geometry. The tangent space at a point is a set whose elements, tangent vectors, arise from calculus in at least two equivalent ways. We have also seen examples of how nonlinear geometric objects can arise from an object defined pointwise at the algebraic level of the tangent spaces. For example, vector fields give rise to more global geometric objects such as integral curves and diffeomorphisms. The passage from the algebraic tangent spaces to the geometric space in these examples was a process of "integration."

Having said that, we have not encountered many questions so far that are "geometric" in the traditional sense of geometry as *measurement*. The goal of this chapter is to introduce the concepts necessary to pose some of the standard measurement problems of modern differential geometry.

Typically, the word "measurement" involves assigning a number to an object that is to describe some quality or aspect of that object. Computing the length, area, or volume of a geometric object, for example, involves assigning numerical measurements (in appropriate units) to the object that describes the extension of the object in the specified number of dimensions.

This chapter introduces differential forms and, more generally, tensors. Linear forms measure collections of vectors, while differential forms measure vector fields, both at the tangent space level and in the way they change. Tensors are more general in the sense that they measure collections of both vector fields and forms.

Differential forms and tensors both emerged at the turn of the twentieth century. Tensors and tensor calculus enjoyed an explosion of interest especially after Einstein formulated his theory of general relativity in the language of Ricci's and Levi-Cività's "absolute calculus." Differential forms, on the other hand, while properly understood as a particular kind of tensor, gained popularity due to the work of Élie Cartan on partial differential equations and transformation groups.

In this chapter, we first return to the setting of linear algebra, reviewing some key facts from Chap. 2 in order to introduce the basic framework for alternating forms and the key algebraic operations unique to forms. We then show how differential

A. McInerney, *First Steps in Differential Geometry: Riemannian, Contact, Symplectic*, Undergraduate Texts in Mathematics, DOI 10.1007/978-1-4614-7732-7_4,

forms arise naturally by applying this algebra to the notions of tangent spaces and vector fields presented in Chap. 3. We introduce the general notion of tensor fields, which includes all the geometric structures that we will introduce in the second part of the text. Finally, we introduce the notion of the Lie derivative, which extends the concept of the directional derivative to the setting of tensors in a way that exploits the interplay between the "infinitesimal" setting of the tangent spaces and the "global" setting of the geometric space.

4.1 The Algebra of Alternating Linear Forms

To set the stage for defining differential forms, we take as a starting point the linear-algebraic foundation developed in Sect. 2.8. Recall that a *linear one-form* α on a vector space V is an element $\alpha \in V^*$, i.e., $\alpha : V \to \mathbf{R}$ is a linear transformation. Starting with a basis $B = \{\mathbf{e}_1, \ldots, \mathbf{e}_n\}$ for V, we define a basis $\{\varepsilon_1, \ldots, \varepsilon_n\}$ for V^* dual to B by

$$\varepsilon_i(\mathbf{e}_j) = \begin{cases} 1 & \text{if } i = j, \\ 0 & \text{if } i \neq j. \end{cases}$$

(From this point forward, we will adopt the convention that forms shall be named using Greek letters, while vectors shall be written using Latin letters.)

Recall also (Definition 2.8.8) that a *multilinear k-form* on a vector space V is a function

$$\alpha : V \times \cdots \times V \to \mathbf{R}$$

whose domain consists of k-tuples of vectors and whose codomain is the set of real numbers, with the additional property that α is linear in *each* of its components. The whole number k is called the *degree* of the multilinear form. A 0-form will be, by definition, a scalar.

For the next several sections we will restrict our attention to *alternating* (or *skew-symmetric*) forms.

Definition 4.1.1. Let V be a vector space. An *alternating k-form* on V ($k \geq 2$) is a multilinear k-form on V that satisfies the additional property that

$$\alpha(\mathbf{v}_1, \ldots, \mathbf{v}_i, \mathbf{v}_{i+1}, \ldots, \mathbf{v}_k) = -\alpha(\mathbf{v}_1, \ldots, \mathbf{v}_{i+1}, \mathbf{v}_i, \ldots, \mathbf{v}_k)$$

for every $i = 1, \ldots, k-1$. In other words, interchanging two consecutive vectors in the domain has the effect of changing the sign of the result. We will adopt the convention that all 0-forms and 1-forms will be considered alternating.

Before presenting some examples of alternating forms, we first illustrate some properties that are immediate consequences of Definition 4.1.1.

Theorem 4.1.2. *Let α be an alternating k-form on a vector space V with $k \geq 2$, and let $\mathbf{v}_1, \ldots, \mathbf{v}_k \in V$. If for some $i \neq j$ we have $\mathbf{v}_i = \mathbf{v}_j$, then $\alpha(\mathbf{v}_1, \ldots, \mathbf{v}_k) = 0$.*

Proof. Exercise. □

In fact, Theorem 4.1.2 is a special case of a more general fact.

Theorem 4.1.3. *Let α be an alternating k-form ($k \geq 2$) and let $\mathbf{v}_1, \ldots, \mathbf{v}_k \in V$. If the set $\{\mathbf{v}_1, \ldots, \mathbf{v}_k\}$ is linearly dependent, then $\alpha(\mathbf{v}_1, \ldots, \mathbf{v}_k) = 0$.*

Proof. Since $k \geq 2$, if the set $\{\mathbf{v}_1, \ldots, \mathbf{v}_k\}$ is linearly dependent, then one of these vectors can be written as a linear combination of the others; see Theorem 2.4.3. The result follows from the fact that α is multilinear and from Theorem 4.1.2. □

We now give some examples of alternating k-forms.

Example 4.1.4. Let (V, ω) be a symplectic vector space; see Sect. 2.10. Then ω by definition is an alternating two-form.

Example 4.1.5 (A fundamental example: the determinant). A fundamental example of an alternating n-form on an n-dimensional vector space V is the determinant. It will also play a key role in our development of alternating forms ahead.

We noted in Example 2.8.10 that the determinant can be viewed as an n-form Ω on an n-dimensional vector space V:

$$\Omega(\mathbf{v}_1, \ldots, \mathbf{v}_n) = \det A,$$

where $A = [\mathbf{v}_1 \cdots \mathbf{v}_n]$ is the matrix whose ith column is given by the components of $\mathbf{v}_i$ relative to some fixed basis for V.

The fact that Ω is also alternating is just a restatement of the fact that switching columns of a matrix has the effect of changing the sign of the determinant.

We now illustrate an important construction of alternating k-forms that will ultimately lead to a basis for the vector space of all alternating k-forms on a vector space V.

Example 4.1.6. Let V be a vector space of dimension n with basis $B = \{\mathbf{e}_1, \ldots, \mathbf{e}_n\}$, and let $B^* = \{\varepsilon_1, \ldots, \varepsilon_n\}$ be the corresponding basis for V^* dual to B. For any choice of k indices $i_1, \ldots, i_k \in \{1, \ldots, n\}$ with $i_1 < \cdots < i_k$, define the function

$$\varepsilon_{i_1 \cdots i_k} : \underbrace{V \times \cdots \times V}_{k \text{ times}} \to \mathbf{R}$$

by

$$\varepsilon_{i_1 \cdots i_k}(\mathbf{v}_1, \ldots, \mathbf{v}_k) = \det \begin{bmatrix} \varepsilon_{i_1}(\mathbf{v}_1) & \ldots & \varepsilon_{i_1}(\mathbf{v}_k) \\ \vdots & & \vdots \\ \varepsilon_{i_k}(\mathbf{v}_1) & \ldots & \varepsilon_{i_k}(\mathbf{v}_k) \end{bmatrix}.$$

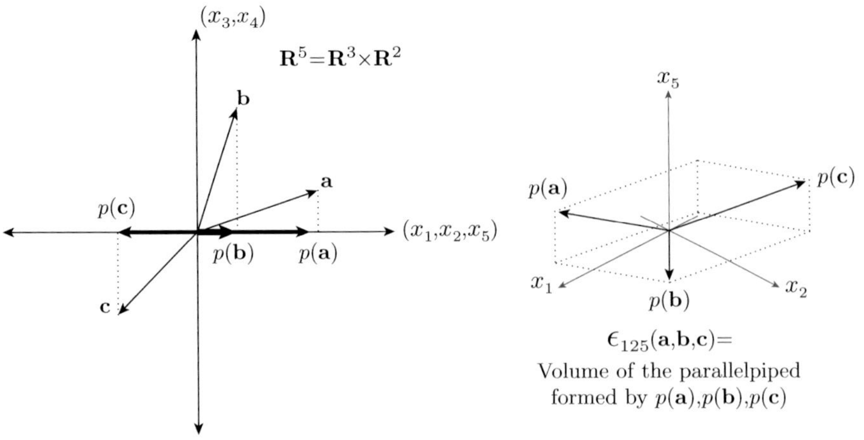

Fig. 4.1 A three-form in $\mathbf{R}^5$.

Then $\varepsilon_{i_1\cdots i_k}$ is an alternating k-form. This follows from properties of the determinant and the fact that the basis forms ε_j are linear.

For example, let $V = \mathbf{R}^5$ and consider the standard basis $\{\varepsilon_1,\ldots,\varepsilon_5\}$ for B^*, where $\varepsilon_i(a_1,a_2,a_3,a_4,a_5) = a_i$ for $i = 1,\ldots,5$. Consider the map $\varepsilon_{125} \in \Lambda_3(\mathbf{R}^5)$,

$$\varepsilon_{125} : \mathbf{R}^5 \times \mathbf{R}^5 \times \mathbf{R}^5 \to \mathbf{R}.$$

Applying this map to the vectors $\mathbf{v}_1 = (2,-3,1,7,4)$, $\mathbf{v}_2 = (1,3,5,7,9)$, $\mathbf{v}_3 = (0,-1,0,1,0)$, we have

$$\begin{aligned}\varepsilon_{125}(\mathbf{v}_1,\mathbf{v}_2,\mathbf{v}_3) &= \det\begin{bmatrix} 2 & 1 & 0 \\ -3 & 3 & -1 \\ 4 & 9 & 0 \end{bmatrix} \\ &= 14.\end{aligned}$$

More generally, let $\mathbf{a} = (a_1,a_2,a_3,a_4,a_5)$, $\mathbf{b} = (b_1,b_2,b_3,b_4,b_5)$, and $\mathbf{c} = (c_1,c_2,c_3,c_4,c_5)$. Then $\varepsilon_{125}(\mathbf{a},\mathbf{b},\mathbf{c})$ represents the (oriented) three-dimensional volume of the parallelepiped formed by the vectors $p(\mathbf{a})$, $p(\mathbf{b})$, and $p(\mathbf{c})$, where $p : \mathbf{R}^5 \to \mathbf{R}^3$ is the projection defined by $p(\mathbf{x}) = (x_1,x_2,x_5)$ for all $\mathbf{x} = (x_1,x_2,x_3,x_4,x_5) \in \mathbf{R}^5$. See Fig. 4.1.

Theorem 4.1.7. *Let V be a vector space. For $k = 1,\ldots,n$, the set $\Lambda_k(V)$ of all alternating k-forms on V, equipped with the standard pointwise addition and scalar multiplication of functions, is a vector space.*

Proof. Exercise. □

By convention, we write $\Lambda_0(V) = \mathbf{R}$.

Theorem 4.1.8. *Let V be a finite-dimensional vector space with basis $\{\mathbf{e}_1, \ldots, \mathbf{e}_n\}$, and let $\{\varepsilon_1, \ldots, \varepsilon_n\}$ be the corresponding dual basis for V^*. Then for $k = 1, \ldots, n$, the set $\mathcal{E}_k = \{\varepsilon_{i_1 \cdots i_k} \mid 1 \leq i_1 < \cdots < i_k \leq n\}$ is a basis for the vector space $\Lambda_k(V)$ of alternating k-forms on V.*

Proof. To show that $\mathcal{E}_k$ spans $\Lambda_k(V)$, choose any $\alpha \in \Lambda_k(V)$. Then for any choice of $i_1, \ldots, i_k \in \{1, \ldots, n\}$ with $i_1 < \cdots < i_k$, define $a_{i_1 \cdots i_k} = \alpha(\mathbf{e}_{i_1}, \ldots, \mathbf{e}_{i_k})$. A calculation shows that $\alpha = \sum a_{i_1 \cdots i_k} \varepsilon_{i_1 \cdots i_k}$, where the summation is taken over all such choices of $i_1, \ldots, i_k$.

To show that $\mathcal{E}_k$ is linearly independent, use the technique of Theorem 2.8.2: For any linear combination of elements of $\mathcal{E}_k$ assumed to be equal to the zero k-form, apply it to ordered k-tuples of basis vectors $(\mathbf{e}_{i_1}, \ldots, \mathbf{e}_{i_k})$ to show that all coefficients must be zero. □

Corollary 4.1.9. *Let V be a vector space with $\dim V = n$. Then*

$$\dim \Lambda_k(V) = \binom{n}{k} = \frac{n!}{k!(n-k)!}.$$

Example 4.1.10. A basis for $\Lambda_2(\mathbf{R}^3)$ is given by the set $\{\varepsilon_{12}, \varepsilon_{13}, \varepsilon_{23}\}$. Explicitly, let $\mathbf{a} = \langle a_1, a_2, a_3 \rangle$ and $\mathbf{b} = \langle b_1, b_2, b_3 \rangle$. Then

$$(\varepsilon_{12})(\mathbf{a}, \mathbf{b}) = \det \begin{bmatrix} \varepsilon_1(\mathbf{a}) & \varepsilon_1(\mathbf{b}) \\ \varepsilon_2(\mathbf{a}) & \varepsilon_2(\mathbf{b}) \end{bmatrix} = a_1 b_2 - a_2 b_1,$$

and likewise,

$$(\varepsilon_{13})(\mathbf{a}, \mathbf{b}) = a_1 b_3 - a_3 b_1,$$

$$(\varepsilon_{23})(\mathbf{a}, \mathbf{b}) = a_2 b_3 - a_3 b_2.$$

4.2 Operations on Linear Forms

The main achievement of the previous section was defining the set $\Lambda_k(V)$ of alternating k-forms on an n-dimensional vector space V. Theorem 4.1.7 shows that this set is in fact a vector space with the operations of addition of functions and multiplication of functions by scalars. In this section we define additional operations on $\Lambda_k(V)$. The first is a kind of product, which assigns to every pair of forms (α, β) a new form $\alpha \wedge \beta$. The second operation is the pullback, which is a generalization of the operation of the same name in the context of the dual space (see Theorem 2.8.5). Finally, we present a kind of contraction operation, which assigns to a k-form α and a vector $\mathbf{v}$ a new form of degree $k - 1$.

We first note that there was nothing essential in Example 4.1.6 about the condition that $i_1 < \cdots < i_k$.

Definition 4.2.1. Let V be an n-dimensional vector space with basis $B = \{\mathbf{e}_1, \dots, \mathbf{e}_n\}$, and let $\{\varepsilon_1, \dots, \varepsilon_n\}$ be the basis of V^* dual to B. Then for every $1 \leq i_1, \dots, i_k \leq n$, define the k-form on V $\varepsilon_{i_1} \wedge \cdots \wedge \varepsilon_{i_k}$ as

$$(\varepsilon_{i_1} \wedge \cdots \wedge \varepsilon_{i_k})(\mathbf{v}_1, \dots, \mathbf{v}_k) = \det \begin{bmatrix} \varepsilon_{i_1}(\mathbf{v}_1) & \cdots & \varepsilon_{i_1}(\mathbf{v}_k) \\ \vdots & & \vdots \\ \varepsilon_{i_k}(\mathbf{v}_1) & \cdots & \varepsilon_{i_k}(\mathbf{v}_k) \end{bmatrix}$$

for all $\mathbf{v}_1, \dots, \mathbf{v}_k \in V$.

Note that in the case $i_1 < \cdots < i_k$, Definition 4.2.1 agrees with the definition in Example 4.1.6, and $\varepsilon_{i_1} \wedge \cdots \wedge \varepsilon_{i_k} = \varepsilon_{i_1 \cdots i_k}$. However, there are some immediate consequences from lifting the restriction that $i_1 < \cdots < i_k$. We leave the proofs as routine exercises involving properties of the determinant.

Proposition 4.2.2. *Let V be an n-dimensional vector space with basis $B = \{\mathbf{e}_1, \dots, \mathbf{e}_n\}$ and let $\{\varepsilon_1, \dots, \varepsilon_n\}$ be the basis of V^* dual to B.*

1. *For $1 \leq k \leq n$, let $1 \leq i_1, \dots, i_k \leq n$. Suppose that for some $s, t \in \{1, \dots, k\}$, $i_s = i_t$. Then*
$$\varepsilon_{i_1} \wedge \cdots \wedge \varepsilon_{i_k} = 0.$$
2. *For all $1 \leq i, j \leq n$,*
$$\varepsilon_i \wedge \varepsilon_j = -\varepsilon_j \wedge \varepsilon_i.$$

Proof. Exercise. □

For the sake of formulating the following proposition dealing with changing the ordering of the indices in the alternating forms $\varepsilon_{i_1} \wedge \cdots \wedge \varepsilon_{i_k}$, we will briefly review permutations and their basic properties. A *permutation* of an ordered set of k elements, denoted for convenience by $S = \{1, \dots, k\}$, is a one-to-one, onto map $\pi : S \to S$ from the set to itself. This is a technical way of saying that it is a reordering of the elements $1, \dots, k$. In this notation, $\pi(1)$ is thought of as the first element in the reordered set, $\pi(2)$ the second, etc.

A specific type of permutation is a *transposition*, which switches the order of two elements and leaves the others unchanged. It is a fact that every permutation can be expressed as a composition of transpositions. The way in which a permutation is written as a composition of transpositions is not at all unique. However, it is another basic fact that the *parity* of two different ways of writing the same permutation as a composition of transpositions must be the same, i.e., for any given permutation, it can be written only as a composition of an even number of transpositions or only as a composition of an odd number. In the former case, the permutation is said to be *even*, while in the latter case, it is *odd*. It is common practice to associate the parity of a transposition with a sign: $\operatorname{sgn} \pi = 1$ or $\operatorname{sgn} \pi = -1$ according to whether π is even or odd.

Proposition 4.2.3. *Let π be a permutation of a set of indices $\{i_1, \dots, i_k\}$, where $1 \leq i_1 < \cdots < i_k \leq n$. Then*

$$\varepsilon_{\pi(i_1)} \wedge \cdots \wedge \varepsilon_{\pi(i_k)} = (\text{sgn } \pi)(\varepsilon_{i_1 \cdots i_k}).$$

Proof. The statement follows directly from Proposition 4.2.2. □

With these basic properties in mind, we are in a position to formulate the definition of the exterior product. We do so in terms of the basis elements $\varepsilon_{i_1 \cdots i_k}$, although it is possible to write the definition without recourse to the basis by means of summation over permutations (see, for example, [3]).

Definition 4.2.4. Let α be a k-form on V and let β be a one-form on V. For any basis $\{\mathbf{e}_1, \ldots, \mathbf{e}_n\}$ for V with dual basis $\{\varepsilon_1, \ldots, \varepsilon_n\}$ for V^*, we can write

$$\alpha = \sum a_{i_1 \cdots i_k} \varepsilon_{i_1 \cdots i_k}$$

and

$$\beta = \sum b_{j_1 \cdots j_l} \varepsilon_{j_1 \cdots j_l},$$

where in both cases the sum is taken over all possible ordered multi-indices $(i_1 < \cdots < i_k)$ and $(j_1 < \cdots < j_l)$. The *exterior product* (or *wedge product*) $\alpha \wedge \beta$ is defined to be the sum

$$\begin{aligned}\alpha \wedge \beta &= \sum a_{i_1 \cdots i_k} b_{j_1 \cdots j_l} \varepsilon_{i_1 \cdots i_k} \wedge \varepsilon_{j_1 \cdots j_l} \\ &\left(= \sum a_{i_1 \cdots i_k} b_{j_1 \cdots j_l} \varepsilon_{i_1} \wedge \cdots \wedge \varepsilon_{i_k} \wedge \varepsilon_{j_1} \wedge \cdots \varepsilon_{j_l}\right),\end{aligned}$$

where the sum this time is over all possible pairs of ordered multi-indices.

The notation may obscure the fact that the definition is essentially just the distributive law in the context of the exterior product. We leave to the reader to confirm that $\alpha \wedge \beta$ so defined is, in fact, a $(k + l)$-form.

Before elaborating basic properties of the exterior product, we illustrate the definition with an example.

Example 4.2.5. Let $V = \mathbf{R}^4$ and let $\{\varepsilon_1, \varepsilon_2, \varepsilon_3, \varepsilon_4\}$ be the standard basis for V^*. Define the one-form $\alpha \in \Lambda_1(V)$ by $\alpha = 3\varepsilon_1 + 2\varepsilon_3$ and the two-form $\beta \in \Lambda_2(V)$ by $\beta = \varepsilon_{14} + 2\varepsilon_{24} - \varepsilon_{34}$. Then

$$\begin{aligned}\alpha \wedge \beta &= (3\varepsilon_1 + 2\varepsilon_3) \wedge (\varepsilon_{14} + 2\varepsilon_{24} - \varepsilon_{34}) \\ &= (3)\varepsilon_1 \wedge \varepsilon_1 \wedge \varepsilon_4 + (6)\varepsilon_1 \wedge \varepsilon_2 \wedge \varepsilon_4 + (-3)\varepsilon_1 \wedge \varepsilon_3 \wedge \varepsilon_4 \\ &\quad + (2)\varepsilon_3 \wedge \varepsilon_1 \wedge \varepsilon_4 + (4)\varepsilon_3 \wedge \varepsilon_2 \wedge \varepsilon_4 + (-2)\varepsilon_3 \wedge \varepsilon_3 \wedge \varepsilon_4 \\ &= 6\varepsilon_1 \wedge \varepsilon_2 \wedge \varepsilon_4 - 3\varepsilon_1 \wedge \varepsilon_3 \wedge \varepsilon_4 - 2\varepsilon_1 \wedge \varepsilon_3 \wedge \varepsilon_4 - 4\varepsilon_2 \wedge \varepsilon_3 \wedge \varepsilon_4 \\ &= 6\varepsilon_{124} - 5\varepsilon_{134} - 4\varepsilon_{234}.\end{aligned}$$

In the course of this computation we have used the results from Propositions 4.2.2 and 4.2.3.

Example 4.2.6. Define the two-form $\omega \in \Lambda_2(\mathbf{R}^4)$ as $\omega = \varepsilon_{12} + \varepsilon_{34}$. Then

$$\begin{aligned}\omega \wedge \omega &= (\varepsilon_{12} + \varepsilon_{34}) \wedge (\varepsilon_{12} + \varepsilon_{34}) \\ &= \varepsilon_1 \wedge \varepsilon_2 \wedge \varepsilon_1 \wedge \varepsilon_2 + \varepsilon_1 \wedge \varepsilon_2 \wedge \varepsilon_3 \wedge \varepsilon_4 \\ &\quad + \varepsilon_3 \wedge \varepsilon_4 \wedge \varepsilon_1 \wedge \varepsilon_2 + \varepsilon_3 \wedge \varepsilon_4 \wedge \varepsilon_3 \wedge \varepsilon_4 \\ &= 2\varepsilon_{1234}.\end{aligned}$$

We list some properties of the exterior product of one-forms, then consider properties of the exterior product in the general context.

Proposition 4.2.7. *Let V be a vector space and let α and β be one-forms on V, i.e., $\alpha, \beta \in V^*$. Then:*

(a) For any vectors $\mathbf{v}, \mathbf{w} \in V$, we have

$$(\alpha \wedge \beta)(\mathbf{v}, \mathbf{w}) = \det \begin{bmatrix} \alpha(\mathbf{v}) & \alpha(\mathbf{w}) \\ \beta(\mathbf{v}) & \beta(\mathbf{w}) \end{bmatrix}.$$

(b) $\alpha \wedge \alpha = 0$.
(c) $\alpha \wedge \beta = -\beta \wedge \alpha$.

Proof. Exercise. □

We emphasize that neither part (b) nor part (c) of Proposition 4.2.7 is true for general k-forms, as can be seen from Example 4.2.6 above.

The results of the previous propositions in fact follow from more general properties of the exterior product. The following properties follow by routine calculations.

Proposition 4.2.8. *Let V be a vector space. If $\alpha \in \Lambda_k(V)$, $\beta \in \Lambda_l(V)$, and $\gamma \in \Lambda_m(V)$, then:*

1. *The exterior product is associative: $(\alpha \wedge \beta) \wedge \gamma = \alpha \wedge (\beta \wedge \gamma)$.*
2. $\beta \wedge \alpha = (-1)^{kl} \alpha \wedge \beta$.
3. *The pairing $(\alpha, \beta) \mapsto \alpha \wedge \beta$ is bilinear.*

Proof. Exercise. □

Turning to our second basic operation on linear forms, we note that the definitions of the pullback of linear one-forms and bilinear forms (see Theorem 2.8.5 and Definition 2.8.15) extend naturally to k-forms.

Definition 4.2.9. Let $T : V \to W$ be a linear transformation between vector spaces V and W, and let α be a k-form on W. The *pullback of α by T*, denoted by $T^*\alpha$, is the k-form on V defined by $(T^*\alpha)(\mathbf{v}_1, \ldots, \mathbf{v}_k) = \alpha(T(\mathbf{v}_1), \ldots, T(\mathbf{v}_k))$ for all $\mathbf{v}_1, \ldots, \mathbf{v}_k \in V$.

The fact that $T^*\alpha$ so defined is in fact a k-form is a consequence of the linearity of T.

Example 4.2.10. Let $T : \mathbf{R}^3 \to \mathbf{R}^2$ be the linear transformation defined by

$$T(a,b,c) = (2a - c, b - 3c).$$

Let $\{\varepsilon_1, \varepsilon_2\}$ be the standard basis for $(\mathbf{R}^2)^*$ and let $\{\eta_1, \eta_2, \eta_3\}$ be the standard basis for $(\mathbf{R}^3)^*$.

Let $\alpha = \varepsilon_1 \wedge \varepsilon_2 = \varepsilon_{12} \in \Lambda_2(\mathbf{R}^2)$; we will compute $T^*\alpha \in \Lambda_2(\mathbf{R}^3)$. To do so, we write arbitrary vectors $\mathbf{v}, \mathbf{w} \in \mathbf{R}^3$ in components relative to the standard basis, $\mathbf{v} = (v_1, v_2, v_3)$ and $\mathbf{w} = (w_1, w_2, w_3)$. Then

$$\begin{aligned}
(T^*\alpha)(\mathbf{v}, \mathbf{w}) &= \alpha\,(T(\mathbf{v}), T(\mathbf{w})) \\
&= \alpha\,((2v_1 - v_3, v_2 - 3v_3), (2w_1 - w_3, w_2 - 3w_3)) \\
&= \varepsilon_1\,((2v_1 - v_3, v_2 - 3v_3))\,\varepsilon_2\,((2w_1 - w_3, w_2 - 3w_3)) \\
&\quad - \varepsilon_2\,((2v_1 - v_3, v_2 - 3v_3))\,\varepsilon_1\,((2w_1 - w_3, w_2 - 3w_3)) \\
&= (2v_1 - v_3)(w_2 - 3w_3) - (v_2 - 3v_3)(2w_1 - w_3) \\
&= (2v_1w_2 - 2w_1v_2) - (6v_1w_3 - 6w_1v_3) \\
&\quad + (v_2w_3 - w_2v_3) + (3v_3w_3 - 3w_3v_3) \\
&= 2(\eta_1 \wedge \eta_2)(\mathbf{v}, \mathbf{w}) - 6(\eta_1 \wedge \eta_3)(\mathbf{v}, \mathbf{w}) + (\eta_2 \wedge \eta_3)(\mathbf{v}, \mathbf{w}) \\
&= [2(\eta_1 \wedge \eta_2) - 6(\eta_1 \wedge \eta_3) + (\eta_2 \wedge \eta_3)]\,(\mathbf{v}, \mathbf{w}).
\end{aligned}$$

Hence $T^*\alpha = 2\eta_1 \wedge \eta_2 - 6\eta_1 \wedge \eta_3 + \eta_2 \wedge \eta_3$.

The reader can verify the following properties of the pullback.

Proposition 4.2.11. *Let U, V, W be vector spaces, and let $T_1 : U \to V$ and $T_2 : V \to W$ be linear transformations.*

1. *Let $\alpha, \beta \in \Lambda_k(V)$. Then*

$$T_1^*(\alpha \wedge \beta) = (T^*\alpha) \wedge (T^*\beta).$$

2. *Let $\alpha \in \Lambda_k(W)$. Then*

$$(T_2 \circ T_1)^*\alpha = T_1^*(T_2^*\alpha).$$

Proof. Exercise. Note that the second statement is in the same spirit as Proposition 2.8.16. □

We illustrate the practical use of Proposition 4.2.11 by revisiting Example 4.2.10.

Example 4.2.12. Let T, ε_1, and ε_2 be as in Example 4.2.10. The reader can verify that $T^*\varepsilon_1 = 2\eta_1 - \eta_3$ and $T^*\varepsilon_2 = \eta_2 - 3\eta_3$. Hence, by Proposition 4.2.11, we have

$$\begin{aligned} T^*(\varepsilon_1 \wedge \varepsilon_2) &= (T^*\varepsilon_1) \wedge (T^*\varepsilon_2) \\ &= (2\eta_1 - \eta_3) \wedge (\eta_2 - 3\eta_3) \\ &= 2\eta_1 \wedge \eta_2 - 6\eta_1 \wedge \eta_3 - \eta_3 \wedge \eta_2 + 3\eta_3 \wedge \eta_3 \\ &= 2\eta_1 \wedge \eta_2 - 6\eta_1 \wedge \eta_3 + \eta_2 \wedge \eta_3, \end{aligned}$$

the same result obtained in Example 4.2.10 directly from the definition.

We conclude this section with one final algebraic operation on linear forms: the interior product.

Definition 4.2.13. Let V be a vector space, and let $\mathbf{v} \in V$ and $\alpha \in \Lambda_k(V)$. The *interior product of α with $\mathbf{v}$* is the $(k-1)$-form, denoted by $i(\mathbf{v})\alpha \in \Lambda_{k-1}(V)$, defined by

$$(i(\mathbf{v})\alpha)(\mathbf{w}_1, \ldots, \mathbf{w}_{k-1}) = \alpha(\mathbf{v}, \mathbf{w}_1, \ldots, \mathbf{w}_{k-1})$$

for every $\mathbf{w}_1, \ldots, \mathbf{w}_{k-1} \in V$.

Again, we leave it as an exercise to verify that $i(\mathbf{v})\alpha$ so defined is multilinear and alternating, so that the result is in fact a $(k-1)$-form.

Example 4.2.14. Let $\omega = \varepsilon_1 \wedge \varepsilon_2 + \varepsilon_3 \wedge \varepsilon_4 \in \Lambda_2(\mathbf{R}^4)$ and let $\mathbf{v} = (v_1, v_2, v_3, v_4)$ be an arbitrary vector in $\mathbf{R}^4$. Then for every $\mathbf{w} = (w_1, w_2, w_3, w_4)$, we have $i(\mathbf{v})\omega \in \Lambda_1(\mathbf{R}^4)$ given by

$$\begin{aligned} (i(\mathbf{v})\omega)(\mathbf{w}) &= \omega(\mathbf{v}, \mathbf{w}) \\ &= (\varepsilon_1 \wedge \varepsilon_2 + \varepsilon_3 \wedge \varepsilon_4)(\mathbf{v}, \mathbf{w}) \\ &= (v_1 w_2 - v_2 w_1) + (v_3 w_4 - v_4 w_3) \\ &= -v_2 \varepsilon_1(\mathbf{w}) + v_1 \varepsilon_2(\mathbf{w}) - v_4 \varepsilon_3(\mathbf{w}) + v_3 \varepsilon_4(\mathbf{w}), \end{aligned}$$

which shows that

$$i(\mathbf{v})\omega = -v_2\varepsilon_1 + v_1\varepsilon_2 - v_4\varepsilon_3 + v_3\varepsilon_4.$$

We conclude by listing properties of how the interior product interacts with the other operations on linear forms we have discussed so far.

Proposition 4.2.15. *Let V and W be vector spaces, $\alpha \in \Lambda_k(V)$, $\beta \in \Lambda_l(V)$, $\gamma \in \Lambda_k(W)$, and let $T : V \to W$ be a linear transformation. Then:*

1. *The pairing $V \times \Lambda_k(V) \to \Lambda_{k-1}(V)$ given by $(\mathbf{v}, \omega) \mapsto i(\mathbf{v})\omega$ is bilinear.*
2. *For all $\mathbf{v} \in V$, we have $i(\mathbf{v})(\alpha \wedge \beta) = (i(\mathbf{v})\alpha) \wedge \beta + (-1)^k \alpha \wedge (i(\mathbf{v})\beta)$.*
3. *For all $\mathbf{v} \in V$, we have $i(\mathbf{v})(T^*\gamma) = T^*(i(T(\mathbf{v}))\gamma)$.*

Proof. Statement (1) follows from the linearity of α and from Definition 4.2.13.

To prove (2), suppose $\{\varepsilon_1, \ldots, \varepsilon_n\}$ is a basis for V^*. We first prove the result for *basis* forms, i.e., $\alpha = \varepsilon_{i_1 \cdots i_k}$ and $\beta = \varepsilon_{j_1 \cdots j_l}$. The general result then follows from statement (1). On the one hand, the reader can verify that

$$i(\mathbf{v})\varepsilon_{i_1\cdots i_k} = \sum_{r=1}^{k}(-1)^{r-1}\varepsilon_{i_r}(\mathbf{v})\varepsilon_{i_1}\wedge\cdots\wedge\widehat{\varepsilon_{i_r}}\wedge\cdots\wedge\varepsilon_{i_k},$$

where the hat indicates that the factor is omitted. On the other hand, note that for $(k+l-1)$ vectors $\mathbf{w}_1,\ldots,\mathbf{w}_{k+l-1}\in V$,

$$\begin{aligned} i(\mathbf{v})\left(\alpha\wedge\beta\right)\left(\mathbf{w}_1,\ldots,\mathbf{w}_{k+l-1}\right) &= \left(\alpha\wedge\beta\right)\left(\mathbf{v},\mathbf{w}_1,\ldots,\mathbf{w}_{k+l-1}\right)\\ &= \det\begin{bmatrix} \varepsilon_{i_1}(\mathbf{v}) & \varepsilon_{i_1}(\mathbf{w}_1) & \cdots & \varepsilon_{i_1}(\mathbf{w}_{k+l-1})\\ \vdots & \vdots & & \vdots\\ \varepsilon_{i_k}(\mathbf{v}) & \varepsilon_{i_k}(\mathbf{w}_1) & \cdots & \varepsilon_{i_k}(\mathbf{w}_{k+l-1})\\ \varepsilon_{j_1}(\mathbf{v}) & \varepsilon_{j_1}(\mathbf{w}_1) & \cdots & \varepsilon_{j_1}(\mathbf{w}_{k+l-1})\\ \vdots & \vdots & & \vdots\\ \varepsilon_{j_l}(\mathbf{v}) & \varepsilon_{j_l}(\mathbf{w}_1) & \cdots & \varepsilon_{j_l}(\mathbf{w}_{k+l-1}) \end{bmatrix}. \end{aligned}$$

Expanding this determinant from the first column and comparing to the right side, using Proposition 4.2.8 as necessary, gives the equality we are aiming to prove. Note that the factor $(-1)^k$ occurs as the sign of the $(k+1)$st term of the determinant expansion.

Finally, to prove (3), let $\mathbf{u}_1,\ldots,\mathbf{u}_{k-1}\in V$. Then for the linear transformation $T: V\to W$, we have

$$\begin{aligned} \Big(i(\mathbf{v})\left(T^*\gamma\right)\Big)(\mathbf{u}_1,\ldots,\mathbf{u}_{k-1}) &= (T^*\gamma)(\mathbf{v},\mathbf{u}_1,\ldots,\mathbf{u}_{k-1})\\ &= \gamma\big(T(\mathbf{v}),T(\mathbf{u}_1),\ldots,T(\mathbf{u}_{k-1})\big)\\ &= \Big(i(T(\mathbf{v}))\gamma\Big)\big(T(\mathbf{u}_1),\ldots,T(\mathbf{u}_{k-1})\big)\\ &= T^*\Big(i(T(\mathbf{v}))\gamma\Big)(\mathbf{u}_1,\ldots,\mathbf{u}_{k-1}). \end{aligned}$$

□

We illustrate the interior product with one example. Others will be seen in later sections.

Example 4.2.16. Let $\Omega=\varepsilon_1\wedge\varepsilon_2\wedge\varepsilon_3\in\Lambda_3(\mathbf{R}^3)$ and let $\mathbf{x}=(x_1,x_2,x_3)$. Then for all $\mathbf{y}=(y_1,y_2,y_3)$, $\mathbf{z}=(z_1,z_2,z_3)$, we have

$$\begin{aligned} \left(i(\mathbf{x})\Omega\right)(\mathbf{y},\mathbf{z}) &= \Omega(\mathbf{x},\mathbf{y},\mathbf{z})\\ &= \varepsilon_1(\mathbf{x})\varepsilon_2(\mathbf{y})\varepsilon_3(\mathbf{z}) - \varepsilon_1(\mathbf{y})\varepsilon_2(\mathbf{x})\varepsilon_3(\mathbf{z})\\ &\quad + \varepsilon_1(\mathbf{y})\varepsilon_2(\mathbf{z})\varepsilon_3(\mathbf{x}) - \varepsilon_1(\mathbf{z})\varepsilon_2(\mathbf{y})\varepsilon_3(\mathbf{x})\\ &\quad + \varepsilon_1(\mathbf{z})\varepsilon_2(\mathbf{x})\varepsilon_3(\mathbf{y}) - \varepsilon_1(\mathbf{x})\varepsilon_2(\mathbf{z})\varepsilon_3(\mathbf{y}) \end{aligned}$$

$$= x_1 (y_2 z_3 - z_2 y_3) - x_2 (y_1 z_3 - y_3 z_1) + x_3 (y_1 z_2 - y_2 z_1)$$
$$= x_1(\varepsilon_2 \wedge \varepsilon_3)(\mathbf{y}, \mathbf{z}) - x_2(\varepsilon_1 \wedge \varepsilon_3)(\mathbf{y}, \mathbf{z})$$
$$+ x_3(\varepsilon_1 \wedge \varepsilon_2)(\mathbf{y}, \mathbf{z}).$$

Hence,

$$i(\mathbf{x})\Omega = x_1 \varepsilon_2 \wedge \varepsilon_3 - x_2 \varepsilon_1 \wedge \varepsilon_3 + x_3 \varepsilon_1 \wedge \varepsilon_2.$$

4.3 Differential Forms

With the previous two sections' introduction of exterior algebra, as the algebra of forms is called, behind us, we now return to the setting of calculus and, more precisely, to the tangent space. Our presentation of differential forms, especially of one-forms, will parallel our presentation of vector fields as smoothly varying families of tangent vectors. Indeed, dual to the notion of a vector field, a differential one-form will be just a smoothly varying family of linear one-forms defined on the tangent spaces.

However, differential calculus has a special impact in the context of forms, which we begin to explore in this chapter. The implications for the development of geometric structures will be the subject of the remaining chapters.

Recall from Chap. 3 that at each point $p \in \mathbf{R}^n$, we associate an n-dimensional vector space, the tangent space $T_p(\mathbf{R}^n)$. Tangent vectors $\mathbf{v}_p \in T_p(\mathbf{R}^n)$ can be thought of as operators assigning to each function f defined in a domain containing p a real number $\mathbf{v}_p[f]$. The standard basis for $T_p(\mathbf{R}^n)$ is then the set of partial differentiation operators

$$\left\{ \left.\frac{\partial}{\partial x_1}\right|_p, \dots, \left.\frac{\partial}{\partial x_n}\right|_p \right\};$$

see Theorem 3.4.5. In this way, a vector field V on $\mathbf{R}^n$ can be written as

$$V = \sum_{i=1}^n v_i \frac{\partial}{\partial x_i},$$

where the $v_i : \mathbf{R}^n \to \mathbf{R}$ are differentiable real-valued functions.

In keeping with this traditional notation, we establish the following definitions, which are really just adaptations of the previous two sections to the notation particular to the vector spaces $T_p(\mathbf{R}^n)$.

Definition 4.3.1. For $i = 1, \dots, n$, define the elements $(dx_i)_p \in (T_p(\mathbf{R}^n))^*$ as the elements dual to the elements $\left.\frac{\partial}{\partial x_1}\right|_p$ in the sense of Example 2.8.4 and the beginning of Sect. 4.1, i.e.,

$$(dx_i)_p \left(\left. \frac{\partial}{\partial x_j} \right|_p \right) = \begin{cases} 1 & \text{if } i = j, \\ 0 & \text{if } i \neq j. \end{cases}$$

In other words, the set $\{(dx_1)_p, \ldots, (dx_n)_p\}$ is the standard basis for the vector space $(T_p(\mathbf{R}^n))^*$.

Using this notation, a (linear) one-form $\alpha_p \in (T_p(\mathbf{R}^n))^*$ will be written using the standard basis as

$$\alpha_p = \sum_{i=1}^{n} a_i \, (dx_i)_p,$$

where for each $i = 1, \ldots, n$, a_i is a constant.

We can then proceed to write a basis for the vector space of alternating k-forms on the vector space $T_p(\mathbf{R}^n)$, according to Definition 4.2.1, as the exterior products of the elements $(dx_i)_p$. Namely, we can consider the set

$$\left\{ (dx_{i_1})_p \wedge \cdots \wedge (dx_{i_k})_p \mid i_1 < \cdots < i_k \right\}.$$

The following proposition is simply Theorem 4.1.8 when $V = T_p(\mathbf{R}^n)$.

Proposition 4.3.2. *The set*

$$\left\{ (dx_{i_1})_p \wedge \cdots \wedge (dx_{i_k})_p \mid i_1 < \cdots < i_k \right\}$$

is a basis for $\Lambda_k \left(T_p(\mathbf{R}^n) \right)$.

All the properties of the exterior product carry over into this setting. For example, for all $i, j = 1, \ldots, n$, we have

$$(dx_i)_p \wedge (dx_j)_p = -(dx_j)_p \wedge (dx_i)_p,$$

and in particular,

$$(dx_i)_p \wedge (dx_i)_p = 0.$$

We now follow the example of Definition 3.7.2 to define the central notion of this chapter. As in Chap. 3, we denote for the sake of the definition the set

$$\Lambda_k \left(T\mathbf{R}^n \right) = \{ (p, \alpha_p) \mid p \in \mathbf{R}^n, \ \alpha_p \in \Lambda_k \left(T_p(\mathbf{R}^n) \right) \}$$

as the vector bundle of k-forms on $T_p(\mathbf{R}^n)$.

Definition 4.3.3. A (smooth) *differential k-form on* $\mathbf{R}^n$ is a function

$$\alpha : \mathbf{R}^n \to \Lambda_k \left(T\mathbf{R}^n \right)$$

such that:

1. For each $p \in \mathbf{R}^n$, $\alpha(p) = (p, \alpha_p)$, where $\alpha_p \in \Lambda_k\left((T_p(\mathbf{R}^n))\right)$.
2. α is differentiable in the following sense: For all $p \in \mathbf{R}^n$ and for all smooth vector fields V_i $(i = 1, \ldots, k)$, the function $F : \mathbf{R}^n \to \mathbf{R}$ given by

$$F(p) = \alpha_p\left(V_1(p), \ldots, V_k(p)\right), \text{ where } \alpha(p) = (p, \alpha_p),$$

is a smooth function of p.

A 0-form will be, by definition, a smooth function $f : \mathbf{R}^n \to \mathbf{R}$.

We are following the practice, mentioned in Chap. 3, of keeping the vector bundle concept in the background and writing a smooth differential k-form as $\alpha(p) = \alpha_p$, where

$$\alpha_p \in \Lambda_k(T_p(\mathbf{R}^n)).$$

More to the point, and keeping in mind Proposition 4.3.2, we can write $\alpha \in \Lambda_k(T\mathbf{R}^n)$ as

$$\alpha(p) = \sum_{i_1 < \cdots < i_k} a_{i_1 \cdots i_k}(p)(dx_{i_1})_p \wedge \cdots \wedge (dx_{i_k})_p,$$

where the $a_{i_1 \cdots i_k} : \mathbf{R}^n \to \mathbf{R}$ are smooth, real-valued functions of p.

4.4 Operations on Differential Forms

All the operations of the exterior algebra of linear k-forms can be defined for differential k-forms by defining the operations pointwise. Namely, let α and β be k-forms, γ be a one-form, and let c be an arbitrary scalar. Then the operations of addition, scalar multiplication, exterior product, and interior product of differential forms are defined as follows:

- $(\alpha + \beta)(p) = \alpha(p) + \beta(p)$.
- $(c\alpha)(p) = c \cdot \alpha(p)$.
- $(\alpha \wedge \gamma)(p) = \alpha(p) \wedge \gamma(p)$.
- Let V be a vector field on $\mathbf{R}^n$. Then $i(V)\alpha$ is a $(k-1)$-form defined by $(i(V)\alpha)(p) = i\left(V(p)\right)\alpha(p)$.

The reader can verify that the operations yield *differential* forms, i.e., that the operations yield smooth forms.

The differential forms endowed with these algebraic operations enjoy the same properties (pointwise) as listed in Propositions 4.2.8 and 4.2.15.

Example 4.4.1. Let

$$\alpha = x\,dx + y\,dy + z\,dz \in \Lambda_1(T\mathbf{R}^3)$$

and

$$\beta = x\,dy \wedge dz - y\,dx \wedge dz + z\,dx \wedge dy \in \Lambda_2(T\mathbf{R}^3).$$

Then $\alpha \wedge \beta \in \Lambda_3(T\mathbf{R}^3)$ is given by:

$$\begin{aligned}\alpha \wedge \beta &= x^2\,dx \wedge dy \wedge dz - xy\,dx \wedge dx \wedge dz + xz\,dx \wedge dx \wedge dy \\ &\quad + yx\,dy \wedge dy \wedge dz - y^2\,dy \wedge dx \wedge dz + yz\,dy \wedge dx \wedge dy \\ &\quad + xz\,dz \wedge dy \wedge dz - zy\,dz \wedge dx \wedge dz + z^2\,dz \wedge dx \wedge dy \\ &= \left(x^2 + y^2 + z^2\right) dx \wedge dy \wedge dz.\end{aligned}$$

Thus far, we have extended linear forms to the setting of differential geometry only in the most obvious way: defining differential forms "pointwise," acting on tangent vectors at a given point $p \in \mathbf{R}^n$, and then requiring that they vary smoothly with the point p. In this sense, they are "fields of linear k-forms." Differential forms, however, are more than simply algebraic objects. In particular, there is a "calculus" associated with them in the form of an intrinsic way of "differentiating" differential forms. The rest of this section is devoted to understanding this operation on forms.

We begin with the case of a 0-form on $\mathbf{R}^n$, i.e., a smooth function.

Definition 4.4.2. Let $f : \mathbf{R}^n \to \mathbf{R}$ be a smooth function. The *exterior derivative* of f is the one-form, denoted by df, defined by

$$df(V) = V\,[f]\,,$$

for every smooth vector field V on $\mathbf{R}^n$.

Note that the linearity of df is a consequence of the way that vector fields act as operators on smooth functions: $df(V_1 + V_2) = (V_1 + V_2)\,[f] = V_1\,[f] + V_2\,[f] = df(V_1) + df(V_2)$ and $df(cV) = (cV)\,[f] = c\,(V\,[f]) = c\,df(V)$.

The first example will justify the notation of Definition 4.3.1 for the basis elements of the dual space $(T\mathbf{R}^n)^* = \Lambda_1(T\mathbf{R}^n)$.

Example 4.4.3. Let $x_i : \mathbf{R}^n \to \mathbf{R}$ be the projection onto the ith coordinate, so that for $p \in \mathbf{R}^n$ with $p = (p_1, \dots, p_n)$,

$$x_i(p) = p_i.$$

Let V be an arbitrary vector field on $\mathbf{R}^n$ defined by

$$V(p) = a_1(p) \left.\frac{\partial}{\partial x_1}\right|_p + \cdots + a_n(p) \left.\frac{\partial}{\partial x_n}\right|_p,$$

where the $a_i : \mathbf{R}^n \to \mathbf{R}$ are smooth functions of p. Then:

$$\begin{aligned} dx_i(V) &= V\,[x_i] \\ &= \left(a_1 \frac{\partial}{\partial x_1} + \cdots + a_n \frac{\partial}{\partial x_n}\right)[x_i] \\ &= a_i. \end{aligned}$$

Written differently, $dx_i\left(\langle a_1, \ldots, a_n \rangle\right) = a_i$. This is precisely the element of $(T\mathbf{R}^n)^*$ that is dual to the standard basis vector field $\dfrac{\partial}{\partial x_i}$.

The following proposition gives precise meaning to a statement that is defined only symbolically in the standard first course in calculus.

Proposition 4.4.4. *Let $f : \mathbf{R}^n \to \mathbf{R}$ be a smooth function. Then, using the standard basis of one-forms $\{dx_1, \ldots, dx_n\}$, we can write*

$$df(p) = \left.\frac{\partial f}{\partial x_1}\right|_p (dx_1)_p + \cdots + \left.\frac{\partial f}{\partial x_n}\right|_p (dx_n)_p.$$

Proof. The statement follows by evaluating the two sides of the equation against a general vector field $V = \sum a_i \dfrac{\partial}{\partial x_i}$, the left side using Definitions 4.4.2 and 3.7.2 and the right side using Definition 4.3.1. □

The reader may notice that the *components* of the one-form df, relative to the standard basis, are the same as the *components* of the usual gradient vector field ∇f. However, the distinction between the vector field ∇f and the differential one-form df is an essential one. The latter notion is defined in terms of the differential structure alone, whereas we will see that the former is dependent on an additional (metric) structure.

We now proceed to use the definition of the exterior derivative of a 0-form, along with the exterior product, to define the exterior derivative of a k-form. In what follows, we shall rely on the standard basis for $\Lambda_k(\mathbf{R}^n)$ given in Proposition 4.3.2.

Definition 4.4.5. Let ω be a k-form on $\mathbf{R}^n$ with smooth component functions $a_{i_1 \cdots i_k} : \mathbf{R}^n \to \mathbf{R}$, i.e., $\omega = \sum a_{i_1 \cdots i_k} dx_{i_1} \wedge \cdots \wedge dx_{i_k}$, where $(i_1 \cdots i_k)$ is a multi-index of the set $\{1, \ldots, n\}$. The *exterior derivative of the k-form ω* is the $(k+1)$-form $d\omega$ defined as follows:

$$d\omega = \sum (da_{i_1 \cdots i_k}) \wedge dx_{i_1} \wedge \cdots \wedge dx_{i_k}.$$

Example 4.4.6. Let $\alpha, \beta \in \Lambda_1(\mathbf{R}^3)$ be defined by $\alpha = x\,dx + y\,dy + z\,dz$, $\beta = x^2 dy - \cos(xz) dz$, and define $\gamma \in \Lambda_2(\mathbf{R}^3)$ as $\gamma = x\,dy \wedge dz + y\,dx \wedge dz + z\,dx \wedge dy$.

Then

$$\begin{aligned} d\alpha &= d(x) \wedge dx + d(y) \wedge dy + d(z) \wedge dz \\ &= dx \wedge dx + dy \wedge dy + dz \wedge dz \\ &= 0, \end{aligned}$$

$$\begin{aligned} d\beta &= d(x^2) \wedge dy + d(-\cos(xz)) \wedge dz \\ &= (2xdx) \wedge dy + \left[z\sin(xz)dx + x\sin(xz)dz\right] \wedge dz \\ &= 2xdx \wedge dy + z\sin(xz)dx \wedge dz, \end{aligned}$$

and

$$\begin{aligned} d\gamma &= d(x) \wedge dy \wedge dz + d(y) \wedge dx \wedge dz + d(z) \wedge dx \wedge dy \\ &= dx \wedge dy \wedge dz + dy \wedge dx \wedge dz + dz \wedge dx \wedge dy \\ &= dx \wedge dy \wedge dz. \end{aligned}$$

Example 4.4.7. Let ω be an n-form on $\mathbf{R}^n$. Then $d\omega = 0$. Indeed, since $\Lambda_n(T_p\mathbf{R}^n)$ is one-dimensional (at each point $p \in \mathbf{R}^n$), we can write $\omega = f dx_1 \wedge \cdots \wedge dx_n$ for some smooth function $f : \mathbf{R}^n \to \mathbf{R}$. Hence

$$d\omega = \left(\frac{\partial f}{\partial x_1}dx_1 + \cdots + \frac{\partial f}{\partial x_n}dx_n\right) \wedge dx_1 \wedge \cdots \wedge dx_n,$$

and each term of the exterior product is zero, since $dx_i \wedge dx_i = 0$.

The following proposition illustrates the relationship between the exterior derivative d and the pointwise operations on differential forms.

Proposition 4.4.8. *Let $\alpha, \beta \in \Lambda_k(T\mathbf{R}^n)$ be smooth k-forms, $\gamma \in \Lambda_l(T\mathbf{R}^n)$ a smooth one-form, and let $s, t \in \mathbf{R}$ be scalars. Then:*

1. $d(s\alpha + t\beta) = sd\alpha + td\beta$; *and*
2. $d(\alpha \wedge \gamma) = (d\alpha) \wedge \gamma + (-1)^k \alpha \wedge (d\gamma)$.

Proof. The equality (1) is immediate. To show (2), we prove the statement for monomial forms, the result then following from (1). So let $\alpha = adx_{i_1} \wedge \cdots \wedge dx_{i_k}$ and $\gamma = bdx_{j_1} \wedge \cdots \wedge dx_{j_l}$, where $a, b : \mathbf{R}^n \to \mathbf{R}$ are smooth real-valued functions. We assume that the multi-indices $(i_1, \ldots, i_k)$ and $(j_1, \ldots, j_l)$ are disjoint, since otherwise, both sides of the equation in (2) will be zero. We can then compare the terms involved. On the one hand,

$$\begin{aligned} d(\alpha \wedge \gamma) &= d\left[(adx_{i_1} \wedge \cdots \wedge dx_{i_k}) \wedge (bdx_{j_1} \wedge \cdots \wedge dx_{j_l})\right] \\ &= d\left[(a \cdot b)dx_{i_1} \wedge \cdots \wedge dx_{i_k} \wedge dx_{j_1} \wedge \cdots \wedge dx_{j_l}\right] \\ &= \left(\sum_{m=1}^{n} \frac{\partial}{\partial x_m}(a \cdot b)dx_m\right) \wedge dx_{i_1} \wedge \cdots \wedge dx_{i_k} \wedge dx_{j_1} \wedge \cdots \wedge dx_{j_l} \\ &= \left(\sum_{m=1}^{n} \left(a\frac{\partial b}{\partial x_m} + b\frac{\partial a}{\partial x_m}\right) dx_m\right) \wedge dx_{i_1} \wedge \cdots \wedge dx_{i_k} \end{aligned}$$

$$\wedge\, dx_{j_1} \wedge \cdots \wedge dx_{j_l}$$

$$= \sum_{m=1}^{n} (a\frac{\partial b}{\partial x_m} + b\frac{\partial a}{\partial x_m}) dx_m \wedge dx_{i_1} \wedge \cdots \wedge dx_{i_k} \wedge dx_{j_1} \wedge \cdots \wedge dx_{j_l}.$$

On the other hand,

$$\begin{aligned}(d\alpha) \wedge \gamma &= \left(\sum_{m=1}^{n} \frac{\partial a}{\partial x_m} dx_m\right) \wedge (dx_{i_1} \wedge \cdots \wedge dx_{i_k}) \wedge (b dx_{j_1} \wedge \cdots \wedge dx_{j_l}) \\ &= \sum_{m=1}^{n} \left(b\frac{\partial a}{\partial x_m}\right) dx_m \wedge dx_{i_1} \wedge \cdots \wedge dx_{i_k} \wedge dx_{j_1} \wedge \cdots \wedge dx_{j_l}\end{aligned}$$

and

$$\begin{aligned}\alpha \wedge (d\gamma) &= (a dx_{i_1} \wedge \cdots \wedge dx_{i_k}) \wedge \left(\sum_{m=1}^{n} (\frac{\partial b}{\partial x_m} dx_m) \wedge dx_{j_1} \wedge \cdots \wedge dx_{j_l}\right) \\ &= \sum_{m=1}^{n} \left(a\frac{\partial b}{\partial x_m}\right) dx_{i_1} \wedge \cdots \wedge dx_{i_k} \wedge dx_m \wedge dx_{j_1} \wedge \cdots \wedge dx_{j_l}.\end{aligned}$$

To add the latter two expressions, the dx_m factor must pass through the k factors $dx_{i_1}, \ldots, dx_{i_k}$, introducing the factor $(-1)^k$ in (2). Comparing the left and right sides shows the desired equality. □

The most important property of the exterior derivative of a smooth differential form is a consequence of the familiar equality of mixed partial derivatives. We will encounter implications of the following theorem throughout the text.

Theorem 4.4.9. *Let α be a k-form on $\mathbf{R}^n$. Then*

$$d\,(d\alpha) = 0.$$

Proof. We will present the proof for a 0-form, i.e., for a smooth function $f : \mathbf{R}^n \to \mathbf{R}$. This is sufficient for the general case in light of Definition 4.4.5 and Proposition 4.4.8. We have

$$\begin{aligned}d\,(df) &= d\left(\frac{\partial f}{\partial x_1} dx_1 + \cdots + \frac{\partial f}{\partial x_n} dx_n\right) \\ &= \left(\frac{\partial^2 f}{\partial x_1^2} dx_1 + \frac{\partial^2 f}{\partial x_2 \partial x_1} dx_2 + \cdots + \frac{\partial^2 f}{\partial x_n \partial x_1} dx_n\right) \wedge dx_1 \\ &\quad + \left(\frac{\partial^2 f}{\partial x_1 \partial x_2} dx_1 + \frac{\partial^2 f}{\partial x_2^2} dx_2 + \cdots + \frac{\partial^2 f}{\partial x_n \partial x_2} dx_n\right) \wedge dx_2 +\end{aligned}$$

$$\vdots$$

$$+\left(\frac{\partial^2 f}{\partial x_1 \partial x_n} dx_1 + \frac{\partial^2 f}{\partial x_2 \partial x_n} dx_2 + \cdots + \frac{\partial^2 f}{\partial x_n^2} dx_n\right) \wedge dx_n$$

$$= \sum_{i<j} \left(\frac{\partial^2 f}{\partial x_i \partial x_j} - \frac{\partial^2 f}{\partial x_j \partial x_i}\right) dx_i \wedge dx_j$$

$$= 0,$$

the last equality due to the equality of mixed partials of f. □

Theorem 4.4.9 prompts the following definitions.

Definition 4.4.10. Let α be a k-form on $\mathbf{R}^n$. Then α is *closed* if $d\alpha = 0$, and α is *exact* if there is a $(k-1)$-form β such that $d\beta = \alpha$.

It follows immediately from Theorem 4.4.9 that every exact form is also closed. In fact, for smooth forms defined on all of $\mathbf{R}^n$, the converse is also true: Every closed form is also exact. This fact is a consequence of the more general Poincaré lemma, which we state here for reference.

Theorem 4.4.11 (Poincaré lemma). *Suppose that $U \subset \mathbf{R}^n$ is a domain that is* star-shaped, *i.e., there is a point $p \in U$ such that for all $q \in U$, the entire line segment with endpoints p and q lies in U. Let α be a closed k-form defined on U. Then there is a $(k-1)$-form β defined on U such that $\alpha = d\beta$.*

For a proof, see, for example, [14] or [37]. We list here two important corollaries of the Poincaré lemma.

Corollary 4.4.12. *If α is a closed k-form defined on all of $\mathbf{R}^n$, then α is exact.*

Proof. $\mathbf{R}^n$ is star-shaped. □

Corollary 4.4.13. *Let α be a closed k-form defined on a domain U. Then for each point $p \in U$, there is a domain $V \subset U$ containing p on which α is exact.*

Proof. By virtue of being an open set, U contains an open ball V centered about p, and an open ball is star-shaped. □

For this reason, a closed form can be said to be *locally exact*.

While we will not pursue the many avenues toward which Definition 4.4.10 leads, we note that there are basically two ways that a closed form can fail to be exact. One way obtains when the domain is a region in $\mathbf{R}^n$ that has "holes." Another occurs when the form itself is either not defined or not differentiable at some point or region in $\mathbf{R}^n$. Both of these obstructions are studied in the field of differential topology.

The pullback of a differential form by a function is also defined pointwise.

Definition 4.4.14. Let $f : \mathbf{R}^n \to \mathbf{R}^m$ be a smooth function and let α be a differential k-form on the codomain $\mathbf{R}^m$. The *pullback of α by f*, denoted by $f^*\alpha$, is a differential k-form on the domain $\mathbf{R}^n$ of f defined by

$$(f^*\alpha)(p)\big((\mathbf{v}_1)_p, \ldots, (\mathbf{v}_k)_p\big) = \alpha(f(p))\big(f_*(\mathbf{v}_1)_p, \ldots, f_*(\mathbf{v}_k)_p\big),$$

where $(\mathbf{v}_1)_p, \ldots, (\mathbf{v}_k)_p \in T_p\mathbf{R}^n$ and f_* is the tangent map of f. The pullback of a 0-form (function) g by f will be the composition

$$f^*g = g \circ f.$$

Note that $f^*\alpha$ is smooth in the sense of Definition 4.3.3, a consequence of the smoothness of f and α.

The following example illustrates the nature of the computations implied by Definition 4.4.14.

Example 4.4.15. Consider $\mathbf{R}^3$ with coordinates (x_1, x_2, x_3) and $\mathbf{R}^2$ with coordinates (y_1, y_2). Let α be the one-form on $\mathbf{R}^2$ given by

$$\alpha_{(y_1,y_2)} = y_1^2 dy_1 + y_2^3 dy_2,$$

and let $f : \mathbf{R}^3 \to \mathbf{R}^2$ be given by

$$f(x_1, x_2, x_3) = \left(x_1^2 + x_2^2, \sin(x_3)\right).$$

Expressed in another way, we have

$$f(\mathbf{R}^3) = \left\{(y_1, y_2) \mid \exists\, (x_1, x_2, x_3) \in \mathbf{R}^3 \text{ such that } y_1 = x_1^2 + x_2^2,\ y_2 = \sin(x_3)\right\}.$$

Consider a general vector field X on $\mathbf{R}^3$,

$$X = X^1 \frac{\partial}{\partial x_1} + X^2 \frac{\partial}{\partial x_2} + X^3 \frac{\partial}{\partial x_3},$$

where $X^1, X^2, X^3 : \mathbf{R}^3 \to \mathbf{R}$ are smooth, real-valued functions. To compute $f^*\alpha$, we first compute the Jacobian matrix of f_*:

$$(f_*)_{(x_1,x_2,x_3)} = \begin{bmatrix} 2x_1 & 2x_2 & 0 \\ 0 & 0 & \cos(x_3) \end{bmatrix}.$$

Hence

$$(f_*)_{(x_1,x_2,x_3)}(X) = \left(2x_1 X^1 + 2x_2 X^2\right) \frac{\partial}{\partial y_1} + \cos(x_3) \cdot X^3 \frac{\partial}{\partial y_2},$$

and so

$$\begin{aligned}\alpha_{f(x_1,x_2,x_3)}(f_*X) &= \alpha_{(x_1^2+x_2^2,\sin(x_3))}(f_*X) \\ &= (x_1^2 + x_2^2)^2(2x_1X^1 + 2x_2X^2) + (\sin(x_3))^3(\cos(x_3) \cdot X^3).\end{aligned}$$

Comparing coefficients shows that

$$f^*\alpha = \left(2x_1(x_1^2+x_2^2)^2\right)dx_1 + \left(2x_2(x_1^2+x_2^2)^2\right)dx_2 + \left(\sin^3(x_3)\cos(x_3)\right)dx_3.$$

We will return to this example after establishing some properties of the pullback of differential forms that yield useful computational techniques.

Proposition 4.4.16. *Let $f : \mathbf{R}^n \to \mathbf{R}^m$ be a smooth function.*

1. *For differential k-forms α, β on $\mathbf{R}^m$ and scalars $s, t \in \mathbf{R}$,*

$$f^*(s\alpha + t\beta) = sf^*\alpha + tf^*\beta.$$

2. *For a differential k-form α and a differential one-form β on $\mathbf{R}^m$,*

$$f^*(\alpha \wedge \beta) = (f^*\alpha) \wedge (f^*\beta).$$

3. *For a differential k-form α on $\mathbf{R}^m$,*

$$f^*(d\alpha) = d(f^*\alpha).$$

Proof. Statements (1) and (2) are proved pointwise in the same manner as their linear counterparts in Propositions 4.2.8 and 4.2.11.

Statement (3) follows from (1) and (2) if we can prove the corresponding statement for 0-forms. For in that case, writing $\alpha = \sum_I a_I dx_I$ using the multi-index I to represent $(i_1, \ldots, i_k)$ and defining $dx_I = dx_{i_1} \wedge \cdots \wedge dx_{i_k}$, we would have

$$\begin{aligned}
f^*(d\alpha) &= f^*\left(d\sum_I a_I dx_I\right) \\
&= f^*\left(\sum_I d(a_I dx_I)\right) \\
&= f^*\left(\sum_I da_I \wedge dx_I\right) \\
&= \sum_I (f^* da_I) \wedge (f^* dx_I) \\
&= \sum_I d(f^* a_I) \wedge (f^* dx_I) \\
&= d\left(\sum_I (f^* a_I)(f^* dx_I)\right) \\
&= d(f^*\alpha).
\end{aligned}$$

To prove the statement for 0-forms, let $a : \mathbf{R}^m \to \mathbf{R}$ be a 0-form (i.e., a real-valued function) on $\mathbf{R}^m$ and write $f : \mathbf{R}^n \to \mathbf{R}^m$ using component functions $f^1, \ldots, f^m$:

$$f(x_1, \ldots, x_n) = \big(f^1(x_1, \ldots, x_n), \ldots, f^m(x_1, \ldots, x_n)\big).$$

Writing the coordinates in $\mathbf{R}^m$ as $(y_1, \ldots, y_m)$, we have

$$da = \sum_{j=1}^{m} \frac{\partial a}{\partial y_j} dy_j.$$

For any vector field X on $\mathbf{R}^n$, we can write $X = \sum_{i=1}^{n} X^i \frac{\partial}{\partial x_i}$, and

$$f_* X = \sum_{i,j} \left(\frac{\partial f^j}{\partial x_i} X^i \right) \frac{\partial}{\partial y_j}.$$

Hence

$$\begin{aligned} (f^* da)(X) &= (da)(f_* X) \\ &= \sum_{i,j} \frac{\partial a}{\partial y_j} \left(\frac{\partial f^j}{\partial x_i} X^i \right). \end{aligned}$$

Examining components, we obtain

$$\begin{aligned} f^*(da) &= \sum_{i,j} \left(\frac{\partial a}{\partial y_j} \frac{\partial f^j}{\partial x_i} \right) dx_i \\ &= d(a \circ f) \\ &= d(f^* a). \quad \square \end{aligned}$$

The preceding proposition has the following computational corollary.

Corollary 4.4.17. *Let $f : \mathbf{R}^n \to \mathbf{R}^m$ be a smooth function with component functions $f^1, \ldots, f^m$. Let $\alpha = \sum a_J dy_{j_1} \wedge \cdots \wedge dy_{j_k}$ be a k-form on $\mathbf{R}^m$, where $a_J = a_{j_1 \cdots j_l} : \mathbf{R}^m \to \mathbf{R}$ are smooth real-valued component functions. Then*

$$f^* \alpha = \sum (f \circ a_J) df^{j_1} \wedge \cdots \wedge df^{j_l}.$$

Example 4.4.18. Returning to Example 4.4.15, consider $f : \mathbf{R}^3 \to \mathbf{R}^2$ given by

$$f(x_1, x_2, x_3) = (x_1^2 + x_2^2, \sin(x_3))$$

and the two-form α on $\mathbf{R}^2$ given by $\alpha = y_1^2 dy_1 + y_2^3 dy_2$. By Corollary 4.4.17,

$$\begin{aligned} f^*\alpha &= (x_1^2 + x_2^2)^2 d(x_1^2 + x_2^2) + (\sin(x_3))^3 d(\sin(x_3)) \\ &= (x_1^2 + x_2^2)^2 \left(2x_1 dx_1 + 2x_2 dx_2\right) + (\sin(x_3))^3 (\cos(x_3)) dx_3 \\ &= 2x_1(x_1^2 + x_2^2)^2 dx_1 + 2x_2(x_1^2 + x_2^2)^2 dx_2 + (\cos(x_3)) \sin^3(x_3) dx_3, \end{aligned}$$

agreeing with the previous computation.

Example 4.4.19. Let $f : \mathbf{R}^2 \to \mathbf{R}^2$ be a smooth function. We write the coordinates of the domain as (x_1, x_2) and the coordinates of the codomain as (y_1, y_2), and we denote the component functions of f by $f^1, f^2 : \mathbf{R}^2 \to \mathbf{R}$. Let $\omega = dy_1 \wedge dy_2$.

Then

$$\begin{aligned} f^*\omega &= f^*(dy_1 \wedge dy_2) \\ &= df^1 \wedge df^2 \\ &= \left(\frac{\partial f^1}{\partial x_1} dx_1 + \frac{\partial f^1}{\partial x_2} dx_2\right) \wedge \left(\frac{\partial f^2}{\partial x_1} dx_1 + \frac{\partial f^2}{\partial x_2} dx_2\right) \\ &= \left(\frac{\partial f^1}{\partial x_1}\frac{\partial f^2}{\partial x_2} - \frac{\partial f^1}{\partial x_2}\frac{\partial f^2}{\partial x_1}\right) dx_1 \wedge dx_2 \\ &= \det(f_*) dx_1 \wedge dx_2. \end{aligned}$$

The preceding example is a special case of a more general fact. Before stating the general result, note that in discussing diffeomorphisms (or, more generally, smooth functions between spaces of the same dimension) there is often an abuse of notation, whereby the same variable names are used for the coordinates of the domain and those of the codomain. This abuse has the advantage of allowing us to write results like that of the previous example in the following nice manner.

Proposition 4.4.20. *Consider the differential n-form on $\mathbf{R}^n$ given by $\Omega = dx_1 \wedge \cdots \wedge dx_n$, where $(x_1, \dots, x_n)$ are coordinates for $\mathbf{R}^n$. Let $f : \mathbf{R}^n \to \mathbf{R}^n$ be a smooth function on $\mathbf{R}^n$. Then*

$$f^*\Omega = (\det(f_*)) \cdot \Omega.$$

Proof. Exercise. □

4.5 Integrating Differential Forms

We began this chapter with the perspective that differential forms give a way of measuring vector fields or collections of vector fields. In this section, we indicate one way that this "local" measurement (i.e., at the level of the tangent space) can be used to measure "global" geometric objects, and in particular the geometric sets of Chap. 2. In essence, this section extends to the nonlinear setting the fact that the determinant of n vectors in $\mathbf{R}^n$ can be interpreted as a kind of "oriented volume" of the parallelpiped formed by the vectors. To do this, we will extend the concept of the multiple integral in such a way that the integral is defined in a coordinate-free way.

As will become clear, integration as presented in this section will not be our main concern in this text. In fact, little of the later exposition will depend on the ideas in this section, especially if the reader has had some prior familiarity with the Gauss-Stokes theorem.

For the same reason, we will not attempt to give a rigorous definition of an integral as a limit of sums, instead relying on familiarity with the standard definitions in a first course in multivariable calculus. Nor will we be overly concerned with the objects over which forms will be integrated. We will simply refer to a *region of integration* as any subset $D \subset \mathbf{R}^n$ on which the multiple integral $\int_D dx_1 \cdots dx_n$ is defined. The simplest such region is a "rectangular" region

$$[a_1, b_1] \times \cdots \times [a_n, b_n] .$$

For a more precise treatment of all the ideas in this section, the reader is encouraged to consult [14] or [37].

We will define the integral of an n-form ω over an n-dimensional region of integration R essentially as a multiple integral. However, from the outset we have the task of reconciling two basic facts. On the one hand, interchanging the order of two of the basis one-forms changes the sign of the alternating form ω. On the other hand, changing the order of integration should not change the result of the integral.

The key to reconciling these two competing facts is the notion of *orientation*. Intuitively, orientation is the notion that allows us to distinguish between left and right on a number line or between clockwise and counterclockwise in the plane.

Definition 4.5.1. Let $U \subset \mathbf{R}^n$ be a domain. An *orientation* of U is a choice of n ordered smooth vector fields $(B_1, \ldots, B_n)$ on U with the property that for each $p \in U$, $\{B_1(p), \ldots, B_n(p)\}$ is a basis for $T_p(\mathbf{R}^n)$.

The key to this definition is that fixing one ordered basis $\mathcal{B}$ in fact partitions the set of *all* ordered n-tuples of basis vector fields into two classes. For any other ordered n-tuple of basis vector fields $\mathcal{B}' = (B_1', \ldots, B_n')$, we have

$$\det \left[B_1'(p) \cdots B_n'(p)\right] \neq 0$$

for all $p \in U$ (the determinant computed by writing $B_i'(p)$ as column vectors relative to the basis $\mathcal{B}$). Since the determinant of a matrix is a continuous function of the entries of the matrix, which are themselves smooth functions, it must be everywhere positive or everywhere negative. If the determinant is positive, we say that $\mathcal{B}'$ determines the *same orientation* as $\mathcal{B}$; otherwise $\mathcal{B}'$ determines the *opposite orientation* from that of $\mathcal{B}$. We sometimes say that $\mathcal{B}'$ is *positive* or *negative* relative to $\mathcal{B}$ according to whether it determines the same or opposite orientation as compared to $\mathcal{B}$.

The *standard orientation* on $U \subset \mathbf{R}^n$ is the one given by selecting as a distinguished basis the standard basis vector fields $\{E_1, \ldots, E_n\}$, where $E_i(p) = \left.\dfrac{\partial}{\partial x_i}\right|_p$.

We are now in a position to define the integral of an n-form on a region of integration in $\mathbf{R}^n$.

Definition 4.5.2. Suppose $U \subset \mathbf{R}^n$ is a domain with the standard orientation, and let $S \subset U$ be a region of integration. Let ω be an n-form on U, so that $\omega = a dx_1 \wedge \cdots \wedge dx_n$ for some smooth function $a : U \to \mathbf{R}$. Then define

$$\int_S \omega = \int \cdots \int_S a dx_1 \cdots dx_n,$$

where the integral on the right is the standard multiple integral.

An important feature of this definition is that it is "almost" invariant with respect to diffeomorphisms. To make this sentence precise, we introduce the following definition.

Definition 4.5.3. Suppose that $U, V \subset \mathbf{R}^n$ are domains oriented by ordered bases $\mathcal{B} = (B_1, \ldots, B_n)$ and $\mathcal{B}' = (B_1', \ldots, B_n')$ respectively, and let $f : U \to V$ be a diffeomorphism. Then f is called *orientation-preserving* (relative to $\mathcal{B}$, $\mathcal{B}$') if the orientation on V induced by the ordered n-tuple $(f_* B_1, \ldots, f_* B_n)$ is the same as that given by $\mathcal{B}'$. Otherwise, f is called *orientation-reversing*.

The reader can check that if U is an oriented domain and $f : U \to U$ is a diffeomorphism of U onto itself, then f is orientation-preserving if and only if $\det[(f_*)_p] > 0$ for all $p \in U$. In Exercise 4.16, we ask the reader to explore behavior typical of an orientation-reversing diffeomorphism in the plane.

The following theorem specifies the extent to which integration of differential forms is "independent of coordinate systems." It underlies several of the definitions that follow later.

Theorem 4.5.4. *Let $f : U \to V$ be a diffeomorphism, where $U, V \subset \mathbf{R}^n$ are oriented domains. Let $D \subset U$ and $S \subset V$ be regions of integration such that $f(D) = S$. Then for any n-form ω on $\mathbf{R}^n$,*

$$\int_D f^* \omega = \pm \int_S \omega,$$

with the sign being positive or negative according to whether f is orientation-preserving or orientation-reversing.

Proof. Writing the pullback $f^*\omega$ in coordinates—essentially the calculation of Proposition 4.4.20—the equality is nothing more than the change of variables formula of multivariable calculus. See, for example, [30]. □

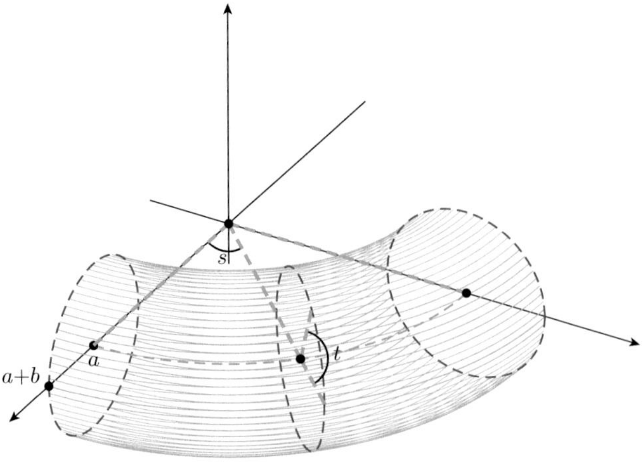

Fig. 4.2 A torus parameterized according to Example 4.5.5.

The following example illustrates this theorem, as well as the ideas of integration we have introduced so far.

Example 4.5.5 (The volume of a solid torus). A solid torus $T \subset \mathbf{R}^3$ can be described as a solid of revolution obtained by revolving a disk in the plane around an axis away from the disk, for example, rotating the disk

$$\left\{(x, 0, z) \mid (x-a)^2 + z^2 \leq b^2\right\},$$

where $0 < b < a$, about the z-axis. Our goal is to compute the volume of T by integrating the standard volume form $\omega = dx \wedge dy \wedge dz$ over T.

However, rather than computing $\int_T \omega$ directly using Definition 4.5.2 (which involves carefully setting up the limits of integration in the appropriate multiple integral), we will instead consider a parameterization of T. For this purpose, consider the parameterization $\phi : \mathbf{R}^3 \to \mathbf{R}^3$ given by

$$\phi(r, s, t) = ((a + r\cos t)\cos s, (a + r\cos t)\sin s, r\sin t),$$

and define the set $Q = [0, b] \times [0, 2\pi] \times [0, 2\pi] \subset \mathbf{R}^3$. The reader should verify that in fact, $\phi(Q) = T$. According to this description, the (r, t) variables correspond to polar coordinates for the disk, while the s variable represents the angle of rotation of the disk around the z-axis. See Fig. 4.2. (There are some technical considerations regarding the parameterization ϕ. For example, it is not one-to-one on all of Q. We will not address these, noting only that removing "singularities" where such problems exist does not change the value of the integral.)

A routine calculation reveals that

$$\det(\phi_*) = r(a + r\cos t),$$

and so, for $r > 0$, ϕ is orientation-preserving on Q, since $r < b < a$.

In order to apply Definition 4.5.6, we resort to Proposition 4.4.20 and the above-referenced computation to note that

$$\begin{aligned}\phi^*\omega &= \det(\phi_*)dr \wedge ds \wedge dt \\ &= r(a + r\cos t)dr \wedge ds \wedge dt.\end{aligned}$$

Then, using Theorem 4.5.4 and Definition 4.5.6, we have

$$\begin{aligned}\int_T \omega &= \int_Q \phi^*\omega \\ &= \int_0^{2\pi}\int_0^{2\pi}\int_0^b r(a + r\cos t)drdsdt \\ &= 2\pi^2 ab^2.\end{aligned}$$

Let $S \subset \mathbf{R}^n$ be a k-dimensional geometric set. An orientation on S is defined exactly as in Definition 4.5.1, i.e., as an ordered k-tuple of vector fields on S that at each point $p \in S$ form a basis for $T_p(S)$. There are several natural ways in which an orientation on S may arise. For example, if $S = \phi(U)$ is a parameterized set, the *orientation on S induced by the parameterization $\phi : U \to S$* is defined by choosing the distinguished ordered basis $\{(\phi_*)_a(E_1(a)), \dots, (\phi_*)_a(E_k(a))\}$, where $a \in U$ and $p = \phi(a) \in S$. Here $(E_1 = \frac{\partial}{\partial x_1}, \dots, E_k = \frac{\partial}{\partial x_k})$ is the standard basis for $U \subset \mathbf{R}^k$.

Another natural way of describing an orientation on a k-dimensional geometric set $S \subset \mathbf{R}^n$ is by specifying a *co-orientation* on $\mathbf{R}^n$. That is, choose an ordered n-tuple of basis vector fields $(B_1, \dots, B_n)$ on $\mathbf{R}^n$ along with an ordered $(n-k)$-tuple of pointwise linearly independent vector fields $(N_1, \dots, N_{n-k})$ having the property that $N_j(p) \notin T_p(S)$ for all $p \in S$ ($j = 1, \dots, n-k$). In that case, an ordered k-tuple of basis vector fields $(X_1, \dots, X_k)$ on S will be given a positive orientation if the orientation $(X_1, \dots, X_k, N_1, \dots, N_{n-k})$ agrees with the given orientation $(B_1, \dots, B_n)$ on $\mathbf{R}^n$, and a negative orientation otherwise. This generalizes the familiar notion of specifying an orientation of a surface in $\mathbf{R}^3$ by means of a nonzero vector field normal to S.

We note that there are sets on which an orientation cannot be defined, the classic example being the Möbius strip in $\mathbf{R}^3$. Deciding whether an orientation can be defined on a set is a topological question that we will not treat here. In what follows, all the sets we will consider will be *orientable* in the sense that an orientation can be defined.

We now define the integral of a k-form over a k-dimensional geometric set in $\mathbf{R}^n$. We especially pay attention to parameterized sets, keeping in mind from Chap. 3 that other geometric sets such as graphs and level sets can be realized as parameterized sets.

Definition 4.5.6. Let ω be a k-form on $\mathbf{R}^n$ and let $S = \phi(U)$ be an oriented k-dimensional geometric set parameterized by a smooth, regular parameterization

$\phi : U \to \mathbf{R}^n$, where $U \subset \mathbf{R}^k$ is a domain. Let $D \subset U$ be a region of integration. Then the *integral of* ω *over* $R = \phi(D)$ is defined to be

$$\int_R \omega = \pm \int_D \phi^* \omega,$$

where the integral on the right is defined according to Definition 4.5.2 and the sign is positive or negative according to whether the given orientation on S agrees with or does not agree with the orientation on S induced by ϕ.

The reader should compare this definition with Theorem 4.5.4. It appears that this definition depends heavily on the parameterization. However, it is another consequence of the change of variables theorem that if $R = \phi_1(D_1) = \phi_2(D_2)$ and if ϕ_1 and ϕ_2 induce the same orientation on R, then $\int_{D_1} \phi_1^* \omega = \int_{D_2} \phi_2^* \omega$.

With this build-up, we finally present some examples of integration of forms on parameterized sets. We will start by reformulating the line integrals from multivariable calculus into the language of differential forms.

Example 4.5.7. Consider $S = \phi(U)$, where $U = (-\pi, \pi) \subset \mathbf{R}^1$ and $\phi : U \to \mathbf{R}^2$ given by

$$\phi(t) = (\cos t, \sin t).$$

Let $D_1 = [0, \pi/2]$ and $D_2 = \phi(D_1)$; D_2 is the part of the unit circle in the first quadrant in $\mathbf{R}^2$.

We will compute $\int_{D_2} \omega_0$, where $\omega_0 = -x_2 dx_1 + x_1 dx_2$. We have

$$\begin{aligned}
\phi^* \omega_0 &= \phi^*(-x_2 dx_1 + x_1 dx_2) \\
&= -(\sin t) d(\cos t) + (\cos t) d(\sin t) \\
&= (\sin^2 t + \cos^2 t) dt \\
&= dt,
\end{aligned}$$

so that

$$\begin{aligned}
\int_{D_2} \omega_0 &= \int_{D_1} \phi^* \omega_0 \\
&= \int_0^{\pi/2} dt \\
&= \pi/2.
\end{aligned}$$

We can repeat this calculation using the one-form $\omega_1 = x_1 dx_1 + x_2 dx_2$. Note that now

$$\begin{aligned}
\phi^* \omega_1 &= \phi^*(x_1 dx_1 + x_2 dx_2) \\
&= (\cos t) d(\cos t) + (\sin t) d(\sin t) \\
&= 0,
\end{aligned}$$

and so

$$\int_{D_2} \omega_1 = \int_{D_1} \phi^* \omega_1 = 0.$$

Example 4.5.8. In $\mathbf{R}^3$, let $\alpha = xdx \wedge dy - ydx \wedge dz + zdy \wedge dz$ and let S be the upper half of the unit sphere described by $x^2 + y^2 + z^2 = 1$ with $z \geq 0$ oriented by means of an inward-pointing normal vector. Then S can be parameterized by means of the map $\phi : R \to \mathbf{R}^3$ defined as $\phi(u, v) = (\cos u \sin v, \sin u \sin v, \cos v)$, where $R = [0, 2\pi] \times [0, \pi/2]$. (We put aside the issue of R being closed and that ϕ is not one-to-one on R.)

We have

$$\begin{aligned} \int_S \alpha &= \int_R \phi^* \alpha \\ &= \int_R [-\sin^2 u \sin^3 v - 2 \cos u \sin^2 v \cos v] du \wedge dv \\ &= \int_0^{\pi/2} \int_0^{2\pi} \left[-\sin^2 u \sin^3 v - 2 \cos u \sin^2 v \cos v\right] du\, dv \\ &= -2\pi/3. \end{aligned}$$

In closing, we mention here the famous Gauss-Stokes theorem, written here in the language of differential forms. This theorem relates the calculus of forms, and particularly the relationship between the exterior derivative and the exterior algebra of forms, with the geometry of the region of integration.

To state the theorem, we let D be an oriented k-dimensional geometric set with boundary in $\mathbf{R}^n$. Defining this precisely would take us too far astray. The reader can picture that if D is the image under a smooth parameterization of the rectangle $[0, 1] \times \cdots \times [0, 1] \subset \mathbf{R}^k$, then the boundary ∂D of D is the image of the points having any coordinate either 0 or 1. Intuitively, the boundary of an interval should be the two endpoints, the boundary of the unit disk in the plane should be the unit circle, and so on. For suitable geometric sets, ∂D can also be considered a geometric set (of dimension $k - 1$) that can be oriented in a way *consistent with* the given orientation of D. We refer the reader to [5], [14], or [37] for a more precise and thorough treatment of these concepts, which are essential to both the statement and the proof of the theorem.

Theorem 4.5.9 (Gauss–Stokes). *Let $D \subset \mathbf{R}^n$ be an oriented k-dimensional region of integration, with boundary ∂D oriented consistently with that of D. Let ω be a smooth $(k-1)$-form defined on an open set containing all of D. Then*

$$\int_{\partial D} \omega = \int_D d\omega.$$

Applying this theorem to problems of integration is a standard goal of a first course in multivariable calculus. We will see further applications later in the text.

4.6 Tensors

In this section, we outline what might be considered a general way of describing differential geometric structures. When we say that differential geometry consists in studying the tangent bundle with some "structure," it will be a tensor that will provide the structure.

A tensor can be thought of as a generalization that encompasses vectors, forms, and more. While the emphasis in this book will be on particular examples rather than the most general setting, we introduce tensors here both for the reader's familiarity and to present a unified treatment of several basic constructions in differential geometry. For a more detailed treatment, the reader may consult [38], [40], or [1].

As usual, we begin in the linear setting. Let V be a vector space with dual space V^*. Recall from Definition 2.8.8 that a multilinear map is one that is linear in each component.

Definition 4.6.1. A *tensor of type* (r, s) *on* V is a multilinear map

$$\alpha : \underbrace{V^* \times \cdots \times V^*}_{r \text{ times}} \times \underbrace{V \times \cdots \times V}_{s \text{ times}} \to \mathbf{R}.$$

The set of all tensors of type (r, s) on V will be denoted by $T^{r,s}(V)$.

We note immediately that the set of all (r, s) tensors has an algebraic structure.

Theorem 4.6.2. *The set $T^{r,s}(V)$ with the standard operations of addition of real-valued functions and multiplication of functions by scalars is a vector space.*

Proof. Since the operations are defined pointwise, the proof is essentially the same as that of Theorem 2.8.1. □

We illustrate the definition with several examples in order to show that it in fact encompasses a number of the objects that we have already encountered.

Example 4.6.3 ($(0,1)$-tensors). A $(0, 1)$-tensor is a linear map $\alpha : V \to \mathbf{R}$. In other words, $\alpha \in V^*$, and so

$$T^{0,1}(V) = V^*.$$

Example 4.6.4 ($(1,0)$-tensors). A $(1, 0)$-tensor is a linear map $a : V^* \to \mathbf{R}$. Note then that $a \in (V^*)^*$.

There is a natural map $I : V \to (V^*)^*$ given by $I(\mathbf{v}) = e_{\mathbf{v}} : V^* \to \mathbf{R}$, where for every $\alpha \in V^*$, $e_{\mathbf{v}}(\alpha) = \alpha(\mathbf{v})$. We leave it as an exercise to show that I is a linear isomorphism under the assumption that V is finite-dimensional. Since we will be concerned with only the finite-dimensional case, we will identify $(V^*)^*$ with V, and so we write

$$T^{1,0}(V) = V.$$

Example 4.6.5 ($(1,1)$-tensors). Let $a : V^* \times V \to \mathbf{R}$ be a $(1, 1)$-tensor on the finite-dimensional vector space V. We follow the thinking of the prior example to interpret the map a in a more familiar way. For every $\mathbf{v} \in V$, the map $A_{\mathbf{v}} : V^* \to \mathbf{R}$ given by $A_{\mathbf{v}}(\alpha) = a(\alpha, \mathbf{v})$ is linear as a consequence of the bilinearity of a. Hence $A_{\mathbf{v}} \in (V^*)^*$, and so by the previous example, there is $\mathbf{w} \in V$ such that $I(\mathbf{w}) = A_{\mathbf{v}}$. Note that we have constructed for each $\mathbf{v} \in V$ a vector $\mathbf{w}$ with the property that for all $\alpha \in V^*$,

$$\alpha(\mathbf{w}) = a(\alpha, \mathbf{v}) = A_{\mathbf{v}}(\alpha).$$

Let A be the map $A : V \to V$ defined by $A(\mathbf{v}) = \mathbf{w}$. It can be shown that A is in fact a linear transformation, and that the correspondence between the set of $(1,1)$-tensors and the vector space $L(V,V)$ of linear transformations from V to itself (with the operations of pointwise addition and scalar multiplication) is a linear isomorphism. So a $(1,1)$-tensor can be thought of as a linear transformation $A : V \to V$.

Example 4.6.6 ($(0,2)$-tensors). A $(0,2)$-tensor on V is nothing other than a bilinear form as described in Sect. 2.8. Recall Proposition 2.8.14, which associates a matrix representation for each bilinear form by means of a basis on V. In particular, the inner products and linear symplectic forms described in Sects. 2.9 and 2.10 are both examples of $(0,2)$-tensors.

Example 4.6.7 (Alternating $(0,k)$-tensors). A $(0,k)$-tensor α is a multilinear map of k-tuples of vectors:

$$\alpha : V \times \cdots \times V \to \mathbf{R}.$$

If α is an *alternating* $(0,k)$-tensor, in the sense of Definition 4.1.1, then α is nothing other than a linear k-form. In other words, we have $\Lambda_k(V) \subset T^{0,k}(V)$.

The final observation we will make in the linear setting concerns the dimension of $T^{r,s}(V)$, where V is a vector space of dimension n. Let $\{\mathbf{e}_1, \ldots, \mathbf{e}_n\}$ be a basis for V and let $\{\varepsilon_1, \ldots, \varepsilon_n\}$ be the corresponding basis for V^*.

Let $\varepsilon_{i_1 \cdots i_r}^{j_1 \cdots j_s}$ be the (r,s)-tensor

$$\varepsilon_{i_1 \cdots i_r}^{j_1 \cdots j_s} : \underbrace{V^* \times \cdots \times V^*}_{r \text{ times}} \times \underbrace{V \times \cdots \times V}_{s \text{ times}} \to \mathbf{R}$$

given by

$$\varepsilon_{i_1 \cdots i_r}^{j_1 \cdots j_s}(\alpha_1, \ldots, \alpha_r, \mathbf{v}_1, \ldots, \mathbf{v}_s) = \alpha_1(\mathbf{e}_{i_1}) \cdots \alpha_r(\mathbf{e}_{i_r}) \cdot \varepsilon_{j_1}(\mathbf{v}_1) \cdots \varepsilon_{j_s}(\mathbf{v}_s).$$

We sometimes use the traditional "tensor product" notation

$$\varepsilon_{i_1 \cdots i_r}^{j_1 \cdots j_s} = \mathbf{e}_{i_1} \otimes \cdots \otimes \mathbf{e}_{i_r} \otimes \varepsilon_{j_1} \otimes \cdots \otimes \varepsilon_{j_s}.$$

In this notation, the exterior product can be expressed in terms of the tensor product. For example,

$$\varepsilon_i \wedge \varepsilon_j = \varepsilon_i \otimes \varepsilon_j - \varepsilon_j \otimes \varepsilon_i.$$

The fact that the functions $\varepsilon_{i_1 \cdots i_r}^{j_1 \cdots j_s}$ are multilinear is a consequence of the linearity of the one-forms α_k and the basis forms ε_l.

Theorem 4.6.8. *The set of all (r, s)-tensors of the form $\varepsilon_{i_1 \cdots i_r}^{j_1 \cdots j_s}$, where*

$$i_1, \ldots, i_r, j_1, \ldots, j_s \in \{1, \ldots, n\},$$

is a basis for $T^{r,s}(V)$.

Proof. Linear independence can be verified by evaluating an arbitrary linear combination of elements having the form in Theorem 4.6.8 against $(r+s)$-tuples of the form $(\varepsilon_{i_1}, \ldots, \varepsilon_{i_r}, \mathbf{e}_{j_i}, \ldots, \mathbf{e}_{j_s})$ to show that in order for the linear combination to be zero, all coefficients of the combination must be zero. (This is essentially the method of proving Theorem 2.8.2.)

To show that the set spans $T^{r,s}(V)$, choose an arbitrary $t \in T^{r,s}(V)$ and define the scalars

$$t_{i_1 \cdots i_r}^{j_1 \cdots j_s} = t(\varepsilon_{i_1}, \ldots, \varepsilon_{i_r}, \mathbf{e}_{j_1}, \ldots, \mathbf{e}_{j_s}).$$

A calculation confirms that

$$t = \sum t_{i_1 \cdots i_r}^{j_1 \cdots j_s} \varepsilon_{i_1 \cdots i_r}^{j_1 \cdots j_s}. \quad \square$$

Corollary 4.6.9. *Suppose V is a vector space with $\dim(V) = n$. Then $\dim T^{r,s}(V) = n^{r+s}$.*

The $\otimes$ notation reflects the fact that there is an operation defined on the set of all tensor fields, which we introduce here although we will not dwell on its corresponding algebra.

Definition 4.6.10. Let S be an (r, s)-tensor and T an (l, m)-tensor on a vector space V. The *tensor product* $S \otimes T$ is defined to be the $(r+l, s+m)$-tensor defined by

$$\begin{aligned}(S \otimes T)(\alpha_1, \ldots, \alpha_r, \alpha_{r+1}, \ldots, \alpha_{r+l}, \mathbf{v}_1, \ldots, \mathbf{v}_s, \mathbf{v}_{s+1}, \ldots, \mathbf{v}_{s+m}) \\ = S(\alpha_1, \ldots, \alpha_r, \mathbf{v}_1, \ldots, \mathbf{v}_s) T(\alpha_{r+1}, \ldots, \alpha_{r+l}, \mathbf{v}_{s+1}, \ldots, \mathbf{v}_{s+m}).\end{aligned}$$

The reader should note that the tensor product is not commutative.

The process of adapting Definition 4.6.1 to the nonlinear setting should by now be a familiar one.

Definition 4.6.11. A *tensor field* S of type (r, s) on $\mathbf{R}^n$ is an assignment of an (r, s)-tensor $S(p)$ to each tangent space $T_p(\mathbf{R}^n)$ that is smooth in the following sense: For any collection of r smooth differential one-forms $\alpha_1, \ldots, \alpha_r$ and s smooth vector fields $X_1, \ldots, X_s$, the function $f : \mathbf{R}^n \to \mathbf{R}$ given by

$$f(p) = \big(S(p)\big) \left(\alpha_1(p), \ldots, \alpha_r(p), X_1(p), \ldots, X_s(p)\right)$$

is a smooth function of p.

The phrase *tensor field* is meant to emphasize the distinction between the smooth Definition 4.6.11 and the purely algebraic notion of a tensor in Definition 4.6.1. However, we will sometimes use the word *tensor* to mean a tensor field when the context is clear.

Keeping in mind Theorem 4.6.8 as well as the standard basis vectors $\left.\frac{\partial}{\partial x_i}\right|_p$ for $T_p(\mathbf{R}^n)$ and the dual basis forms $dx_i(p)$ for $(T_p(\mathbf{R}^n))^*$, we can write a typical (r,s)-tensor field on $\mathbf{R}^n$ as a sum of terms having the form

$$a_{i_1\cdots i_r}^{j_1\cdots j_s}(p)\left.\frac{\partial}{\partial x_{i_1}}\right|_p \otimes\cdots\otimes \left.\frac{\partial}{\partial x_{i_r}}\right|_p \otimes dx_{j_1}(p)\otimes\cdots\otimes dx_{j_s}(p),$$

where the $a_{i_1\cdots i_r}^{j_1\cdots j_s}:\mathbf{R}^n\to\mathbf{R}$ are smooth functions.

Examples of tensor fields can be produced by extending each of Examples 4.6.3 through 4.6.7 to the nonlinear setting. In particular, we note that a $(0,1)$-tensor field is a differential one-form, a $(1,0)$-tensor field is a vector field, and an *alternating* $(0,k)$-tensor field is a differential k-form. As mentioned earlier in the linear setting, the exterior product of differential k-forms can be expressed in terms of the tensor product. For example,

$$dx\wedge dy = dx\otimes dy - dy\otimes dx.$$

Much of the latter part of the text will be devoted to an in-depth presentation of some of these examples of tensor fields.

One of the essential features that distinguishes the theory of linear tensors and that of nonlinear tensor fields is the behavior of tensor fields under "change of variables," or in other words, how tensor fields transform under diffeomorphisms. While the model for this study will be the previously encountered pullback of differential forms, the issue is somewhat more complicated owing to the "mixed" nature of (r,s)-tensor fields acting on both differential forms and vector fields. That is why it is essential to consider diffeomorphisms instead of just smooth maps (compare to Definition 4.4.14).

Definition 4.6.12. Let $U,V\subset\mathbf{R}^n$ be domains and let $\phi:U\to V$ be a diffeomorphism with inverse $\psi=\phi^{-1}:V\to U$. Let S be an (r,s)-tensor field on V. The *pullback of* S, denoted by ϕ^*S, is defined to be the (r,s)-tensor field on U given by

$$(\phi^*S)(\alpha_1,\ldots,\alpha_r,X_1,\ldots,X_s) = S(\psi^*\alpha_1,\ldots,\psi^*\alpha_r,\phi_*X_1,\ldots,\phi_*X_s),$$

where the α_i are smooth one-forms and the X_j are smooth vector fields on U.

We illustrate the essence of this definition with two elementary examples. In both cases, we write the coordinates of the domain as $(x_1,\ldots,x_n)\in\mathbf{R}^n$ and the coordinates of the codomain as $(y_1,\ldots,y_n)\in\mathbf{R}^n$. We write the component functions of $\phi:\mathbf{R}^n\to\mathbf{R}^n$ as $\phi^i:\mathbf{R}^n\to\mathbf{R}$, i.e.,

$$\phi(x_1, \dots, x_n) = \left(\phi^1(x_1, \dots, x_n), \dots, \phi^n(x_1, \dots, x_n)\right),$$

and likewise for the inverse $\psi = (\psi^1, \dots, \psi^n)$. This allows us to use the traditional notation

$$y_i = \phi^i(x_1, \dots, x_n),$$
$$x_j = \psi^j(y_1, \dots, y_n).$$

Example 4.6.13 (The pullback of a basis $(0,1)$*-tensor).* Let $T = dy_j$. Then for any vector field $X = \sum a_i \dfrac{\partial}{\partial x_i}$, we have

$$\begin{aligned}(\phi^* T)(X) &= T(\phi_* X) \\ &= dy_j \left(\sum_{k=1}^n \left(a_1 \frac{\partial y_k}{\partial x_1} + \dots + a_n \frac{\partial y_k}{\partial x_n} \right) \frac{\partial}{\partial y_k} \right) \\ &= a_1 \frac{\partial y_j}{\partial x_1} + \dots + a_n \frac{\partial y_j}{\partial x_n} \\ &(= X[y_j]) .\end{aligned}$$

Example 4.6.14 (The pullback of a basis $(1,0)$*-tensor field).* Let $T = \frac{\partial}{\partial y_j}$. Then for any differential one-form $\alpha = \sum a_i dx_i$, we have

$$\begin{aligned}(\phi^* T)(\alpha) &= T(\psi^* \alpha) \\ &= \left(\frac{\partial}{\partial y_j}\right) \left(\sum_{k=1}^n (a_k \circ \psi) \left(\frac{\partial x_k}{\partial y_1} dy_1 + \dots + \frac{\partial x_k}{\partial y_n} dy_n \right) \right) \\ &= \left(\frac{\partial}{\partial y_j}\right) \left(\sum_{i=1}^n \left((a_1 \circ \psi) \frac{\partial x_1}{\partial y_i} + \dots + (a_n \circ \psi) \frac{\partial x_n}{\partial y_i} \right) dy_i \right) \\ &= (a_1 \circ \psi) \frac{\partial x_1}{\partial y_j} + \dots + (a_n \circ \psi) \frac{\partial x_n}{\partial y_j} \\ &\left(= \alpha \left(\psi^* \frac{\partial}{\partial y_j} \right) \right) .\end{aligned}$$

The difference in how the basis tensors transform under the pullback operation, or change of variables, can be seen in the role of the "summation variables" x_i. In the first example they play the role of independent variables that are "operating on" the dependent variables y_j, while in the second example they play the role of dependent variables "operated on" by the independent variables y_j.

This difference distinguishes between what are called "contravariant" and "covariant" tensors, as the first and second examples, respectively, are known. In classical tensor analysis and differential geometry, the behavior of objects under change of variables was in fact used as the key feature in defining tensors. The interested reader may refer, for example, to [38].

We list here a few properties of the tensor product that we will encounter in the future.

Proposition 4.6.15. *Let S be an (r,s)-tensor field, T a (k,l)-tensor field on $\mathbf{R}^n$, and $\phi : \mathbf{R}^n \to \mathbf{R}^n$ a diffeomorphism. Then:*

1. *The mapping $T^{r,s} \times T^{k,l} \to T^{r+k,s+l}$ given by $(S,T) \mapsto S \otimes T$ is bilinear.*
2. $\phi^*(S \otimes T) = (\phi^* S) \otimes (\phi^* T)$.

Proof. We leave the verification of (1) to the reader. The equality in (2) is verified by applying both $\phi^*(S \otimes T)$ and $(\phi^* S) \otimes (\phi^* T)$ to an arbitrary $(r+k+s+l)$-tuple of $r+k$ one-forms and $s+l$ vector fields:

$$(\alpha_1, \ldots, \alpha_r, \alpha_{r+1}, \ldots, \alpha_{r+k}, X_1, \ldots, X_s, X_{s+1}, \ldots, X_{s+l}). \quad \square$$

There is a rich algebra of tensors and tensor fields that is outside the scope of this text. However, we will encounter the following operation on tensor fields, which we present only in the specific context of $(0,k)$-tensors.

Definition 4.6.16. Let T be a $(0,k)$-tensor field ($k \geq 1$) and let X be a smooth vector field. Then the *interior product* of T with X, denoted by $i(X)T$, is the $(0,k-1)$-tensor field defined by

$$(i(X)T)(Y_1, \ldots, Y_{k-1}) = T(X, Y_1, \ldots, Y_{k-1}).$$

The following examples give some feel for the computations with the interior product, as well as some of the common contexts in which it will be encountered in the following chapters. The reader should compare the following examples to the computations of Sect. 4.2.

Example 4.6.17. Let $g \in T^{0,2}(\mathbf{R}^2)$ be defined by

$$g = dx_1 \otimes dx_1 + dx_2 \otimes dx_2.$$

For any smooth $f : \mathbf{R}^2 \to \mathbf{R}$, let

$$X = \frac{\partial f}{\partial x_1}\frac{\partial}{\partial x_1} + \frac{\partial f}{\partial x_2}\frac{\partial}{\partial x_2}.$$

Then for any $Y = Y^1 \frac{\partial}{\partial x_1} + Y^2 \frac{\partial}{\partial x_2}$, we have

$$(i(X)g)(Y) = g(X,Y)$$

$$
\begin{aligned}
&= dx_1(X)dx_1(Y) + dx_2(X)dx_2(Y) \\
&= \frac{\partial f}{\partial x_1}Y^1 + \frac{\partial f}{\partial x_2}Y^2 \\
&= df(Y).
\end{aligned}
$$

Hence $i(X)g = df$. We will return to this example in Chap. 6.

Example 4.6.18. Let

$$\omega = dx_1 \wedge dx_2 + dx_3 \wedge dx_4 \in T^{0,2}(\mathbf{R}^4),$$

and let

$$X = X^1\frac{\partial}{\partial x_1} + X^2\frac{\partial}{\partial x_2} + X^3\frac{\partial}{\partial x_3} + X^4\frac{\partial}{\partial x_4}$$

be a given smooth vector field. Then the reader may verify that

$$i(X)\omega = X^1dx_2 - X^2dx_1 + X^3dx_4 - X^4dx_3.$$

Many of the cases in which we will encounter the interior product later in the text will be in the context of differential forms—*alternating* $(0, k)$-tensors. We conclude this section with a proposition that relates the interior product with the operations we have seen earlier in this section.

Proposition 4.6.19. *Let α be a differential k-form and X a smooth vector field on $\mathbf{R}^n$. Then:*

- *If f is a smooth function, then $i(fX)\alpha = fi(X)\alpha$.*
- *If Y is a smooth vector field, then $i(X+Y)\alpha = i(X)\alpha + i(Y)\alpha$.*
- *If f_1, f_2 are smooth functions and β is another k-form, then*

$$i(X)\left(f_1\alpha + f_2\beta\right) = f_1 i(X)\alpha + f_2 i(X)\beta.$$

- *If β is a one-form, then*

$$i(X)(\alpha\wedge\beta) = (i(X)\alpha)\wedge\beta + (-1)^k\alpha\wedge(i(X)\beta).$$

- *If $\phi : \mathbf{R}^n \to \mathbf{R}^n$ is a diffeomorphism, then*

$$i(X)\left(\phi^*\alpha\right) = \phi^*\left(i(Y)\alpha\right),$$

where $Y = \phi_* X$.

Proof. All statements are proved pointwise. See Proposition 4.2.15. □

Note that we have not addressed the interaction of the interior product with the exterior differentiation operation d. Here the relationship is more complex, and will be addressed in the next section.

4.7 The Lie Derivative

Given the nonlinear nature of differential geometry, with a "structure" tensor varying smoothly from point to point, it is natural to try to measure how the structure tensor changes in a given direction. A basic tool for this purpose is the Lie derivative, named after one of the pioneers of modern differential geometry, Sophus Lie. The notion of the directional derivative of a function will provide the motivation for defining this key concept (Fig. 4.3).

We recall the basic framework from Chap. 3. Let $f : \mathbf{R}^n \to \mathbf{R}$ be a smooth function and let X be a smooth vector field on $\mathbf{R}^n$. Let ϕ_t be the flow generated by X. For each $p \in \mathbf{R}^n$, there exist a domain $U_p \subset \mathbf{R}^n$ containing p and an interval $I_p \subset \mathbf{R}$ containing 0 such that for each $t \in I_p$,

$$\phi_t : U_p \to \phi_t(U_p)$$

is a diffeomorphism. Moreover, for all $p \in U_p$, $c_p : I_p \to \mathbf{R}^n$ given by $c_p(t) = \phi_t(p)$ is an integral curve for X through p, i.e., $c_p(0) = p$ and $c_p'(t) = X(c_p(t))$. In this framework, interpreting X analytically as operating on functions f, with $X[f]$ being the directional derivative of f along X, we can write

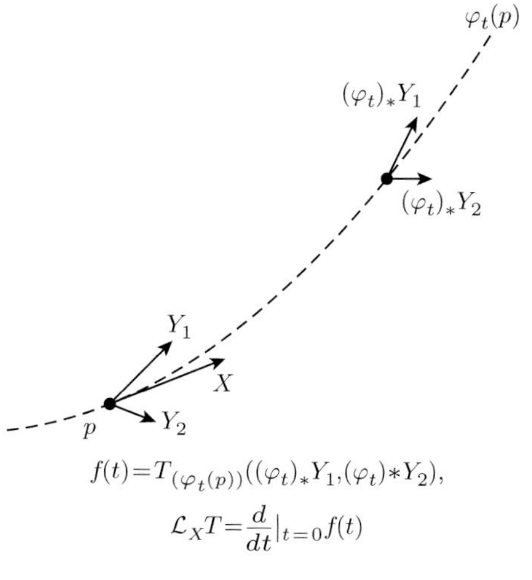

Fig. 4.3 The Lie derivative of a $(0, 2)$-tensor T with respect to a vector field X with flow φ_t.

$$\begin{aligned} X[f](p) &= \left.\frac{d}{dt}\right|_{t=0} (f(c_p(t))) \\ &= \left.\frac{d}{dt}\right|_{t=0} (f \circ \phi_t(p)) \\ &= \left.\frac{d}{dt}\right|_{t=0} ((\phi_t^* f)(p)) \,. \end{aligned}$$

Keeping in mind the definition of the pullback of a tensor field (Definition 4.6.12), the preceding discussion prompts the following:

Definition 4.7.1. Let T be a tensor field of type (r, s) on $\mathbf{R}^n$ and let X be a vector field with flow $\phi_t : \mathbf{R}^n \to \mathbf{R}^n$. The *Lie derivative* of T with respect to X, denoted by $\mathcal{L}_X T$, is the (r, s)-tensor defined as

$$\mathcal{L}_X T = \left.\frac{d}{dt}\right|_{t=0} (\phi_t^* T) \,.$$

We leave it as an exercise to show that $\mathcal{L}_X T$ is in fact an (r, s)-tensor. As the definition suggests, it measures the rate of change of the tensor T along the flow lines of X.

The definition can be extended immediately to the derivative of a pullback for any value of t, not just at $t = 0$.

Proposition 4.7.2. *Let T, X, and ϕ_t be as in Definition 4.7.1. Then for every t_0 in the interval where ϕ_t is defined, we have*

$$\left.\frac{d}{dt}\right|_{t=t_0} (\phi_t^* T) = \phi_{t_0}^* \left(\mathcal{L}_X T\right) .$$

Proof. We rely on the properties of the flow from Theorem 3.9.2:

$$\begin{aligned} \left.\frac{d}{dt}\right|_{t=t_0} (\phi_t^* T) &= \left.\frac{d}{ds}\right|_{s=0} (\phi_{s+t_0}^* T) \\ &= \left.\frac{d}{ds}\right|_{s=0} ((\phi_s \circ \phi_{t_0})^* T) \\ &= \left.\frac{d}{ds}\right|_{s=0} ((\phi_{t_0})^* (\phi_s)^* T) \\ &= \phi_{t_0}^* \left(\left.\frac{d}{ds}\right|_{s=0} (\phi_s^* T) \right) \\ &= \phi_{t_0}^* (\mathcal{L}_X T). \end{aligned}$$

□

Before narrowing our attention to the cases that we will encounter most frequently in this text, we list here some useful properties of the Lie derivative acting on tensors. Of course, these apply to differential forms as a special case.

Proposition 4.7.3. *Let S and T be tensors and let X be a smooth vector field on $\mathbf{R}^n$ (with the degrees of the tensors implicit and depending on the context below). Then:*

1. $\mathcal{L}_X(S+T) = \mathcal{L}_X S + \mathcal{L}_X T$.
2. *For any constant* $c \in \mathbf{R}$, $\mathcal{L}_X(cT) = c(\mathcal{L}_X T)$.
3. $\mathcal{L}_X(S \otimes T) = (\mathcal{L}_X S) \otimes T + S \otimes (\mathcal{L}_X T)$.

Proof. We leave the verifications of (1) and (2) to the reader. Statement (3) follows by resorting to the same tactic used to prove the Leibniz product rule in a first calculus course, along with the properties of the tensor product in Proposition 4.6.15. To see this, let ϕ_t be the flow generated by X:

$$\begin{aligned}
\mathcal{L}_X(S \otimes T) &= \left.\frac{d}{dt}\right|_{t=0} (\phi_t^*(S \otimes T)) \\
&= \lim_{h \to 0} \frac{\phi_h^*(S \otimes T) - \phi_0^*(S \otimes T)}{h} \\
&= \lim_{h \to 0} \frac{(\phi_h^* S) \otimes (\phi_h^* T) - (S \otimes T)}{h} \\
&= \lim_{h \to 0} \frac{(\phi_h^* S) \otimes (\phi_h^* T) - \phi_h^* S \otimes T + \phi_h^* S \otimes T - (S \otimes T)}{h} \\
&= \lim_{h \to 0} \frac{(\phi_h^* S) \otimes (\phi_h^* T - T) + (\phi_h^* S - S) \otimes T}{h} \\
&= \left(\lim_{h \to 0} \phi_h^* S\right) \otimes \left(\lim_{h \to 0} \frac{\phi_h^* T - T}{h}\right) + \left(\lim_{h \to 0} \frac{\phi_h^* S - S}{h}\right) \otimes T \\
&= S \otimes \mathcal{L}_X T + \mathcal{L}_X S \otimes T.
\end{aligned}$$

□

Although the definition and the preceding proposition have been phrased in full generality, this section is really devoted to presenting the Lie derivative of those objects that we will encounter most frequently: vector fields, differential forms, and $(0,2)$-tensors. We will do this by expressing the Lie derivative in coordinates.

Let $X = \sum X^j \frac{\partial}{\partial x_j}$ be a vector field on $\mathbf{R}^n$ with flow $\phi_t : \mathbf{R}^n \to \mathbf{R}^n$, which we write in components as $\phi_t = (\phi_t^1, \ldots, \phi_t^n)$. In particular, $\phi_0(x) = x$ for all $x \in \mathbf{R}^n$ and

$$\left.\frac{d}{dt}\right|_{t=0} (\phi_t(x)) = X(x).$$

Finally, we recall that $\phi_t^{-1} = \phi_{-t}$.

Proposition 4.7.4. *For each* $i = 1, \dots, n$, *let* $\partial_i = \dfrac{\partial}{\partial x_i}$ *be the standard basis vector fields on* $\mathbf{R}^n$ *with coordinates* $(x_1, \dots, x_n)$. *Let* $X = \sum_{j=1}^n X^j \partial_j$ *be a smooth vector field on* $\mathbf{R}^n$. *Then for all* $i = 1, \dots, n$,

$$\mathcal{L}_X \partial_i = -\sum_{j=1}^n \frac{\partial X^j}{\partial x_i} \partial_j.$$

Proof. The proof relies essentially on the chain rule and the product rule from multivariable calculus. We calculate the effect of the operator $\mathcal{L}_X \partial_i$ on an arbitrary smooth function $f : \mathbf{R}^n \to \mathbf{R}$ at a point $x \in \mathbf{R}^n$:

$$\begin{aligned}
(\mathcal{L}_X \partial_i)[f](x) &= \frac{d}{dt}\bigg|_{t=0} (\phi_t^* \partial_i)[f](x) \quad \text{by Definition 4.7.1} \\
&= \frac{d}{dt}\bigg|_{t=0} [(\phi_{-t})_* \partial_i][f](x) \quad \text{by Definition 4.6.12} \\
&= \frac{d}{dt}\bigg|_{t=0} \partial_i\bigg|_{\phi_t(x)} [f \circ \phi_{-t}] \quad \text{by Definition 3.7.6} \\
&= \frac{d}{dt}\bigg|_{t=0} \sum_j \frac{\partial f}{\partial x_j}(x) \cdot \frac{\partial \phi_{-t}^j}{\partial x_i}(\phi_t(x)) \quad \text{by the chain rule} \\
&= \sum_j \frac{\partial f}{\partial x_j}(x) \frac{d}{dt}\bigg|_{t=0} \frac{\partial \phi_{-t}^j}{\partial x_i}(\phi_t(x)) \quad \text{since } f \text{ is independent of } t \\
&= \sum_j \frac{\partial f}{\partial x_j}(x) \frac{\partial}{\partial x_i}\bigg|_x \frac{d\phi_{-t}^j}{dt}(\phi_t(x))\bigg|_{t=0} \\
&= \sum_j \frac{\partial f}{\partial x_j}(x) \frac{\partial}{\partial x_i}\bigg|_x (-X^j(x)) \quad \text{by the definition of } \phi_t \\
&= -\sum_j \frac{\partial X^j}{\partial x_i}(x) \frac{\partial f}{\partial x_j}(x),
\end{aligned}$$

so $\mathcal{L}_X \partial_i = -\sum \dfrac{\partial X^j}{\partial x_i} \partial_j$ as desired. Note that we have relied on the equality of mixed partials for the sixth equality. □

This basic proposition in turn gives the more general statement.

Proposition 4.7.5. *Let*

$$X = \sum X^i \frac{\partial}{\partial x_i} \quad \text{and} \quad Y = \sum Y^i \frac{\partial}{\partial x_i}$$

be smooth vector fields on $\mathbf{R}^n$. *Then the Lie derivative of* Y *with respect to* X *is given by*

$$\mathcal{L}_X Y = \sum_{j=1}^{n} A^j \frac{\partial}{\partial x_j},$$

where the functions $A^j : \mathbf{R}^n \to \mathbf{R}$ $(j = 1, \dots, n)$ *are given by*

$$A^j = \sum_{k=1}^{n} \left(X^k \frac{\partial Y^j}{\partial x_k} - Y^k \frac{\partial X^j}{\partial x_k} \right).$$

Expressed more compactly, $\mathcal{L}_X Y = XY - YX$.

Proof. Apply the results of Proposition 4.7.3 to the result of Proposition 4.7.4 for the basis vector fields. Here we consider the component functions Y^i as $(0,0)$-tensors for the purpose of applying Proposition 4.7.3(3). □

The following proposition follows immediately from Proposition 4.7.5; we leave the details to the reader.

Proposition 4.7.6. *Let* X *and* Y *be vector fields on* $\mathbf{R}^n$. *Then:*

1. $\mathcal{L}_X X = 0$.
2. $\mathcal{L}_X Y = -\mathcal{L}_Y X$.
3. *For all* $p \in \mathbf{R}^n$, *the map* $\mathcal{L}_p : T_p(\mathbf{R}^n) \times T_p(\mathbf{R}^n) \to T_p(\mathbf{R}^n)$ *given by*

$$(X_p, Y_p) \mapsto (\mathcal{L}_p)(X_p, Y_p) = (\mathcal{L}_X Y)_p$$

is bilinear.

Proof. Exercise. □

The first statement of Proposition 4.7.6 says that a vector field is "constant along its own flow lines." The second statement, on the surface merely a statement of anticommutativity, has deeper implications. According to the definition of the Lie derivative, the left-hand side of the equality, $\mathcal{L}_X Y$, appears to depend on the vector field Y and the *flow* of the vector field X. Conversely, the right-hand side, involving $\mathcal{L}_Y X$, depends on the vector field X and the flow of Y.

More precisely, let ϕ_t, ψ_t be the flows generated by X and Y, respectively. Consider the curve $c : I \to \mathbf{R}^n$ through the point $x \in \mathbf{R}^n$ given by

$$c(t) = c_t(x) = (\psi_{-t} \circ \phi_{-t} \circ \psi_t \circ \phi_t)(x).$$

Here I is an interval containing 0 on which all flows are defined. This curve can be thought of as taking a point $p = c_0(p)$ and following a "loop" formed by the flow lines of X, then Y, then $-X$, and then $-Y$ to reach the point $q = c_t(p)$. In general, $c_t(p)$ may or may not be equal to p, as the following two examples illustrate.

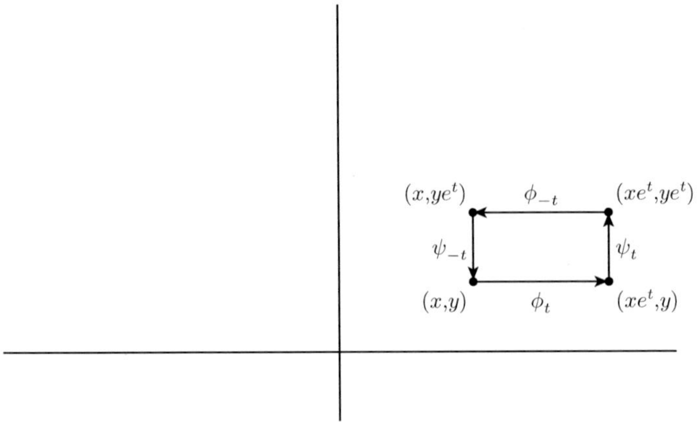

Fig. 4.4 The flow loop of the vector fields in Example 4.7.7.

Example 4.7.7. On $\mathbf{R}^2$, let $X = x\dfrac{\partial}{\partial x}$ and let $Y = y\dfrac{\partial}{\partial y}$. The reader can verify that the corresponding flows ϕ_t for X and ψ_t for Y are given by $\phi_t(x,y) = (xe^t, y)$ and $\psi_t(x,y) = (x, ye^t)$. Both are defined for all $t \in \mathbf{R}$. Hence we have

$$\begin{aligned}
c_t(x,y) &= \psi_{-t}\left(\phi_{-t}\left(\psi_t\left(\phi_t(x,y)\right)\right)\right)\\
&= \psi_{-t}\left(\phi_{-t}\left(\psi_t(xe^t,y)\right)\right)\\
&= \psi_{-t}\left(\phi_{-t}(xe^t,ye^t)\right)\\
&= \psi_{-t}\left((xe^t)\cdot e^{-t}, ye^t\right)\\
&= \left(x, (ye^t)\cdot e^{-t}\right)\\
&= (x,y).
\end{aligned}$$

See Fig. 4.4.

Example 4.7.8. On $\mathbf{R}^2$, let $X = y\dfrac{\partial}{\partial x}$ and let $Y = x\dfrac{\partial}{\partial y}$, with corresponding flows given by $\phi_t(x,y) = (x+ty, y)$ and $\psi_t(x,y) = (x, y+tx)$. Then

$$\begin{aligned}
c_t(x,y) &= \psi_{-t}\left(\phi_{-t}\left(\psi_t\left(\phi_t(x,y)\right)\right)\right)\\
&= \psi_{-t}\left(\phi_{-t}\left(\psi_t(x+ty,y)\right)\right)\\
&= \psi_{-t}\left(\phi_{-t}\left(x+ty, y+t(x+ty)\right)\right)\\
&= \psi_{-t}\left(x+ty-t(y+tx+t^2y), y+tx+t^2y\right)\\
&= \left(x-t^2x-t^3y, y+t^2y+t^3x+t^4y\right).
\end{aligned}$$

For example, for every $(x,y) \in \mathbf{R}^2$, $c_1(x,y) = (-y, 3y+x)$. See Fig. 4.5.

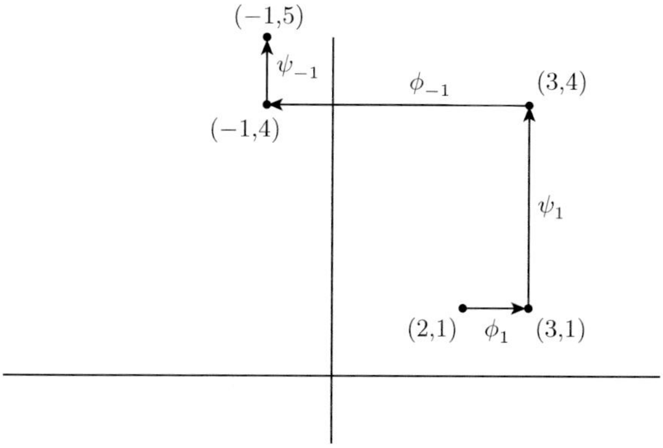

Fig. 4.5 The flow "loop" for the vector fields in Example 4.7.8, with $c_1(2,1) = (-1,5)$.

What is the difference between these two examples? Simply put, it is the Lie derivative. A computation reveals that in Example 4.7.7, $\mathcal{L}_X Y = 0$, whereas in Example 4.7.8,

$$\mathcal{L}_X Y = -x\frac{\partial}{\partial x} + y\frac{\partial}{\partial y}.$$

At the heart of the previous discussion lies the following theorem. It is actually a consequence of more general results; we refer the interested reader to [38, pp. 218–225].

Theorem 4.7.9. *Let X and Y be smooth vector fields on $\mathbf{R}^n$ with corresponding flows ϕ_t and ψ_t. Then*

$$c_t(p) = (\psi_{-t} \circ \phi_{-t} \circ \psi_t \circ \phi_t)(p) = p$$

for all p if and only if $\mathcal{L}_X Y = 0$.

Corollary 4.7.10. *$\mathcal{L}_X Y = 0$ if and only if for all t on which the corresponding flows are defined, $\psi_t \circ \phi_t = \phi_t \circ \psi_t$.*

Proof (of Corollary 4.7.10). The statement is really a reformulation of the theorem, keeping in mind the important property $(\phi_t)^{-1} = \phi_{-t}$ of one-parameter flows (and likewise for ψ_t). The statement about c_t in Theorem 4.7.9 really states, then, that

$$(\psi_t)^{-1} \circ (\phi_t)^{-1} \circ \psi_t \circ \phi_t = \mathrm{Id},$$

and the result follows by composing both sides on the left by $\phi_t \circ \psi_t$. □

In light of Theorem 4.7.9 and its corollary, the Lie derivative of vector fields measures the degree to which the flows of vector fields do not *commute*, where the flows commute (in the sense of Corollary 4.7.10) if and only if $\mathcal{L}_X Y = 0$.

Before turning away from the Lie derivative of vector fields, we introduce one definition (really in this presentation a notation) that is in common usage.

Definition 4.7.11. Let X and Y be smooth vector fields on $\mathbf{R}^n$. Then the *Lie bracket* of X and Y, denoted by $[X,Y]$, is defined to be

$$[X,Y] = \mathcal{L}_X Y.$$

Certain authors vary the definition by a sign. The notation comes from the terminology of Lie groups and Lie algebras, which are lurking beneath the surface of our discussion of vector fields. Exploring these ideas—even listing the algebraic properties of the Lie bracket—would lead us too far astray. The interested reader is invited to read Warner's exposition in [40].

One algebraic property of the Lie derivative of vector fields that appears often in calculations is the *Jacobi identity*, which is usually written in the notation of Lie brackets.

Proposition 4.7.12 (Jacobi identity). *Let X, Y, and Z be vector fields on a domain $U \subset \mathbf{R}^n$. Then*

$$[X,[Y,Z]] + [Y,[Z,X]] + [Z,[X,Y]] = 0.$$

Proof. The identity follows by applying Proposition 4.7.5. For every smooth function f on U, we have

$$\begin{aligned}
\Big([X,[Y,Z]] + [Y,[Z,X]] + [Z,[X,Y]]\Big)[f] &= X[[Y,Z][f]] - [Y,Z][X[f]] \\
&\quad + Y[[Z,X][f]] - [Z,X][Y[f]] \\
&\quad + Z[[X,Y][f]] - [X,Y][Z[f]] \\
&= X\Big[Y[Z[f]] - Z[Y[f]]\Big] \\
&\quad - Y[Z[X[f]]] - Z[Y[X[f]]] \\
&\quad + Y\Big[Z[X[f]] - X[Z[f]]\Big] \\
&\quad - Z[X[Y[f]]] - X[Z[Y[f]]] \\
&\quad + Z\Big[X[Y[f]] - Y[X[f]]\Big] \\
&\quad - X[Y[Z[f]]] - Y[X[Z[f]]] \\
&= 0. \quad \square
\end{aligned}$$

The following proposition shows that diffeomorphisms preserve the Lie derivative of vector fields.

Proposition 4.7.13. *Let $\phi : U \to V$ be a diffeomorphism between domains $U, V \subset \mathbf{R}^n$. Then for all vector fields X, Y on U,*

$$\phi_* (\mathcal{L}_X Y) = \mathcal{L}_{\phi_* X} \phi_* Y.$$

Proof. We again rely on Proposition 4.7.5 as well as Definition 3.7.6. Let $f : V \to \mathbf{R}$ be a smooth function on V. Then

$$\begin{aligned}
\phi_* (\mathcal{L}_X Y) [f] &= (\mathcal{L}_X Y) [f \circ \phi] \\
&= (XY - YX) [f \circ \phi] \\
&= X [Y [f \circ \phi]] - Y [X [f \circ \phi]] \\
&= X [(\phi_* Y) [f] \circ \phi] - Y [(\phi_* X) [f] \circ \phi] \\
&= \phi_* X [\phi_* Y [f]] - \phi_* Y [\phi_* X [f]] \\
&= (\phi_* X \phi_* Y - \phi_* Y \phi_* X) [f] \\
&= \left(\mathcal{L}_{\phi_* X} \phi_* Y\right) [f]. \quad \square
\end{aligned}$$

Written in the language of Lie brackets, Proposition 4.7.13 says that

$$\phi_* ([X, Y]) = [\phi_* X, \phi_* Y].$$

We now turn to the Lie derivative of differential forms. As before, we begin with a coordinate calculation for the Lie derivative of a basis one-form, followed by a coordinate expression for the Lie derivative of any one-form on $\mathbf{R}^n$. The more general case of the Lie derivative of a k-form on $\mathbf{R}^n$ is best handled by means of the Cartan formula, which we present following the initial coordinate calculations and some basic properties.

Proposition 4.7.14. *Let $X = \sum X^j \frac{\partial}{\partial x_j}$ be a smooth vector field on $\mathbf{R}^n$, and let dx_i be a standard basis one-form on $\mathbf{R}^n$. Then*

$$\mathcal{L}_X (dx_i) = \sum_{j=1}^{n} \frac{\partial X^i}{\partial x_j} dx_j \; (= dX^i).$$

Proof. Exercise. See Proposition 4.7.4. $\square$

Proposition 4.7.15. *Let $X = \sum_{i=1}^{n} X^i \frac{\partial}{\partial x_i}$ be a smooth vector field on $\mathbf{R}^n$ and let $\alpha = \sum_{i=1}^{n} a_i dx_i$ be a differential one-form on $\mathbf{R}^n$. Then*

$$\mathcal{L}_X \alpha = \sum_{j=1}^{n} b_j dx_j,$$

where the functions $b_j : \mathbf{R}^n \to \mathbf{R}$ *(*$j = 1, \dots, n$*) are given by*

$$b_j = \sum_{k=1}^{n} \left(\frac{\partial a_j}{\partial x_k} \cdot X^k + a_k \cdot \frac{\partial X^k}{\partial x_j} \right).$$

Proof. As in the proof of Proposition 4.7.5, the result is an application of Proposition 4.7.3 to the coordinate calculation for the basis forms in Proposition 4.7.14. □

We now present a useful property that expresses the Lie derivative of $(0,k)$-tensor fields in terms of the Lie derivative of vector fields.

Proposition 4.7.16. *Let* T *be a* $(0,k)$*-tensor field on* $\mathbf{R}^n$*, and let* X *and* $Y_1, \dots, Y_k$ *be vector fields on* $\mathbf{R}^n$*. Then*

$$\begin{aligned}(\mathcal{L}_X T)(Y_1, \dots, Y_k) &= X\left[T(Y_1, \dots, Y_k)\right] \\ &\quad - \sum_{i=1}^{k} T(Y_1, \dots, Y_{i-1}, \mathcal{L}_X Y_i, Y_{i+1}, \dots, Y_k).\end{aligned}$$

Proof. Consider the standard bases $\left\{ \frac{\partial}{\partial x_1}, \dots, \frac{\partial}{\partial x_n} \right\}$ for $T(\mathbf{R}^n)$ and $\{dx_1, \dots, dx_n\}$ for $T^*(\mathbf{R}^n)$. We write $X = \sum_{i=1}^{n} X^i \frac{\partial}{\partial x_i}$ and $Y_j = \sum_{i=1}^{n} Y_j^i \frac{\partial}{\partial x_i}$ for $j = 1, \dots, k$. Furthermore, the corresponding standard basis elements for the set of $(0,k)$-tensors have the form $dx_{i_1} \otimes \cdots \otimes dx_{i_k}$, where $i_1, \dots, i_k \in \{1, \dots, n\}$.

Using these coordinates, we first prove the statement for monomial tensors of the form $A = a dx_{i_1} \otimes \cdots \otimes dx_{i_k}$, where $a : \mathbf{R}^n \to \mathbf{R}$ is a smooth function. The calculations will rely extensively on Proposition 4.7.3(3) as well as on Proposition 4.7.5.

As a consequence of the usual Leibniz product rule, we have

$$\begin{aligned}X\left[A(Y_1, \dots, Y_k)\right] &= X\left[a \cdot Y_1^{i_1} \cdots Y_k^{i_k}\right] \\ &= \sum_{j=1}^{n} X^j \frac{\partial}{\partial x_j}\left[a \cdot Y_1^{i_1} \cdots Y_k^{i_k}\right] \\ &= \sum_{j=1}^{n} X^j \left(\frac{\partial a}{\partial x_j} \cdot Y_1^{i_1} \cdots Y_k^{i_k} + a \cdot \frac{\partial Y_1^{i_1}}{\partial x_j} Y_2^{i_2} \cdots Y_k^{i_k} \right. \\ &\quad \left. + \cdots + a \cdot Y_1^{i_1} \cdots Y_{k-1}^{i_{k-1}} \cdot \frac{\partial Y_k^{i_k}}{\partial x_j} \right).\end{aligned}$$

Using Proposition 4.7.3(3) along with Proposition 4.7.14, we have

$$\begin{aligned}
\mathcal{L}_X A &= X\,[a]\, dx_{i_1} \otimes \cdots \otimes dx_{i_k} \\
&\quad + a \cdot (\mathcal{L}_X dx_{i_1}) \otimes dx_{i_2} \otimes \cdots \otimes dx_{i_k} \\
&\quad + \cdots + a \cdot dx_{i_1} \otimes \cdots \otimes dx_{i_{k-1}} \otimes (\mathcal{L}_X dx_{i_k}) \\
&= \left(\sum_{j=1}^{n} X^j \frac{\partial a}{\partial x_j} \right) \cdot dx_{i_1} \otimes \cdots \otimes dx_{i_k} \\
&\quad + a \cdot \left(\sum_{j=1}^{n} \frac{\partial X^{i_1}}{\partial x_j} dx_j \right) \otimes dx_{i_2} \otimes \cdots dx_{i_k} \\
&\quad + \cdots \\
&\quad + a \cdot dx_{i_1} \otimes \cdots \otimes dx_{i_{k-1}} \otimes \left(\sum_{j=1}^{n} \frac{\partial X^{i_k}}{\partial x_j} dx_j \right),
\end{aligned}$$

and so

$$\begin{aligned}
(\mathcal{L}_X A)(Y_1, \ldots, Y_k) &= \sum_{j=1}^{n} X^j \cdot \frac{\partial a}{\partial x_j} \cdot Y_1^{i_1} \cdots Y_k^{i_k} \\
&\quad + a \cdot \left(\sum_{j=1}^{n} \frac{\partial X^{i_1}}{\partial x_j} Y_1^j \right) \cdot Y_2^{i_2} \cdots Y_k^{i_k} \\
&\quad + \cdots \\
&\quad + a \cdot Y_1^{i_1} \cdots Y_{k-1}^{i_{k-1}} \cdot \left(\sum_{j=1}^{n} \frac{\partial X^{i_k}}{\partial x_j} Y_k^j \right).
\end{aligned}$$

Finally, using Proposition 4.7.5, we have

$$\begin{aligned}
A\Big((\mathcal{L}_X Y_1), Y_2, \cdots, Y_k \Big) &= a \cdot (\mathcal{L}_X Y_1)^{i_1}\, Y_2^{i_2} \cdots Y_r^{i_r} \\
&= a \cdot \left(\sum_{j=1}^{n} \left(X^j \frac{\partial Y_1^{i_1}}{\partial x_j} - Y_1^j \frac{\partial X^{i_1}}{\partial x_j} \right) \right) \cdot Y_2^{i_2} \cdots Y_r^{i_r},
\end{aligned}$$

$$\begin{aligned}
&A\Big(Y_1, (\mathcal{L}_X Y_2), Y_3, \cdots, Y_k \Big) \\
&\quad = a \cdot Y_1^{i_1} \cdot \left(\sum_{j=1}^{n} \left(X^j \frac{\partial Y_2^{i_2}}{\partial x_j} - Y_2^j \frac{\partial X^{i_2}}{\partial x_j} \right) \right) \cdot Y_3^{i_3} \cdots Y_r^{i_r}, \\
&\qquad \vdots
\end{aligned}$$

$$
\begin{aligned}
A\Big(Y_1,\cdots,Y_{k-1},(\mathcal{L}_X Y_k)\Big) \\
&= a\cdot Y_1^{i_1}\cdots Y_{k-1}^{i_{k-1}}\cdot\left(\sum_{j=1}^{n}\left(X^j\frac{\partial Y_k^{i_k}}{\partial x_j}-Y_k^j\frac{\partial X^{i_k}}{\partial x_j}\right)\right).
\end{aligned}
$$

Here $(\mathcal{L}_X Y_j)^i$ represents the i component of the vector field $\mathcal{L}_X Y_j$.

Putting all these calculations together shows that

$$
X\left[A(Y_1,\dots,Y_k)\right]=(\mathcal{L}_X A)(Y_1,\dots,Y_k)+\sum_{i=1}^{k}A(Y_1,\dots,\mathcal{L}_X Y_i,\dots,Y_k),
$$

as was to be proven.

Since a general $(0,k)$-tensor field is a sum of tensors of the form A above, applying Proposition 4.7.3(1) proves the proposition. □

In addition to the properties of the Lie derivative on general tensors listed in Proposition 4.7.3, there are several additional properties of how the Lie derivative acts on differential forms, and in particular how it interacts with the exterior product and the exterior derivative.

Proposition 4.7.17. *Let α and β be differential forms and let X be a smooth vector field on $\mathbf{R}^n$ (with the degrees of the forms implicit and depending on the context below). Then:*

1. $\mathcal{L}_X(\alpha\wedge\beta)=(\mathcal{L}_X\alpha)\wedge\beta+\alpha\wedge(\mathcal{L}_X\beta)$.
2. $\mathcal{L}_X d\alpha=d(\mathcal{L}_X\alpha)$.

Proof. Statement (1) follows from Proposition 4.7.3 and the fact that differential forms are a specific type of tensor, namely $(0,k)$-tensors that have the additional property of being alternating. Moreover, the reader should verify that the exterior product $\wedge$ can be written completely in terms of the tensor product $\otimes$.

Statement (2) follows by applying Proposition 4.4.16 to the definition of the Lie derivative:

$$
\begin{aligned}
\mathcal{L}_X d\alpha &= \left.\frac{d}{dt}\right|_{t=0}(\phi_t^*(d\alpha)) \\
&= \left.\frac{d}{dt}\right|_{t=0} d(\phi_t^*\alpha) \\
&= d\left(\left.\frac{d}{dt}\right|_{t=0}(\phi_t^*\alpha)\right) \\
&= d(\mathcal{L}_X\alpha),
\end{aligned}
$$

where ϕ_t is the flow generated by X. However, there are important technical details associated with the second-to-last equality, "passing the d-operator through the derivative." See [40, pp. 71–72] for those details. □

We will frequently rely on the following theorem in calculating Lie derivatives of differential forms.

Theorem 4.7.18 (Cartan's formula). *Let X be a smooth vector field on $\mathbf{R}^n$ and let ω be a differential k-form. Then*

$$\mathcal{L}_X\omega = i(X)d\omega + d\left(i(X)\omega\right).$$

Proof. We verify the equality in coordinates with ω a one-form; the result then follows from Propositions 4.7.15, 4.6.19, and induction. Let $X = \sum X^i \frac{\partial}{\partial x_i}$, and let $\omega = \sum a_i dx_i$. Then $d\omega = \sum da_i \wedge dx_i$, and so

$$i(X)d\omega = \sum \left(\left(i(X)da_i\right)dx_i - X^i da_i\right).$$

Also

$$\begin{aligned} d\left(i(X)\omega\right) &= d\left(\sum X^i a_i\right) \\ &= \sum d(X^i a_i) \\ &= \sum \left(X^i da_i + a_i dX^i\right). \end{aligned}$$

Together, this means that

$$i(X)d\omega + d(i(X)\omega) = \sum \left(i(X)da_i + a_i dX^i\right).$$

Comparing this with Proposition 4.7.15 proves the statement. □

We can now return to the comment at the end of the last section, relating the interior product to the exterior differential.

Corollary 4.7.19. *Suppose Ω is a k-form and X is a vector field satisfying $\mathcal{L}_X\Omega = 0$. Then*

$$i(X)d\Omega = -d\left(i(X)\Omega\right).$$

In other words, the Lie derivative describes the degree to which the interior product of a k-form fails to anticommute with the exterior product.

We are now in a position to illustrate some calculations of Lie derivatives for differential forms that we will encounter in the future.

Example 4.7.20. On $\mathbf{R}^3$ with coordinates (x, y, z), define the differential one-form $\alpha_0 = xdy + dz$. Let

$$X_1 = \frac{\partial}{\partial x}, \qquad X_2 = x\frac{\partial}{\partial x} + y\frac{\partial}{\partial y} + 2z\frac{\partial}{\partial z}, \qquad X_3 = \frac{\partial}{\partial y}.$$

Then we have

$$\begin{aligned}\mathcal{L}_{X_1}\alpha_0 &= i(X_1)d\alpha_0 + d\left(i(X_1)\alpha_0\right)\\ &= i(X_1)\left(dx \wedge dy\right) + d(0)\\ &= dy,\end{aligned}$$

$$\begin{aligned}\mathcal{L}_{X_2}\alpha_0 &= i(X_2)d\alpha_0 + d\left(i(X_2)\alpha_0\right)\\ &= i(X_2)\left(dx \wedge dy\right) + d(xy + 2z)\\ &= (xdy - ydx) + (xdy + ydx + 2dz)\\ &= 2xdy + 2dz\\ &= 2\alpha_0,\end{aligned}$$

and

$$\begin{aligned}\mathcal{L}_{X_3}\alpha_0 &= i(X_3)d\alpha_0 + d\left(i(X_3)\alpha_0\right)\\ &= i(X_3)\left(dx \wedge dy\right) + d(x)\\ &= -dx + dx\\ &= 0.\end{aligned}$$

Example 4.7.21. On $\mathbf{R}^4$ with coordinates (x, y, z, w), consider the differential two-form $\omega_0 = dx \wedge dy + dz \wedge dw$. Note that $d\omega_0 = 0$, so Cartan's formula is just $\mathcal{L}_X\omega_0 = d\left(i(X)\omega_0\right)$.

Let

$$X_1 = x\frac{\partial}{\partial x} + z\frac{\partial}{\partial z}, \qquad X_2 = y\frac{\partial}{\partial x} + w\frac{\partial}{\partial z}.$$

Then

$$\begin{aligned}\mathcal{L}_{X_1}\omega_0 &= d\left(i(X_1)\omega_0\right)\\ &= d\left(xdy + zdw\right)\\ &= dx \wedge dy + dz \wedge dw\\ &= \omega_0,\end{aligned}$$

and

$$\begin{aligned}\mathcal{L}_{X_2}\omega_0 &= d\left(i(X_2)\omega_0\right)\\ &= d\left(ydy + wdw\right)\\ &= 0.\end{aligned}$$

There is an analogy to Proposition 4.7.16 for differential forms, which states that the exterior derivative can be expressed completely in terms of the Lie derivative.

Proposition 4.7.22. *Let α be a smooth k-form on $\mathbf{R}^n$. Let $Y_0, Y_1, \dots, Y_k$ be smooth vector fields. Then*

$$\begin{aligned} d\alpha(Y_0, Y_1, \dots, Y_k) &= \sum_{i=0}^{k} (-1)^i Y_i \left[\alpha(Y_0, Y_1, \dots, \widehat{Y_i}, \dots, Y_k)\right] \\ &\quad + \sum_{i<j} (-1)^{i+j} \alpha(\mathcal{L}_{Y_i} Y_j, X, Y_1, \dots, \widehat{Y_i}, \dots, \widehat{Y_j}, \dots, Y_k). \end{aligned}$$

Here the hat notation $\widehat{Y_j}$ signifies omitting Y_j from the list.

In particular, for $k = 1$, we have

$$d\alpha(X, Y) = X\left[\alpha(Y)\right] - Y\left[\alpha(X)\right] - \alpha(\mathcal{L}_X Y).$$

Proof. The full proof can be accomplished by induction on k, relying repeatedly on Proposition 4.7.16 and Theorem 4.7.18. We will prove the basis step $k = 1$ here (which is in fact the case that we will encounter most in the future), and then indicate how to prove the inductive step, leaving the details to the reader.

Let α be a one-form and let X, Y be vector fields. By Proposition 4.7.16, we have

$$(\mathcal{L}_X \alpha)(Y) = X\left[\alpha(Y)\right] - \alpha(\mathcal{L}_X Y). \tag{A}$$

By Cartan's formula, we have

$$\begin{aligned} (\mathcal{L}_X \alpha)(Y) &= \left(d(\alpha(X))\right)(Y) + \left(i(X) d\alpha\right)(Y) \\ &= Y\left[\alpha(X)\right] + d\alpha(X, Y). \end{aligned} \tag{B}$$

Substituting (B) into (A) yields the result.

For the induction step, assume that the statement holds for all $(k-1)$-forms and let α be a k-form. Relying again on Proposition 4.7.16 and Theorem 4.7.18, we have

$$(\mathcal{L}_X \alpha)(Y_1, \dots, Y_k) = X\left[\alpha(Y_1, \dots, Y_k)\right] + \sum_{i=1}^{n} (-1)^i \alpha(Y_1, \dots, \mathcal{L}_X Y_i, \dots, Y_k)$$

and

$$\mathcal{L}_X \alpha = d(i(X)\alpha) + i(X)(d\alpha).$$

Then the induction hypothesis applies to the $(k-1)$-form

$$\beta = i(X)\alpha.$$

Substituting, keeping careful track of the signs, yields the result. □

We close this section with an important interpretation of the Lie derivative of a differential form that will be discussed in more detail in the subsequent chapters. Although we will most often see this in the special case of differential forms, we state it for a general tensor field.

Theorem 4.7.23. *Let S be a tensor field on $\mathbf{R}^n$. Let X be a smooth vector field on $\mathbf{R}^n$ with flow ϕ_t. If $\mathcal{L}_X S = 0$, then for all t for which the flow is defined we have*

$$\phi_t^* S = S.$$

Proof. In essence, this statement reduces to the observation that the definition of the Lie derivative,

$$\mathcal{L}_X S = \left.\frac{d}{dt}\right|_{t=0} (\phi_t^* S),$$

implies that if the Lie derivative is 0, then $\phi_t^* S$ must be constant with respect to t. Hence

$$\phi_t^* S = \phi_0^* S = (\mathrm{Id})^* S = S. \quad \square$$

Read literally, Theorem 4.7.23 states that if $\mathcal{L}_X S = 0$, then S is constant along the flow generated by X. In the geometric spirit that we will be developing in the next chapters, where we consider a geometric "structure" defined by a differential form or tensor, the theorem states that diffeomorphisms generated by vector fields with the property that $\mathcal{L}_X S = 0$ "preserve the structure" defined by S.

4.8 For Further Reading

There are two standard ways of presenting tensors. Historically, tensors received the most attention from physicists, whose interest in calculation and coordinate systems placed the distinct emphasis on the behavior of these objects under change of coordinates. This emphasis led to an elaborate index and summation system. A classical presentation of the subject from this perspective can be found, for example, in [15]. The tedious notation and component calculations has often led to a kind of disdain, best typified by Élie Cartan's characterization of tensor calculus as "the debauch of indices."

The reformulation of tensor calculus and differential forms in particular was a consequence of the coordinate-free ideal that drove the development of differential geometry within pure mathematics. This was done in the context of manifolds, and can best be appreciated in the classic texts of Spivak [38] and Warner [40]. Most of the results in this chapter can be found in those two sources.

Integration of forms has a slightly different historical trajectory. Standard references for this topic include Spivak [37] and do Carmo [14]. Our treatment most closely follows the latter.

We also note that the coordinate-free approach of mathematicians found its way back to physicists through texts like [1, 3], written by mathematicians for an audience with applied and physical tastes.

The effort to bring differential geometry to students at an earlier point in their formal mathematical training—a primary goal of this text—has inspired several texts whose goal is to make differential forms accessible to students with a background in calculus and linear algebra. The elegant presentation of differential geometry by O'Neill [33], first in 1966 and then again in 1997, is a case in point, although forms appear only in passing. Weintraub's 1997 text [41] set itself the task of presenting a first course in vector calculus through the lens of differential forms; its only drawback is its somewhat unorthodox notation. Finally, we mention the 2006 text by Bachman [5], which shares many of the goals of the present chapter and has the additional benefit of introducing contact structures, a topic we will visit in detail in Chap. 7.

4.9 Exercises

4.1. Consider the linear two-form $\gamma \in \Lambda_2(\mathbf{R}^4)$ given by $\gamma = 3\epsilon_{12} - 2\epsilon_{34}$. For arbitrary vectors $\mathbf{a}, \mathbf{b} \in \mathbf{R}^4$ with components $\mathbf{a} = (a_1, a_2, a_3, a_4)$ and $\mathbf{b} = (b_1, b_2, b_3, b_4)$, evaluate $\gamma(\mathbf{a}, \mathbf{b})$. Then find nonzero vectors $\mathbf{a}$ and $\mathbf{b}$ such that $\gamma(\mathbf{a}, \mathbf{b}) = 0$.

4.2. Prove Proposition 4.2.7.

4.3. Prove Proposition 4.2.8.

4.4. Prove Proposition 4.2.11.

4.5. Let $\alpha = 2\epsilon_2 + \epsilon_3 \in \Lambda_1(\mathbf{R}^3)$. Let $T_1 : \mathbf{R}^3 \to \mathbf{R}^3$ be given by

$$T_1(x, y, z) = (3y + 2z, -x - y + z, x - 4z)$$

and let $T_2 : \mathbf{R}^2 \to \mathbf{R}^3$ be given by

$$T_2(x, y) = (12x - 13y, x + 2y, -x - y).$$

Compute $T_1^*\alpha$ and $T_2^*\alpha$ in two ways, first using the method of Example 4.2.10 and then using the method of Example 4.2.12.

4.6. Prove the following converse of Theorem 4.1.2: Suppose that α is a k-form with the property that for $i \neq j$, $\mathbf{v}_i = \mathbf{v}_j$ implies that $\alpha(\mathbf{v}_1, \dots, \mathbf{v}_k) = 0$. Then α is an alternating form.

4.7. Let $\alpha = c\epsilon_2 + \epsilon_3 \in \Lambda_1(\mathbf{R}^3)$, where $c \in \mathbf{R}$, and let $T : \mathbf{R}^3 \to \mathbf{R}^3$ be a linear transformation represented by the matrix $A = [T] = [a_{ij}]$. Find necessary and sufficient conditions on the entries a_{ij} such that $T^*\alpha = k\alpha$ for some constant k. Write these conditions in the specific cases that $k = 1$ and $k = 0$.

4.8. Let $\alpha = a_1 dx + a_2 dy + a_3 dz$ be a smooth differential one-form defined on all of $\mathbf{R}^3$. Define a function $f : \mathbf{R}^3 \to \mathbf{R}$ by

$$f(x, y, z) = \int_0^1 \big[x \cdot a_1(tx, ty, tz) \\ + y \cdot a_2(tx, ty, tz) + z \cdot a_3(tx, ty, tz)\big]\, dt.$$

Show that $d\alpha = 0$ if and only if $\alpha = df$. (This proves that every closed one-form defined on all of $\mathbf{R}^3$ is exact.)

For the following Exercises 4.9–4.11, let $\alpha, \beta \in \Lambda_1(\mathbf{R}^3)$ and $\gamma \in \Lambda_2(\mathbf{R}^3)$ be given by

$$\alpha = x dx + y dy + z dz,$$
$$\beta = x dy - y dx + dz,$$
$$\gamma = z^2 dx \wedge dy + x^2 dy \wedge dz + y^2 dx \wedge dz.$$

4.9. Compute the following:

(a) $d\alpha,\ d\beta,\ d\gamma$.
(b) $\alpha \wedge \beta,\ \alpha \wedge \gamma,\ \beta \wedge \gamma$.
(c) $\alpha \wedge d\alpha,\ \beta \wedge d\beta,\ \alpha \wedge d\beta$.

4.10. Let $X = y\frac{\partial}{\partial x} - x\frac{\partial}{\partial y} + \frac{\partial}{\partial z}$.

(a) Compute $i(X)\alpha,\ i(X)\beta,\ i(X)\gamma$.
(b) Compute $i(X)d\alpha,\ i(X)d\beta,\ i(X)d\gamma$.
(c) Compute $d(i(X)\alpha),\ d(i(X)\beta),\ d(i(X)\gamma)$.

4.11. Let $\phi : \mathbf{R}^3 \to \mathbf{R}^3$ be given by

$$\phi(u, v, w) = (e^w \sin u, e^w \cos u, v^3 - u^3)$$

and let $\psi : \mathbf{R}^2 \to \mathbf{R}^3$ be given by

$$\psi(s, t) = (s \cos t, s \sin t, t).$$

(a) Compute $\phi^*\alpha,\ \phi^*\beta,\ \phi^*\gamma$.
(b) Compute $\psi^*\alpha,\ \psi^*\beta,\ \psi^*\gamma$.
(c) Compute $\phi^*(d\alpha),\ \phi^*(d\beta),\ \phi^*(d\gamma)$.
(d) Compute $\psi^*(d\alpha),\ \psi^*(d\beta),\ \psi^*(d\gamma)$.

4.12. Prove Proposition 4.4.20.

4.13. Let $\{E_1, \ldots, E_n\}$ be a set of vector fields defined on a domain $U \subset \mathbf{R}^n$ such that for all $p \in U$, $\{E_1(p), \ldots, E_n(p)\}$ is an *orthonormal* basis for T_pU in the sense that

$$E_i(p) \centerdot E_j(p) = \delta_{ij}$$

using the notation of the "dot product" on $\mathbf{R}^n$. Such a set is said to be an *orthonormal frame* relative to the standard inner product. Note that by making the identification $T_pU = \mathbf{R}^n$, we can consider each of the vector fields E_i to be a differentiable map

$$E_i : U \to \mathbf{R}^n,$$

and in particular we can consider the tangent map $(E_i)_* : TU \to T\mathbf{R}^n$.
For $i, j = 1, \ldots, n$, define for all vector fields V on U,

$$\omega_{ij}(V) = (E_i)_*(V) \centerdot E_j.$$

(a) Show that for all $i, j = 1, \ldots, n$, ω_{ij} is a differential one-form, i.e., that it is smoothly varying and linear on each tangent space.
(b) Show that $\omega_{ji} = -\omega_{ij}$. [Hint: Differentiate $E_i \centerdot E_j = \delta_{ij}$.]
(c) Let $\{\varepsilon_1, \ldots, \varepsilon_n\}$ be the set of one-forms dual to the vector fields $\{E_1, \ldots, E_n\}$. Show that

$$d\varepsilon_i = \sum_k \varepsilon_k \wedge \omega_{ki}$$

and

$$d\omega_{ij} = \sum_k \omega_{ik} \wedge \omega_{kj}.$$

These two sets of equations are called *Cartan's structural equations.*

4.14. This exercise is an example of the constructions in Exercise 4.13. Consider the vector fields E_1, E_2 on $U = \mathbf{R}^2 \setminus \{(0,0)\}$ given by

$$E_1(x,y) = \frac{x}{\sqrt{x^2+y^2}} \frac{\partial}{\partial x} + \frac{y}{\sqrt{x^2+y^2}} \frac{\partial}{\partial y},$$

$$E_2(x,y) = \frac{-y}{\sqrt{x^2+y^2}} \frac{\partial}{\partial x} + \frac{x}{\sqrt{x^2+y^2}} \frac{\partial}{\partial y}.$$

(a) Verify that for all $p \in U$, $\{E_1(p), E_2(p)\}$ is an orthonormal basis for T_pU.
(b) Compute ε_1, ε_2.
(c) Verify Cartan's structure equations by computing each side of both sets of equations in Exercise 4.13(c).

4.15. For each of the diffeomorphisms in Example 3.6.10, determine whether they are orientation-preserving or orientation-reversing.

4.16. Let U and f be as in Example 3.6.8 and let $0 < r < 1$.

(a) Show that if $p = (p_1, p_2) \in U$ satisfies $p_1^2 + p_2^2 < r^2$, then $f(p) = (q_1, q_2)$ satisfies $q_1^2 + q_2^2 > (1/r)^2$. That is to say, show that if C_r is the circle of radius r, then points inside C_r are mapped to points outside $C_{1/r} = f(C_r)$.
(b) Show that f is orientation-reversing (relative to any fixed orientation of U).

4.17. Suppose that ω_1 is a closed differential form and ω_2 is an exact differential form, both defined on some open set $U \subset \mathbf{R}^n$. Show that $\omega_1 \wedge \omega_2$ is exact.

4.18. Let $\phi : U \to \mathbf{R}^3$ be the regular parameterization of a paraboloid given in Example 3.3.4, i.e.,

$$U = \{(r, \theta) \mid 0 < r < 2, -\pi < \theta < \pi\} \subset \mathbf{R}^2$$

and

$$\phi(r, \theta) = (r\cos\theta, r\sin\theta, r^2).$$

Let $\omega = xdy \wedge dz - ydx \wedge dz + zdx \wedge dy \in \Lambda_2(\mathbf{R}^3)$. Compute $\int_S \omega$, where $S = \phi(R)$ has the orientation induced by ϕ and $R = [1/2, 3/2] \times [0, \pi/2]$.

4.19. Given a parameterized surface S with the standard orientation induced by the parameterization and the differential two-form ω below, compute $\int_S \omega$.

(a) $S = \phi_1(R)$ is the part of the helicoid of Exercise 3.14(a), where $R = [\pi/2, \pi] \times [0, 1]$, and $\omega = xdy \wedge dz - ydx \wedge dz + zdx \wedge dy$.
(b) $S = \phi_2(R)$ is the part of the catenoid of Exercise 3.14(b), where $R = [\pi/2, \pi] \times [0, 1]$, and $\omega = dx \wedge dy$.

4.20. Suppose that V is a finite-dimensional vector space. Show that the map $I : V \to (V^*)^*$ described in Example 4.6.4 is an isomorphism.

4.21. Let $\epsilon_{12}^{23} \in T^{2,2}(\mathbf{R}^3)$ be as in Theorem 4.1.8 and let $\Phi : \mathbf{R}^3 \to \mathbf{R}^3$ be defined by

$$\Phi(x, y, z) = (x + y, x - y - z, y + z).$$

Compute $\Phi^* \epsilon_{12}^{23}$.

4.22. Let V, W be two vector fields on $\mathbf{R}^2$, and define $g : T\mathbf{R}^2 \times T\mathbf{R}^2 \to \mathbf{R}$ by

$$\begin{aligned} g(V, W) = (1 + 4x^2)V^1W^1 + (4xy)V^1W^2 \\ + (4xy)V^2W^1 + (1 + 4y^2)V^2W^2, \end{aligned}$$

where (x, y) are coordinates on $\mathbf{R}^2$ and $V = \langle V^1, V^2 \rangle$, $W = \langle W^1, W^2 \rangle$ are the corresponding coordinate expressions for the vector fields V and W. In particular, V^1, V^2, W^1, W^2 are smooth functions of (x, y).

(a) Show directly from Definition 4.6.11 that g is a $(0, 2)$-tensor field.
(b) Let $X(x, y) = x\frac{\partial}{\partial x} + y\frac{\partial}{\partial y}$. Compute $i(X)g$.
(c) Compute $\phi^* g$, where $\phi : \mathbf{R}^2 \to \mathbf{R}^2$ is given by

$$\phi(x, y) = (2x, xy).$$

4.23. Repeat Exercise 4.22 for $g(V, W) = V^1W^1 + V^2W^2$, $X = x\frac{\partial}{\partial y} - y\frac{\partial}{\partial x}$, and $\phi(x, y) = (x^2 + y^2, 2xy)$.

4.24. Let $\phi : \mathbf{R}^n \to \mathbf{R}^m$ be a smooth map. Let X be a smooth vector field on $\mathbf{R}^n$ and let Y be a smooth vector field on $\mathbf{R}^m$. We say that X and Y are *ϕ-related* if $\phi_* X = Y$. Show that if X is ϕ-related to Y and X' is ϕ-related to Y', then $[X, X']$ is ϕ-related to $[Y, Y']$.

4.25. Prove Proposition 4.7.6.

4.26. Consider the vector fields X, Y on $\mathbf{R}^3$ given by

$$X = y\frac{\partial}{\partial x} - xz\frac{\partial}{\partial y} + y^2\frac{\partial}{\partial z}, \quad Y = e^{-x}\frac{\partial}{\partial x} + e^{-y}\frac{\partial}{\partial y} + e^{-z}\frac{\partial}{\partial z}.$$

Compute $\mathcal{L}_X Y$.

4.27. Consider the vector fields X, Y on $\mathbf{R}^2$ given by

$$X(u, v) = u\frac{\partial}{\partial u} + v\frac{\partial}{\partial v}, \quad Y(u, v) = -v\frac{\partial}{\partial u} + u\frac{\partial}{\partial v}.$$

(a) Compute the flows ϕ_t of X and ψ_t of Y. Use these to compute the curve $c(t)$ through $c(0) = (1, 0)$ given by

$$c(t) = (\psi_{-t} \circ \phi_{-t} \circ \psi_t \circ \phi_t)(1, 0).$$

(b) Compute $\mathcal{L}_X Y$. How does your result correspond to the result of part (a)?

4.28. Repeat Exercise 4.27 using

$$X(u, v) = 4u\frac{\partial}{\partial u} + v\frac{\partial}{\partial v}, \quad Y(u, v) = v\frac{\partial}{\partial u} - 4u\frac{\partial}{\partial v}.$$

4.29. Following the outline of Proposition 4.7.4, prove Proposition 4.7.14.

4.30. Let X, Y be vector fields on $\mathbf{R}^3$ given by

$$X = y\frac{\partial}{\partial x} - z\frac{\partial}{\partial y} + x^2\frac{\partial}{\partial z}, \quad Y = (xy - z)\frac{\partial}{\partial x} + yz\frac{\partial}{\partial z}.$$

Let $\alpha = xdy + dz$, $\beta = yz\, dx + xz\, dy + xy\, dz$, and $\omega = dx \wedge dy$ be differential forms on $\mathbf{R}^3$.

(a) Compute $\mathcal{L}_X\alpha$, $\mathcal{L}_X\beta$, and $\mathcal{L}_X\omega$.
(b) Compute $\mathcal{L}_Y\alpha$, $\mathcal{L}_Y\beta$, and $\mathcal{L}_Y\omega$.

4.31. Let X, Y be vector fields on $\mathbf{R}^4$ with coordinates (x_1, y_1, x_2, y_2) given by

$$X = -x_1\frac{\partial}{\partial x_1} + y_1\frac{\partial}{\partial y_1} - x_2\frac{\partial}{\partial x_2} + y_2\frac{\partial}{\partial y_2},$$

$$Y = y_1\frac{\partial}{\partial x_1} + x_1\frac{\partial}{\partial y_1} + y_2\frac{\partial}{\partial x_2} + x_2\frac{\partial}{\partial y_2}.$$

Let $\alpha = x_1dy_1 - y_1dx_1 + x_2dy_2 - y_2dx_2$ and $\omega = dx_1 \wedge dy_1 + dx_2 \wedge dy_2$ be differential forms on $\mathbf{R}^4$.

(a) Compute $\mathcal{L}_X\alpha$ and $\mathcal{L}_X\omega$.
(b) Compute $\mathcal{L}_Y\alpha$ and $\mathcal{L}_Y\omega$.

4.32. Consider the differential one-form $\alpha = xdy + dz$ on $\mathbf{R}^3$.

(a) Find conditions for a vector field X to satisfy $\mathcal{L}_X\alpha = 0$. Give three examples of such vector fields.
(b) For each of the examples you constructed in part (a), verify the conclusion of Theorem 4.7.23 by computing the flow ϕ_t of X and then $\phi_t^*\alpha$.

4.33. Consider the differential two-form $\omega_0 = dx_1 \wedge dy_1 + dx_2 \wedge dy_2$ on $\mathbf{R}^4$ with coordinates (x_1, y_1, x_2, y_2).

(a) Find conditions on a smooth vector field X so that

$$\mathcal{L}_X\omega_0 = 0.$$

Give three examples of such vector fields.
(b) For each of the examples you constructed in part (a), verify the conclusion of Theorem 4.7.23 by computing the flow ϕ_t of X and then $\phi_t^*\omega_0$.

4.34. For a smooth vector field $X = \sum X^i \frac{\partial}{\partial x_i}$ on $\mathbf{R}^n$, use the techniques used to prove Proposition 4.7.4 to show that for every $j, k = 1, \ldots, n$,

$$\mathcal{L}_X(dx_j \otimes dx_k) = \sum_{r=1}^{n}\left(\frac{\partial X^j}{\partial x_r}dx_r \otimes dx_k + \frac{\partial X^k}{\partial x_r}dx_j \otimes dx_r\right).$$

4.35. Let $g = \sum_{i,j} g_{ij}dx_i \otimes dx_j$ be a smooth $(0, 2)$-tensor. Use the result of Exercise 4.34 to show that for a smooth vector field $X = \sum X^i \frac{\partial}{\partial x_i}$,

$$\mathcal{L}_X g = \sum_{i,j} h_{ij} dx_i \otimes dx_j,$$

where

$$h_{ij} = \sum_{k=1}^{n} \left(X^k \frac{\partial g_{ij}}{\partial x_k} + g_{kj} \frac{\partial X^k}{\partial x_i} + g_{ik} \frac{\partial X^k}{\partial x_j} \right).$$

(The tensor $h = \mathcal{L}_X g$ occurs in physics as the *strain tensor for small deformations*; cf. [15, p. 210].)

Chapter 5
Riemannian Geometry

We begin our presentation of differential-geometric structures with the one whose origins are most closely tied to the way the subject was developed by Gauss and Riemann, and that carries the name of the latter. The concepts of Riemannian geometry are familiar: length, angle, distance, and curvature, among others. Historically tied to the origins of differential geometry, and with such familiar concepts, Riemannian geometry is often presented in textbooks as being synonymous with differential geometry itself, instead of as one differential-geometric structure among many.

In our presentation, Riemannian geometry will be the study of a domain $U \subset \mathbf{R}^n$ equipped with a $(0,2)$-tensor with particular properties, called the Riemannian metric tensor. In this way, we will foreshadow many concepts that will appear again in the other geometric structures that we will consider. We will also see later that key concepts in Riemannian geometry do not carry over to the contact and symplectic settings of the later chapters.

It is worth noting at the outset that this point of view will sacrifice the historical development of Riemannian geometry in favor of a more axiomatic approach. This approach involves a definite loss. The classical, more historic, presentation beginning with curves and surfaces in $\mathbf{R}^3$ has both concreteness and a sense of motivation for the ideas to be generalized. We hope to compensate for this loss later by the ease of presenting contact and symplectic structures in a parallel manner.

5.1 Basic Concepts

The central object of Riemannian geometry is the metric tensor. This object should be seen as the analogue in the smooth setting of the inner product from linear algebra presented in Chap. 2. As with the inner product, the metric tensor will ultimately give rise to notions of length and distance, as well as concepts related to angles such as parallelism and orthogonality.

A. McInerney, *First Steps in Differential Geometry: Riemannian, Contact, Symplectic*, Undergraduate Texts in Mathematics, DOI 10.1007/978-1-4614-7732-7_5,

As in earlier chapters, we work mainly in the setting of a *domain* U in $\mathbf{R}^n$. In particular, a domain is *open* in the sense that for every point $p \in U$, there is an open ball centered at p that is completely contained in U.

Definition 5.1.1. Let $U \subset \mathbf{R}^n$ be a domain. A *Riemannian metric* on U is a smooth $(0,2)$-tensor field g satisfying the following two properties:

1. g is symmetric: for all $p \in U$ and all tangent vectors $X_p, Y_p \in T_pU$,

$$g_p(X_p, Y_p) = g_p(Y_p, X_p);$$

2. g is positive definite: for all $p \in U$ and all tangent vectors $X_p \in T_pU$,

$$g_p(X_p, X_p) \geq 0,$$

with $g_p(X_p, X_p) = 0$ if and only if $X_p = \mathbf{0}_p$.

In other words, a Riemannian metric is a smooth assignment of an inner product to each tangent space.

We sometimes simply refer to a Riemannian metric as a *metric tensor*.

Definition 5.1.2. A *Riemannian space* is a domain $U \subset \mathbf{R}^n$ equipped with a Riemannian metric g, and is denoted by (U, g).

In the notation of Chap. 4 (see Theorem 4.6.8), we can write

$$g = \sum_{i,j=1}^{n} g_{ij} dx_i \otimes dx_j,$$

where the $g_{ij} : \mathbf{R}^n \to \mathbf{R}$ are smooth functions and the dx_i are the standard basis one-forms for T^*U. Symmetry is then expressed by saying that

$$g_{ij} = g_{ji}.$$

It is also convenient at times to express the metric tensor g by means of the matrix $G = [g_{ij}]$. In this notation,

$$g(X, Y) = \mathbf{y}^T G \mathbf{x},$$

where $\mathbf{x}$ and $\mathbf{y}$ are column vector representations of vector fields X, Y using the standard basis for TU. Then G is a symmetric, positive definite matrix, with positive definiteness ensuring that $\det G > 0$.

Before proceeding to examples, we immediately present the most basic geometric measurement associated with a Riemannian metric: the length of a smooth curve. This will allow us to appreciate the difference between Riemannian metrics once we turn to examples.

Definition 5.1.3. Let (U, g) be a Riemannian space and let $c : [a, b] \to U$ be a smooth, regular parameterized curve (in particular, $\dot{c}(t) \neq \mathbf{0}$ for all $t \in I$). The *length of* c is defined to be

$$\ell(c) = \int_a^b \left[g_{c(t)}\left(\dot{c}(t), \dot{c}(t)\right)\right]^{1/2} dt.$$

It is a straightforward application of the chain rule and the bilinearity of g to see that the length is in fact *independent of parameterization*, although the definition is written in terms of a particular parameterization. A precise statement and proof of this fact can be found in [25, p. 8].

Example 5.1.4. Let $U = \mathbf{R}^2$ with coordinates (x, y) and let

$$g_0 = dx \otimes dx + dy \otimes dy,$$

or, using matrix notation,

$$G_0 = \begin{bmatrix} 1 & 0 \\ 0 & 1 \end{bmatrix}.$$

This Riemannian metric is known as the *standard Euclidean metric* on $\mathbf{R}^2$ for reasons that will be seen immediately and throughout. Note that g_0 is constant, acting in the same way on all tangent spaces.

For a regular parameterized curve $c : [a, b] \to \mathbf{R}^2$ described by $c(t) = (x(t), y(t))$, we have

$$\begin{aligned}(g_0)_{c(t)}\left(\dot{c}(t), \dot{c}(t)\right) &= (g_0)_{c(t)}\Big(\left\langle \dot{x}(t), \dot{y}(t)\right\rangle, \left\langle \dot{x}(t), \dot{y}(t)\right\rangle\Big) \\ &= (\dot{x}(t))^2 + (\dot{y}(t))^2,\end{aligned}$$

and so

$$\ell_0(c) = \int_a^b \sqrt{\left(\frac{dx}{dt}\right)^2 + \left(\frac{dy}{dt}\right)^2}\, dt.$$

The reader will recognize this as the standard calculus formula for the arc length of a parameterized curve.

Example 5.1.5. Let $U = \mathbf{R}^2$ with coordinates (u, v), and let

$$\begin{aligned} g_1 = (1 + 4u^2) du \otimes du &+ (4uv) du \otimes dv \\ &+ (4uv) dv \otimes du + (1 + 4v^2) dv \otimes dv, \end{aligned}$$

or, in matrix notation,

$$G_1 = \begin{bmatrix} 1 + 4u^2 & 4uv \\ 4uv & 1 + 4v^2 \end{bmatrix}.$$

The reader should verify that g_1 as a $(0, 2)$-tensor satisfies the properties of a metric tensor.

The length of a curve $c : [a, b] \to \mathbf{R}^2$, $c(t) = (u(t), v(t))$, according to the metric g_1 is given by

$$\ell_1(c) = \int_a^b \left[(1+4u^2)\left(\frac{du}{dt}\right)^2 + (8uv)\left(\frac{du}{dt}\right)\left(\frac{dv}{dt}\right) \right.$$
$$\left. + (1+4v^2)\left(\frac{dv}{dt}\right)^2 \right]^{1/2} dt.$$

This length is different from the standard Euclidean metric structure of Example 5.1.4. However, we note that for a circle with center $(0,0)$ and radius a parameterized as $c(t) = (a\cos t, a\sin t)$, $0 \le t \le 2\pi$, we have

$$\begin{aligned} \ell_1(c) &= \int_0^{2\pi} \left[(1+4a^2\cos^2 t)(-a\sin t)^2 \right. \\ &\qquad + 8(a\cos t)(a\sin t)(-a\sin t)(a\cos t) \\ &\qquad \left. + (1+4a^2\sin^2 t)(a\cos t)^2 \right]^{1/2} dt \\ &= \int_0^{2\pi} a\, dt \\ &= 2\pi a, \end{aligned}$$

which agrees with the length of the same circle using the Euclidean metric g_0. We leave to the reader to list examples of curves whose ℓ_1-length is different from the ℓ_0-length: any line segment, for example.

Example 5.1.6. Let $U = \mathbf{R}^2$, with coordinates again denoted by (u,v), and let g_2 be given by

$$\begin{aligned} g_2 &= (1+4u^2)du \otimes du - (4uv)du \otimes dv \\ &\qquad - (4uv)dv \otimes du + (1+4v^2)dv \otimes dv, \end{aligned}$$

or, in matrix notation,

$$G_2 = \begin{bmatrix} 1+4u^2 & -4uv \\ -4uv & 1+4v^2 \end{bmatrix}.$$

Proceeding in the same way as in the previous example, the reader may verify that

$$\ell_2(c) = \int_a^b \left[(1+4u^2)\left(\frac{du}{dt}\right)^2 - (8uv)\left(\frac{du}{dt}\right)\left(\frac{dv}{dt}\right) \right.$$
$$\left. + (1+4v^2)\left(\frac{dv}{dt}\right)^2 \right]^{1/2} dt.$$

While the differences between this metric and the previous one look minor, they are in fact quite significant. For example, the reader can verify that the length of the same circle parameterized as above by $c(t) = (a\cos t, a\sin t)$ with $0 \le t \le 2\pi$ is now, according to the metric g_2, given by

$$\ell_2(c) = \int_0^{2\pi} a\left[1 + 4a^2\sin^2(2t)\right]^{1/2} dt,$$

which is an elliptic integral that cannot be evaluated by means of elementary functions. Numerical methods show that for a circle of radius $a = 1$, we have

$$\ell_2(c) \approx 7.64,$$

compared to

$$\ell_1(c) = 2\pi \approx 6.28.$$

Example 5.1.7. Let $U = \{(x, y) \mid y > 0\} \subset \mathbf{R}^2$ be the upper half-plane, and define the metric g_3 by

$$g_3 = \frac{1}{y^2} dx \otimes dx + \frac{1}{y^2} dy \otimes dy,$$

or

$$G_3 = \begin{bmatrix} 1/y^2 & 0 \\ 0 & 1/y^2 \end{bmatrix}.$$

The half-plane U equipped with the metric g_3 is sometimes called the *Poincaré upper half-plane* and g_3 the *Poincaré metric*, after the renowned mathematician Henri Poincaré.

Notice that g_3 is obtained by scaling the standard Euclidean metric g_0 by a factor of $1/y^2$. We now illustrate this scaling's dramatic impact.

First note that the length of a parameterized curve $c : [a, b] \to U$ according to g_3 is given by

$$\begin{aligned} \ell_3(c) &= \int_a^b \left[\left(\frac{1}{y^2}\right)\left(\frac{dx}{dt}\right)^2 + \left(\frac{1}{y^2}\right)\left(\frac{dy}{dt}\right)^2\right]^{1/2} dt \\ &= \int_a^b \frac{1}{y}\left[\left(\frac{dx}{dt}\right)^2 + \left(\frac{dy}{dt}\right)^2\right]^{1/2} dt. \end{aligned}$$

For curves parallel to the x-axis, for example parameterized as $c(t) = (x(t), y_0)$, we have $\ell_3(c) = (1/y_0)\ell_0(c)$, i.e., horizontal curves with fixed g_3-length "get longer" (relative to the standard Euclidean metric) farther away from the x-axis. See Fig. 5.1.

Consider the parameterized curves $c : \mathbf{R} \to \mathbf{R}^2$ with initial point (x_0, y_0) having the form $c(t) = (x_0, y_0 + at)$ with $0 \le t \le 1$ and $a > 0, y_0 > 0$. These are curves parallel to the y-axis whose Euclidean length is $\ell_0(c) = a$. If we compute the length of c with respect to g_3, we see that

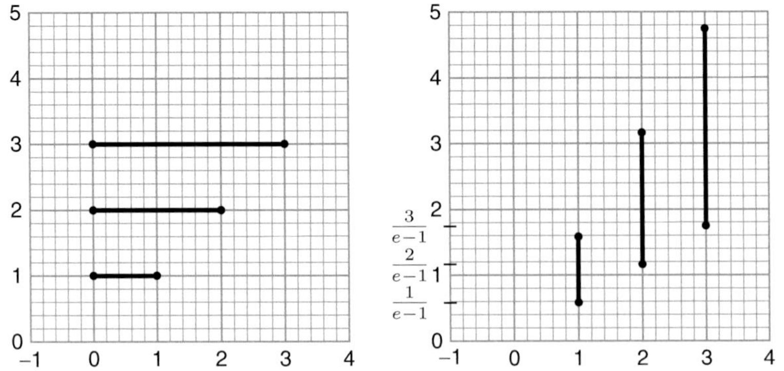

Fig. 5.1 Various curves whose g_3-length is 1.

$$\ell_3(c) = \int_0^1 \frac{a}{y_0 + at}\,dt$$
$$= \ln\left(\frac{y_0 + a}{y_0}\right).$$

Again, vertical curves with this parameterization having fixed g_3-length "get longer" relative to the standard Euclidean metric with increasing y_0.

Example 5.1.8. We give an example of a non-Euclidean metric structure on $U = \mathbf{R}^3$, one that arises naturally in the setting of contact geometry. Using the standard coordinates (x, y, z) on U, let g_4 be described by the matrix

$$G_4 = \begin{bmatrix} 1 & 0 & 0 \\ 0 & 1 + x^2 & x \\ 0 & x & 1 \end{bmatrix}.$$

The matrix G_4 can be seen to be symmetric and positive definite, with $\det(G_4) = 1$ for all (x, y, z); in this way the reader can see that g_4 is a Riemannian metric.

Rather than write out the length integral $\ell_4(c)$ with respect to g_4 in general (which is not complicated, but messy), we will illustrate it for a few specific curves. To that end, let $c_1 : [0, 2\pi] \to \mathbf{R}^3$ be defined by $c_1(t) = (a\cos t, a\sin t, 0)$, a parameterization of a circle of radius a in the xy-plane. In this case, relying on matrix notation for computational purposes, we have

$$g_4(\dot{c_1}(t), \dot{c_1}(t)) = \begin{bmatrix} -a\sin t & a\cos t & 0 \end{bmatrix} \begin{bmatrix} 1 & 0 & 0 \\ 0 & 1 + (a\cos t)^2 & a\cos t \\ 0 & a\cos t & 1 \end{bmatrix} \begin{bmatrix} -a\sin t \\ a\cos t \\ 0 \end{bmatrix}$$
$$= a^2 + a^4\cos^4 t,$$

and so

$$\ell_4(c_1) = \int_0^{2\pi} a\left[1 + a^2\cos^4 t\right]^{1/2}\, dt,$$

an elliptic integral similar to the one encountered in Example 5.1.6.

On the other hand, for $c_2 : [0, 2\pi] \to \mathbf{R}^3$ a circle of the same radius but in the xz-plane parameterized by $c_2(t) = (a\cos t, 0, a\sin t)$, we have

$$g_4(\dot{c_2}(t), \dot{c_2}(t)) = a^2,$$

yielding

$$\ell_4(c_2) = \int_0^{2\pi} a\, dt = 2\pi a,$$

which agrees with the Euclidean length of a circle.

Length is not the only metric quantity that carries over from an inner product space to its smooth, Riemannian counterpart. Translating the notion of the angle between two vectors defined by an inner product (Definition 2.9.7) to the context of the tangent space, the metric tensor allows us to define the notion of the *angle between (nonzero) tangent vectors* $X_p, Y_p \in T_pU$, which we will denote by $\angle_g$:

$$\angle_g(X_p, Y_p) = \cos^{-1}\left(\frac{g_p(X_p, Y_p)}{\left[g_p(X_p, X_p)\right]^{1/2}\left[g_p(Y_p, Y_p)\right]^{1/2}}\right).$$

As before, we say that tangent vectors X_p, Y_p are *orthogonal* if $g_p(X_p, Y_p) = 0$.

The final concept we present in this section is one that we will encounter again in other geometric settings. The following proposition, a smooth analogue of Theorem 2.9.22, shows that the metric tensor induces a smoothly varying isomorphism between the vector space of tangent vectors and the vector space of one-forms. The isomorphism will allow us to define the gradient of a smooth function.

Proposition 5.1.9. *Let (U, g) be a Riemannian space, $\mathcal{X}(U)$ the set of smooth vector fields on U, and $\Lambda_1(U)$ the set of smooth one-forms on U. Then the map $\gamma : \mathcal{X}(U) \to \Lambda_1(U)$ given by $\gamma(X) = i(X)g$ for $X \in \mathcal{X}(U)$, i.e., $\gamma(X)$ is the differential one-form such that for any vector field Y on U,*

$$\big(\gamma(X)\big)(Y) = g(X, Y),$$

induces a vector space isomorphism $\gamma_p : T_pU \to T_p^*U$ for all $p \in U$.

Proof. The fact that $\gamma(p) : T_pU \to T_p^*U$ is a vector space isomorphism for each $p \in U$ is just Theorem 2.9.22. We only point out that our proof in Theorem 2.9.22

that γ_p is onto required the choice of an orthonormal basis. In the present setting, beginning with any basis of smooth vector fields, the Gram–Schmidt process for creating an orthonormal basis is in fact smooth, yielding a basis of vector fields that is orthonormal at each point $p \in U$.

Note that $\gamma(X)$ varies smoothly with p, since both g and X vary smoothly with p. Hence $\gamma(X)$ is a smooth one-form on U. □

We note that *every* smooth $(0, 2)$-tensor A induces a smooth map between $\mathcal{X}(U)$ and $\Lambda_1(U)$ that is linear at the tangent space level. Recall that when the map induced by A is one-to-one for each $p \in U$ (and so, as a linear map between vector spaces of the same dimension, A in fact induces an isomorphism), A is said to be *nondegenerate*. The fact that a Riemannian metric tensor is nondegenerate is a consequence of positive definiteness. We will encounter the concept of nondegeneracy and its consequences many times in the chapters ahead.

We are now in a position to define the gradient of a smooth function, which will depend essentially on the Riemannian metric.

Definition 5.1.10. Let $f : U \to \mathbf{R}$ be a smooth function defined on a Riemannian space (U, g). Let $\gamma : \mathcal{X}(U) \to \Lambda_1(U)$ be the map induced by g from Proposition 5.1.9. Then the *gradient of* f, denoted by ∇f, is the vector field defined by

$$\nabla f = \gamma^{-1}(df).$$

Using matrix notation with coordinates $(x_1, \ldots, x_n)$, we can write

$$\nabla f = G^{-1} \begin{bmatrix} \frac{\partial f}{\partial x_1} \\ \vdots \\ \frac{\partial f}{\partial x_n} \end{bmatrix},$$

where as usual, G is the matrix representation of g.

Example 5.1.11. Let g_0 be the standard Euclidean metric on $\mathbf{R}^n$, i.e.,

$$g_0 = dx_1 \otimes dx_1 + \cdots + dx_n \otimes dx_n,$$

whose matrix representation is the identity matrix,

$$G_0 = I_n = \begin{bmatrix} 1 & 0 & \cdots & 0 \\ & \ddots & & \\ & & \ddots & \\ 0 & \cdots & 0 & 1 \end{bmatrix}.$$

In this case, the gradient takes the familiar form

$$\nabla f = \frac{\partial f}{\partial x_1}\frac{\partial}{\partial x_1} + \cdots + \frac{\partial f}{\partial x_n}\frac{\partial}{\partial x_n},$$

which is the definition presented in a first course in multivariable calculus.

Example 5.1.12. Let g_4 be the metric on $\mathbf{R}^3$ given in Example 5.1.8. The reader can verify that

$$G_4^{-1} = \begin{bmatrix} 1 & 0 & 0 \\ 0 & 1 & -x \\ 0 & -x & 1+x^2 \end{bmatrix}.$$

Hence, using coordinates (x, y, z), the gradient according to g_4 is given by

$$\nabla f = \frac{\partial f}{\partial x}\frac{\partial}{\partial x} + \left(\frac{\partial f}{\partial y} - x\frac{\partial f}{\partial z}\right)\frac{\partial}{\partial y} + \left((1+x^2)\frac{\partial f}{\partial z} - x\frac{\partial f}{\partial y}\right)\frac{\partial}{\partial z}.$$

We close this section by mentioning one property of the gradient of a scalar function; others may be found, for instance, in [1].

Proposition 5.1.13. *Let $f : U \to \mathbf{R}$ be a smooth function on a Riemannian space (U, g). Let ∇f be the gradient of f induced by g, and let ϕ_t be the flow generated by ∇f, so that $\phi_0(p) = p$ and $\frac{d}{dt}\phi_t(p) = \nabla f(\phi_t(p))$. Then f is nondecreasing along the flow lines of ∇f:*

$$f(\phi_t(p)) \geq f(\phi_s(p)) \text{ when } t \geq s.$$

Proof. Using the standard basis for TU, we write

$$\nabla f = X^1\frac{\partial}{\partial x_1} + \cdots + X^n\frac{\partial}{\partial x_n},$$

where $X^i : \mathbf{R}^n \to \mathbf{R}$ are smooth functions. Let

$$h(t) = f(\phi_t(p)) = f(\phi_t^1(p), \ldots, \phi_t^n(p)).$$

Relying on the chain rule, we have

$$\begin{aligned} \frac{dh}{dt} &= \frac{\partial f}{\partial x_1}\frac{d\phi_t^1}{dt} + \cdots + \frac{\partial f}{\partial x_n}\frac{d\phi_t^n}{dt} \\ &= \frac{\partial f}{\partial x_1}X^1 + \cdots + \frac{\partial f}{\partial x_n}X^n \\ &= df(\nabla f) \\ &= g(\nabla f, \nabla f) \geq 0, \end{aligned}$$

where the final inequality is just the positive definiteness of the metric tensor g. Hence h is nondecreasing, which proves the proposition. □

5.2 Constructing Metrics; Metrics on Geometric Sets

In this section, we illustrate two methods of constructing new metric tensors from a given one, both based on constructions outlined in Chap. 4. We then will show how to define Riemannian metrics on geometric sets, in particular on parameterized sets.

The most straightforward way to create a new metric is to multiply an existing one pointwise by a smooth, positive function.

Proposition 5.2.1. *Let (U, g) be a Riemannian space, and let $h : U \to \mathbf{R}$ be a smooth, positive function on U, i.e., $h(p) > 0$ for all $p \in U$. Then $g_h = h \cdot g$ is also a Riemannian metric on U.*

Proof. Exercise. □

For example, the Poincaré metric of Example 5.1.7 is obtained by scaling the standard Euclidean metric g_0 by the positive function $h(x, y) = 1/y^2$. The length computations in that example begin to show the drastic geometric impact that the scaling factor may have. We will see more evidence of this phenomenon in the coming sections.

Certain geometric properties do remain unchanged for two metric tensors related by scaling.

Proposition 5.2.2. *Suppose (U, g) is a Riemannian space. Let $h : U \to \mathbf{R}$ be a smooth positive function, and let $g_h = h \cdot g$ be the Riemannian metric on U as in Proposition 5.2.1. Then for all $p \in U$ and for any two nonzero tangent vectors $X_p, Y_p \in T_pU$, the angle between X_p and Y_p relative to g is the same as the angle between them relative to g_h:*

$$\angle_{g_h}(X_p, Y_p) = \angle_g(X_p, Y_p).$$

In particular, X_p and Y_p are orthogonal relative to g if and only if they are orthogonal relative to g_h.

Proof. The proof is a routine calculation using the definition above:

$$\begin{aligned}
&\frac{(g_h)_p(X_p, Y_p)}{\left[(g_h)_p(X_p, X_p)\right]^{1/2}\left[(g_h)_p(Y_p, Y_p)\right]^{1/2}} \\
&\qquad = \frac{h(p) \cdot g_p(X_p, Y_p)}{\left[h(p) \cdot g_p(X_p, X_p)\right]^{1/2}\left[h(p) \cdot g_p(Y_p, Y_p)\right]^{1/2}} \\
&\qquad = \frac{g_p(X_p, Y_p)}{\left[g_p(X_p, X_p)\right]^{1/2}\left[g_p(Y_p, Y_p)\right]^{1/2}}.
\end{aligned}$$ □

The second method for constructing metrics involves important techniques in differential geometry. We will discuss its importance after introducing the method in the following proposition.

Proposition 5.2.3. *Let (U, g) be a Riemannian space, where $U \subset \mathbf{R}^n$ is a domain. Suppose that for a domain $V \subset \mathbf{R}^k$ ($1 \leq k \leq n$), we are given a smooth, one-to-one function $\phi : V \to U$ that is* regular *in the sense that for all $p \in V$, $(\phi_*)_p$ is one-to-one. Then the pullback $g_\phi = \phi^* g$ of g by ϕ is a Riemannian metric defined on V.*

Proof. The only property of a metric tensor that does not follow immediately from the properties of the pullback is the positive definiteness of g_ϕ. Suppose that for $p \in V$, $X_p \in T_p V$ is such that $g_\phi(X_p, X_p) = 0$. Then, by definition,

$$\begin{aligned} 0 &= g_\phi(X_p, X_p) \\ &= (\phi^* g)(X_p, X_p) \\ &= g(\phi_* X_p, \phi_* X_p), \end{aligned}$$

and so $\phi_* X_p = \mathbf{0}_{\phi(p)}$, since g is positive definite. But since $(\phi_*)_p$ is one-to-one, $X_p = \mathbf{0}_p$. Hence g_ϕ is positive definite, as desired. □

Several of the examples of Riemannian metrics from the preceding section in fact arise naturally from applying Proposition 5.2.3 to the standard Euclidean metric on $\mathbf{R}^3$. We leave it to the reader to check that the maps given in the following two examples are regular, along with the details of the pullback calculations.

Example 5.2.4. Let $\phi_1 : \mathbf{R}^2 \to \mathbf{R}^3$ be defined by

$$\phi_1(u, v) = (u, v, u^2 + v^2),$$

and let g_0 be the Euclidean metric tensor on $\mathbf{R}^3$. Then

$$\begin{aligned} g_1 &= \phi_1^* g_0 \\ &= (1 + 4u^2) du \otimes du + (4uv)\, du \otimes dv + (4uv)\, dv \otimes du + (1 + 4v^2) dv \otimes dv \end{aligned}$$

is a metric tensor on $\mathbf{R}^2$. See Example 5.1.5.

Example 5.2.5. Let $\phi_2 : \mathbf{R}^2 \to \mathbf{R}^3$ be given by

$$\phi_2(u, v) = (u, v, u^2 - v^2),$$

and let g_0 be the standard Euclidean metric tensor on $\mathbf{R}^3$. Then

$$\begin{aligned} g_2 &= \phi_2^* g_0 \\ &= (1 + 4u^2)\, du \otimes du + (-4uv)\, du \otimes dv \\ &\quad + (-4uv)\, dv \otimes du + (1 + 4v^2)\, dv \otimes dv \end{aligned}$$

is a metric tensor on $\mathbf{R}^2$. See Example 5.1.6.

These examples give an insight into a dominant theme in modern differential geometry: the goal of describing geometric structures "intrinsically." The reader might have recognized in Example 5.2.4 a parameterization of the paraboloid

$$S = \{(x, y, z) \in \mathbf{R}^3 \mid z = x^2 + y^2\}.$$

There are two conceivable ways to perform geometric measurements on S and subsets of S. The first approach would be to perform measurements on S as a subset of the "ambient space" $\mathbf{R}^3$, in particular by means of the metric tensor g_0 on $\mathbf{R}^3$. This is the approach of classical differential geometry.

The second approach is to regard S as an object in itself with its "own" geometric structure, without explicitly resorting to the structure of the ambient $\mathbf{R}^3$. This is the "intrinsic" approach mentioned above.

This discussion motivates the following definition, which extends the notion of a metric tensor from domains in Euclidean space to geometric sets. The reader is encouraged to review Corollary 3.3.11.

Definition 5.2.6. Let $S \subset \mathbf{R}^n$ be a parameterized set as in Definition 3.3.2, i.e., there exist a domain $U \subset \mathbf{R}^k$ and a one-to-one, smooth, regular function $\phi : U \to \mathbf{R}^n$ such that $\phi(U) = S$. A Riemannian metric g_S on S will be defined in terms of a Riemannian metric g_U on U: For any two vectors $\mathbf{v}_1, \mathbf{v}_2 \in T_pS$, with $p = \phi(a) \in S$ and $a \in U$, there are unique vectors $\mathbf{u}_1, \mathbf{u}_2 \in T_aU$ such that $(\phi_*)_a(\mathbf{u}_1) = \mathbf{v}_1$ and $(\phi_*)_a(\mathbf{u}_2) = \mathbf{v}_2$. Define

$$(g_S)_p(\mathbf{v}_1, \mathbf{v}_2) = (g_U)_a(\mathbf{u}_1, \mathbf{u}_2).$$

The essence of this definition is that calculations involving the tensor g_S on S are performed not in the ambient space $\mathbf{R}^n$, but in the parameter domain U by means of g_U. We will sometimes blur the distinction between g_S and g_U, referring to them both simply as g.

There is a price to this approach. By shifting the emphasis so decisively to the parameterization of S, it would be easy to conclude that we are not really studying the geometry of S at all, but rather particularities of ϕ or of U. In particular, given two different parameterizations of S, say $\phi_1 : U_1 \to \mathbf{R}^n$ and $\phi_2 : U_2 \to \mathbf{R}^n$ with $\phi_1(U_1) = \phi_2(U_2) = S$, the corresponding metrics g_1 and g_2 might conceivably have very different properties. One of the hallmarks of modern differential geometry, however, is to define geometric properties in such a way that they do not depend on the parameterization, which will in fact be the case for Definition 5.2.6 above. We will return to this subject in Sect. 5.6.

Example 5.2.7. Let $S_1 \subset \mathbf{R}^3$ be the paraboloid defined by $z = x^2 + y^2$. Then the Riemannian metric g_1 from Examples 5.1.5 and 5.2.4 can be considered a Riemannian metric on the paraboloid S_1 itself. Note that circles centered at the origin in the (u, v) parameterization space, given by $u^2 + v^2 = a^2$, correspond to circles of radius a in the plane $z = a^2$, and so their Euclidean length $2\pi a$ matches their g_1-length.

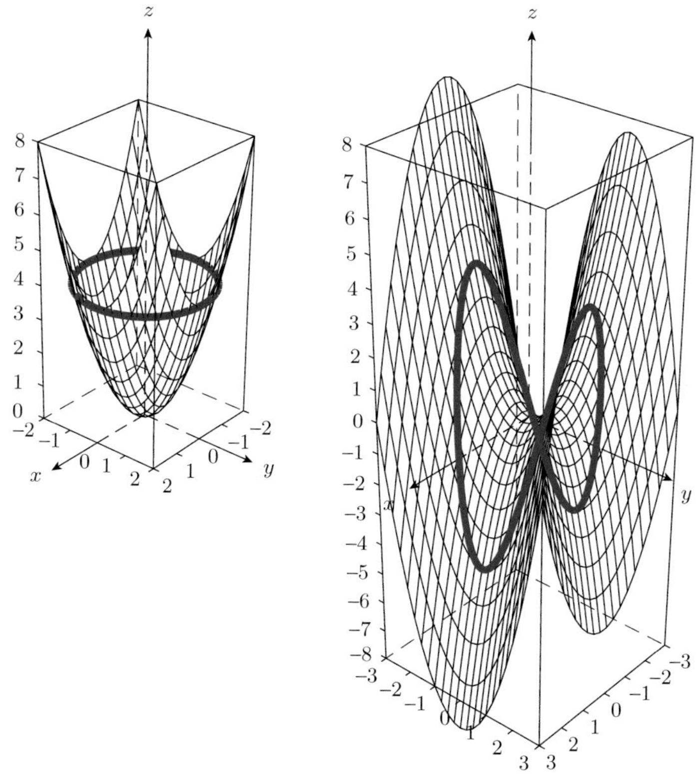

Fig. 5.2 The "circle of radius 2" on the paraboloid and the parabolic hyperboloid, according to the parameterizations of Examples 5.2.4 and 5.2.5.

Example 5.2.8. Let $S_2 \subset \mathbf{R}^3$ be the hyperbolic paraboloid defined by $z = x^2 - y^2$. Then the Riemannian metric g_2 from Examples 5.1.6 and 5.2.5 can be considered to be a metric on S_2 itself. Notice that the g_2-length of the circle $u^2 + v^2 = a^2$ in the parameterization space represents the Euclidean length of a deformed circle, the intersection of the hyperbolic paraboloid with the cylinder $x^2 + y^2 = a^2$ in $\mathbf{R}^3$ whose z-coordinate "height" ranges from $-a^2$ to a^2. See Fig. 5.2.

5.3 The Riemannian Connection

In Riemannian geometry, the covariant derivative is second in importance only to the metric tensor. While the unfolding of this concept is presented in great detail in treatments like [25] or [13], we will try to give a brief sense of the development of this concept, followed by a modern definition and some of the more important properties of this differential-geometric concept.

The notion of the covariant derivative emerges from the following problem: For a vector field X on a surface $S \subset \mathbf{R}^3$, find the "derivative of X in a given direction from the point of view of the surface S." The operating intuition here is that all the geometry defined on S should be defined relative to the tangent space T_pS.

We first try to approach this problem "classically," using the standard Euclidean metric on the "ambient" $\mathbf{R}^3$. We will carry this out by means of an example on the paraboloid described by $z = x^2 + y^2$. Following this example, we will formulate a modern definition of the covariant derivative. We will then show that, at least in the case of the paraboloid, this definition coincides with the more cumbersome classical approach.

Example 5.3.1. Consider the paraboloid $S \subset \mathbf{R}^3$ described by

$$S = \left\{(x, y, z) \mid z = x^2 + y^2\right\},$$

which we describe as a geometric set by means of the parameterization $\phi : \mathbf{R}^2 \to \mathbf{R}^3$ given by

$$\phi(u, v) = (u, v, u^2 + v^2).$$

A vector field X on S will then be given, following Corollary 3.3.11, as the image under ϕ_* of a vector field $\tilde{X}$ on $\mathbf{R}^2$. In other words, for $p = \phi(a) \in S$ with $a \in \mathbf{R}^2$, we have $X(p) \in T_pS$ if and only if there is $\tilde{X}(a) \in T_a\mathbf{R}^2$ such that $(\phi_*)_a(\tilde{X}(a)) = X(p)$.

In more detail, let $\tilde{X}$ be a vector field on $\mathbf{R}^2$. Using matrix notation relative to the standard basis for $T\mathbf{R}^2$, we can write

$$\tilde{X}(u, v) = \begin{bmatrix} X^1(u, v) \\ X^2(u, v) \end{bmatrix} \in T_{(u,v)}\mathbf{R}^2,$$

where $X^1, X^2 : \mathbf{R}^2 \to \mathbf{R}$ are smooth functions. Thus vector fields X on S must have the form

$$\begin{aligned} X(x, y, z) &= (\phi_*)_{(u,v)} \begin{bmatrix} X^1 \\ X^2 \end{bmatrix} \\ &= \begin{bmatrix} 1 & 0 \\ 0 & 1 \\ 2u & 2v \end{bmatrix} \begin{bmatrix} X^1 \\ X^2 \end{bmatrix} \\ &= \begin{bmatrix} X^1 \\ X^2 \\ 2xX^1 + 2yX^2 \end{bmatrix} \in T_{(x,y,z)}\mathbf{R}^3, \end{aligned} \tag{5.1}$$

where $(x, y, z) = \phi(u, v)$, and in particular, $z = u^2 + v^2$.

Consider, for example, two vector fields X, Y on S defined by

$$X(x, y, z) = (\phi_*) \begin{bmatrix} 1 \\ 0 \end{bmatrix} = \begin{bmatrix} 1 \\ 0 \\ 2x \end{bmatrix}$$

and

$$Y(x, y, z) = (\phi_*) \begin{bmatrix} 1 \\ 1 \end{bmatrix} = \begin{bmatrix} 1 \\ 1 \\ 2x + 2y \end{bmatrix}.$$

We are going to imitate somewhat naively the definition of the directional derivative of a function in order to try to define "the derivative of the vector field Y in the direction X."

To that end, let us define $D_X Y$ as the componentwise derivative of Y in the direction X:

$$D_X Y = \begin{bmatrix} X[Y^1] \\ X[Y^2] \\ X[Y^3] \end{bmatrix},$$

which in the example we are considering is a vector field on $\mathbf{R}^3$. In our case,

$$D_X Y = \begin{bmatrix} \left(\frac{\partial}{\partial x} + 2x\frac{\partial}{\partial z}\right)[1] \\ \left(\frac{\partial}{\partial x} + 2x\frac{\partial}{\partial z}\right)[1] \\ \left(\frac{\partial}{\partial x} + 2x\frac{\partial}{\partial z}\right)[2x + 2y] \end{bmatrix}$$

$$= \begin{bmatrix} 0 \\ 0 \\ 2 \end{bmatrix}.$$

The problem with this naive definition is that the resulting vector field $D_X Y$ is not a vector field on S; it does not have the form (5.1) above (Fig.5.3).

In order to adapt this definition, we consider instead the *tangential component* of $D_X Y$, which we denote by $(D_X Y)_T$. To do this, we will compute a unit normal vector to S by the standard methods of vector calculus:

$$\mathbf{n}(x, y, z) = \frac{1}{(1 + 4x^2 + 4y^2)^{1/2}} \begin{bmatrix} -2x \\ -2y \\ 1 \end{bmatrix}.$$

Note that we are relying heavily on the standard Euclidean metric tensor on $\mathbf{R}^3$, since both unit length and orthogonality are metric concepts.

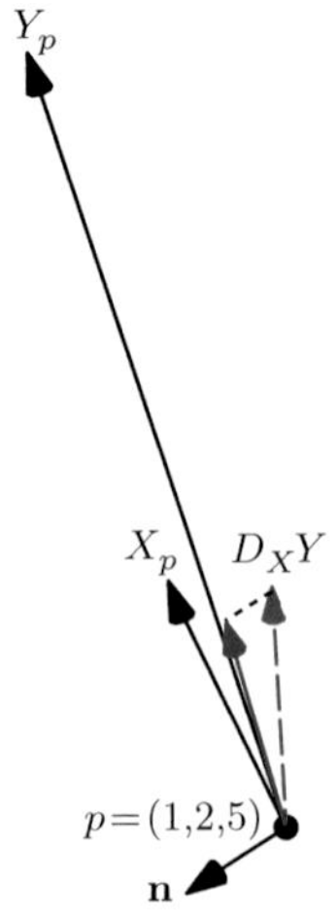

Fig. 5.3 The covariant derivative as a tangential directional derivative.

We then compute the tangential component by subtracting from D_XY its normal component, using "dot product" notation (really the standard Euclidean metric again!):

$$\begin{aligned}(D_XY)_T &= D_XY - (D_XY \cdot \mathbf{n})\,\mathbf{n} \\ &= \begin{bmatrix}0\\0\\2\end{bmatrix} - \left(\frac{2}{(1+4x^2+4y^2)^{1/2}}\right)\begin{bmatrix}\frac{-2x}{(1+4x^2+4y^2)^{1/2}}\\ \frac{-2y}{(1+4x^2+4y^2)^{1/2}}\\ \frac{1}{(1+4x^2+4y^2)^{1/2}}\end{bmatrix} \\ &= \frac{1}{1+4z}\begin{bmatrix}4x\\4y\\8z\end{bmatrix},\end{aligned}$$

where $(x, y, z) \in S$, and so $z = x^2 + y^2$. Note that

$$2x\left(\frac{4x}{1+4z}\right) + 2y\left(\frac{4y}{1+4z}\right) = \frac{8z}{1+4z},$$

i.e., $(D_XY)_T$ satisfies (5.1) above and so is a vector field on S. We call $(D_XY)_T$ the *tangential directional derivative of* Y *in the direction* X.

We have presented this example in detail to illustrate the concept of the "directional derivative of a vector field Y in the direction X from the perspective of a surface" in familiar terms from vector calculus. It also illustrates several features that are awkward from a modern point of view. The first, which we noted in the course of the discussion, was the heavy reliance on the standard Euclidean metric structure in the "ambient" $\mathbf{R}^3$. This flies in the face of the goal of differential geometry to present concepts intrinsically.

There is another important way that the procedure above relies on special features of the ambient $\mathbf{R}^3$. Implicit in the construction above was the reliance on the standard basis $\{\mathbf{e}_1, \mathbf{e}_2, \mathbf{e}_3\}$ of $T_p\mathbf{R}^3$ for $p \in \mathbf{R}^3$, where as usual,

$$(\mathbf{e}_1)_p = \begin{bmatrix} 1 \\ 0 \\ 0 \end{bmatrix}_p, \quad (\mathbf{e}_2)_p = \begin{bmatrix} 0 \\ 1 \\ 0 \end{bmatrix}_p, \quad (\mathbf{e}_3)_p = \begin{bmatrix} 0 \\ 0 \\ 1 \end{bmatrix}_p.$$

Writing $Y = Y^1\mathbf{e}_1 + Y^2\mathbf{e}_2 + Y^3\mathbf{e}_3$, our naive definition of D_XY took the form

$$D_XY = X[Y^1]\mathbf{e}_1 + X[Y^2]\mathbf{e}_2 + X[Y^3]\mathbf{e}_3.$$

On the other hand, if we expect this naive concept to compare favorably to an ordinary differential operator, we would expect a product rule to be at work, where each component would have the form

$$D_X(Y^i\mathbf{e}_i) = X[Y^i\mathbf{e}_i] = X[Y^i]\mathbf{e}_i + Y^iX[\mathbf{e}_i].$$

The two expressions yield the same result as long as we assume that the "directional derivatives" of the basis vector fields $\mathbf{e}_i$ are 0, i.e., that these basis fields are "constant" with respect to every vector field X. It is what would allow us to say that $(\mathbf{e}_i)_p$ is "parallel to" $(\mathbf{e}_i)_q$ when $p \neq q$, despite our insistence in Chap. 3 that $T_p\mathbf{R}^3$ and $T_q\mathbf{R}^3$ are formally different vector spaces.

With this motivation, we now proceed to the central definition of this section.

Definition 5.3.2. Suppose X and Y are smooth vector fields on a Riemannian space (U, g). Let $\theta_Y = \gamma(Y)$ be the one-form corresponding to the vector field Y under the isomorphism γ induced by g defined in Proposition 5.1.9. Construct a new one-form $\theta_{Y,X}$ as follows:

$$\theta_{Y,X} = \frac{1}{2} i(X) \left[\mathcal{L}_Y g + d\theta_Y \right].$$

The *covariant derivative of Y with respect to X* (relative to the metric tensor g), denoted by $\nabla_X Y$, is the vector field

$$\nabla_X Y = \gamma^{-1}(\theta_{Y,X}).$$

The assignment $\nabla : (X, Y) \mapsto \nabla_X Y$ is also known as the *Riemannian connection* corresponding to g.

In other words, $\nabla_X Y$ is the unique vector field satisfying

$$2g(\nabla_X Y, Z) = (\mathcal{L}_Y g)(X, Z) + (d\theta_Y)(X, Z), \tag{5.2}$$

for all smooth vector fields Z.

We will use the phrase "Riemannian connection" interchangeably with the phrase "covariant derivative adapted to the metric g," although we will generally refer to the operator ∇ as the Riemannian connection and the vector field $\nabla_X Y$ as the covariant derivative of Y with respect to X.

Equation (5.2) is often written in an equivalent form, known as *Koszul's formula*:

$$\begin{aligned} 2g(\nabla_X Y, Z) &= X \cdot g(Y, Z) + Y \cdot g(X, Z) - Z \cdot g(X, Y) \\ &\quad - g(\mathcal{L}_X Z, Y) - g(\mathcal{L}_Y Z, X) + g(\mathcal{L}_X Y, Z). \end{aligned} \tag{5.3}$$

We leave the proof of the equivalence of Koszul's formula and Definition 5.3.2 as an exercise for the reader.

Before outlining some properties of this admittedly strange-looking definition, we give a coordinate expression for the covariant derivative relative to a coordinate system $(x_1, \dots, x_n)$ on a domain $U \subset \mathbf{R}^n$. In particular, we write $\partial_i = \frac{\partial}{\partial x_i}$ to denote the standard basis vector fields for TU, and we express the metric g in matrix form $G = [g_{ij}]$ with inverse matrix $G^{-1} = [g^{ij}]$.

Proposition 5.3.3. *Let $X = \sum_{i=1}^{n} X^i \partial_i$ and $Y = \sum_{j=1}^{n} Y^j \partial_j$ be smooth vector fields on U, so that the component functions X^i and Y^j are smooth for all i and j. Then*

$$\nabla_X Y = \sum_{k=1}^{n} \left(\sum_i X^i \frac{\partial Y^k}{\partial x_i} + \sum_{i,j} \Gamma_{ij}^k X^i Y^j \right) \partial_k,$$

where the n^3 functions $\Gamma_{ij}^k : \mathbf{R}^n \to \mathbf{R}$ are given by

$$\Gamma_{ij}^k = \frac{1}{2} \left[\sum_{m=1}^{n} \left(\frac{\partial g_{jm}}{\partial x_i} + \frac{\partial g_{im}}{\partial x_j} - \frac{\partial g_{ij}}{\partial x_m} \right) g^{mk} \right]. \tag{5.4}$$

The functions Γ_{ij}^k are called the *Christoffel symbols* for the Riemannian connection ∇.

We take advantage of the proof of this proposition in order to introduce a basic convention of tensor calculus known as the *Einstein summation convention*. According to this convention, the components of k-forms are denoted with "lowered" indices, while components of vector fields are denoted with "raised" indices. Moreover, when the same index appears "above" and "below" in the same term, it is assumed that there is a summation on the repeated variable. For example, we write

$$X = X^i \partial_i$$

to represent

$$X = \sum_{i=1}^{n} X^i \partial_i.$$

The action of a one-form $\alpha = \alpha_i dx^i$ (i.e., $\alpha = \sum_i \alpha_i dx^i$) on a vector field $X = X^j \partial_j$ is given by

$$\alpha(X) = \alpha_i X^i.$$

As a final example, matrix multiplication can be expressed in Einstein summation notation as follows: Let $A = [a^i_j]$ be an $m \times r$ matrix, so that a^i_j is the entry of the ith row and jth column, and let $B = [b^r_s]$ be an $r \times n$ matrix. Then $C = AB = [c^k_l]$ is given by

$$c^k_l = a^k_t \cdot b^t_l,$$

where $k = 1, \ldots, m$, $t = 1, \ldots, r$, and $l = 1, \ldots, n$.

We also note that in Einstein summation notation, summed indices are named (and renamed) arbitrarily, so that $X^i \partial_i$ and $X^j \partial_j$ have exactly the same meaning.

Proof (of Proposition 5.3.3).

In Exercise 4.35 of Chap. 4, we asked the reader to show that if the metric g is expressed as

$$g = g_{ij} dx^i \otimes dx^j,$$

then $\mathcal{L}_Y g = h_{ij} dx^i \otimes dx^j$, where

$$h_{ij} = Y^k \frac{\partial g_{ij}}{\partial x_k} + g_{kj} \frac{\partial Y^k}{\partial x_i} + g_{ik} \frac{\partial Y^k}{\partial x_j}.$$

This gives, on the one hand,

$$\begin{aligned} i(X)\mathcal{L}_Y g &= (h_{ij} X^i) dx^j \\ &= \left(X^i Y^k \frac{\partial g_{ij}}{\partial x_k} + X^i g_{kj} \frac{\partial Y^k}{\partial x_i} + X^i g_{ik} \frac{\partial Y^k}{\partial x_j} \right) dx^j. \end{aligned} \tag{5.5}$$

Turning to the other term in (5.2), we have

$$\theta_Y = i(Y)g = g_{ij} Y^i dx^j,$$

and so

$$d\theta_Y = \left(\frac{\partial g_{ij}}{\partial x^k} Y^i + g_{ij} \frac{\partial Y^i}{\partial x^k} \right) dx^k \wedge dx^j.$$

Hence

$$
\begin{aligned}
i(X)d\theta_Y &= \left(\frac{\partial g_{ij}}{\partial x^k}Y^i + g_{ij}\frac{\partial Y^i}{\partial x^k}\right)\left(X^k dx^j - X^j dx^k\right) \\
&= X^k\left(\frac{\partial g_{ij}}{\partial x^k}Y^i + g_{ij}\frac{\partial Y^i}{\partial x^k}\right)dx^j \\
&\quad - X^j\left(\frac{\partial g_{ij}}{\partial x^k}Y^i + g_{ij}\frac{\partial Y^i}{\partial x^k}\right)dx^k.
\end{aligned}
$$

We can now reindex the summation indices in the second term by interchanging the indices k and j,

$$
\begin{aligned}
i(X)d\theta_Y &= X^k\left(\frac{\partial g_{ij}}{\partial x^k}Y^i + g_{ij}\frac{\partial Y^i}{\partial x^k}\right)dx^j \\
&\quad - X^k\left(\frac{\partial g_{ik}}{\partial x^j}Y^i + g_{ik}\frac{\partial Y^i}{\partial x^j}\right)dx^j \\
&= \left(X^k\frac{\partial Y^i}{\partial x_k}g_{ij} + Y^i X^k\frac{\partial g_{ij}}{\partial x_k}\right. \qquad (5.6) \\
&\quad \left. - X^k\frac{\partial Y^i}{\partial x_j}g_{ik} - Y^i X^k\frac{\partial g_{ik}}{\partial x_j}\right)dx^j.
\end{aligned}
$$

We now compare the expressions in (5.5) and (5.6) in order to evaluate the expression (5.2) defining the covariant derivative, combining terms that are linear in the X^k as well as the quadratic terms of the form X^iY^k, both after reindexing appropriately. We obtain

$$
\begin{aligned}
\theta_{Y,X} &= \frac{1}{2}\left[i(X)\left(\mathcal{L}_Y g\right) + i(X)d\theta_Y\right] \\
&= \frac{1}{2}\left(2X^k\frac{\partial Y^i}{\partial x_k}g_{ij} + 2\Gamma_{ik,j}X^iY^k\right)dx^j \\
&= \left(X^k\frac{\partial Y^i}{\partial x_k}g_{ij} + \Gamma_{ik,j}X^iY^k\right)dx^j,
\end{aligned}
$$

where

$$
\Gamma_{ik,j} = \frac{1}{2}\left[\frac{\partial g_{jk}}{\partial x_i} + \frac{\partial g_{ij}}{\partial x_k} - \frac{\partial g_{ik}}{\partial x_j}\right].
$$

Finally, note that

$$
\gamma(Z^i\partial_i) = Z^i g_{ij}dx^j,
$$

and so

$$
\gamma^{-1}(\alpha_j dx^j) = \left(g^{lj}\alpha_j\right)\partial_l.
$$

Hence

$$
\begin{aligned}
\nabla_X Y &= \gamma^{-1}(\theta_{Y,X}) \\
&= \gamma^{-1}\left[\left(X^k \frac{\partial Y^i}{\partial x_k} g_{ij} + \Gamma_{ik,j} X^i Y^k\right) dx^j\right] \\
&= \left[g^{lj}\left(X^k \frac{\partial Y^i}{\partial x_k} g_{ij} + \Gamma_{ik,j} X^i Y^k\right)\right] \partial_l \\
&= \left[X^k \frac{\partial Y^i}{\partial x_k} \delta_i^l + \left(g^{lj}\Gamma_{ik,j}\right) X^i Y^k\right] \partial_l \\
&= \left[X^k \frac{\partial Y^l}{\partial x_k} + \Gamma_{ik}^l X^i Y^k\right] \partial_l,
\end{aligned}
$$

where

$$\Gamma_{ik}^l = g^{lj}\Gamma_{ik,l}.$$

Here we have used the Kronecker delta notation $\delta_i^l = 0$ when $i \neq l$, $\delta_i^l = 1$ when $i = l$; in other words, δ_i^l is the (l, i) entry of the identity matrix I. Hence the equality we have used,

$$g^{lj} g_{ji} = \delta_i^l,$$

is just the Einstein notation for $G^{-1}G = I$. Moreover, the reader can verify the identity, which is also encountered frequently in Einstein notation calculations, that

$$\frac{\partial Y^i}{\partial x_k}\delta_i^l = \frac{\partial Y^l}{\partial x_k}.$$

This proves the proposition up to reindexing. □

We will often rely on the following corollary, which states that the Christoffel symbols are the components of the vector field obtained by taking the covariant derivative of a basis vector field with respect to another basis vector field.

Corollary 5.3.4. *For the standard basis vector fields* $\partial_i = \dfrac{\partial}{\partial x_i}$ *corresponding to coordinates* $(x_1, \ldots, x_n)$ *on* $\mathbf{R}^n$,

$$\nabla_{\partial_i}\partial_j = \sum_k \Gamma_{ij}^k \partial_k.$$

Note also that the Riemannian connection ∇ adapted to g is defined completely in terms of the Riemannian metric. As the coordinate calculation shows, the vector field $\nabla_X Y$ depends on the components of X at the point p, but depends on the components of Y in a neighborhood around p, since the partial derivatives of the Y^j appear in the coordinate expression. The is the first indication of a point we shall make later: The Riemannian connection is not a tensor.

We proceed to illustrate the definition with a number of examples.

Example 5.3.5. Let $\mathbf{R}^n$ be equipped with the standard Euclidean metric tensor, i.e., $[g_{ij}] = I_n$. Since in particular all the components g_{ij} of g are constant, we have, by Eq. (5.4) of Proposition 5.3.3, that

$$\Gamma_{ij}^k = 0$$

for all choices of indices i, j, k. In this case, for smooth vector fields

$$X = \sum X^i \partial_i, \quad Y = \sum Y^j \partial_j,$$

we have

$$\nabla_X Y = \sum_k X[Y^k] \partial_k.$$

Hence in the most basic case of the standard Euclidean metric structure on $\mathbf{R}^n$, the covariant derivative coincides with the "naive" definition of $D_X Y$ that we put forward in Example 5.3.1.

Instead of using the coordinate expression in Proposition 5.3.3, we can compute the Christoffel symbols directly from Definition 5.3.2 along with Corollary 5.3.4, as we illustrate in the following example.

Example 5.3.6. Let g be the metric tensor on $\mathbf{R}^2$ induced by pulling back the standard Euclidean metric tensor on $\mathbf{R}^3$ from the paraboloid, as in Example 5.2.4, i.e.,

$$G = [g_{ij}] = \begin{bmatrix} 1+4u^2 & 4uv \\ 4uv & 1+4v^2 \end{bmatrix}.$$

In the coordinates (u, v), we will denote the basis vector fields

$$U = \frac{\partial}{\partial u}, \quad V = \frac{\partial}{\partial v}.$$

In this notation, Corollary 5.3.4 can be written as

$$\nabla_U U = \Gamma_{11}^1 U + \Gamma_{11}^2 V, \; \nabla_U V = \Gamma_{12}^1 U + \Gamma_{12}^2 V$$

$$\nabla_V U = \Gamma_{21}^1 U + \Gamma_{21}^2 V, \; \nabla_V V = \Gamma_{22}^1 U + \Gamma_{22}^2 V.$$

In order to find Γ_{11}^1 and Γ_{11}^2, we first evaluate $g(\nabla_U U, U)$:

$$\begin{aligned} g(\nabla_U U, U) &= g(\Gamma_{11}^1 U + \Gamma_{11}^2 V, U) \\ &= g(\Gamma_{11}^1 U, U) + g(\Gamma_{11}^2 V, U) \\ &= \Gamma_{11}^1 g(U, U) + \Gamma_{11}^2 g(V, U) \\ &= \Gamma_{11}^1 g_{11} + \Gamma_{11}^2 g_{21} \\ &= \Gamma_{11}^1 (1+4u^2) + \Gamma_{11}^2 (4uv). \end{aligned}$$

On the other hand, turning to Definition 5.3.2, we have

$$g(\nabla_U U, U) = \frac{1}{2}\left[(\mathcal{L}_U g)(U,U) + (d\theta_U)(U,U)\right].$$

Recalling again the result of Exercise 4.35, we have

$$\mathcal{L}_U g = (8u)du \otimes du + (4v)du \otimes dv + (4v)dv \otimes du + (0)dv \otimes dv.$$

Also,

$$\begin{aligned}\theta_U &= i(U)g \\ &= (1+4u^2)du + (4uv)dv,\end{aligned}$$

and so

$$d\theta_U = 4v du \wedge dv.$$

Hence,

$$\begin{aligned}g(\nabla_U U, U) &= \frac{1}{2}\left[(8u) + 0\right] \\ &= 4u.\end{aligned}$$

We have shown so far that

$$\Gamma^1_{11}(1+4u^2) + \Gamma^2_{11}(4uv) = 4u.$$

We invite the reader to perform the analogous calculations for $g(\nabla_U U, V)$ to obtain a second equation involving Γ^1_{11} and Γ^2_{11}, namely

$$\Gamma^1_{11}(4uv) + \Gamma^2_{11}(1+4v^2) = 4v.$$

Note that this system is in fact linear in Γ^1_{11} and Γ^2_{11}:

$$\begin{cases}\Gamma^1_{11}(1+4u^2) + \Gamma^2_{11}(4uv) = 4u, \\ \Gamma^1_{11}(4uv) + \Gamma^2_{11}(1+4v^2) = 4v,\end{cases}$$

which can be written in matrix form as

$$G\begin{bmatrix}\Gamma^1_{11} \\ \Gamma^2_{11}\end{bmatrix} = \begin{bmatrix}4u \\ 4v\end{bmatrix}.$$

Since

$$G^{-1} = \left(\frac{1}{1+4u^2+4v^2}\right)\begin{bmatrix}1+4v^2 & -4uv \\ -4uv & 1+4u^2\end{bmatrix},$$

we can solve the system to obtain

$$\Gamma_{11}^{1} = \frac{4u}{1+4u^2+4v^2}, \qquad \Gamma_{11}^{2} = \frac{4v}{1+4u^2+4v^2}.$$

In a similar way, the reader can verify that $\Gamma_{12}^{1} = \Gamma_{12}^{2} = 0$, $\Gamma_{21}^{1} = \Gamma_{21}^{2} = 0$, and

$$\Gamma_{22}^{1} = \frac{4u}{1+4u^2+4v^2}, \qquad \Gamma_{22}^{2} = \frac{4v}{1+4u^2+4v^2}.$$

We now turn to some of the essential properties of the Riemannian connection.

Theorem 5.3.7. *Let (U,g) be a Riemannian space with corresponding Riemannian connection ∇. Let X and Y be smooth vector fields on U. Then:*

1. *$\nabla_X Y$ is a smooth vector field.*
2. *The assignment*

$$(X,Y) \mapsto \nabla_X Y$$

is linear in the first argument. In other words, given smooth vector fields X_1, X_2 and smooth functions f_1, f_2 on U, we have

$$\nabla_{f_1 X_1 + f_2 X_2} Y = f_1 \nabla_{X_1} Y + f_2 \nabla_{X_2} Y.$$

3. *The assignment*

$$(X,Y) \mapsto \nabla_X Y$$

is additive in the second argument: For smooth vector fields Y_1, Y_2 on U, we have

$$\nabla_X (Y_1 + Y_2) = \nabla_X Y_1 + \nabla_X Y_2.$$

4. *The assignment*

$$(X,Y) \mapsto \nabla_X Y$$

is a derivation (i.e., obeys the Leibniz "product rule") in the second argument: Given a smooth function $f: U \to \mathbf{R}$, we have

$$\nabla_X (fY) = (X[f]) \cdot Y + f \cdot \nabla_X Y.$$

5. *The Riemannian connection is* symmetric *(or* torsion-free*), by which we mean that*

$$\nabla_X Y - \nabla_Y X = \mathcal{L}_X Y.$$

In particular, for basis vector fields ∂_i and ∂_j, we have $\nabla_{\partial_i} \partial_j = \nabla_{\partial_j} \partial_i$, and so $\Gamma_{ij}^{k} = \Gamma_{ji}^{k}$ for all k.

6. *The Riemannian connection is* compatible with the metric *g, by which we mean that*

$$Z[g(X,Y)] = g(\nabla_Z X, Y) + g(X, \nabla_Z Y)$$

for all smooth vector fields X, Y, Z on U.

Proof. Statements (1)–(3) are immediate consequences of Definition 5.3.2. Statement (4) is really a local statement that must be verified for all $p \in U$, and so can be verified using the coordinate expression in Proposition 5.3.3.

Statements (5) and (6) can best be seen using the Koszul formula (5.3). The calculations are similar, so we illustrate only the proof of (5).

For any smooth vector field Z on U, we have

$$\begin{aligned}
2g(\nabla_X Y - \nabla_Y X, Z) &= 2g(\nabla_X Y, Z) - 2g(\nabla_Y X, Z) \\
&= X \cdot g(Y,Z) + Y \cdot g(X,Z) - Z \cdot g(X,Y) \\
&\qquad - g(\mathcal{L}_X Z, Y) - g(\mathcal{L}_Y Z, X) + g(\mathcal{L}_X Y, Z) \\
&\quad - \big[Y \cdot g(X,Z) + X \cdot g(Y,Z) - Z \cdot g(Y,X) \\
&\qquad - g(\mathcal{L}_Y Z, X) - g(\mathcal{L}_X Z, Y) + g(\mathcal{L}_Y X, Z)\big] \\
&= 2g(\mathcal{L}_X Y, Z),
\end{aligned}$$

where we have relied on the symmetry and bilinearity of g and the fact that

$$\mathcal{L}_X Y = -\mathcal{L}_Y X.$$

Hence, by the nondegeneracy of g (see the comment after Proposition 5.1.9),

$$\nabla_X Y - \nabla_Y X = \mathcal{L}_X Y. \qquad \square$$

We note here that many modern differential geometry texts present the covariant derivative in an axiomatic way. Namely, a Riemannian connection is *defined* as an assignment

$$(X,Y) \mapsto \nabla_X Y$$

satisfying properties (1)–(6) of Theorem 5.3.7. The existence of such an assignment is then proved by showing that the formula of Definition 5.3.2 does in fact satisfy the required properties. One of the advantages of the more axiomatic definition is that it points to a more general mathematical structure on a domain, an *affine connection*, which is independent of the metric structure. An affine connection is defined to be an assignment that satisfies statements (1)–(4) of Theorem 5.3.7. Doing so provides a general framework for developing a range of differential-geometric structures. This approach is epitomized in the classic (but advanced) work of Kobayashi and Nomizu [24].

In the more general context of an affine connection, one defines a $(1,2)$-tensor quantity

$$T(X,Y) = \nabla_X Y - \nabla_Y X - \mathcal{L}_X Y,$$

called the *torsion*. (Note that the assignment $(X,Y) \mapsto \nabla_X Y$ is not tensorial, i.e., multilinear, in the Y component due to statement (4) of Theorem 5.3.7.) It is then traditional to prove that while there are many possible affine connections defined on a domain U, there is a *unique* affine connection that is torsion-free and compatible with a given metric tensor g, i.e., satisfying statements (5) and (6) of Theorem 5.3.7.

Let us now return to Example 5.3.1 from the beginning of the section, which served to motivate the concepts involved in terms of a tangential directional derivative of a vector field.

Example 5.3.8. Consider $\mathbf{R}^2$ with the metric g obtained by the pullback of the standard Euclidean metric tensor on $\mathbf{R}^3$ from the paraboloid, discussed in Example 5.2.4. We have

$$[g_{ij}] = \begin{bmatrix} 1+4u^2 & 4uv \\ 4uv & 1+4v^2 \end{bmatrix}.$$

Consider the constant vector fields on $\mathbf{R}^2$,

$$\tilde{X} = \frac{\partial}{\partial u} = \begin{bmatrix} 1 \\ 0 \end{bmatrix}$$

and

$$\tilde{Y} = \frac{\partial}{\partial u} + \frac{\partial}{\partial v} = \begin{bmatrix} 1 \\ 1 \end{bmatrix}.$$

We will now use the results of Proposition 5.3.3 and Example 5.3.6 to compute $\nabla_{\tilde{X}}\tilde{Y}$. We have

$$\begin{aligned} \nabla_{\tilde{X}}\tilde{Y} = & \Big[\tilde{X}[\tilde{Y}^1] + \Gamma^1_{11}(\tilde{X}^1)(\tilde{Y}^1) + \Gamma^1_{12}(\tilde{X}^1)(\tilde{Y}^2) \\ & + \Gamma^1_{21}(\tilde{X}^2)(\tilde{Y}^1) + \Gamma^1_{22}(\tilde{X}^2)(\tilde{Y}^2)\Big]\frac{\partial}{\partial u} \\ & + \Big[\tilde{X}[\tilde{Y}^2] + \Gamma^2_{11}(\tilde{X}^1)(\tilde{Y}^1) + \Gamma^2_{12}(\tilde{X}^1)(\tilde{Y}^2) \\ & + \Gamma^2_{21}(\tilde{X}^2)(\tilde{Y}^1) + \Gamma^2_{22}(\tilde{X}^2)(\tilde{Y}^2)\Big]\frac{\partial}{\partial v} \\ = & \left(\frac{4u}{1+4u^2+4v^2}\right)\frac{\partial}{\partial u} + \left(\frac{4v}{1+4u^2+4v^2}\right)\frac{\partial}{\partial v} \\ = & \frac{1}{1+4u^2+4v^2}\begin{bmatrix} 4u \\ 4v \end{bmatrix}. \end{aligned}$$

Recall from Example 5.2.4 that $g = \phi_1^* g_0$, where g_0 is the standard Euclidean metric on $\mathbf{R}^3$ and

$$\phi_1(u, v) = (u, v, u^2 + v^2)$$

is a parameterization of the paraboloid. We have deliberately chosen $\tilde{X}$ and $\tilde{Y}$ so that

$$X = (\phi_1)_*(\tilde{X}) = \begin{bmatrix} 1 \\ 0 \\ 2u \end{bmatrix}, \quad Y = (\phi_1)_*(\tilde{Y}) = \begin{bmatrix} 1 \\ 1 \\ 2u + 2v \end{bmatrix},$$

correspond to the vector fields we used for illustration in Example 5.3.1.

We have

$$\begin{aligned}(\phi_1)_*\left(\nabla_{\tilde{X}}\tilde{Y}\right) &= \begin{bmatrix} 1 & 0 \\ 0 & 1 \\ 2u & 2v \end{bmatrix} \cdot \begin{bmatrix} \frac{4u}{1+4u^2+4v^2} \\ \frac{4v}{1+4u^2+4v^2} \end{bmatrix} \\ &= \frac{1}{1+4u^2+4v^2}\begin{bmatrix} 4u \\ 4v \\ 8u^2+8v^2 \end{bmatrix} \\ &= \frac{1}{1+4z}\begin{bmatrix} 4x \\ 4y \\ 8z \end{bmatrix},\end{aligned}$$

where $x = u$, $y = v$, and $z = u^2 + v^2$. Note that

$$(\phi_1)_*\left(\nabla_{\tilde{X}}\tilde{Y}\right) = (D_X Y)_T,$$

using the notation of Example 5.3.1. We will return to this observation in Sect. 5.6.

At least for the chosen example, the covariant derivative corresponds to the tangential directional derivative defined earlier.

In Exercise 5.13, we ask the reader to show that the previous example is a special case of calculations for a surface described as the graph of a function $f : \mathbf{R}^2 \to \mathbf{R}$.

We conclude this section by extending the construction of the covariant derivative as the tangential directional derivative of a vector field, with a vector field being understood as a $(1,0)$-tensor, to a broader class of tensor fields. Although it is possible to define the covariant derivative of a general (r,s)-tensor, we present here only the definition of the covariant derivative of a $(0,s)$-tensor with respect to a vector field, which will cover the most important cases we will encounter, namely metric tensors and differential forms.

Definition 5.3.9. Let T be a $(0,s)$-tensor field and let X be a vector field, both defined on a domain $U \subset \mathbf{R}^n$. We define the *covariant derivative of* T, denoted by $\nabla_X T$, as the $(0,s)$-tensor field given by

$$(\nabla_X T)(Y_1, \dots, Y_s) = X\left[T(Y_1, \dots, Y_s)\right] - \sum_{i=1}^{s} T(Y_1, \dots, \nabla_X Y_i, \dots, Y_s)$$

for smooth vector fields $Y_1, \dots, Y_s$. Note that we interpret $T(Y_1, \dots, Y_s)$ as a smooth real-valued function $f : U \to \mathbf{R}$ given by

$$f(p) = T_p(Y_1(p), \ldots, Y_s(p)).$$

We leave to the reader to show that $\nabla_X T$ so defined is in fact a $(0,s)$-tensor field.

Example 5.3.10. Considering a smooth function f as a $(0,0)$-tensor, Definition 5.3.9 takes the form

$$\nabla_X f = X[f],$$

showing that ∇_X agrees with the standard directional derivative as well as with the Lie derivative $\mathcal{L}_X$ of a smooth function.

Example 5.3.11. Let α be a one-form. Then Definition 5.3.9 takes the form

$$(\nabla_X \alpha)(Y) = X[\alpha(Y)] - \alpha(\nabla_X Y).$$

Example 5.3.12. In the particular case of the metric $(0,2)$-tensor g to which ∇ corresponds, the compatibility condition (6) of Theorem 5.3.7 implies that

$$\begin{aligned}(\nabla_X g)(Y,Z) &= X[g(Y,Z)] - g(\nabla_X Y, Z) - g(Y, \nabla_X Z)\\ &= 0.\end{aligned}$$

Like any good quantity with the name "derivative," the covariant derivative of tensor fields obeys a kind of product rule with respect to the tensor product.

Proposition 5.3.13. *Let S be a $(0,r)$-tensor and T a $(0,s)$-tensor. Then for a smooth vector field X,*

$$\nabla_X(S \otimes T) = (\nabla_X S) \otimes T + S \otimes (\nabla_X T).$$

Proof. Consider the r vector fields $Y_1, \ldots, Y_r$ and the s vector fields $Z_1, \ldots, Z_s$. Unraveling the definition, we have

$$\begin{aligned}(\nabla_X(S\otimes T))\,(Y_1,\ldots,Y_r,Z_1,\ldots,Z_s) &=\\ X\left[(S\otimes T)(Y_1,\ldots,Y_r,Z_1,\ldots,Z_s)\right]&\\ -\sum_{i=1}^{r}(S\otimes T)(Y_1,\ldots,\nabla_X Y_i,\ldots,Y_r,Z_1,\ldots,Z_s)&\\ -\sum_{j=1}^{s}(S\otimes T)(Y_1,\ldots,Y_r,Z_1,\ldots,\nabla_X Z_j,\ldots,Z_s)&\\ = X\left[S(Y_1,\ldots,Y_r)\cdot T(Z_1,\ldots,Z_s)\right]&\\ -\sum_{i=1}^{r} S(Y_1,\ldots,\nabla_X Y_i,\ldots,Y_r)T(Z_1,\ldots,Z_s)&\\ -\sum_{j=1}^{s} S(Y_1,\ldots,Y_r)T(Z_1,\ldots,\nabla_X Z_j,\ldots,Z_s)&\end{aligned}$$

$$
\begin{aligned}
&= X\left[S(Y_1,\ldots,Y_r)\right]\cdot T(Z_1,\ldots,Z_s)\\
&\quad + S(Y_1,\ldots,Y_r)\cdot X\left[T(Z_1,\ldots,Z_s)\right]\\
&\quad - \sum_{i=1}^{r} S(Y_1,\ldots,\nabla_X Y_i,\ldots,Y_r)T(Z_1,\ldots,Z_s)\\
&\quad - \sum_{j=1}^{s} S(Y_1,\ldots,Y_r)T(Z_1,\ldots,\nabla_X Z_j,\ldots,Z_s)\\
&= \Bigg(X\left[S(Y_1,\ldots,Y_r)\right]\\
&\quad - \sum_{i=1}^{r} S(Y_1,\ldots,\nabla_X Y_i,\ldots,Y_r)\Bigg)\cdot T(Z_1,\ldots,Z_s)\\
&\quad + S(Y_1,\ldots,Y_r)\cdot\Bigg(X\left[T(Z_1,\ldots,Z_s)\right]\\
&\quad - \sum_{j=1}^{s} T(Z_1,\ldots,\nabla_X Z_j,\ldots,Z_s)\Bigg)\\
&= \left[(\nabla_X S)\otimes T + S\otimes(\nabla_X T)\right](Y_1,\ldots,Y_r,Z_1,\ldots,Z_s). \qquad \square
\end{aligned}
$$

The following theorem, which we state for reference, relates the covariant derivative to the two other types of derivative we have seen so far, namely the Lie derivative and the exterior derivative. It is really just a version of Propositions 4.7.16 and 4.7.22 in light of the symmetry condition $\nabla_X Y - \nabla_Y X = \mathcal{L}_X Y$.

Theorem 5.3.14. *Let (U, g) be a Riemannian space with Riemannian connection ∇. Let X be a smooth vector field, T a $(0, r)$-tensor, and ω a differential r-form. Then for any vector fields $Y_1, \ldots, Y_r$:*

1. *The Lie derivative can be expressed in terms of the covariant derivative:*

$$(\mathcal{L}_X T)(Y_1,\ldots,Y_r) = (\nabla_X T)(Y_1,\ldots,Y_r) + \sum_{i=1}^{r} T(Y_1,\ldots,\nabla_{Y_i} X,\ldots,Y_r).$$

2. *The exterior derivative can be expressed in terms of the covariant derivative:*

$$d\omega(Y_0,Y_1,\ldots,Y_r) = \sum_{i=0}^{r}(-1)^i(\nabla_{Y_i}\omega)(Y_0,\ldots,\hat{Y}_i,\ldots,Y_r).$$

Proof. Exercise. $\square$

In fact, the definition of the Riemannian connection, depending heavily on the Riemannian metric, has been made in such a way to ensure compatibility with the Lie derivative and the exterior derivative, two concepts whose definitions do not depend on the metric tensor at all. The compatibility that is expressed in Theorem 5.3.14 can be considered a major consequence of the compatibility conditions (5) and (6) of Theorem 5.3.7.

5.4 Parallelism and Geodesics

Having introduced two fundamental notions of Riemannian geometry, the Riemannian metric tensor and its associated Riemannian connection, we are now in a position to examine some geometric notions beyond those of length and angle introduced in Sect. 5.1. In particular, we will define geodesics, the analogue of "straight lines" in the context of "curved space." This concept relies heavily on the notion of the covariant derivative introduced in the previous section.

In order to generalize the notion of a line to the setting of a Riemannian space, we first need to have a sense of what qualities of a line we hope to generalize. In fact, as with many fundamental notions, the concept of a line unites a number of seemingly distinct properties. In Euclid's axiomatic geometry, a line is completely described by two distinct points. In this setting, two lines are parallel if they have no point of intersection. In the analytic description of Euclidean geometry, a (nonvertical) line is described by one point and a number, the slope; two distinct lines are parallel if they have the same slope.

In the vector geometry of $\mathbf{R}^3$, a line can be characterized by one point (represented by a position vector) and a unit "direction" vector. Two lines can then be said to be parallel if their direction vectors are the same, up to sign. Key to this notion is the ability to compare the direction vectors at different points in $\mathbf{R}^3$.

Another property of a Euclidean line might be offered by physics. A particle's motion is linear if its acceleration at each point is zero. This is essentially Newton's first law.

Finally, there is another metric property of lines in Euclidean space that seems to be outside all the above conceptions: The shortest distance between two points is the length of the line segment between them. Phrased differently, the shortest path from a point P to a point Q is a line segment.

In this section, we show how several of these different properties can be formulated in a non-Euclidean setting using the Riemannian connection associated to a given Riemannian metric. The Euclidean case, when $[g_{ij}]$ is the identity matrix and $\Gamma_{ij}^k = 0$, will still be the reference case.

For the notions introduced in this section, we will consider vector fields V defined along a curve. Let $c : I \to \mathbf{R}^n$ be a smooth parameterized curve. A *vector field along* c is a map $V : I \to T\mathbf{R}^n$, given by

$$t \mapsto V(t) \in T_{c(t)}\mathbf{R}^n,$$

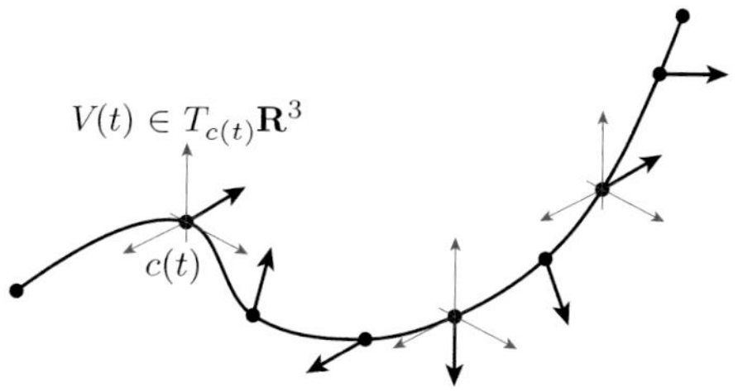

Fig. 5.4 A vector field V along the curve c.

which is smooth in any of the obvious senses from Chap. 3, for example that the component functions $V^i(t)$ relative to a coordinate basis of $T_{c(t)}\mathbf{R}^n$ are smooth functions of t. See Fig. 5.4.

Such might be the case, for example, for a vector field V defined on $\mathbf{R}^n$ and then restricted to a parameterized curve $c : I \to \mathbf{R}^n$, i.e., $V(t) = V(c(t))$ for all $t \in I$. However, vector fields along a curve need not arise in this way. For a smooth parameterized curve $c : I \to \mathbf{R}^n$, the *velocity vector field* $\dot{c}(t) = (c_*)\left(\frac{d}{dt}\right)$ (also denoted by $\dfrac{dc}{dt}$), where $\frac{d}{dt}$ is the standard basis vector field on $T\mathbf{R}^1$, is a case in point. In coordinates,

$$\dot{c}(t) = \sum_{i=1}^{n} \frac{dc_i}{dt}\frac{\partial}{\partial x_i},$$

where $c(t) = (c_1(t), \ldots, c_n(t))$. This vector field is not defined for points not on $c(I)$.

It is not hard to see that the definition of the covariant derivative $\nabla_X Y$ extends directly to vector fields X, Y along a curve c. With this in mind, we define the *derivative of a vector field V along the curve c* to be

$$\frac{DV}{dt} = \nabla_{\dot{c}(t)} V. \tag{5.7}$$

Using Proposition 5.3.3 and taking into account the chain rule, we obtain

$$\frac{DV}{dt} = \sum_{k=1}^{n} \left(\frac{dV^k}{dt} + \sum_{i,j} \left(\Gamma_{ij}^k \cdot \frac{dc_i}{dt} \cdot V^j \right) \right) \frac{\partial}{\partial x_k},$$

where $V(t) = \sum V^i \frac{\partial}{\partial x_i}$. Note that when V is a vector field on $\mathbf{R}^n$ restricted to a curve $c : I \to \mathbf{R}^n$, we have $\dfrac{dV^k}{dt} = \sum \dfrac{\partial V^k}{\partial x_i}\dfrac{dx_i}{dt}$,

Definition 5.4.1. Let V be a vector field along a parameterized curve $c : I \to \mathbf{R}^n$. Then V is *parallel along c* if $\dfrac{DV}{dt} = 0$ for all $t \in I$.

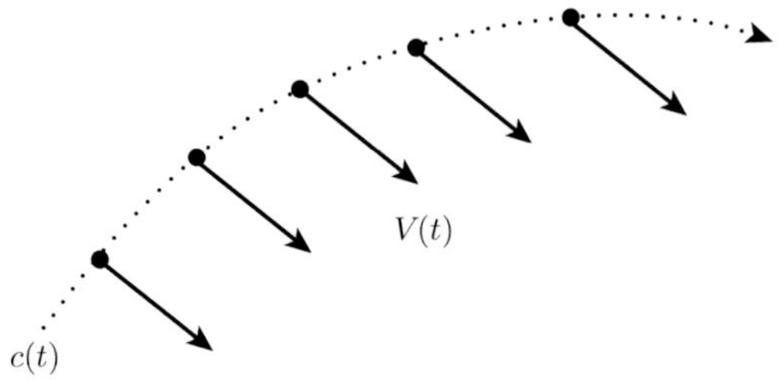

Fig. 5.5 A parallel vector field V along a curve c with the standard Euclidean metric.

The terminology is suggestive here: V is parallel along c if it is "constant along c." We will see shortly that this version of "constant" is closely related to the geometry and, specifically, to the metric tensor.

The condition for V to be parallel along c can be expressed, using coordinates, by saying that the components V^i of V are solutions to the first-order system of differential equations

$$\begin{cases} \frac{dV^1}{dt} + \sum_{i,j} \Gamma^1_{ij} \frac{dc_i}{dt} V^j = 0, \\ \vdots \\ \frac{dV^n}{dt} + \sum_{i,j} \Gamma^n_{ij} \frac{dc_i}{dt} V^j = 0. \end{cases} \tag{5.8}$$

We illustrate the impact of the metric tensor on this notion of parallelism with two examples.

Example 5.4.2. Let $\mathbf{R}^2$ be endowed with the standard Euclidean metric g_0 with the corresponding Riemannian connection $\Gamma^k_{ij} = 0$ for $i, j, k = 1, 2$. Then the system (5.8) takes the form

$$\begin{cases} \frac{dV^1}{dt} = 0, \\ \frac{dV^2}{dt} = 0, \end{cases}$$

i.e., V is a constant vector field in the standard sense that

$$V(t) = k^1 \frac{\partial}{\partial x_1} + k^2 \frac{\partial}{\partial x_2} \quad \text{for all } t \in \mathbf{R}.$$

Note that in the Euclidean case, the notion of parallelism is really independent of the curve c: For any parameterized curve c, V is parallel along c if and only if V is constant. See Fig.5.5.

Example 5.4.3. Let $U = \{(x, y) \mid y > 0\} \subset \mathbf{R}^2$ be the upper half-plane endowed with the Poincaré metric tensor g_3 described by the matrix $\begin{bmatrix} 1/y^2 & 0 \\ 0 & 1/y^2 \end{bmatrix}$. Let $c : I \to \mathrm{U}$ given by $c(t) = (c_1(t), c_2(t))$ be a parameterized curve. Then using the

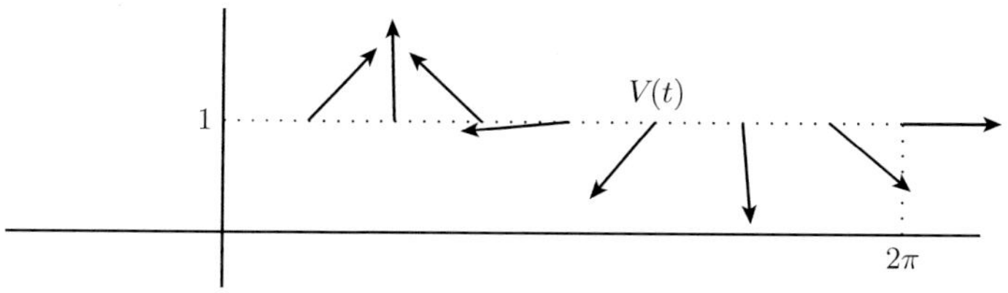

Fig. 5.6 A parallel vector field along a line segment, relative to the Poincaré metric on the upper half-plane.

calculation of the Christoffel symbols in Exercise 5.9, the system (5.8) takes the form

$$\begin{cases} \frac{dV^1}{dt} - \frac{1}{c_2(t)} V^1 \frac{dc_2}{dt} - \frac{1}{c_2(t)} V^2 \frac{dc_1}{dt} = 0, \\ \\ \frac{dV^2}{dt} + \frac{1}{c_2(t)} V^1 \frac{dc_1}{dt} - \frac{1}{c_2(t)} V^2 \frac{dc_2}{dt} = 0. \end{cases}$$

The reader can verify, for example, that along the vertical line segment $c : [0,1] \to U$ given by $c(t) = (0, 1+t)$, the constant vector field $V_0 = \frac{\partial}{\partial x}$ is *not* parallel, but that the vector field $V_1 : I \to TU$ given by $V_1(t) = (1+t)\frac{\partial}{\partial x}$ is parallel.

Along the horizontal line segment $c : [0, 2\pi] \to U$ parameterized by $c(t) = (t, 1)$, the reader can verify that the vector field $V : I \to TU$ given by

$$V(t) = (\cos t)\frac{\partial}{\partial x} + (\sin t)\frac{\partial}{\partial y}$$

is parallel. See Fig. 5.6.

There are several consequences of the close relationship between the Riemannian connection and the metric tensor in the context of parallel vector fields.

Proposition 5.4.4. *For a Riemannian space (U, g) with associated Riemannian connection ∇, let V be a parallel vector field along the parameterized curve $c : I \to U$. Then V has constant magnitude along c.*

Proof. Define $h : I \to \mathbf{R}$ by $h(t) = g_{c(t)}(V(t), V(t))$. We will show that $h'(t) = 0$. Let $X : I \to TU$ be defined by

$$X(t) = \frac{dc}{dt}(c(t)) = \sum_{i=1}^{n} \frac{dc_i}{dt}\frac{\partial}{\partial x_i},$$

the tangential vector field along c. By the chain rule, we have

$$\frac{dh}{dt} = X\left[g_{c(t)}(V(t), V(t))\right].$$

However, by property (6) of the Riemannian connection in Theorem 5.3.7,

$$\begin{aligned} X\left[g(V,V)\right] &= g(\nabla_X V, V) + g(V, \nabla_X V) \\ &= 2g(\nabla_X V, V) \\ &= 0, \end{aligned}$$

since V is parallel along c. □

The proof of the following proposition is nearly identical to the previous one.

Proposition 5.4.5. *Let (U,g) be a Riemannian space, and suppose that V and W are two parallel vector fields along the parameterized curve $c : I \to U$. Then the angle between V and W is constant along c.*

The following is a consequence of the existence and uniqueness theorem for systems of first-order linear differential equations. Details of the proof can be found in [13, p. 52].

Theorem 5.4.6. *Let (U,g) be a Riemannian space with corresponding Riemannian connection ∇. Let $c : [0,1] \to U$ be a smooth curve with $c(0) = p$. Let $V_0 \in T_pU$. Then there is a unique parallel vector field V along c such that $V(0) = V_0$. The vector field V is called the* parallel transport *of V_0 along c.*

As is typical in the context of differential equations, explicit computations of parallel transport are often difficult. We illustrate a manageable example here.

Example 5.4.7. Consider $\mathbf{R}^2$ with the metric g_2 given in matrix form by

$$[(g_2)_{ij}] = \begin{bmatrix} 1+4u^2 & 4uv \\ 4uv & 1+4v^2 \end{bmatrix}$$

with the corresponding Christoffel symbols as computed in Example 5.3.6:

$$\begin{aligned} \Gamma^1_{11} &= \Gamma^1_{22} = \frac{4u}{1+4u^2+4v^2}, \\ \Gamma^2_{11} &= \Gamma^2_{22} = \frac{4v}{1+4u^2+4v^2}, \\ \Gamma^1_{12} &= \Gamma^2_{12} = \Gamma^1_{21} = \Gamma^2_{21} = 0. \end{aligned}$$

We will compute the parallel transport V of the tangent vector $V_0 = \left.\frac{\partial}{\partial u}\right|_p$ at the point $p = (0,0)$ along the curve described by $c(t) = (t,t)$ for $t \in [0,1]$.

The system of differential equations in this case takes the form

$$\begin{cases} \dot{V^1} + \frac{4t}{1+8t^2}\left(V^1 + V^2\right) = 0, \\ \dot{V^2} + \frac{4t}{1+8t^2}\left(V^1 + V^2\right) = 0. \end{cases} \tag{5.9}$$

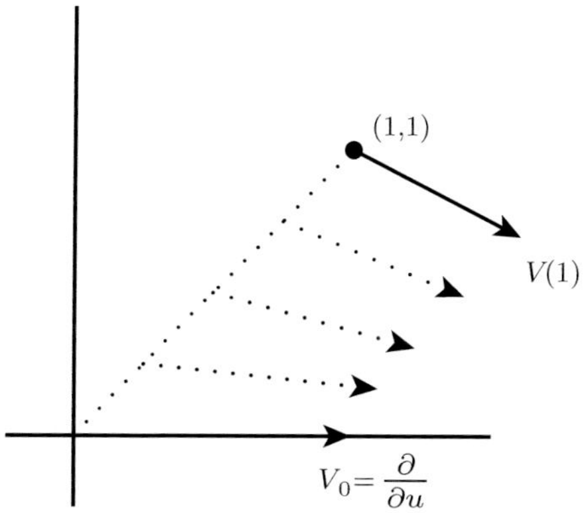

Fig. 5.7 The parallel transport of the tangent vector V_0 along the curve c of Example 5.4.7, relative to the parabolic metric.

Adding these equations gives the separable equation

$$(V^1 + V^2)^{\cdot} = \frac{-8t}{1+8t^2}(V^1 + V^2),$$

which can be solved using initial conditions

$$V^1(0) = 1, \quad V^2(0) = 0$$

to yield

$$V^1 + V^2 = \frac{1}{(1+8t^2)^{1/2}}.$$

Substituting into the system (5.9) gives

$$\begin{cases} \dot{V}^1 + \frac{4t}{(1+8t^2)^{3/2}} = 0, \\ \dot{V}^2 + \frac{4t}{(1+8t^2)^{3/2}} = 0. \end{cases}$$

Integrating and again using the initial conditions gives

$$V(t) = \frac{1}{2}\left(\frac{1}{(1+8t^2)^{1/2}} + 1\right)\frac{\partial}{\partial u} + \frac{1}{2}\left(\frac{1}{(1+8t^2)^{1/2}} - 1\right)\frac{\partial}{\partial v}.$$

See Fig. 5.7.

In the above example, we might abuse terminology by saying that $V(0) = \frac{\partial}{\partial u} \in T_{(0,0)}\mathbf{R}^2$ is parallel to $V(1) = \frac{2}{3}\frac{\partial}{\partial u} - \frac{1}{3}\frac{\partial}{\partial v} \in T_{(1,1)}\mathbf{R}^2$ along c, instead of the more cumbersome but correct "$V(1)$ is obtained by parallel translation of $V(0)$ along the curve c." It is exactly in this sense that the Riemannian connection gives a way of comparing ("connecting") tangent vectors at one point with tangent vectors at another as mentioned in the introduction to this section. In particular, in the case of the Euclidean connection, parallel transport is what allows us to freely transport tangent vectors from one tangent space to another—since the result will be the same constant vector regardless of the path.

Thus far in this section, our primary object of attention has been a vector field defined on a curve. To say that V is parallel along c is understood as a condition on V, with the curve playing a secondary (although necessary) reference role.

We now turn our attention to certain curves that are well adapted to the Riemannian geometry as determined by the metric tensor. These are the curves that are "self-parallel" in the sense of the preceding definitions of the section.

Definition 5.4.8. Let (U, g) be a Riemannian space with adapted Riemannian connection ∇. A parameterized curve $c : I \to \mathbf{R}^n$ is a *geodesic* if

$$\nabla_{\dot{c}}\dot{c} = 0.$$

Using the notation introduced earlier in this section, c is a geodesic if

$$\frac{D}{dt}(\dot{c}) = 0.$$

This notation suggests that geodesics are curves that have zero "acceleration" in some sense, which should remind the reader of the properties of Euclidean lines discussed at the outset of the section.

As with parallel vector fields along a curve, the condition for $c : I \to \mathbf{R}^n$ to be a geodesic amounts to saying that the component functions $(c_1, \dots, c_n)$, $c_i : I \to \mathbf{R}$, satisfy a system of differential equations—this time, a system of second-order, nonlinear differential equations. Explicitly, the system (5.8) applied to the tangential vector field

$$V = \frac{dc}{dt} = \frac{dc_1}{dt}\frac{\partial}{\partial x_1} + \cdots + \frac{dc_n}{dt}\frac{\partial}{\partial x_n}$$

takes the form

$$\begin{cases} \frac{d^2 c_1}{dt^2} + \sum_{i,j} \Gamma^1_{ij} \cdot \frac{dc_i}{dt}\frac{dc_j}{dt} = 0, \\ \quad\vdots \\ \frac{d^2 c_n}{dt^2} + \sum_{i,j} \Gamma^n_{ij} \cdot \frac{dc_i}{dt}\frac{dc_j}{dt} = 0. \end{cases} \tag{5.10}$$

Let us examine geodesics for several of the metric tensors we have encountered so far.

Example 5.4.9. Consider $\mathbf{R}^2$ with the standard Euclidean metric g_0 and Riemannian connection defined by $\Gamma^k_{ij} = 0$ for all i, j, k. Then the system (5.10) takes the form

$$\begin{cases} \frac{d^2 c_1}{dt^2} = 0, \\ \frac{d^2 c_2}{dt^2} = 0, \end{cases}$$

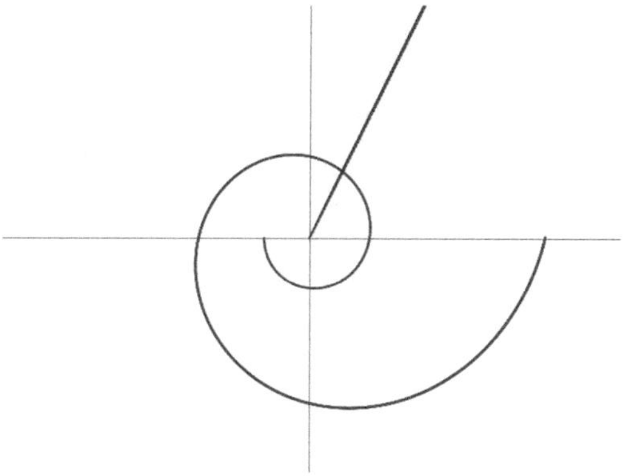

Fig. 5.8 Graphs of geodesics in $\mathbf{R}^2$ with the parabolic metric, created with the computer software 3D-XplorMath-J.

whose solutions all have the form

$$c(t) = (x_0 + at, y_0 + bt)$$

for constants a, b, x_0, y_0. Hence in the Euclidean case, geodesics are lines with the usual (linear) parameterization.

Example 5.4.10. Let us again turn to the metric g_2 on $\mathbf{R}^2$ obtained by pulling back the standard Euclidean metric tensor on $\mathbf{R}^3$ from the paraboloid, as in Examples 5.2.4 and 5.3.6:

$$[(g_2)_{ij}] = \begin{bmatrix} 1+4u^2 & 4uv \\ 4uv & 1+4v^2 \end{bmatrix}$$

with the corresponding Christoffel symbols as computed in Example 5.3.6.

In this case, the system (5.10) takes the form

$$\begin{cases} \ddot{c}_1 + \left(\frac{4c_1}{1+4c_1^2+4c_2^2}\right)(\dot{c_1}^2 + \dot{c_2}^2) = 0 \\ \ddot{c}_2 + \left(\frac{4c_2}{1+4c_1^2+4c_2^2}\right)(\dot{c_1}^2 + \dot{c_2}^2) = 0. \end{cases}$$

Needless to say, producing explicit solutions for this system by means of integration will not be easy. Computer software can create image curves for these geodesics. See Fig. 5.8.

For example, there are geodesics that satisfy the relationship $u = kv$, for any constant k. These lines through the origin in the (u, v)-plane correspond under the parameterization of the paraboloid to the intersection of $z = x^2 + y^2$ with the plane $x = ky$; these are called "meridians" of the paraboloid. Even for these relatively simple geodesics (there are others!), the parameterization is quite complicated.

The reader can verify, for example, that the parameterization for the geodesic corresponding to $u = 0$ is given by $c(t) = (0, c_2(t))$, where c_2 is given implicitly by

$$2c_2\sqrt{1+4(c_2)^2} + \ln\left|2c_2 + \sqrt{1+4(c_2)^2}\right| = k_1 t + k_2,$$

and k_1, k_2 are constants of integration obtained from the initial conditions.

Example 5.4.10, along with Exercise 5.23, shows that the parameterization is essential to the way a geodesic is defined, in the sense that the same image curve may be the image of a geodesic under one parameterization and not under another.

We close this section by listing several geometric properties of geodesics. Precise statements and proofs of these properties would take us too far from the introductory scope of this presentation; the interested reader is encouraged to refer to [13] or [25].

In the following, let (U, g) be a Riemannian space. Let $I = [a, b]$, and recall that for a curve $c : I \to U$, the length of c is defined as

$$\ell(c) = \int_a^b [g(\dot{c}, \dot{c})]^{1/2}\, dt.$$

The first theorem states that geodesics are the shortest paths between two points—as long as the points are "close enough" together. In this further sense, geodesics are the generalization of lines to the context of "curved space" that we discussed in the introduction to this section.

Theorem 5.4.11. *Let (U, g) be a Riemannian space. For each point $p \in U$, there is a domain $V_p \subset U$ containing p that has the following property: If $c_0 : [a, b] \to V_p$ is a geodesic such that $c_0(a) = p$ and $c_0(b) = q$, and if $c_1 : [a, b] \to V_p$ is any other smooth curve satisfying $c_1(a) = p$ and $c_1(b) = q$, then*

$$\ell(c_0) \leq \ell(c_1).$$

In light of this property of geodesics as "length-minimizing" curves, questions of existence and uniqueness are all the more important. The two final theorems of this section address these questions in slightly different ways. The first says that geodesics locally exist in every direction.

Theorem 5.4.12. *Let (U, g) be a Riemannian space. Then for all $p \in U$ and all $\mathbf{v}_0 \in T_pU$, there exist an $\epsilon > 0$ and a unique geodesic $c : (-\epsilon, \epsilon) \to U$ such that $c(0) = p$ and $c'(0) = \mathbf{v}_0$.*

The second theorem is a less straightforward application of the usual existence and uniqueness theorems for ordinary differential equations. It says that for any two points "close enough together," there is a geodesic connecting them. This is closely related to the definition and properties of the "exponential map," an important construction that lies at the heart of many proofs of more advanced theorems in Riemannian geometry. For more details, see [13].

Theorem 5.4.13. *Let (U, g) be a Riemannian space. For every $p \in U$, there exists a domain $V_p \subset U$ containing p such that for every $q \in V_p$, there is a geodesic $c : [a, b] \to V_p$ with $c(a) = p$ and $c(b) = q$.*

5.5 Curvature

A theme of the previous sections has been the extent to which the Riemannian metric tensor is able to intrinsically capture ideas of shape that had in classical differential geometry been described extrinsically, referring to the ambient space with its "natural" or "absolute" metric structure. We have seen how the Riemannian connection, viewed as covariant differentiation, can be seen to measure the extent to which a metric differs from the "flat" standard Euclidean metric (where the Christoffel symbols are identically zero). The adjective "flat" here describes such facts as that geodesics are lines and that parallel translation consists in "translating the base point" of a tangent vector.

The concept that has been lurking beneath all of this is curvature. Since the first efforts to measure curvature numerically in the classical context of curves and surfaces, curvature has come to be measured by means of a tensor. The curvature tensor has the disadvantage of being a complicated quantity. Mikhail Gromov, a prominent figure in late twentieth-/early twenty-first-century differential geometry, called it "a little monster of multilinear algebra whose full geometric meaning remains obscure."[1] It is complicated because it is rich. Much active research in Riemannian geometry centers on the ways in which the curvature tensor determines, and is determined by, the metric (and topological) properties of a geometric space.

In keeping with our theme of presenting geometry as the study of tensor structures on the tangent space, we will not follow the more historical (and more concrete) route of presenting the geometric problems that gave rise to the various concepts associated with the curvature tensor. Rather, we will define the curvature tensor in terms of the Riemannian connection, present some elementary examples, and then outline some of the many geometric measurements contained within the little monster of the curvature tensor.

Up to now, we have generally avoided the use of the Lie bracket notation $[X, Y]$ in favor of the Lie derivative notation $\mathcal{L}_X Y$. However, the Lie bracket notation is standard in the presentation of curvature, so we will use it here. Recall that for smooth vector fields X and Y on a domain $U \subset \mathbf{R}^n$,

$$([X, Y])[f] = (\mathcal{L}_X Y)[f] = X[Y[f]] - Y[X[f]]. \tag{5.11}$$

[1]Quoted in [[10], p. 15].

Furthermore, if $(x_1, \ldots, x_n)$ is a coordinate system for U with corresponding basis vector fields $\partial_i = \dfrac{\partial}{\partial x_i}$ for the tangent space TU, we have

$$[\partial_i, \partial_j] = 0$$

for all i, j. The reader is encouraged to review Sect. 4.7 for the other relevant properties of the Lie bracket (i.e., Lie derivative) of vector fields, which we will use often below.

Definition 5.5.1. Let (U, g) be a Riemannian space with associated Riemannian connection ∇. Let $\mathcal{X}(U)$ denote the set of smooth vector fields on U. The *curvature* of (U, g) is a map R that associates to each pair of vector fields $X, Y \in \mathcal{X}(U)$ the map

$$R(X, Y) : \mathcal{X}(U) \to \mathcal{X}(U)$$

defined by

$$R(X, Y)(Z) = \nabla_X(\nabla_Y Z) - \nabla_Y(\nabla_X Z) - \nabla_{[X,Y]} Z.$$

Key to the definition is the following proposition, stating that R so defined is a tensor, even though the Riemannian connection ∇ is not. In the same sense that a linear map from a vector space V to itself is considered a $(1, 1)$-tensor (see Example 4.6.5), a multilinear map

$$V \times V \times V \to V$$

can be considered as a $(1, 3)$-tensor on the vector space V.

Proposition 5.5.2. *For smooth vector fields X, Y, Z on U, the assignment*

$$(X_p, Y_p, Z_p) \mapsto R(X_p, Y_p)(Z_p)$$

is multilinear for all $p \in U$, and depends smoothly on p. Hence R is a $(1, 3)$-tensor field.

Proof. The smoothness with respect to p follows from the smoothness of X, Y, Z as well as that of ∇ (in the sense of Theorem 5.3.7 (1)). In the remainder of the proof we will suppress the dependance on p, with the understanding that all vector fields in the calculations are being considered at a point p.

The linearity in the first and second components are consequences of Theorem 5.3.7 (2) and Proposition 4.7.6 (3). Hence the only component in which we need to check linearity is the third. The additivity in the third component follows from Theorem 5.3.7 (3). So the only remaining property to check is that

$$R(X, Y)(fZ) = fR(X, Y)(Z)$$

for all smooth functions $f : U \to \mathbf{R}$.

Relying then on Theorem 5.3.7 (4) and (5.11), we have

$$\begin{aligned}
R(X,Y)(fZ) &= \nabla_X(\nabla_Y(fZ)) - \nabla_Y(\nabla_X(fZ)) - \nabla_{[X,Y]}(fZ) \\
&= \nabla_X\left((Y[f])Z + f\nabla_Y Z\right) \\
&\quad - \nabla_Y\left((X[f])Z + f\nabla_X Z\right) \\
&\quad - \Big([X,Y]\,[f]Z + f\nabla_{[X,Y]}Z\Big)
\end{aligned}$$

$$\begin{aligned}
&= (X\,[Y[f]])\,Z + (Y[f])\,\nabla_X Z + (X[f])\,\nabla_Y Z + f\,(\nabla_X(\nabla_Y Z)) \\
&\quad - (Y\,[X[f]])\,Z - (X[f])\,\nabla_Y Z - (Y[f])\,\nabla_X Z \\
&\quad - f\,(\nabla_Y(\nabla_X Z)) - (X\,[Y[f]] - Y\,[X[f]])\,Z \\
&\quad - f\nabla_{[X,Y]}Z \\
&= f\Big(\nabla_X(\nabla_Y Z) - \nabla_Y(\nabla_X Z) - \nabla_{[X,Y]}Z\Big) \\
&= fR(X,Y)(Z). \qquad \square
\end{aligned}$$

It is also common to consider a related tensor quantity, which is obtained from Definition 5.5.1 by means of the Riemannian metric g.

Definition 5.5.3. The *Riemannian curvature tensor* R is the $(0,4)$-tensor defined by

$$R(X,Y,Z,W) = g(R(X,Y)(Z),W)$$

for all vector fields X, Y, Z, W on U.

We leave it as an exercise to verify that R so defined is smooth and multilinear.

There is an inherent confusion of notation in using not only the same notation R, but also the same "curvature" terminology to describe two related but different objects. We will try to adhere to the convention of using the term *curvature* for the assignment of a linear map to pairs of vector fields, which will often be written with two arguments $R(X,Y)$, while the *curvature tensor* will be the assignment of a scalar quantity to a quadruple of vector fields, and it will be written with four arguments $R(X,Y,Z,W)$. It will also be clearer in context using the tensor notation for coordinate calculations, which we present now.

Let $U \subset \mathbf{R}^n$ be a domain with coordinates $(x_1,\ldots,x_n)$ and basis vector fields $\partial_i = \dfrac{\partial}{\partial x_i}$. Let g be a Riemannian metric on U described by the matrix $[g_{ij}]$, with associated Riemannian connection ∇ given by Christoffel symbols Γ_{ij}^k.

Note that for vector fields $X = \sum X^i\partial_i$, $Y = \sum Y^j\partial_j$, and $Z = \sum Z^k\partial_k$, Proposition 5.5.2 implies that

$$\begin{aligned}R(X,Y)(Z) &= \sum_{i,j,k} X^i Y^j Z^k \left(R(\partial_i,\partial_j)(\partial_k)\right)\\&= \sum_{l=1}^{n}\left(\sum_{i,j,k} X^i Y^j Z^k R^l_{ijk}\right)\partial_l,\end{aligned}$$

where by definition, R^l_{ijk} is the l-component of the vector field $R(\partial_i,\partial_j)(\partial_k)$:

$$R(\partial_i,\partial_j)(\partial_k) = \sum_{l=1}^{n} R^l_{ijk}\partial_l.$$

Proposition 5.5.4. *Let (U,g) be a Riemannian space whose associated connection is described by the Christoffel symbols Γ^k_{ij}. Then the components of the $(1,3)$ curvature tensor are given by*

$$R^l_{ijk} = \frac{\partial}{\partial x_i}\left[\Gamma^l_{jk}\right] - \frac{\partial}{\partial x_j}\left[\Gamma^l_{ik}\right] + \sum_{s=1}^{n}\left(\Gamma^s_{jk}\Gamma^l_{is} - \Gamma^s_{ik}\Gamma^l_{js}\right).$$

Furthermore, for $R_{ijkl} = R(\partial_i,\partial_j,\partial_k,\partial_l)$, we have

$$R_{ijkl} = \sum_{t=1}^{n} R^t_{ijk} g_{tl}.$$

Proof. We apply Theorem 5.3.7 and the fact that $[\partial_i,\partial_j] = 0$ for all i,j to the curvature tensor defined in Definition 5.5.1. As in the proof of Proposition 5.3.3, we rely on the Einstein summation convention of summing over repeated upper–lower indices. In particular, we will alternate freely between summation indices s and l below:

$$\begin{aligned}R(\partial_i,\partial_j)(\partial_k) &= \nabla_{\partial_i}\left(\nabla_{\partial_j}\partial_k\right) - \nabla_{\partial_j}\left(\nabla_{\partial_i}\partial_k\right) - \nabla_{[\partial_i,\partial_j]}\partial_k\\&= \nabla_{\partial_i}\left(\Gamma^s_{jk}\partial_s\right) - \nabla_{\partial_j}\left(\Gamma^s_{ik}\partial_s\right)\\&= \left(\frac{\partial}{\partial x_i}\left[\Gamma^l_{jk}\right]\partial_l + \Gamma^s_{jk}\nabla_{\partial_i}\partial_s\right) - \left(\frac{\partial}{\partial x_j}\left[\Gamma^l_{ik}\right]\partial_l + \Gamma^s_{ik}\nabla_{\partial_j}\partial_s\right)\end{aligned}$$

$$\begin{aligned}&= \frac{\partial}{\partial x_i}\left[\Gamma^l_{jk}\right]\partial_l + \Gamma^s_{jk}\Gamma^l_{is}\partial_l - \frac{\partial}{\partial x_j}\left[\Gamma^l_{ik}\right]\partial_l - \Gamma^s_{ik}\Gamma^l_{js}\partial_l\\&= \left(\frac{\partial}{\partial x_i}\left[\Gamma^l_{jk}\right] - \frac{\partial}{\partial x_j}\left[\Gamma^l_{ik}\right] + \Gamma^s_{jk}\Gamma^l_{is} - \Gamma^s_{ik}\Gamma^l_{js}\right)\partial_l.\end{aligned}$$

Hence

$$R^l_{ijk} = \frac{\partial}{\partial x_i}\left[\Gamma^l_{jk}\right] - \frac{\partial}{\partial x_j}\left[\Gamma^l_{ik}\right] + \Gamma^s_{jk}\Gamma^l_{is} - \Gamma^s_{ik}\Gamma^l_{js}.$$

The components of the curvature tensor follow immediately from the bilinearity of g:

$$\begin{aligned} R(\partial_i, \partial_j, \partial_k, \partial_l) &= g(R_{ijk}^t \partial_t, \partial_l) \\ &= R_{ijk}^t g(\partial_t, \partial_l) \\ &= R_{ijk}^t g_{tl}. \qquad \square \end{aligned}$$

Before turning to examples, we point out that on its face, computing the curvature tensor is even more formidable than computing the covariant derivative. In general, the curvature requires specifying n^4 components, so even in $\mathbf{R}^2$, Proposition 5.5.4 requires 16 computations (compared to the 8 required to compute the Christoffel symbols for the Riemannian connection). These difficulties in computation might be overwhelming were it not for the many symmetries and relations between components that are concealed in the definition. We list some of them now.

Proposition 5.5.5. *Let R be the curvature associated with a Riemannian metric g and Riemannian connection ∇. Let X, Y, Z, W be smooth vector fields. Then:*

1. $R(X, X) \equiv 0$. *In particular,* $R_{iik}^l = 0$.
2. $R(X, X, Z, W) = 0$. *In particular,* $R_{iikl} = 0$.
3. $R(X, Y, Z, Z) = 0$. *In particular,* $R_{ijkk} = 0$.
4. $R(X, Y)Z + R(Y, Z)X + R(Z, X)Y = 0$. *In particular,*

$$R_{ijk}^l + R_{jki}^l + R_{kij}^l = 0.$$

We note that (1) is an equality of linear maps, (2) and (3) are scalar equalities, and (4) is an equality of vector fields.

Proof. Statements (1) and (2), which are in fact equivalent, follow immediately from Definition 5.5.1, keeping in mind that $[X, X] = 0$ for all vector fields X.

To prove (3), we rely on the compatibility of ∇ with g, expressed in Theorem 5.3.7(6). We have

$$\begin{aligned} X\left[Y\left[g(Z, Z)\right]\right] &= X\left[g(\nabla_Y Z, Z) + g(Z, \nabla_Y Z)\right] \\ &= X\left[2g(\nabla_Y Z, Z)\right] \\ &= 2\left(g(\nabla_X(\nabla_Y Z), Z) + g(\nabla_Y Z, \nabla_X Z)\right). \end{aligned} \tag{5.12}$$

Likewise, we have

$$Y\left[X\left[g(Z, Z)\right]\right] = 2\left(g(\nabla_Y(\nabla_X Z), Z) + g(\nabla_X Z, \nabla_Y Z)\right) \tag{5.13}$$

and

$$[X, Y]\left(g(Z, Z)\right) = 2g\left(\nabla_{[X,Y]} Z, Z\right). \tag{5.14}$$

Using (5.12), (5.13), and (5.14) together with the definition of the Lie bracket, we have

$$\begin{aligned} R(X,Y,Z,Z) &= g(R(X,Y)(Z),Z) \\ &= g(\nabla_X\nabla_Y Z - \nabla_Y\nabla_X Z - \nabla_{[X,Y]}Z, Z) \\ &= g(\nabla_X\nabla_Y Z, Z) - g(\nabla_Y\nabla_X Z, Z) - g(\nabla_{[X,Y]}Z, Z) \\ &= \frac{1}{2}\Big(X\left[Y\left[g(Z,Z)\right]\right] - Y\left[X\left[g(Z,Z)\right]\right] - [X,Y]\left(g(Z,Z)\right)\Big) \\ &= \frac{1}{2}\left(XY - YX - [X,Y]\right)\left[g(Z,Z)\right] \\ &= 0. \end{aligned}$$

To prove (4), we rely repeatedly on the symmetry of ∇, expressed in Theorem 5.3.7 (5), i.e.,

$$\nabla_X Y - \nabla_Y X = [X,Y].$$

We have

$$R(X,Y)(Z) = \nabla_X\nabla_Y Z - \nabla_Y\nabla_X Z - \nabla_{[X,Y]}Z, \tag{5.15}$$

$$\begin{aligned} R(Y,Z)(X) &= \nabla_Y\nabla_Z X - \nabla_Z\nabla_Y X - \nabla_{[Y,Z]}X \\ &= \nabla_Y\left(\nabla_X Z + [Z,X]\right) - \nabla_Z\nabla_Y X - \nabla_{[Y,Z]}X \\ &= \nabla_Y\nabla_X Z + \nabla_Y\left[Z,X\right] - \nabla_Z\nabla_Y X - \nabla_{[Y,Z]}X \\ &= \nabla_Y\nabla_X Z + \nabla_{[Z,X]}Y + [Y,[Z,X]] \\ &\qquad - \nabla_Z\nabla_Y X - \nabla_{[Y,Z]}X, \end{aligned} \tag{5.16}$$

and

$$\begin{aligned} R(Z,X)(Y) &= \nabla_Z\nabla_X Y - \nabla_X\nabla_Z Y - \nabla_{[Z,X]}Y \\ &= \nabla_Z\left(\nabla_Y X + [X,Y]\right) - \nabla_X\left(\nabla_Y Z + [Z,Y]\right) - \nabla_{[Z,X]}Y \\ &= \nabla_Z(\nabla_Y X) + \nabla_Z([X,Y]) - \nabla_X(\nabla_Y Z) - \nabla_X([Z,Y]) - \nabla_{[Z,X]}Y \\ &= \nabla_Z\nabla_Y X + \nabla_{[X,Y]}Z + [Z,[X,Y]] \\ &\qquad - \nabla_X\nabla_Y Z - \nabla_{[Z,Y]}X - [X,[Z,Y]] - \nabla_{[Z,X]}Y. \end{aligned} \tag{5.17}$$

Adding Eqs. (5.15), (5.16), and (5.17) and using the anticommutativity of the Lie bracket $[X,Y] = -[Y,X]$ gives

$$\begin{aligned} R(X,Y)(Z) + R(Y,Z)(X) + R(Z,X)(Y) &= [X,[Y,Z]] + [Y,[Z,X]] + [Z,[X,Y]] \\ &= 0 \end{aligned}$$

by the Jacobi identity 4.7.12. □

The first three properties of Proposition 5.5.5 are often presented in an equivalent form, emphasizing various anticommutativity relations within the curvature tensor. These all flow from the observation that if α is a bilinear form, then the condition that $\alpha(\mathbf{v},\mathbf{v}) = 0$ for all vectors $\mathbf{v}$ is equivalent to the condition that $\alpha(\mathbf{v},\mathbf{w}) = -\alpha(\mathbf{w},\mathbf{v})$ for all pairs of vectors $\mathbf{v},\mathbf{w}$ (see Theorem 4.1.2 and Exercise 4.6).

Corollary 5.5.6. *Let R, X, Y, Z, W be as in Proposition 5.5.5. Then*

1. $R(Y,X)(Z) = -R(X,Y)(Z)$. *In particular,* $R^l_{jik} = -R^l_{ijk}$.
2. $R(Y,X,Z,W) = -R(X,Y,Z,W)$. *In particular,* $R_{jikl} = -R_{ijkl}$.
3. $R(X,Y,W,Z) = -R(X,Y,Z,W)$. *In particular,* $R_{ijlk} = -R_{ijkl}$.

The following symmetry is a consequence of those already presented.

Proposition 5.5.7. *For any vector fields X,Y,Z,W, the curvature tensor satisfies*

$$R(Z,W,X,Y) = R(X,Y,Z,W).$$

In particular, $R_{klij} = R_{ijkl}$.

Proof. Using Proposition 5.5.5 (4), the reader can verify that for any four vector fields X_1, X_2, X_3 and Y, we have

$$R(X_1,X_2,X_3,Y) + R(X_2,X_3,X_1,Y) + R(X_3,X_1,X_2,Y) = 0.$$

In particular, the following four identities hold:

$$\begin{aligned} R(X,Y,Z,W) + R(Y,Z,X,W) + R(Z,X,Y,W) &= 0, \\ R(W,X,Z,Y) + R(X,Z,W,Y) + R(Z,W,X,Y) &= 0, \\ R(Y,W,X,Z) + R(W,X,Y,Z) + R(X,Y,W,Z) &= 0, \\ R(Z,Y,W,X) + R(Y,W,Z,X) + R(W,Z,Y,X) &= 0. \end{aligned}$$

Adding the four and using Corollary 5.5.6 (2) and (3) gives

$$2R(Y,Z,X,W) + 2R(Z,X,Y,W) + 2R(W,Z,Y,X) = 0.$$

Since

$$R(Y,Z,X,W) + R(Z,X,Y,W) = -R(X,Y,Z,W)$$

and since Corollary 5.5.6 implies

$$R(W,Z,Y,X) = R(Z,W,X,Y),$$

we have

$$R(X,Y,Z,W) = R(Z,W,X,Y). \qquad \square$$

All of these symmetries and relations drastically reduce the computations required to compute the n^4 components of the curvature tensor. For example, the reader can verify that for $n = 2$, the only independent component is R_{1212}.

With these observations, we now proceed to some examples.

Example 5.5.8. Let $\mathbf{R}^n$ be equipped with the standard Euclidean metric tensor g_0 with the associated Riemannian connection defined by the Christoffel symbols $\Gamma_{ij}^k = 0$ for all i, j, k. Proposition 5.5.4 then gives

$$R_{ijk}^l = R_{ijkl} = 0$$

for all $i, j, k, l = 1, \ldots, n$. This further justifies the terminology "flat," in the sense of "having no curvature," for the standard Euclidean structure on $\mathbf{R}^n$.

Notice that this implies that

$$\nabla_X \nabla_Y Z - \nabla_Y \nabla_X Z - \nabla_{[X,Y]} Z = 0$$

for all vector fields X, Y, Z. Symbolically, this can be written

$$\nabla_X \nabla_Y - \nabla_Y \nabla_X = \nabla_{[X,Y]},$$

and in particular, for basis vector fields ∂_1, ∂_j, we have

$$\nabla_{\partial_i} \nabla_{\partial_j} = \nabla_{\partial_j} \nabla_{\partial_i},$$

a kind of "equality of mixed partials."

In the following two examples of metrics on domains $U \subset \mathbf{R}^2$, we compute the independent component R_{1212} of the curvature tensor in two different ways: the first using Definitions 5.5.1 and 5.5.3 and the second using the coordinate formula of Proposition 5.5.4.

Example 5.5.9. Let $\mathbf{R}^2$ be endowed with the metric g of Example 5.3.6, i.e.,

$$[g_{ij}] = \begin{bmatrix} 1 + 4u^2 & 4uv \\ 4uv & 1 + 4v^2 \end{bmatrix},$$

with the associated Riemannian connection defined by the Christoffel symbols

$$\begin{aligned} \Gamma_{11}^1 = \Gamma_{22}^1 &= \frac{4u}{1 + 4u^2 + 4v^2}, \\ \Gamma_{11}^2 = \Gamma_{22}^2 &= \frac{4v}{1 + 4u^2 + 4v^2}, \\ \Gamma_{12}^1 = \Gamma_{12}^2 &= \Gamma_{21}^1 = \Gamma_{21}^2 = 0. \end{aligned}$$

Let $\partial_1 = \frac{\partial}{\partial u}$ and $\partial_2 = \frac{\partial}{\partial v}$; as usual, $[\partial_1, \partial_2] = 0$. Using Definition 5.5.1, we have

$$\begin{aligned} R(\partial_1, \partial_2)(\partial_1) &= \nabla_{\partial_1} \nabla_{\partial_2} \partial_1 - \nabla_{\partial_2} \nabla_{\partial_1} \partial_1 \\ &= \nabla_{\partial_1} \left(\Gamma_{21}^1 \partial_1 + \Gamma_{21}^2 \partial_2 \right) - \nabla_{\partial_2} \left(\Gamma_{11}^1 \partial_1 + \Gamma_{11}^2 \partial_2 \right) \end{aligned}$$

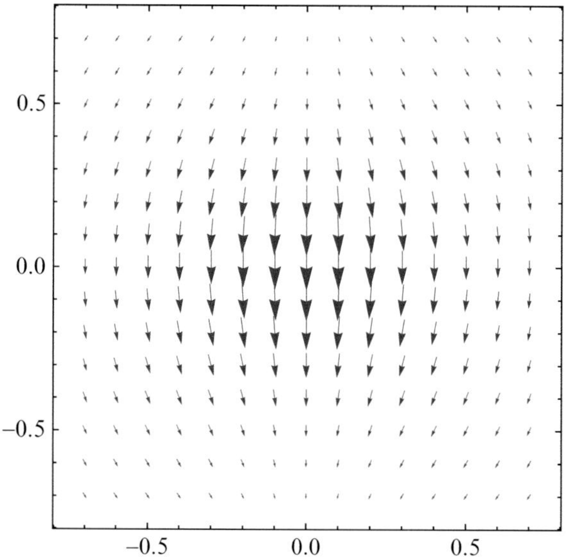

Fig. 5.9 A Mathematica plot of the vector field $R(\partial_1, \partial_2)\partial_1$ with the parabolic metric.

$$
\begin{aligned}
&= \nabla_{\partial_1}(0) - \nabla_{\partial_2}\left(\frac{4u}{1+4u^2+4v^2}\partial_1 + \frac{4v}{1+4u^2+4v^2}\partial_2\right) \\
&= -\left(\partial_2\left[\frac{4u}{1+4u^2+4v^2}\right]\partial_1 + \frac{4u}{1+4u^2+4v^2}\nabla_{\partial_2}\partial_1\right. \\
&\quad \left. + \partial_2\left[\frac{4v}{1+4u^2+4v^2}\right]\partial_2 + \frac{4v}{1+4u^2+4v^2}\nabla_{\partial_2}\partial_2\right) \\
&= \frac{16uv}{(1+4u^2+4v^2)^2}\partial_1 - \frac{4(1+4u^2)}{(1+4u^2+4v^2)^2}\partial_2.
\end{aligned}
$$

See Fig. 5.9.

Then by Definition 5.5.3,

$$
\begin{aligned}
R_{1212} &= g\left(R(\partial_1, \partial_2)(\partial_1), \partial_2\right) \\
&= g\left(\frac{16uv}{(1+4u^2+4v^2)^2}\partial_1 - \frac{4(1+4u^2)}{(1+4u^2+4v^2)^2}\partial_2, \partial_2\right) \\
&= g_{12}\cdot\frac{16uv}{(1+4u^2+4v^2)^2} + g_{22}\cdot\frac{-4(1+4u^2)}{(1+4u^2+4v^2)^2} \\
&= \frac{-4}{1+4u^2+4v^2}.
\end{aligned}
$$

Example 5.5.10. Let U be the upper half-plane with the Poincaré metric

$$[g_{ij}] = \begin{bmatrix} 1/y^2 & 0 \\ 0 & 1/y^2 \end{bmatrix}$$

and associated Riemannian connection given by the Christoffel symbols

$$\Gamma_{11}^2 = 1/y, \quad \Gamma_{12}^1 = \Gamma_{21}^1 = \Gamma_{22}^2 = -1/y, \quad \Gamma_{11}^1 = \Gamma_{12}^2 = \Gamma_{21}^2 = \Gamma_{22}^1 = 0,$$

as in Exercise 5.9.

Applying Proposition 5.5.4, we have

$$\begin{aligned} R_{1212} &= R_{121}^1 \cdot g_{12} + R_{121}^2 \cdot g_{22} \\ &= R_{121}^2 (1/y^2) \\ &= \frac{1}{y^2}\left[\frac{\partial}{\partial x}\left[\Gamma_{21}^2\right] - \frac{\partial}{\partial y}\left[\Gamma_{11}^2\right] + \Gamma_{21}^1\Gamma_{11}^2 + \Gamma_{21}^2\Gamma_{21}^2 - \Gamma_{11}^1\Gamma_{21}^2 - \Gamma_{11}^2\Gamma_{22}^2\right] \\ &= \frac{1}{y^4}. \end{aligned}$$

So far, we have essentially defined differential geometry as the study of tensors defined on domains of $\mathbf{R}^n$. Our examples have been chosen to illustrate explicit calculations with such tensors.

Further development of Riemannian geometry proceeds by defining these tensors on geometric sets such as spheres, surfaces, and especially their generalizations to higher dimensions, then using these tensors to make measurements.

While this goal is outside the introductory framework that we are presenting here, we will exhibit several of the more specialized curvature concepts that appear in the development of Riemannian geometry, as well as the statements of some theorems to give the reader a flavor of the subject.

For the following, consider two vector fields X and Y on a domain $U \subset \mathbf{R}^n$ equipped with metric tensor g and curvature tensor R. Suppose that for all $p \in U$, $X(p)$ and $Y(p)$ are linearly independent. Let $\sigma_p \subset T_pU$ be the two-dimensional vector subspace of T_pU spanned by $X(p)$ and $Y(p)$. In analogy with the transition from tangent vectors to vector fields, we write $\sigma \subset TU$ to be $\sigma(p) = \sigma_p$. The field of subspaces σ is called a *plane field* on U.

Definition 5.5.11. For the plane field σ defined by linearly independent vector fields X and Y, the *sectional curvature with respect to* σ is the scalar quantity

$$K(\sigma) = K(X,Y) = \frac{R(X,Y,Y,X)}{g(X,X) \cdot g(Y,Y) - [g(X,Y)]^2}.$$

Since the denominator can be interpreted as the (square of the) area of the parallelogram formed by vectors X and Y, the sectional curvature might be thought of as a "curvature per unit area."

Implicit in this definition is the following:

Proposition 5.5.12. *Suppose that $\{X_1, Y_1\}$ and $\{X_2, Y_2\}$ are two pairs of (pointwise) linearly independent vector fields that span the same plane field σ. Then*

$$K(X_1, Y_1) = K(X_2, Y_2).$$

Hence the sectional curvature is independent of the vector fields defining σ.

Proof. Writing one basis in terms of the other,

$$X_2 = aX_1 + bY_1, \qquad Y_2 = cX_1 + dY_1,$$

the reader can verify that

$$R(X_2, Y_2, Y_2, X_2) = (ad - bc)^2 R(X_1, Y_1, Y_1, X_1)$$

and

$$\begin{aligned} &g(X_2, X_2) \cdot g(Y_2, Y_2) - [g(X_2, Y_2)]^2 \\ &\quad = (ad - bc)^2 \left(g(X_1, X_1) \cdot g(Y_1, Y_1) - [g(X_1, Y_1)]^2 \right). \end{aligned}$$ □

Definition 5.5.11 takes a special form in the two-dimensional case, i.e., when $U \subset \mathbf{R}^2$. The only two-dimensional subspace $\sigma \subset TU$ is TU itself. Hence K is a scalar function of $p \in U$ alone. In this case, the quantity K is known as the *Gaussian curvature.* It can be expressed explicitly using the coordinate basis fields $\partial_1 = \dfrac{\partial}{\partial x_1}$ and $\partial_2 = \dfrac{\partial}{\partial x_2}$ using the components of the curvature tensor:

$$K = \frac{R_{1221}}{g_{11}g_{22} - (g_{12})^2} = \frac{-R_{1212}}{g_{11}g_{22} - (g_{12})^2}. \tag{5.18}$$

Example 5.5.13. Let g be the metric on $\mathbf{R}^2$ corresponding to the pullback of the standard Euclidean metric from the paraboloid, and let R be the corresponding curvature tensor (see Example 5.5.9). In this case, the reader can verify that

$$g_{11}g_{22} - (g_{12})^2 = 1 + 4u^2 + 4v^2,$$

and so the Gaussian curvature is given by

$$K(u, v) = \frac{-R_{1212}}{g_{11}g_{22} - (g_{12})^2} = \frac{4}{(1 + 4u^2 + 4v^2)^2}.$$

Example 5.5.14. Let U be the upper half-plane with the Poincaré metric g and corresponding curvature tensor R, as in Example 5.5.10. The Gaussian curvature is then

$$K(x, y) = \frac{-1/y^4}{(1/y^2)\,(1/y^2) - (0)(0)} = -1.$$

Example 5.5.14 shows that the Poincaré metric is special in the sense that the Gaussian curvature is constant. Metrics whose sectional curvature is the same constant for all plane fields σ are called *metrics of constant curvature*. Similarly, a parameterized set $S = \phi(U) \subset \mathbf{R}^3$ with regular parameterization (U, ϕ) is called a *space of constant curvature* if $\phi^* g_0$ is a metric of constant curvature, where g_0 is the standard Euclidean metric on $\mathbf{R}^3$. They are studied extensively; the reader may refer to [13, 25] or any number of other texts on Riemannian geometry for a sample of the results that exist about constant-curvature metrics.

We present one final example of a sectional curvature computation, this time when $n = 3$.

Example 5.5.15. Consider $\mathbf{R}^3$ with the metric g given by

$$[g_{ij}] = \begin{bmatrix} 1 & 0 & 0 \\ 0 & 1+x^2 & x \\ 0 & x & 1 \end{bmatrix}.$$

In Exercise 5.10, we ask the reader to show that

$$\Gamma^1_{11} = \Gamma^2_{11} = \Gamma^3_{11} = 0,$$

$$\Gamma^1_{12} = \Gamma^1_{21} = 0, \quad \Gamma^2_{12} = \Gamma^2_{21} = \frac{x}{2}, \quad \Gamma^3_{12} = \Gamma^3_{21} = \frac{1-x^2}{2},$$

$$\Gamma^1_{13} = \Gamma^1_{31} = 0, \quad \Gamma^2_{13} = \Gamma^2_{31} = \frac{1}{2}, \quad \Gamma^3_{13} = \Gamma^3_{31} = -\frac{x}{2},$$

$$\Gamma^1_{22} = -x, \quad \Gamma^2_{22} = \Gamma^3_{22} = 0,$$

$$\Gamma^1_{23} = \Gamma^1_{32} = -\frac{1}{2}, \quad \Gamma^2_{23} = \Gamma^2_{32} = \Gamma^3_{23} = \Gamma^3_{32} = 0,$$

$$\Gamma^1_{33} = \Gamma^2_{33} = \Gamma^3_{33} = 0.$$

We will first compute the sectional curvature $K(\sigma_{12})$ corresponding to the coordinate plane field σ_{12} spanned by $\frac{\partial}{\partial x}$ and $\frac{\partial}{\partial y}$.

We have

$$\begin{aligned} R_{1221} &= R^1_{122} g_{11} + R^2_{122} g_{21} + R^3_{122} g_{31} \\ &= \Bigg[\left(\frac{\partial}{\partial x} \Gamma^1_{22} \right) - \left(\frac{\partial}{\partial y} \Gamma^1_{12} \right) \\ &\qquad + \Gamma^1_{22}\Gamma^1_{11} + \Gamma^2_{22}\Gamma^1_{21} + \Gamma^3_{22}\Gamma^1_{31} - \Gamma^1_{12}\Gamma^1_{12} - \Gamma^2_{12}\Gamma^1_{22} - \Gamma^3_{12}\Gamma^1_{32} \Bigg] \\ &\qquad + R^2_{122}(0) + R^3_{122}(0) \\ &= \frac{1}{4}\left(-3 + x^2\right). \end{aligned}$$

Another computation shows that $g_{11} \cdot g_{22} - (g_{12})^2 = 1 + x^2$, and so

$$K(\sigma_{12}) = \frac{R_{1221}}{g_{11} \cdot g_{22} - (g_{12})^2} = \frac{-3 + x^2}{4(1 + x^2)}.$$

We compute one more sectional curvature. Let σ_H be the plane field spanned by the vector fields

$$X_1 = \frac{\partial}{\partial x}, \quad X_2 = \frac{\partial}{\partial y} - x\frac{\partial}{\partial z}.$$

The reader can verify that

$$g(X_1, X_1) = g(X_2, X_2) = 1, \quad g(X_1, X_2) = 0,$$

so that

$$K(\sigma_H) = R(X_1, X_2, X_2, X_1).$$

Computing, using the fact that

$$g(\frac{\partial}{\partial x}, \frac{\partial}{\partial y}) = g(\frac{\partial}{\partial x}, \frac{\partial}{\partial z}) = 0,$$

$$\begin{aligned} R(X_1, X_2, X_2, X_1) &= g(R(X_1, X_2)X_2, X_1) \\ &= (R(X_1, X_2)X_2)^1 \\ &= \sum_{i,j,k} (X_1)^i (X_2)^j (X_2)^k R^1_{ijk} \\ &= R^1_{122} - xR^1_{123} - xR^1_{132} + x^2 R^1_{133}. \end{aligned}$$

Further calculations show that $R^1_{122} = \frac{1}{4}(-3 + x^2)$, $R^1_{123} = \frac{x}{4}$, $R^1_{132} = \frac{x}{4}$, and $R^1_{133} = \frac{1}{4}$. Combining all this gives

$$\begin{aligned} K(\sigma_H) &= \frac{1}{4}(-3 + x^2) - x\left(\frac{x}{4}\right) - x\left(\frac{x}{4}\right) + x^2\left(\frac{1}{4}\right) \\ &= -\frac{3}{4}. \end{aligned}$$

This illustrates the fact that the sectional curvature may in general be constant for certain plane fields but not for others.

We close this section by defining two other curvature-related quantities. We will not explore them in detail, however. We include them only for the sake of completeness.

Definition 5.5.16. Let (U, g) be a Riemannian space. Let $\{E_1, \ldots, E_n\}$ be a set of vector fields such that at each point $p \in U$, the set of tangent vectors $\{E_1(p), \ldots, E_n(p)\}$ is an orthonormal basis for T_pU. The *Ricci curvature tensor* Ric is the $(0,2)$-tensor defined by

$$Ric(X, Y) = \frac{1}{n-1} \sum_{i=1}^{n} R(X, E_i, Y, E_i),$$

for any vector fields X, Y. The *Ricci curvature* at $p \in U$ in the direction X, where X is a unit vector field, is defined to be

$$Ric(X) = Ric(X, X).$$

We leave it as an exercise to show that these definitions are independent of the choice of orthonormal basis.

We follow do Carmo [13] in including the factor of $\frac{1}{n-1}$ so that for every $j = 1, \ldots, n$, $Ric(E_j)$ is the average of the sectional curvatures of plane fields σ_{ij}, $i \neq j$, where σ_{ij} is the plane field spanned by $\{E_i, E_j\}$.

In addition, we have the following fact in the case that $n = 2$.

Proposition 5.5.17. *If* $U \subset \mathbf{R}^2$, *then for every unit vector field* E_1,

$$Ric(E_1) = K,$$

i.e., the Ricci curvature agrees with the Gaussian curvature.

Proof. Choose a unit vector field E_2 such that for all $p \in U$, $\{E_1(p), E_2(p)\}$ is a basis for T_pU. The results then follows by comparing Definition 5.5.16 with Eq. (5.18). □

In fact, for metrics of constant curvature, the Ricci curvature agrees with the sectional curvature in higher dimensions as well.

The components of the Ricci curvature tensor are computed by applying Definition 5.5.16 to compute $(Ric)_{jk} = Ric(E_j, E_k)$:

$$(Ric)_{jk} = \frac{1}{n-1} \sum_{i=1}^{n} R^i_{jik}. \tag{5.19}$$

The final curvature quantity that we present here is the same sort of construction.

Definition 5.5.18. Let (U, g) be a Riemannian space, and let $\{E_1, \ldots, E_n\}$ be a set of orthonormal basis fields for TU. The *scalar curvature* S is defined as

$$S = \frac{1}{n} \sum_{k=1}^{n} Ric(E_k).$$

As above, it is an exercise to verify that the definition is independent of the choice of orthonormal basis. Again, the normalization factor ensures that in the case of $n = 2$, the scalar curvature agrees with the Gaussian curvature.

5.6 Isometries

The previous sections provide the first illustration of a major theme of this text, differential geometry as the study of smooth structures—tensors—defined on domains in $\mathbf{R}^n$. Indeed, the reader should have begun to appreciate the wealth of concepts that arise starting "just" from the Riemannian metric tensor and the calculus required to take derivatives.

Alongside the study of sets with structure is the study of those functions between such sets that "preserve the structure" in a sense we will make precise below. Such functions will allow us to talk about when two sets with structure are "equivalent," for example in the sense of congruence in Euclidean geometry.

In addition, such structure-preserving functions single out those subsidiary structures and concepts that belong within the confines of the geometry in question, as opposed to incidental or "extrinsic" structures. One could even try, once a family of structure-preserving functions is identified, to define a geometry as the study of all properties that are preserved by the given family.

Definition 5.6.1. Suppose $\phi : U \to V$ is a diffeomorphism between two Riemannian spaces (U, g_1) and (V, g_2). Then ϕ is said to be an *isometry* if $\phi^* g_2 = g_1$. In other words, ϕ is an isometry if for all $p \in U$ and tangent vectors $X_p, Y_p \in T_pU$, we have

$$g_1(p)\left(X_p, Y_p\right) = g_2(\phi(p))\left(\phi_* X_p, \phi_* Y_p\right).$$

In this sense, we say that ϕ preserves the metric structure. Under the pairing ϕ between U and V, which in turn induces the linear isomorphisms $(\phi_*)_p : T_pU \to T_{\phi(p)}V$ at each point $p \in U$, we can perform measurements on U (with g_1) or on V (with g_2) and obtain the same results.

The isometry condition is a stringent one. As we will see in the subsequent examples, it amounts to saying that ϕ satisfies a system of partial differential equations. Needless to say, the project of trying to find all isometries for a given pair of metrics by solving the system of PDEs explicitly in terms of elementary functions is usually beyond hope.

We begin our examples with the "test case" of isometries of the standard Euclidean plane.

Example 5.6.2. Let $U = V = \mathbf{R}^2$ and $g_1 = g_2 = dx \otimes dx + dy \otimes dy$. Let $\phi : \mathbf{R}^2 \to \mathbf{R}^2$ be a smooth map, written in components as

$$\phi(x, y) = \left(\phi^1(x, y), \phi^2(x, y)\right).$$

Then computing the pullback using Proposition 4.6.15, we have

$$
\begin{aligned}
\phi^* g_2 &= d\phi^1 \otimes d\phi^1 + d\phi^2 \otimes d\phi^2 \\
&= \left(\left(\frac{\partial \phi^1}{\partial x} \right)^2 + \left(\frac{\partial \phi^2}{\partial x} \right)^2 \right) dx \otimes dx + \left(\frac{\partial \phi^1}{\partial x} \frac{\partial \phi^1}{\partial y} + \frac{\partial \phi^2}{\partial x} \frac{\partial \phi^2}{\partial y} \right) dx \otimes dy \\
&\quad + \left(\frac{\partial \phi^1}{\partial y} \frac{\partial \phi^1}{\partial x} + \frac{\partial \phi^2}{\partial y} \frac{\partial \phi^2}{\partial x} \right) dy \otimes dx \\
&\quad + \left(\left(\frac{\partial \phi^1}{\partial y} \right)^2 + \left(\frac{\partial \phi^2}{\partial y} \right)^2 \right) dy \otimes dy.
\end{aligned}
$$

Hence the condition that $\phi^* g_2 = g_1$ is equivalent to the components of ϕ satisfying the PDEs

$$
\begin{cases}
\left(\frac{\partial \phi^1}{\partial x} \right)^2 + \left(\frac{\partial \phi^2}{\partial x} \right)^2 = 1, \\
\left(\frac{\partial \phi^1}{\partial x} \right) \left(\frac{\partial \phi^1}{\partial y} \right) + \left(\frac{\partial \phi^2}{\partial x} \right) \left(\frac{\partial \phi^2}{\partial y} \right) = 0, \\
\left(\frac{\partial \phi^1}{\partial y} \right)^2 + \left(\frac{\partial \phi^2}{\partial y} \right)^2 = 1.
\end{cases}
$$

Example 5.6.3. Consider again $\mathbf{R}^2$ with the standard Euclidean metric g_0, and suppose that $\phi : \mathbf{R}^2 \to \mathbf{R}^2$ is a *linear* map, in the sense that $\phi(x, y) = A\mathbf{x} = (ax + by, cx + dy)$, where $A = \begin{bmatrix} a & b \\ c & d \end{bmatrix}$. Then the system of PDEs in Example 5.6.2 takes the form

$$
\begin{cases}
a^2 + c^2 = 1, \\
ab + cd = 0, \\
b^2 + d^2 = 1.
\end{cases}
$$

The reader can verify that these equations can be written as $A^T A = I$, i.e., that A is an orthogonal matrix. Hence the system of PDEs that defines the Euclidean isometries agrees in the linear case with the classical fact of linear algebra that the orthogonal matrices are exactly the linear isometries of $\mathbf{R}^2$ as described in Sect. 2.9.

Example 5.6.4. Let $U = \{(x, y) \mid y > 0\}$ be the upper half-plane in $\mathbf{R}^2$ with the Poincaré metric $g = \frac{1}{y^2} (dx \otimes dx + dy \otimes dy)$. We will show explicitly that the map $\phi : U \to U$ given by

$$
\phi(x, y) = \left(\frac{-x}{x^2 + y^2}, \frac{y}{x^2 + y^2} \right)
$$

is an isometry of g.

First, the fact that ϕ is a diffeomorphism can be seen from the observation that ϕ is the composition of a (linear) reflection about the y-axis with the diffeomorphism of Example 3.6.8.

To show that $\phi = (\phi^1, \phi^2)$ preserves the Poincaré metric tensor, we first compute

$$d\phi^1 = \frac{x^2 - y^2}{(x^2 + y^2)^2} dx + \frac{2xy}{(x^2 + y^2)^2} dy$$

and

$$d\phi^2 = \frac{-2xy}{(x^2 + y^2)^2} dx + \frac{x^2 - y^2}{(x^2 + y^2)^2} dy.$$

Hence

$$\begin{aligned}
\phi^* g &= \frac{1}{(\phi^2)^2} \left(d\phi^1 \otimes d\phi^1 + d\phi^2 \otimes d\phi^2 \right) \\
&= \frac{1}{y^2 (x^2 + y^2)^2} \left[(x^2 + y^2)^2 dx \otimes dx + (x^2 + y^2)^2 dy \otimes dy \right] \\
&= \frac{1}{y^2} \left[dx \otimes dx + dy \otimes dy \right] \\
&= g.
\end{aligned}$$

This shows that ϕ is an isometry.

We made note earlier of the fact that the project of describing all isometries of a given metric tensor by means of solving a system of nonlinear PDEs is usually impractical or impossible. At the end of the section, we will give some examples of theorems that describe all isometries of a given metric tensor. In fact, Examples 5.6.3 and 5.6.4 are both special instances of those theorems.

There is a method of constructing *local isometries*, meaning isometries defined in a (possibly small) domain around each point. This is essentially Theorem 4.7.23, which we restate here in the specific instance of a Riemannian metric tensor.

Theorem 5.6.5. *Let (U, g) be a Riemannian space, and let X be a vector field such that $\mathcal{L}_X g = 0$. Then for each $p \in U$, the flow $\phi_t : U_p \to \phi_t(U_p)$ generated by X, satisfying $\phi_0(p) = p$ and $\frac{d}{dt}(\phi_t(p)) = X(\phi_t(p))$, is a family of isometries, i.e.,*

$$\phi_t^* g = g$$

for all t such that ϕ_t is defined.

Theorem 5.6.5 prompts the following definition, named for the German mathematician Wilhelm Killing. Killing's work on geometry and transformation groups paralleled that of Sophus Lie and presaged that of Élie Cartan.

Definition 5.6.6. Let (U, g) be a Riemannian space. A *Killing vector field* is a vector field X satisfying the condition that

$$\mathcal{L}_X g = 0.$$

Because of Theorem 5.6.5, Killing vector fields are also known as *infinitesimal isometries*, a terminology that arises from the idea of "integrating" vector fields to obtain isometries.

We now illustrate the Killing vector field condition in local coordinates, expressing it as a system of first-order, linear partial differential equations.

Proposition 5.6.7. *Let $(x_1, \dots, x_n)$ be a system of coordinates on a domain U with corresponding basis $\left\{(\partial_i)_p = \frac{\partial}{\partial x_i}\big|_p\right\}$ of $T_p(\mathbf{R}^n)$. Let g be a Riemannian metric tensor with components $[g_{ij}]$. Then $X = \sum X^i \partial_i$ is a Killing vector field if and only if the components X^i satisfy the $\frac{n(n+1)}{2}$ partial differential equations*

$$\sum_{k=1}^{n} \left(X^k \frac{\partial g_{ij}}{\partial x_k} + g_{jk} \frac{\partial X^k}{\partial x_i} + g_{ik} \frac{\partial X^k}{\partial x_j} \right) = 0, \quad i, j = 1, \dots, n, \quad i \le j.$$

Proof. This follows immediately from Exercise 4.35. □

Example 5.6.8. Let g_0 be the standard Euclidean metric on $\mathbf{R}^2$ with coordinates (x, y). Let $X = X^1 \frac{\partial}{\partial x} + X^2 \frac{\partial}{\partial y}$ be a vector field on $\mathbf{R}^2$. Our goal is to first describe Killing vector fields in the Euclidean context in $\mathbf{R}^2$, and then to integrate them to obtain (local) Euclidean isometries.

Applying Proposition 5.6.7 yields, after the appropriate calculations,

$$\frac{\partial X^1}{\partial x} = 0, \tag{5.20}$$

$$\frac{\partial X^1}{\partial y} + \frac{\partial X^2}{\partial x} = 0, \tag{5.21}$$

$$\frac{\partial X^2}{\partial y} = 0. \tag{5.22}$$

By (5.20) and (5.22), we see that $X^1 = X^1(y)$ and $X^2 = X^2(x)$. Equation (5.21) then implies that

$$\frac{\partial X^1}{\partial y} = -\frac{\partial X^2}{\partial x} = a$$

for some constant a, and so the Euclidean Killing vector fields must have the form

$$X(x, y) = (ay + b)\frac{\partial}{\partial x} + (-ax + c)\frac{\partial}{\partial y},$$

for some constants a, b, c.

The fact that Euclidean Killing vector fields have such a simple form allows them to be integrated in order to obtain isometries. The flow ϕ_t given by $\phi_t(x, y) = (x(t), y(t))$ generated by a Euclidean Killing vector field must satisfy the system of first-order ordinary differential equations

$$\begin{cases} \dot{x} = ay + b, \\ \dot{y} = -ax + c. \end{cases}$$

Solving this system can be accomplished by employing the toolbox of a first course in differential equations to obtain the general solution

$$x(t) = c_1 \cos(at) + c_2 \sin(at) + c/a,$$
$$y(t) = c_2 \cos(at) - c_1 \sin(at) - b/a,$$

where c_1 and c_2 are constants depending (smoothly) on the initial position

$$(x(0), y(0)) = (x_0, y_0).$$

This shows, incidentally, that the local isometries arising as integrals of Euclidean Killing vector fields are in fact defined globally, and can be expressed as compositions of rotations and translations.

Example 5.6.9. Let $U = \{(x, y) \mid y > 0\}$ be the upper half-plane with the Poincaré metric $[g_{ij}] = \begin{bmatrix} 1/y^2 & 0 \\ 0 & 1/y^2 \end{bmatrix}$. Then the components of a vector field

$$X = X^1 \frac{\partial}{\partial x} + X^2 \frac{\partial}{\partial y}$$

must satisfy, by Proposition 5.6.7,

$$-\frac{1}{y^3}\left(X^2 - y\frac{\partial X^1}{\partial x}\right) = 0, \tag{5.23}$$

$$\frac{1}{y^2}\left(\frac{\partial X^1}{\partial y} + \frac{\partial X^2}{\partial x}\right) = 0, \tag{5.24}$$

$$-\frac{1}{y^3}\left(X^2 - y\frac{\partial X^2}{\partial y}\right) = 0. \tag{5.25}$$

This system implies, noting that $y \neq 0$ and by subtracting (5.25) from (5.23), that

$$\begin{cases} \frac{\partial X^1}{\partial y} + \frac{\partial X^2}{\partial x} = 0, \\ \frac{\partial X^1}{\partial x} - \frac{\partial X^2}{\partial y} = 0. \end{cases} \tag{5.26}$$

The relations between component functions X^1 and X^2 expressed in the system (5.26) arise naturally in a number of settings. For example, in the study of functions of one complex variable, Eqs. (5.26) are the celebrated Cauchy–Riemann equations. The functions X^1, X^2 are said to be *harmonic conjugates*. Both X^1 and X^2 individually are said to be *harmonic* functions in that they both satisfy *Laplace's equation*:

$$\frac{\partial^2 X^1}{\partial x^2} + \frac{\partial^2 X^1}{\partial y^2} = 0,$$
$$\frac{\partial^2 X^2}{\partial x^2} + \frac{\partial^2 X^2}{\partial y^2} = 0.$$

This is a direct consequence (with the equality of mixed partials) of the system (5.26).

We summarize this discussion as follows:

The component functions X^1 and X^2 of a Killing vector field for the Poincaré metric on the upper half-plane are harmonic conjugates. In particular, they are both harmonic functions.

The reader may verify, however, by way of example that the following vector field is a Killing vector field for the Poincaré metric on U:

$$X(x,y) = \left(x^2 - y^2\right)\frac{\partial}{\partial x} + (2xy)\frac{\partial}{\partial y}.$$

Integrating such vector fields to come up with a local isometry in closed form is impractical. However, a wide variety of computer software can illustrate flow lines in these cases. See Fig. 5.10.

So far, the examples of isometries we have encountered in Examples 5.6.3, 5.6.4, and 5.6.8 have all been examples of what might be called "self-isometries," in the sense of being differentiable maps between a domain equipped with a Riemannian metric and itself. In this case, we can use the notation

$$\phi : (U,g) \to (U,g),$$

where the geometric condition for ϕ to be an isometry is then $\phi^* g = g$. Before turning to the more general case, we point out that there is an algebra associated with self-isometries expressed in the following theorem.

Theorem 5.6.10. *Let (U,g) be a Riemannian space, and let $Iso(U,g)$ be the set of all self-isometries of (U,g), i.e., $\phi \in Iso(U,g)$ if and only if $\phi : U \to U$ is a diffeomorphism satisfying $\phi^* g = g$. Then:*

1. $\mathrm{Id} \in Iso(U,g)$, *where* $\mathrm{Id}(p) = p$ *is the identity map.*
2. *If* $\phi_1, \phi_2 \in Iso(U,g)$, *then* $\phi_1 \circ \phi_2 \in Iso(U,g)$.
3. *If* $\phi \in Iso(U,g)$, *then* $\phi^{-1} \in Iso(U,g)$.

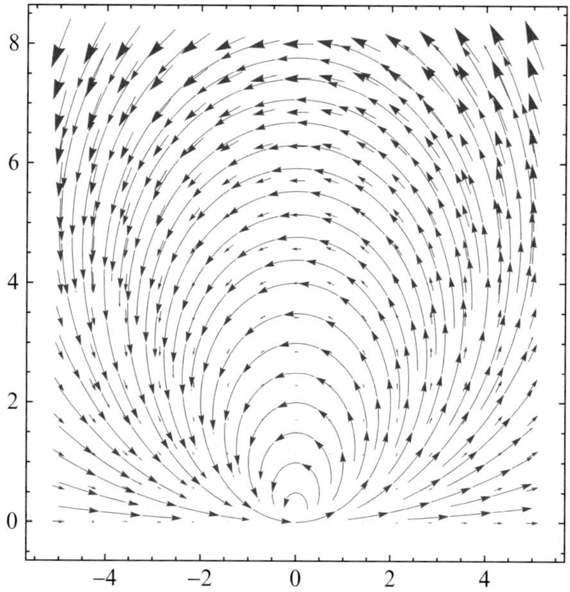

Fig. 5.10 A Mathematica plot of integral curves of the vector field $X = \left(x^2 - y^2\right) \frac{\partial}{\partial x} + (2xy) \frac{\partial}{\partial y}$, which is a Killing vector field for the Poincaré metric.

Proof. All of the statements follow from Theorem 3.6.11 and properties of the pullback operation. We illustrate the proof of (2) as an example. Assume $\phi_1^* g = g$ and $\phi_2^* g = g$, where ϕ_1 and ϕ_2 are diffeomorphisms. Then $\phi_1 \circ \phi_2$ is a diffeomorphism by Theorem 3.6.11. Further,

$$\begin{aligned} (\phi_1 \circ \phi_2)^* g &= \phi_2^* (\phi_1^* g) \\ &= \phi_2^* g \\ &= g. \quad \square \end{aligned}$$

Theorem 5.6.10 is summarized in mathematical terminology by saying that the set $Iso(U, g)$ is a group under the operation of function composition.

We now present one method for constructing examples of isometries between domains with different Riemannian metrics. We start with two different parameterizations of the same geometric set $S \subset \mathbf{R}^n$, where $\mathbf{R}^n$ is equipped with a Riemannian metric g. Let $U_1, U_2 \subset \mathbf{R}^k$ be domains and let

$$\phi_1 : U_1 \to S, \quad \phi_2 : U_2 \to S$$

be two parameterizations of S. Recall that ϕ_1, ϕ_2 should be regular, i.e., one-to-one, smooth, with one-to-one tangent maps, and so have inverses on S.

Theorem 5.6.11. *Suppose that $S \subset \mathbf{R}^n$ is a geometric set parameterized by two regular parameterizations $\phi_1 : U_1 \to S$ and $\phi_2 : U_2 \to S$, where $U_1, U_2 \subset \mathbf{R}^k$ ($k \leq n$) are domains. For a Riemannian metric g on $\mathbf{R}^n$, define $g_i = \phi_i^* g$ ($i = 1, 2$). Then $\phi : (U_1, g_1) \to (U_2, g_2)$ defined by $\phi = \phi_2^{-1} \circ \phi_1$ is an isometry.*

Proof. We leave it as an exercise to show that g_1 and g_2 are Riemannian metrics, which relies heavily on the fact that $(\phi_1)_*$, resp. $(\phi_2)_*$, is a vector space isomorphism between TU_1, resp. TU_2, and TS. Hence all the properties of the Riemannian metric can be checked by referring to the given Riemannian metric g.

To show that ϕ is a diffeomorphism is a routine exercise in the differentiability of inverses and compositions. Hence to show that ϕ is an isometry, we need to show only that $\phi^* g_2 = g_1$. This is again an application of the properties of pullbacks:

$$\begin{aligned}
\phi^* g_2 &= \left(\phi_2^{-1} \circ \phi_1\right)^* g_2 \\
&= \phi_1^* \left((\phi_2^{-1})^*\right) g_2 \\
&= \phi_1^* (\phi_2^{-1})^* (\phi_2^* g) \\
&= \phi_1^* \left(\phi_2 \circ \phi_2^{-1}\right)^* g \\
&= \phi_1^* (\mathrm{Id})^* g \\
&= \phi_1^* g \\
&= g_1. \qquad \square
\end{aligned}$$

We illustrate this theorem with one example.

Example 5.6.12. Let $S \subset \mathbf{R}^3$ be the piece of the paraboloid

$$S = \left\{(x, y, z) \mid z = x^2 + y^2,\ x > 0,\ 0 < z < 1\right\}$$

and $g_0 = dx \otimes dx + dy \otimes dy + dz \otimes dz$ the standard Euclidean metric on $\mathbf{R}^3$. Let (U_1, ϕ_1) be the standard parameterization of S that we have relied on thus far: $U_1 = \left\{(u, v) \mid u > 0,\ 0 < u^2 + v^2 < 1\right\}$ and $\phi_1(u, v) = (u, v, u^2 + v^2)$. In this case, we have seen in Example 5.2.4 that

$$\begin{aligned}
g_1 &= \phi_1^* g_0 \\
&= (1 + 4u^2) du \otimes du + (4uv) du \otimes dv + (4uv) dv \otimes du + (1 + 4v^2) dv \otimes dv.
\end{aligned}$$

Let (U_2, ϕ_2) be the parameterization of S by polar coordinates: $U_2 = \{(r, \theta) \mid 0 < r < 1, -\pi/2 < \theta < \pi/2\}$, $\phi_2(r, \theta) = (r\cos\theta, r\sin\theta, r^2)$; note then that on S,

$$\phi_2^{-1}(x, y, z) = \left(\sqrt{z}, \tan^{-1}(y/x)\right).$$

The metric defined by the pullback of g_0 by ϕ_2 is given by

$$
\begin{aligned}
g_2 &= \phi_2^* g_0 \\
&= d(r\cos\theta) \otimes d(r\cos\theta) + d(r\sin\theta) \otimes d(r\sin\theta) + d(r^2) \otimes d(r^2) \\
&= (1 + 4r^2) dr \otimes dr + r^2 d\theta \otimes d\theta.
\end{aligned}
$$

In this case, Theorem 5.6.11 shows that $\phi : U_1 \to U_2$ given by

$$
\begin{aligned}
\phi(u, v) &= \phi_2^{-1} \circ \phi_1(u, v) \\
&= \phi_2^{-1}\left(u, v, u^2 + v^2\right) \\
&= \left(\sqrt{u^2 + v^2}, \tan^{-1}(v/u)\right)
\end{aligned}
$$

is an isometry between (U_1, g_1) and (U_2, g_2).

Two Riemannian spaces (U_1, g_1) and (U_2, g_2) can be considered equivalent if there exists an isometry $\phi : U_1 \to U_2$. (The previous sentence can be made precise by saying that this notion defines an *equivalence relation* on the set of Riemannian spaces.) In that case, the two spaces can be considered two different "models" for the same geometric object.

We now show that isometries preserve not only the Riemannian metric, but all the key geometric constructions we have introduced, such as geodesics and curvature. For that reason, we say that the constructions are *invariants* of Riemannian geometry, in the sense that they are independent of a particular model of the space in question.

Before stating some of these results, we first prove two lemmas that technically belong in the context of tensor calculus.

The first lemma is a property of diffeomorphisms in general. The reader may compare this to Proposition 4.7.13.

Lemma 5.6.13. *Let $\phi : U \to V$ be a diffeomorphism between domains $U, V \subset \mathbf{R}^n$. If X is a vector field on U, let $Y = \phi_* X$ to be the corresponding vector field on V. Then for any tensor field T defined on V,*

$$\phi^*\left(\mathcal{L}_Y T\right) = \mathcal{L}_X(\phi^* T).$$

Note that in fact, every vector field Y on V also determines a unique vector field X on U such that $\phi_* X = Y$, since ϕ is a diffeomorphism.

Proof. Let $\psi_t : V \to \psi_t(V)$ be the flow generated by Y. We use the fact (see Exercise 3.31) that the condition $\phi_* X = Y$ implies that the flow generated by X is given by

$$\phi^{-1} \circ \psi_t \circ \phi : U \to \phi^{-1}\left(\psi_t(V)\right).$$

Then, relying on Definition 4.7.1,

$$\begin{aligned}
\phi^* (\mathcal{L}_Y T) &= \phi^* \left(\frac{d}{dt}\bigg|_{t=0} (\psi_t^* T) \right) \\
&= \frac{d}{dt}\bigg|_{t=0} (\phi^* \psi_t^* T) \\
&= \frac{d}{dt}\bigg|_{t=0} \left(\phi^* \psi_t^* \left[(\phi^{-1})^* \phi^* \right] T\right) \\
&= \frac{d}{dt}\bigg|_{t=0} \left[(\phi^{-1} \circ \psi_t \circ \phi)^* (\phi^* T) \right] \\
&= \mathcal{L}_X (\phi^* T). \qquad \square
\end{aligned}$$

The second lemma is a property of isometries. We first introduce some notation. Let X be a vector field on a Riemannian space (U, g). We define $\theta_{g,X} = i(X)g$ as the unique one-form on U satisfying $\theta_{g,X}(Y) = g(X, Y)$ for all vector fields Y on U. In the notation of Proposition 5.1.9, $\theta_{g,X} = \gamma(X)$.

Lemma 5.6.14. *Let $\phi : (U, g) \to (V, h)$ be an isometry between Riemannian spaces (U, g) and (V, h). Then for any vector field X on U and corresponding vector field $Y = \phi_* X$ on V, we have*

$$\phi^* (\theta_{h,Y}) = \theta_{g,X}.$$

Proof. Let Z be any vector field on U. Then

$$\begin{aligned}
(\phi^* \theta_{h,Y}) (Z) &= \theta_{h,Y} (\phi_* Z) \\
&= h(Y, \phi_* Z) \\
&= h(\phi_* X, \phi_* Z) \\
&= (\phi^* h)(X, Z) \\
&= g(X, Z) \qquad \text{since } \phi \text{ is an isometry} \\
&= \theta_{g,X}(Z).
\end{aligned}$$

Hence $\phi^* \theta_{h,Y} = \theta_{g,X}$. □

We now put these technical lemmas to use in proving theorems to show how isometries preserve the geometric constructions of the previous sections.

Theorem 5.6.15. *Let $\phi : (U, g) \to (V, h)$ be an isometry between Riemannian spaces. Let $\nabla, \overline{\nabla}$ be the Riemannian connections associated to g, h respectively. Then for all vector fields X, Y on U,*

$$\phi_* (\nabla_X Y) = \overline{\nabla}_{\phi_* X} (\phi_* Y).$$

Proof. We will use Lemmas 5.6.13 and 5.6.14 along with Definition 5.3.2 of the Riemannian connection. Let W be a vector field on V, and let Z be the vector field on U such that $\phi_* Z = W$.

On the one hand, we have

$$\begin{aligned}
2h\left(\overline{\nabla}_{\phi_* X}(\phi_* Y), W\right) &= 2h\left(\overline{\nabla}_{\phi_* X}(\phi_* Y), \phi_* Z\right) \\
&= \left(\mathcal{L}_{\phi_* Y} h\right)(\phi_* X, \phi_* Z) + d\theta_{h,\phi_* Y}\left(\phi_* X, \phi_* Z\right) \\
&= \left(\phi^*\left(\mathcal{L}_{\phi_* Y} h\right)\right)(X, Z) + \left(\phi^*\left(d\theta_{h,\phi_* Y}\right)\right)(X, Z) \\
&= \left(\mathcal{L}_Y(\phi^* h)\right)(X, Z) + \left(d\phi^*\theta_{h,\phi_* Y}\right)(X, Z) \\
&= \left(\mathcal{L}_Y g\right)(X, Z) + d\theta_{g,Y}(X, Z) \\
&= 2g\left(\nabla_X Y, Z\right),
\end{aligned}$$

where we have also used the fact from Proposition 4.4.16 that the exterior derivative commutes with pullback.

On the other hand,

$$\begin{aligned}
2h\left(\phi_*(\nabla_X Y), W\right) &= 2h\left(\phi_*(\nabla_X Y), \phi_* Z\right) \\
&= 2\left(\phi^* h\right)\left(\nabla_X Y, Z\right) \\
&= 2g\left(\nabla_X Y, Z\right).
\end{aligned}$$

Hence

$$h\left(\overline{\nabla}_{\phi_* X}(\phi_* Y), W\right) = h\left(\phi_*(\nabla_X Y), W\right)$$

for all vector fields W, and since h is positive definite (hence nondegenerate),

$$\overline{\nabla}_{\phi_* X}(\phi_* Y) = \phi_*(\nabla_X Y). \qquad \square$$

The reader should refer back to Example 5.3.6 in light of Theorem 5.6.15.

Corollary 5.6.16. *Let (U_1, g_1) and (U_2, g_2) be Riemannian spaces with corresponding Riemannian connections ∇ and $\overline{\nabla}$. Suppose $\phi : (U_1, g_1) \to (U_2, g_2)$ is an isometry. If $c_1 : I \to U_1$ is a geodesic with respect to g_1, then*

$$c_2 = \phi \circ c_1 : I \to U_2$$

is a geodesic with respect to g_2.

Proof. By the chain rule, $\dot{c_2} = \phi_* \dot{c_1}$. So by Theorem 5.6.15,

$$\begin{aligned}
\overline{\nabla}_{\dot{c_2}} \dot{c_2} &= \overline{\nabla}_{\phi_* \dot{c_1}}(\phi_* \dot{c_1}) \\
&= \phi_*\left(\nabla_{\dot{c_1}} \dot{c_1}\right) \\
&= 0 \qquad \text{since } c_1 \text{ is a geodesic.}
\end{aligned}$$

Hence c_2 is a geodesic with respect to $\overline{\nabla}$. $\square$

This corollary, in addition to confirming that geodesics are really an invariant of the metric structure, also gives a way of using an isometry (or set of isometries) to build new geodesics out of old ones.

Example 5.6.17. Let U be the upper half-plane with the Poincaré metric g. We ask the reader to show in Exercise 5.23 that the parameterized curve $c : I \to U$ given by

$$c(t) = (1, e^t),$$

a parameterization of a line parallel to the y-axis, is a geodesic.

In Example 5.6.4, we showed that the map

$$\phi(x, y) = \left(\frac{-x}{x^2 + y^2}, \frac{y}{x^2 + y^2} \right)$$

is an isometry of (U, g). Hence, by Corollary 5.6.16,

$$\tilde{c} = \phi \circ c : I \to U$$

given explicitly as

$$\tilde{c}(t) = \left(\frac{-1}{1 + e^{2t}}, \frac{e^t}{1 + e^{2t}} \right)$$

is also a geodesic. Geometrically, $\tilde{c}$ can be seen to be a parameterization of the upper half $(y > 0)$ of the circle

$$(x + 1/2)^2 + y^2 = 1/4$$

with center $(-1/2, 0)$ and radius $1/2$. See Fig. 5.11.

The following statement, which is also essentially a consequence of Theorem 5.6.15, is important enough to be stated as a theorem in its own right.

Theorem 5.6.18. *Let (U_1, g_1) and (U_2, g_2) be Riemannian spaces with corresponding curvature tensors R_1, R_2. Suppose $\phi : (U_1, g_1) \to (U_2, g_2)$ is an isometry. Then*

$$\phi^* R_2 = R_1.$$

In other words, for all vector fields X, Y, Z on U_1,

$$\phi_* \left(R_1(X, Y)(Z) \right) = R_2(\phi_* X, \phi_* Y)(\phi_* Z).$$

Proof. According to the definition

$$R(X, Y)(Z) = \nabla_X(\nabla_Y Z) - \nabla_Y(\nabla_X Z) - \nabla_{[X,Y]} Z,$$

the statement follows from Theorem 5.6.15 and Proposition 4.7.13 (using here the notation of the Lie bracket). □

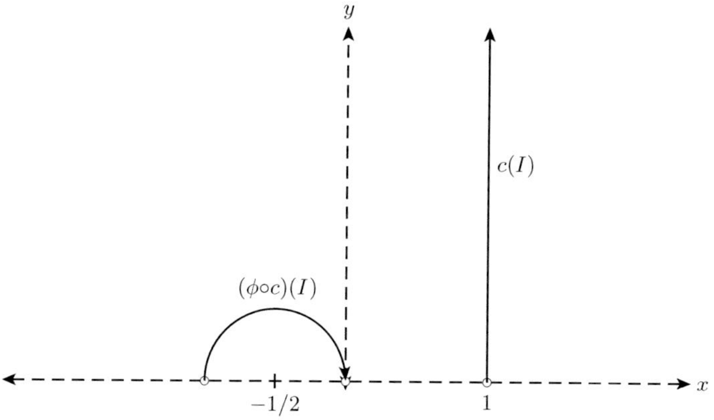

Fig. 5.11 Graphs of some geodesics in the Poincaré upper half-plane.

Corollary 5.6.19. *The following curvature-related tensors are preserved by isometries:*

- *the* $(0,4)$ *curvature tensor* R*;*
- *the Ricci curvature tensor* Ric*;*
- *the scalar curvature* S*.*

Proof. We leave the proof as an exercise in writing the statements precisely and then applying Theorem 5.6.18. □

Incidentally, read in the context of Theorem 5.6.11, the preceding statement says that for a geometric set, all of the curvature concepts are *independent of parameterization.*

We mention in particular one instance of Corollary 5.6.19, the "remarkable theorem" (theorem egregium) proved by Gauss, recalling that the Gaussian curvature for two-dimensional domains coincides with sectional, Ricci, and scalar curvatures.

Corollary 5.6.20 (Theorem egregium). *Let* $U \subset \mathbf{R}^2$ *be a domain, and let* $f : U \to \mathbf{R}^3$ *be a regular parameterization of the surface* $S = f(U)$*. Let* $g = f^* g_0$ *be the metric tensor on* U *obtained by pulling back the standard Euclidean metric on* $\mathbf{R}^3$*. Then for all isometries* $\phi : (U,g) \to (U,g)$*, the Gaussian curvature* K *of* g *is invariant:*

$$K \circ \phi = K.$$

In order to give the reader a taste of the kind of theorems that are much sought after in Riemannian geometry, we state without proof a kind of local converse to Theorem 5.6.18.

Theorem 5.6.21. *Let* (U_1, g_1) *and* (U_2, g_2) *be Riemannian spaces of the same dimension and the same constant sectional curvature. Then they are locally* *isometric: For each pair of points $p \in U_1$ and $q \in U_2$, there exist a domain $V_1 \subset U_1$ containing p, a domain $V_2 \subset U_2$ containing q, and an isometry $\phi : V_1 \to V_2$ such that $\phi(p) = q$.*

For a proof, the reader may consult do Carmo [13, p. 158].

We close this section by giving some examples of the types of theorems characterizing the isometries of a given Riemannian metric tensor. Proofs can be found, for example, in [15].

Theorem 5.6.22. *Let ϕ be an isometry of $\mathbf{R}^2$ with the standard Euclidean metric g_0. Then ϕ can be expressed as a composition*

$$\phi = \phi_3 \circ \phi_2 \circ \phi_1,$$

where ϕ_1 is a translation, ϕ_2 is a rotation, and ϕ_3 is either the identity (if $\det(\phi_*) > 0$) or a reflection (if $\det(\phi_*) < 0$).

In particular, all isometries of the Euclidean metric are linear. This gives an indication of how rigid the partial differential equations defining the isometry really are. Also note that the flows of Killing vector fields (see Example 5.6.8) yield all possible isometries, up to reflection.

By way of comparison, we state the corresponding result for the Poincaré metric. For ease of notation, we will use complex notation

$$(x, y) = z = x + iy.$$

Theorem 5.6.23. *Let U be the upper half-plane with the Poincaré metric g. Let $\phi : (U, g) \to (U, g)$ be an isometry. Then ϕ can be expressed as a composition*

$$\phi = \phi_2 \circ \phi_1,$$

where

$$\phi_1(z) = \frac{az + b}{cz + d}, \qquad a, b, c, d \in \mathbf{R},\ ad - bc > 0,$$

and ϕ_2 is either the identity (if $\det(\phi_*) > 0$) or $\phi_2(z) = -\bar{z}$ (if $\det(\phi_*) < 0$).

Functions of the form $\phi(z) = \dfrac{az + b}{cz + d}$ have been studied extensively, and are known as *Möbius transformations*. The interested reader may consult, among many texts, the presentation by Schwerdtfeger [36].

Finally, while the two preceding theorems give an inkling of how restrictive the isometry condition is, it should be noted that for general Riemannian spaces, the *only* isometry might be the identity map. Riemannian spaces that have (relatively) many isometries, usually due to the presence of built-in symmetries, are hence of particular interest.

5.7 For Further Reading

The number of texts on Riemannian geometry, often in the guise of texts on differential geometry in general, is so large that it would be impossible to list even the high-quality treatments. In such cases, the author's tastes can hardly be disguised.

As we have already noted, most of the classical treatments of Riemannian geometry begin with a treatment of curves and surfaces. Dirk Struik's *Lectures on Classical Differential Geometry* [39], first published in 1950, is representative of how the subject was treated for decades, both in style of presentation and choice of topics.

Michael Spivak's *Comprehensive Introduction to Differential Geometry* [38] can be seen as an effort to bridge the classical approach with the modern. In the second volume, Spivak includes key writings of Riemann, followed by a discussion of how to understand them in modern notation and terminology. Spivak's volumes remain standard reference guides in differential geometry, next to the advanced volumes of Kobayashi and Nomizu [24].

Two important contributions paved the way to a wider undergraduate audience for Riemannian geometry. Do Carmo's *Differential Geometry of Curves and Surfaces* [14], first translated from Portuguese into English in 1976, set a gold standard in terms of exposition, and can be viewed as an introduction to his more advanced but perhaps even more readable *Riemannian Geometry* [13]. The latter has influenced the presentation in this chapter.

O'Neill's *Elementary Differential Geometry* [33] shares the readable style of do Carmo, and the notation is at this point more conventional. A more recent text, Kühnel's 2006 Differential Geometry: Curves–Surfaces–Manifolds [25], will likely become a standard, combining a strong motivation with a clear and rapid development into the heart of Riemannian geometry. Frank Morgan's 1998 text [32] deserves special mention, having a very similar audience and style as the current text, and features a chapter of examples of undergraduate research projects in the subject.

The more advanced treatments are characterized by two important features that are absent from our introductory presentation here. First, they present results in the context of manifolds, which generalize our use of domains in $\mathbf{R}^n$. This opens up questions about the relationship of topology to Riemannian geometry, which is a major subject of research that might be best illustrated by the (Riemannian-geometric) proof of the famous Poincaré conjecture. Second, they present in far greater depth the extent to which curvature determines, and is determined by, other geometric properties of a Riemannian space. This line of study makes it possible to make a strong claim that Riemannian geometry is the study of curvature, even more so than the study of the Riemannian metric tensor.

We will not even list the many more advanced texts that are available. Beyond do Carmo [13], we mention only the more recent *Riemannian Geometry* by Peterson [34], which includes important and illustrative modern examples, and also presents, to this author's knowledge, the first systematic use of the formulation of the

Riemannian connection in terms of the Lie and exterior derivatives that we have used as Definition 5.3.2.

5.8 Exercises

5.1. Given the metrics g below defined on a subset $U \subset \mathbf{R}^2$, along with a parameterized curve $c : I \to U$: (i) Write an explicit coordinate expression for g; and (ii) set up integrals representing the g-lengths of the curve c, evaluating the integral if possible. In these examples, g_0 represents the standard Euclidean metric structure on $\mathbf{R}^2$ or $\mathbf{R}^3$, as appropriate.

(a) $U = \mathbf{R}^2$, $g = \left(\frac{1}{1+x^2+y^2}\right) g_0$, $c : [0, 2\pi] \to U$ given by $c(t) = (2\cos t, 2\sin t)$;
(b) $U = \{(x, y) \mid x^2 + y^2 < r^2\}$, $g = \phi_1^* g_0$, where $\phi_1 : U \to \mathbf{R}^3$ is given by

$$\phi_1(u, v) = (u, v, \sqrt{r^2 - u^2 - v^2}),$$

$c : [0, 2\pi] \to U$ given by $c(t) = (\frac{r}{2}\cos t, \frac{r}{2}\sin t)$;
(c) $U = \{(u, v) \mid 0 < u < 2\pi, 0 < v < \pi\}$, $g = \phi_2^* g_0$, where $\phi_2 : U \to \mathbf{R}^3$ is given by

$$\phi_2(u, v) = (r\cos u \sin v, r \sin u \sin v, r\cos v),$$

$c : (0, \pi) \to U$ given by $c(t) = (t, t)$.

5.2. Prove Proposition 5.2.1.

5.3. Let g_0 be the standard Euclidean metric tensor on $\mathbf{R}^3$, and let

$$\phi(u, v) = \left(\frac{2u}{1 + u^2 + v^2}, \frac{2v}{1 + u^2 + v^2}, \frac{u^2 + v^2 - 1}{1 + u^2 + v^2}\right).$$

Give an explicit coordinate expression for $\phi^* g_0$. (The map ϕ is an example of a geometric construction known as *stereographic projection*; see Exercise 3.13. Note that $\phi(\mathbf{R}^2) = S \setminus \{(0, 0, 1)\}$, where $S \subset \mathbf{R}^3$ is given by

$$S = \{(x, y, z) \mid x^2 + y^2 + z^2 = 1\},$$

is the sphere of radius one. The metric $g = \phi^* g_0$ can then be considered a metric on $\phi(\mathbf{R}^2)$.)

5.4. Let $X_1, \ldots, X_n$ be smooth vector fields on a domain $U \subset \mathbf{R}^n$ such that at each point $p \in U$, the set $\{X_1(p), \ldots, X_n(p)\}$ forms a basis for T_pU. Show that there is a metric tensor g such that for $p \in U$,

$$g(p)(X_i(p), X_j(p)) = \delta_{ij}.$$

5.5. Write the coordinate expressions for the gradient of a smooth function according to the metrics given in Examples 5.1.5 through 5.1.8.

5.6. Write the coordinate expressions for the gradient of a smooth function according to the metrics on $\mathbf{R}^2$ given in Exercise 5.1. For each, include an explicit description of the isomorphism in Proposition 5.1.9.

5.7. Use Propositions 4.7.16 and 4.7.22 to show that Koszul's formula (5.3) is equivalent to Eq. 5.2 of Definition 5.3.2.

5.8. Use the method of Example 5.3.6 to show that the Christoffel symbols for the metric g of Example 5.2.5,

$$[g_{ij}] = \begin{bmatrix} 1+4u^2 & -4uv \\ -4uv & 1+4v^2 \end{bmatrix},$$

are given by

$$\Gamma^1_{11} = \frac{4u}{1+4u^2+4v^2}, \quad \Gamma^2_{11} = \frac{-4v}{1+4u^2+4v^2},$$
$$\Gamma^1_{12} = \Gamma^2_{12} = \Gamma^1_{21} = \Gamma^2_{21} = 0,$$
$$\Gamma^1_{22} = \frac{-4u}{1+4u^2+4v^2}, \quad \Gamma^2_{22} = \frac{4v}{1+4u^2+4v^2}.$$

Then repeat the verification using the coordinate calculation of Proposition 5.3.3.

5.9. Show that the Christoffel symbols for the Poincaré metric g on the upper half-plane $U = \{(x, y) \mid y > 0\}$ described in Example 5.1.7,

$$[g_{ij}] = \begin{bmatrix} 1/y^2 & 0 \\ 0 & 1/y^2 \end{bmatrix},$$

are given by

$$\Gamma^1_{11} = 0, \quad \Gamma^2_{11} = 1/y,$$
$$\Gamma^1_{12} = \Gamma^1_{21} = -1/y, \quad \Gamma^2_{12} = \Gamma^2_{21} = 0,$$
$$\Gamma^1_{22} = 0, \quad \Gamma^2_{22} = -1/y.$$

5.10. Let g be the metric tensor on $\mathbf{R}^3$ given in Example 5.1.8:

$$[g_{ij}] = \begin{bmatrix} 1 & 0 & 0 \\ 0 & 1+x^2 & x \\ 0 & x & 1 \end{bmatrix}.$$

Show that the Christoffel symbols for g are given by

$$\Gamma_{11}^1 = \Gamma_{11}^2 = \Gamma_{11}^3 = 0,$$

$$\Gamma_{12}^1 = \Gamma_{21}^1 = 0, \quad \Gamma_{12}^2 = \Gamma_{21}^2 = \frac{x}{2}, \quad \Gamma_{12}^3 = \Gamma_{21}^3 = \frac{1-x^2}{2},$$

$$\Gamma_{13}^1 = \Gamma_{31}^1 = 0, \quad \Gamma_{13}^2 = \Gamma_{31}^2 = \frac{1}{2}, \quad \Gamma_{13}^3 = \Gamma_{31}^3 = -\frac{x}{2},$$

$$\Gamma_{22}^1 = -x, \quad \Gamma_{22}^2 = \Gamma_{22}^3 = 0,$$

$$\Gamma_{23}^1 = \Gamma_{32}^1 = -\frac{1}{2}, \quad \Gamma_{23}^2 = \Gamma_{32}^2 = \Gamma_{23}^3 = \Gamma_{32}^3 = 0,$$

$$\Gamma_{33}^1 = \Gamma_{33}^2 = \Gamma_{33}^3 = 0.$$

5.11. For each of the metrics in Exercise 5.1, write explicitly the isomorphism $\theta_{Y,X}$ of Definition 5.3.2 for arbitrary vector fields $X = X^1\frac{\partial}{\partial x} + X^2\frac{\partial}{\partial y}$ and $Y = Y^1\frac{\partial}{\partial x} + Y^2\frac{\partial}{\partial y}$. Then find $\nabla_X Y$ and write the corresponding Christoffel symbols.

5.12. Let $E_1, \ldots, E_n$ be vector fields on the standard Riemannian space $(\mathbf{R}^n, g_0)$ such that at each point $p \in \mathbf{R}^n$, $\{E_1(p), \ldots, E_n(p)\}$ forms an orthonormal basis for $T_p\mathbf{R}^n$ relative to g_0. Show that the one-forms ω_{ij} defined in Exercise 4.13 satisfy, for all vector fields V,

$$\omega_{ij}(V) = g_0(\nabla_V E_i, E_j).$$

In the notation of Exercise 4.13,

$$\nabla_V E_i = (E_i)_*(V).$$

5.13. Let $U \subset \mathbf{R}^2$ be a domain and let $f : U \to \mathbf{R}$ be a smooth function. Let $S = \phi(U)$ be the geometric set described by the parameterization $\phi : U \to \mathbf{R}^3$ given by $\phi(u,v) = (u, v, f(u,v))$. Define $g = \phi^* g_0$ as the pullback of the standard Euclidean metric g_0 on $\mathbf{R}^3$.

(a) Show that g is given by

$$g = (1+f_u^2)\, du \otimes du + (f_u f_v)\, du \otimes dv + (f_u f_v)\, dv \otimes du + (1+f_v^2)\, dv \otimes dv,$$

where f_u and f_v are the respective partial derivatives of f.

(b) Let ∇ be the Riemannian connection on U corresponding to g and suppose that

$$\tilde{X} = \tilde{X}^1\frac{\partial}{\partial u} + \tilde{X}^2\frac{\partial}{\partial v}, \quad \tilde{Y} = \tilde{Y}^1\frac{\partial}{\partial u} + \tilde{Y}^2\frac{\partial}{\partial v}$$

are smooth vector fields on U. Show that

$$\nabla_{\tilde{X}}\tilde{Y} = \begin{bmatrix} \tilde{X}[\tilde{Y}^1] \\ \tilde{X}[\tilde{Y}^2] \end{bmatrix} + \left(\frac{H_f(\tilde{X},\tilde{Y})}{N^2}\right)\begin{bmatrix} f_u \\ f_v \end{bmatrix}, \tag{5.27}$$

where $N^2 = 1 + f_u^2 + f_v^2$ and

$$H_f(\tilde{X},\tilde{Y}) = f_{uu}\tilde{X}^1\tilde{Y}^1 + f_{uv}\tilde{X}^1\tilde{Y}^2 + f_{vu}\tilde{X}^2\tilde{Y}^1 + f_{vv}\tilde{X}^2\tilde{Y}^2$$

is the Hessian of f interpreted as a quadratic form.

(c) Show that

$$\begin{aligned} \Gamma_{11}^1 &= \frac{f_u \cdot f_{uu}}{N^2}, & \Gamma_{11}^2 &= \frac{f_v \cdot f_{uu}}{N^2}, \\ \Gamma_{12}^1 = \Gamma_{21}^1 &= \frac{f_u \cdot f_{uv}}{N^2}, & \Gamma_{12}^2 = \Gamma_{21}^2 &= \frac{f_v \cdot f_{uv}}{N^2}, \\ \Gamma_{22}^1 &= \frac{f_u \cdot f_{vv}}{N^2}, & \Gamma_{22}^2 &= \frac{f_v \cdot f_{vv}}{N^2}. \end{aligned} \tag{5.28}$$

(d) Show that if $X = \phi_*(\tilde{X})$ and $Y = \phi_*(\tilde{Y})$, then the tangential directional derivative $(D_X Y)_T$ on S defined in Example 5.3.1 coincides with the image of $\nabla_{\tilde{X}}\tilde{Y}$ under ϕ_*:

$$(D_X Y)_T = \phi_*(\nabla_{\tilde{X}}\tilde{Y}). \tag{5.29}$$

5.14. Use Proposition 5.13 to compute the Christoffel symbols for the standard metric $\phi^* g_0$ of the graphs in $\mathbf{R}^3$ of the following smooth functions (on appropriate domains):

(a) $f(u,v) = 2u + 3v$;
(b) $f(u,v) = \sqrt{u^2 + v^2}$;
(c) $f(u,v) = \sqrt{1 - u^2 - v^2}$.

5.15. Consider a parameterized curve $c : I \to \mathbf{R}^2$ written as $c(t) = (x(t), z(t))$, whereby we envision $\mathbf{R}^2$ as the xz-plane in $\mathbf{R}^3$. The *surface of revolution* obtained by revolving the image of c around the y-axis in $\mathbf{R}^3$ is parameterized by

$$\phi(t,u) = (x(t)\cos u, x(t)\sin u, z(t)).$$

(a) Write the coordinate expression for the metric $\phi^* g_0$, where g_0 is the standard Euclidean metric on $\mathbf{R}^3$.
(b) Compute the Christoffel symbols for the metric g in part (a).

5.16. Prove Theorem 5.3.14. Hint: Use Propositions 4.7.16, 4.7.22, and the fact that ∇ is torsion-free.

5.17. On the standard Riemannian space $(\mathbf{R}^3, g_0)$, let $X = \langle x, y, x\rangle$. Compute $\dfrac{DX}{dt}$ along the helix $c : \mathbf{R} \to \mathbf{R}^3$ given by $c(t) = (\cos t, \sin t, t)$.

5.18. On the Poincaré upper half-plane (U, g) of Example 5.4.3, let

$$X = -y\frac{\partial}{\partial x} + x\frac{\partial}{\partial y}.$$

Compute $\dfrac{DX}{dt}$ along the curve $c : [0,1] \to U$ given by:

(a) $c(t) = (t, 1)$;
(b) $c(t) = (t, 1 + t)$;
(c) $c(t) = (\cos(\pi t), 2 + \sin(\pi t))$.

5.19. Let (U, g) be the Riemannian space of Example 5.3.6. For the vector field $X = \langle -v, u \rangle$ on U, compute $\dfrac{DX}{dt}$ along the curve $c : [0, \pi] \to U$ given by $c(t) = (\cos t, \sin t)$.

5.20. For the Riemannian space (U, g) of Exercise 5.19, compute the parallel transport V of the tangent vector $\mathbf{v}_p = \langle 0, 1 \rangle_p \in T_pU$ along the curve $c : [0, 2\pi] \to \mathbf{R}^2$ given by $c(t) = (\cos t, \sin t)$, where $p = c(0) = (1, 0)$. Hint: To solve the system of differential equations for $V(t) = \langle V^1(t), V^2(t) \rangle$, use the substitution

$$u = (\cos t)V^1 + (\sin t)V^2.$$

5.21. Let (U, g) be the Poincaré upper half-plane, and for $p = (1, 0)$, consider the tangent vector $\mathbf{v}_p = \langle 0, 1 \rangle_p \in T_pU$. Find the parallel transport of $\mathbf{v}_p$ along $c : [0, 1] \to \mathbf{R}^2$ given by $c(t) = (t, t + 1)$.

5.22. Consider the unit sphere $S^2 \subset \mathbf{R}^3$ with the metric obtained by restricting the metric g_0 on $\mathbf{R}^3$. In other words, for $\mathbf{v}_p, \mathbf{w}_p \in T_p(S^2) \subset T_p(\mathbf{R}^3)$, define $g(\mathbf{v}_p, \mathbf{w}_p) = g_0(\mathbf{v}_p, \mathbf{w}_p)$, Let $c : I \to S^2$ be a parameterization of a great circle with constant speed, i.e., $g_0(\dot{c}(t), \dot{c}(t)) = k$ for all $t \in I$. Show that c is a geodesic on S by computing $\nabla_{\dot{c}}\dot{c}$ as the tangential directional derivative as in Example 5.3.1.

5.23. Show that for the upper half-plane with the Poincaré metric of Example 5.1.7, the vertical lines parameterized as $c(t) = (x_0, y_0e^{kt})$ are geodesics (where k is any constant), but the vertical lines parameterized by $c_1(t) = (x_0, y_0 + t)$ are not.

5.24. For each of the metrics in Exercise 5.1, compute R_{1212}.

5.25. Compute the six independent components of curvature tensor

$$R_{1212}, R_{1313}, R_{2323}, R_{1223}, R_{1323}, R_{1213}$$

for the metric of Example 5.1.8. Refer to Exercise 5.10.

5.26. Consider the Riemannian space $(\mathbf{R}^3, g)$, where g is the metric described in Example 5.5.15. As in that example, let $X_2 = \frac{\partial}{\partial y} - x\frac{\partial}{\partial z}$. Compute the sectional curvatures corresponding to the planes spanned by:

(a) $\left\{\frac{\partial}{\partial x}, \frac{\partial}{\partial z}\right\}$;

(b) $\left\{\frac{\partial}{\partial y}, \frac{\partial}{\partial z}\right\}$;

(c) $\left\{X_2, \frac{\partial}{\partial z}\right\}$.

5.27. Show that the Ricci curvature tensor is symmetric: For all vector fields X, Y, $Ric(X,Y) = Ric(Y,X)$.

5.28. All of the definitions of this chapter related to the covariant derivative, and the different curvature tensors carry over to the case of *pseudo-Riemannian metrics*, which are symmetric $(0,2)$-tensor fields that are nondegenerate in the sense that if a tangent vector X_p satisfies $g(X_p, Y_p) = 0$ for all tangent vectors Y_p, then $X_p = 0_p$. Pseudo-Riemannian metrics may not be positive definite. Such tensors provide the natural mathematical setting for general relativity. (For more information on pseudo-Riemannian metrics, the reader may consult [25] or [32].)

Let $U \subset \mathbf{R}^4$ be defined as

$$U = \left\{(r, \phi, \theta, t) \in \mathbf{R}^4 \mid 0 < r < 2G_0M,\ 0 < \phi < \pi,\ 0 < \theta < 2\pi\right\},$$

where G_0 and M are some physical constants. Consider the *Schwarzschild metric* tensor

$$g = -\left(1 - \frac{2G_0M}{r}\right)^{-1} dr \otimes dr - r^2 \left(d\phi \otimes d\phi + \sin^2 \phi d\theta \otimes d\theta\right) + \left(1 - \frac{2G_0M}{r}\right) dt \otimes dt.$$

Verify that the Ricci tensor Ric and the scalar curvature S are both identically zero.

5.29. Verify the result of Example 5.5.9 using the method of Proposition 5.5.4.

5.30. Verify the result of Example 5.5.10 using Definitions 5.5.1 and 5.5.3.

5.31. For each of the metrics in Exercise 5.1, write the differential equations that describe the Killing vector fields for the metric.

5.32. Show that X is a Killing vector field for the Riemannian space (U, g) if and only if for all vector fields Y, Z on U,

$$g(\nabla_Y X, Z) + g(\nabla_Z X, Y) = 0.$$

5.33. Let (U, g_1) be the Riemannian space of Example 5.6.12, i.e., $U \subset \mathbf{R}^2$ is given by

$$U = \left\{(u, v) \mid 0 < u^2 + v^2 < 1\right\},$$

and $g_1 = \phi_1^* g_0$, where g_0 is the standard Euclidean metric on $\mathbf{R}^3$ and $\phi_1 : U \to \mathbf{R}^3$ is given by

$$\phi_1(u,v) = (u, v, u^2 + v^2).$$

For fixed $\theta \in [0, 2\pi]$, let $\psi : U \to U$ be given by

$$\psi(u,v) = (u\cos\theta + v\sin\theta, -u\sin\theta + v\cos\theta).$$

Show that ψ is an isometry of (U, g_1). Interpret this result geometrically, with respect to both U and $S = \phi(U) \subset \mathbf{R}^3$.

5.34. For a domain $U \subset \mathbf{R}^2$, let $\phi : U \to \mathbf{R}^3$ be a regular parameterization with component functions $\phi(u,v) = (x(u,v), y(u,v), z(u,v))$. Let $\mathbf{x}_u$ and $\mathbf{x}_v$ be the vectors fields on $\mathbf{R}^3$ defined by

$$\mathbf{x}_u = \phi_*(\partial_u) = \left\langle \frac{\partial x}{\partial u}, \frac{\partial y}{\partial u}, \frac{\partial z}{\partial u} \right\rangle,$$

$$\mathbf{x}_v = \phi_*(\partial_v) = \left\langle \frac{\partial x}{\partial v}, \frac{\partial y}{\partial v}, \frac{\partial z}{\partial v} \right\rangle,$$

where $\partial_u = \frac{\partial}{\partial u}$ and $\partial_v = \frac{\partial}{\partial v}$ are the standard basis vector fields for $T\mathbf{R}^2$. For the standard Euclidean metric tensor g_0 on $\mathbf{R}^3$, define functions $E, F, G : S \to \mathbf{R}$ on $S = \mathbf{x}(U)$ by

$$E = (\phi^* g_0)(\partial_u, \partial_u) = g_0(\mathbf{x}_u, \mathbf{x}_u),$$

$$F = (\phi^* g_0)(\partial_u, \partial_v) = g_0(\mathbf{x}_u, \mathbf{x}_v),$$

$$G = (\phi^* g_0)(\partial_v, \partial_v) = g_0(\mathbf{x}_v . \mathbf{x}_v).$$

Finally, define vector fields E_1, E_2 on U by

$$E_1 = \frac{1}{\sqrt{E}}\partial_u, \quad E_2 = \frac{1}{\sqrt{E}\sqrt{D}}\left(-F\partial_u + E\partial_v\right),$$

where $D = EG - F^2$.

(a) Show that $\{E_1, E_2\}$ is an orthonormal frame of vector fields on U (in the sense of Exercise 4.13; see also Exercise 5.12) relative to the metric tensor $g = \phi^* g_0$.
(b) Again referring to Exercise 4.13, compute the one-forms ε_1, ε_2, and ω_{12}. In this context, ω_{12} is defined to be the one-form on U

$$\omega_{12}(V) = g_0((E_1)_*(V), E_2),$$

where $(E_1)_*$ is the tangent map of $\phi_* E_1 : U \to \mathbf{R}^3$.
(c) Let ∇ be the Riemannian connection on U associated to the metric tensor $g = \phi^* g_0$. Show that for all vector fields X on U,

$$g(\nabla_X E_1, E_2) = \omega_{12}(X).$$

(d) Show that the Gaussian curvature K of $g = \phi^* g_0$ is given by

$$K = \frac{1}{2\sqrt{D}}\left[\frac{\partial}{\partial u}\left(\frac{FE_v - EG_u}{E\sqrt{D}}\right) + \frac{\partial}{\partial v}\left(\frac{2EF_u - FE_u - EE_v}{E\sqrt{D}}\right)\right],$$

where E_u represents $\partial_u[E]$, etc.

(e) Show that $d\omega_{12} = -K\varepsilon_1 \wedge \varepsilon_2$, where K is the Gaussian curvature described in part (d).

5.35. Continuing Exercise 5.34, define $\tilde{E}_1 = \phi_* E_1$, $\tilde{E}_2 = \phi_* E_2$ as vector fields on the surface $S = \phi(U)$. Define $\tilde{E}_3$ to be a vector field on $\mathbf{R}^3$ for which at each point $p \in S$, $\left\{\tilde{E}_1(p), \tilde{E}_2(p), \tilde{E}_3(p)\right\}$ is an orthonormal basis for $T_p\mathbf{R}^3$ relative to the standard Euclidean metric g_0. For example, using the standard cross product in $\mathbf{R}^3$, we could define $\tilde{E}_3 = \tilde{E}_1 \times \tilde{E}_2$.

(a) Let $\varepsilon_1, \varepsilon_2, \varepsilon_3$ be the corresponding one-forms on $S \subset \mathbf{R}^3$, i.e., $\varepsilon_i = i(\tilde{E}_i)g_0$. Show that $\mathbf{v}_p \in T_pS$ if and only if $\varepsilon_3(p)(\mathbf{v}_p) = 0$. In particular, $\phi^*\varepsilon_3 = 0$.

(b) Define the *attitude matrix* $A = [a_{ij}]$ relative to the orthonormal frame $\left\{\tilde{E}_1, \tilde{E}_2, \tilde{E}_3\right\}$ to be the 3×3 matrix of smooth functions $a_{ij} : U \to \mathbf{R}$, where

$$\tilde{E}_1 = a_{11}\frac{\partial}{\partial x} + a_{12}\frac{\partial}{\partial y} + a_{13}\frac{\partial}{\partial z},$$

$$\tilde{E}_2 = a_{21}\frac{\partial}{\partial x} + a_{22}\frac{\partial}{\partial y} + a_{23}\frac{\partial}{\partial z},$$

$$\tilde{E}_3 = a_{31}\frac{\partial}{\partial x} + a_{32}\frac{\partial}{\partial y} + a_{33}\frac{\partial}{\partial z}.$$

Further, let dA be the 3×3 matrix of one-forms on U, $[dA]_{ij} = da_{ij}$. Let ω be defined as the formal matrix product $\omega = (dA)A^T$ and let $\omega_{ij} = [\omega]_{ij}$. Show that $\omega_{ji} = -\omega_{ij}$ and that ω_{ij} satisfy Cartan's structural equations in a manner exactly as in Exercise 4.13.

(c) Verify that ω_{12} agrees with the form of the same name in the previous exercise.

5.36. Let $\left\{\tilde{E}_1, \tilde{E}_2, \tilde{E}_3\right\}$ be the orthonormal frame adapted to a parameterized surface $S = \phi(U) \subset \mathbf{R}^3$ as in the previous example, with associated one-forms $\{\varepsilon_1, \varepsilon_2, \varepsilon_3\}$ and ω_{ij} for $i, j = 1, 2, 3$. Show that for all vector fields V on S,

$$\nabla_V \tilde{E}_i = \sum_{j=1}^{3} \omega_{ij}(V)\tilde{E}_j,$$

where ∇ is the Riemannian connection associated with g_0. (For this reason, the forms ω_{ij} are called the *connection forms* associated to the orthonormal frame $\left\{\tilde{E}_1, \tilde{E}_2, \tilde{E}_3\right\}$.)

5.37. For a parameterized surface $S = \phi(U) \subset \mathbf{R}^3$ with the adapted orthonormal frame $\left\{\tilde{E}_1, \tilde{E}_2, \tilde{E}_3\right\}$, define the map $N : TS \to TS$ by

$$N(V) = \nabla_V \tilde{E}_3 = \omega_{13}(V)\tilde{E}_1 + \omega_{23}(V)\tilde{E}_2,$$

where V is a vector field on S. The map N is known as the *shape map* associated to the parameterization of S.

(a) Show that N is a linear transformation.
(b) Show that for vector fields V, W on S,

$$g_0(N(V), W) = g_0(V, N(W)).$$

(c) Show that $\det(N) = K$, where as in Exercise 5.34, K is the Gaussian curvature of S.

Chapter 6
Contact Geometry

Despite the effort that might be required to master the constructions of the previous chapter, the core concerns of Riemannian geometry are traditional ones: length, angles, curvature, etc. Indeed, the way we view the underlying set of spatial "points" in Riemannian geometry is not much different from the way that Euclid described them over one thousand years ago.

The development of projective geometry in the nineteenth century challenged that view of spatial points. Ideas like "points at infinity" and "lines as points" ultimately put an end to the centuries-long efforts to deduce Euclid's fifth postulate from the other axioms. The freedom to consider "points" as objects more general than spatial objects was one of the main conceptual developments that opened the door to modern geometries, and to contact geometry in particular.

Another development that paved the way for new, non-Euclidean geometries like contact geometry was the line of work initiated by Felix Klein in his 1872 Erlangen program. According to Klein, the proper objects and relations of geometry were those that remained invariant under a given collection of transformations. Euclidean geometry, for example, was the study of those objects that are preserved under the set of translations, rotations, and reflections. Klein's perspective, combined with Sophus Lie's 1890 use of "contact transformations" for the study of partial differential equations, gave further impetus to the modern theory of contact geometry.

Contact geometry slowly emerged as its own distinct field of study beginning in the 1950s. It experienced a surge of attention in the 1990s thanks to developments in other areas of mathematical research. The first textbook on contact geometry at the graduate level, by Hansjörg Geiges [19], was published only in 2008. Nevertheless, key ideas of contact geometry can be seen as early as Huygens' 1690 formulation of geometric optics. Enthusiasts cite geometer V.I. Arnold's claim, "Contact geometry is all geometry."[1]

[1]Quoted in H. Geiges, *An Introduction to Contact Topology*, Cambridge University Press, New York, 2008, p. ix.

A. McInerney, *First Steps in Differential Geometry: Riemannian, Contact, Symplectic*, Undergraduate Texts in Mathematics, DOI 10.1007/978-1-4614-7732-7_6,

We begin this chapter with two expositions of topics in which contact geometry provides a natural setting, the first in physics and the second in mathematics. We do this with the understanding that the reader will undoubtedly be less familiar with the basic concepts in play, and we hope that the expositions will motivate and justify later definitions.

Our presentation of contact geometry then proceeds in a manner parallel to the previous chapter's presentation of Riemannian geometry. The structure tensor of contact geometry, the contact form, will be a nondegenerate differential one-form. After introducing this form, we identify the objects that arise naturally from the definition of the contact form. We note from the outset, however, that the proper object of study in contact geometry is not the contact form itself, but the associated contact hyperplane. The consequences of this sentence will be seen throughout the chapter, most notably in the definition of the structure-preserving transformations, the contact diffeomorphisms.

One important difference between contact geometry and Riemannian geometry is the fact that all contact structures are "locally the same." This is the content of Darboux's theorem, which we present in its own section. For the field as a whole, this has the consequence of putting a far greater weight on topology than on calculus-based techniques like those presented in the previous chapter. Natural questions include, for example, establishing which topological spaces (or manifolds) can support a contact structure. These questions are outside the scope of this text. We will try to compensate, however, by introducing special geometric sets, Legendre sets, which are interesting in their own right.

Finally, we note that for ease of exposition, most of this chapter will be situated in $\mathbf{R}^3$, although, as we have already indicated, we might not always be envisioning $\mathbf{R}^3$ in the familiar way as a set of points in space. In the last section of this chapter, we will indicate how the definitions translate to higher (odd!) dimensions.

6.1 Motivation I: Huygens' Principle and Contact Elements

One of the fundamental goals of classical physics is to describe a physical theory of light. The competing "particle theories" and "wave theories" of light, and the way in which these conflicting theories were harmonized by Maxwell's theory of electromagnetism (and later further refined by theories of quantum electrodynamics) is a dramatic story that is properly told in a first course in electricity and magnetism. One of the early attempts to formulate a theory of light and the associated physical phenomena of reflection and refraction was carried out by Christiaan Huygens in his 1690 text *Treatise on Light*. In this section, we give a brief outline of his model, although in modern terminology. This is an instance of what is generally known as geometric optics. In so doing, we show how basic concepts of contact geometry have a natural role in setting up a mathematical model for Huygens' principle.

Huygens' description of the propagation of light begins with a model of a point source emitting light rays that travel through the "ether" at finite (but very high)

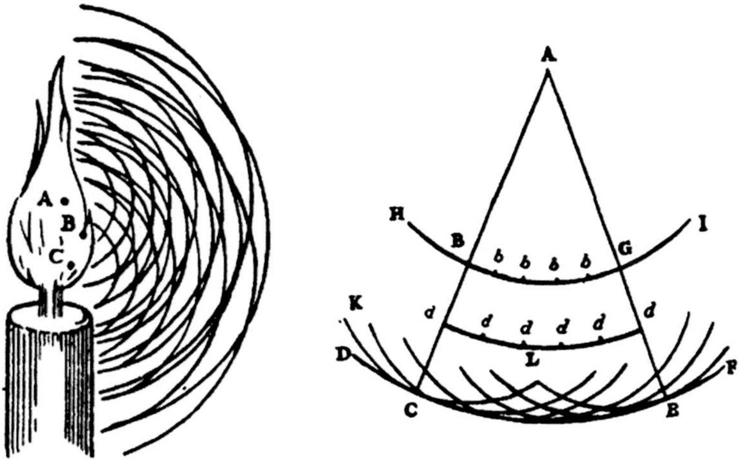

Fig. 6.1 Illustrations from a 1912 edition of Huygens' *Treatise on Light* (University of Chicago Press, tr. Silvanus Thompson).

speed. An "elementary wave" consists of all points that would be reached by all rays emitted from a single point source in some fixed time. Geometrically, assuming that the medium permitted constant speed for light rays, an elementary ray would be represented by a sphere whose center was the point source.

More important for his later explanations of reflection and refraction, Huygens made the further assumption that each point on the wave becomes, in turn, a new point source for further elementary waves. His original rationale for this assumption was based on the analogy of a billiard ball transmitting momentum to neighboring balls.

Moreover, since the speed of the light rays from the secondary source wave was the same as that of the original elementary wave, the secondary wave should, in Huygens' terminology, be "touching" (i.e., tangent to) the original elementary wave. In geometric language that we will make more precise below, a light wave propagates in such a way that the resultant wave at time $t_0 + \Delta t$ is obtained as the *envelope* of elementary waves obtained from points on the wave at time t_0. See Fig. 6.1.

Here is a summary of Huygens' ideas in geometric terminology, following the exposition by Geiges [17, 18].

> HUYGENS' PRINCIPLE: Every point of a wave is the source of an elementary wave. The wavefront at a later time is the envelope of these elementary waves.

Our goal in the remainder of this section is to provide a mathematical framework for Huygens' principle. We begin using the language of vector calculus. Then, defining "points" differently, we will rework the formulation in a way that introduces some key elements of contact geometry. Throughout this section, we illustrate the

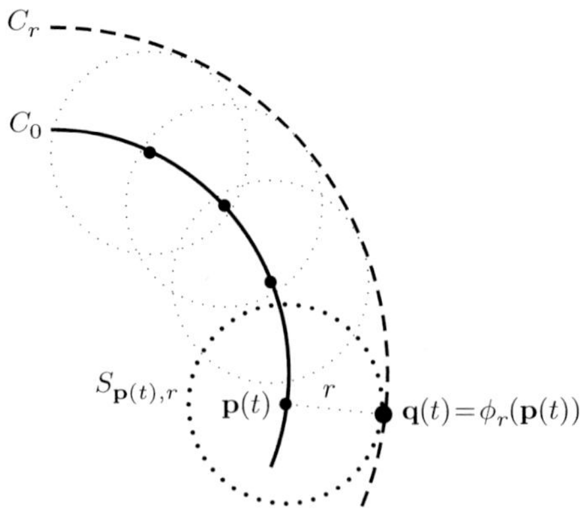

Fig. 6.2 The wavefront map ϕ_r.

concepts by describing *plane* waves, which we describe as curves in $\mathbf{R}^2$. Elementary waves, then, will be circles instead of spheres.

Let the initial wavefront be described by a curve $C_0 \subset \mathbf{R}^2$, which we will assume to be smoothly parameterized by $\mathbf{p} : I \to \mathbf{R}^2$ with components $\mathbf{p}(t) = (p_1(t), p_2(t))$. Moreover, for each $\mathbf{p}(t) \in C_0$, let $S_{\mathbf{p}(t),r}$ be a circle ("elementary wave") centered at $\mathbf{p}(t)$ with radius r, with r representing the distance traveled by a light ray in some fixed time. Let $\mathcal{W}_r$ represent the collection of all such circles $S_{\mathbf{p}(t),r}$, where $t \in I$.

Definition 6.1.1. The *envelope* C_r of the set $\mathcal{W}_r$ described in the preceding paragraph is a curve in $\mathbf{R}^2$ satisfying the following properties:

- Each point $\mathbf{q} \in C_r$ is a point on exactly one circle $S_{\mathbf{p}(t),r}$, i.e., for each $\mathbf{q} \in C_r$, there exists a unique $\mathbf{p}(t) \in C_0$ such that $\mathbf{q} \in S_{\mathbf{p}(t),r}$.
- The tangent line to C_r at $\mathbf{q}$ coincides with the tangent line to $S_{\mathbf{p}(t),r}$ at $\mathbf{q}$, where $\mathbf{q} \in S_{\mathbf{p}(t),r}$.

We will assume the existence of such envelopes for the remainder of this discussion.

Our goal is to express Huygens' principle by means of a diffeomorphism

$$\phi_r : \mathbf{R}^2 \to \mathbf{R}^2,$$

where $r \geq 0$ is a given distance. This diffeomorphism will describe the evolution of the wavefront in the following sense: For a wavefront C_0, the evolution of C_0 through a distance r will be given by $C_r = \phi_r(C_0)$, and points $\mathbf{p}(t) \in C_0$ are mapped to the corresponding point $\mathbf{q} = \phi_r(\mathbf{p}(t)) \in C_r$. To do this, note that the same parameter t used to parameterize C_0 can be used to parameterize C_r, writing $\mathbf{q}(t) = (q_1(t), q_2(t))$ as the unique point on C_r such that $\mathbf{q}(t) \in S_{\mathbf{p}(t),r}$. See Fig. 6.2. We make the nontrivial assumption that the parameterization $\mathbf{q}(t)$ is smooth—an assumption that can easily be false in general envelope constructions.

In this framework, then, the two conditions of Definition 6.1.1 can be formulated using vector calculus notation as

$$(\mathbf{q}(t) - \mathbf{p}(t)) \cdot (\mathbf{q}(t) - \mathbf{p}(t)) = r^2, \tag{6.1}$$

$$\mathbf{q}'(t) \cdot (\mathbf{q}(t) - \mathbf{p}(t)) = 0. \tag{6.2}$$

Equation (6.1) expresses the fact that $\mathbf{q}(t)$ lies on the circle with center $\mathbf{p}(t)$ and radius r. Equation (6.2) expresses the fact that the tangent vector $\mathbf{q}'(t)$ to C_r is perpendicular to the radial vector $\mathbf{q}(t) - \mathbf{p}(t)$ for all t, and hence tangent also to the circle $S_{\mathbf{p}(t),r}$.

Differentiating Eq. (6.1) yields

$$(\mathbf{q}'(t) - \mathbf{p}'(t)) \cdot (\mathbf{q}(t) - \mathbf{p}(t)) = 0,$$

which together with Eq. (6.2) implies that

$$\mathbf{p}'(t) \cdot (\mathbf{q}(t) - \mathbf{p}(t)) = 0.$$

In components, this is written

$$p_1'(q_1 - p_1) + p_2'(q_2 - p_2) = 0.$$

For convenience, we assume for the moment that $p_1'(t) \neq 0$ for all t, so we can write

$$q_1 - p_1 = -m(q_2 - p_2), \tag{6.3}$$

where $m(t) = p_2'(t)/p_1'(t)$ is the slope of the tangent line to C_0 at $\mathbf{p}(t)$.

Substituting Eq. (6.3) into (6.1) and solving for $(q_2 - p_2)^2$ yields

$$(q_2 - p_2)^2 = \frac{r^2}{1 + m^2},$$

or, choosing the positive square root for specificity,

$$q_2 = p_2 + \frac{r}{\sqrt{1 + m^2}}.$$

This along with Eq. (6.3) implies that

$$q_1 = p_1 - \frac{mr}{\sqrt{1 + m^2}}.$$

The preceding discussion shows how the conditions of Huygens' principle translate into the fact that the map

$$\phi_r(p_1, p_2) = \left(p_1 - \frac{mr}{\sqrt{1+m^2}}, p_2 + \frac{r}{\sqrt{1+m^2}} \right) \tag{6.4}$$

maps points $\mathbf{p} = (p_1, p_2) \in C_0$ to points $\mathbf{q} = \phi_r(\mathbf{p}) \in C_r$.

The problem with the map ϕ_r so defined is that it does not depend on $\mathbf{p} = (p_1, p_2)$ alone. The quantity we have labeled m depends not just on the point $\mathbf{p}(t)$ but also on the parameterization $\mathbf{p} : I \to \mathbf{R}^2$. In order to address this problem, we introduce a new setting for describing wavefronts: the set of "contact elements."

Definition 6.1.2. A *contact element in* $\mathbf{R}^2$ is a pair (P, ℓ) consisting of a point $P \in \mathbf{R}^2$ along with a nonvertical line $\ell \subset \mathbf{R}^2$ such that $P \in \ell$. The set of all contact elements in $\mathbf{R}^2$ will be denoted by $C\mathbf{R}^2$.

There is a standard system of coordinates for $C\mathbf{R}^2$, namely,

$$C\mathbf{R}^2 = \left\{ (x, y, m) \mid (x, y) \in \mathbf{R}^2,\ m \in \mathbf{R} \right\},$$

where the ordered triple (x, y, m) corresponds to the contact element (P, ℓ), with m the slope of the line ℓ through $P = (x, y)$. Hence as a set, $C\mathbf{R}^2$ is the same as $\mathbf{R}^3$. In particular, we can consider $T_p(C\mathbf{R}^2) = T_p(\mathbf{R}^3)$ for all $p = (x, y, m)$. We maintain the notational distinction in order to emphasize that we are not envisioning "points" in $C\mathbf{R}^2$ as spatial points in the same way that we traditionally view points in $\mathbf{R}^3$. See Fig. 6.3.

Every smooth curve $c : I \to \mathbf{R}^2$ with nowhere-vertical tangent line gives rise to a curve $\tilde{c} : I \to C\mathbf{R}^2$ as follows: For all $t \in I$, define

$$\tilde{c}(t) = (x(t), y(t), m(t)),$$

where $c(t) = (x(t), y(t))$ and

$$m(t) = \frac{\dot{y}(t)}{\dot{x}(t)}$$

is the slope of the tangent line to $c(I)$ at the point $c(t) = (x(t), y(t))$. We will call the curve $\tilde{c}$ the *lift of* c *to* $C\mathbf{R}^2$.

Suppose that the images of two curves $c_1 : I_1 \to \mathbf{R}^2$ and $c_2 : I_2 \to \mathbf{R}^2$ intersect at a point $P \in \mathbf{R}^2$. The images of the corresponding lifts $\tilde{c_1}$, $\tilde{c_2}$ intersect in $C\mathbf{R}^2$ only when the images of c_1 and c_2 have a common tangent line ℓ at P. In that case, c_1 and c_2 are said to be "in contact" (to first order) at p, whence the name "contact elements".

Not every curve in $C\mathbf{R}^2$ can be considered the lift of a curve in $\mathbf{R}^2$. In fact, a necessary and sufficient condition for a curve $\gamma : I \to C\mathbf{R}^2$ given by components $\gamma(t) = (x(t), y(t), m(t))$ to be the lift of a curve $c : I \to \mathbf{R}^2$ is that for all t,

$$m(t) = \frac{\dot{y}(t)}{\dot{x}(t)},$$

again under the assumption that $\dot{x}(t) \neq 0$ for all t.

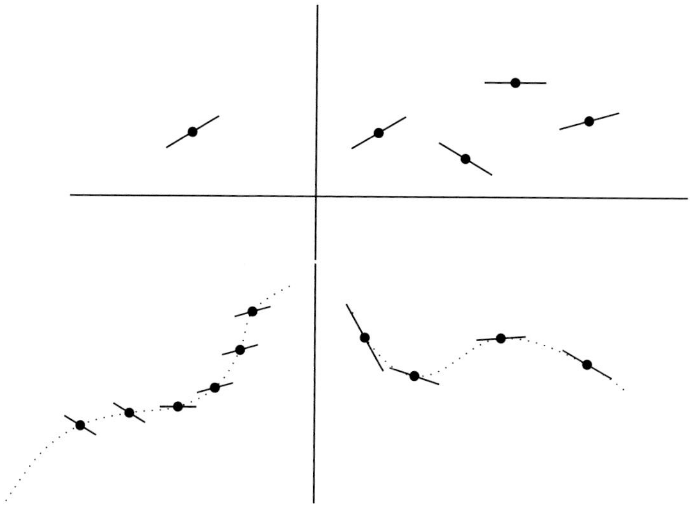

Fig. 6.3 Contact elements and curves in $C\mathbf{R}^2$.

The above discussion can be framed nicely in the language of differential forms. With the standard coordinates (x, y, m) on $C\mathbf{R}^2$, consider the special differential one-form

$$\alpha = dy - m dx.$$

At each point $p = (x, y, m) \in C\mathbf{R}^2$, $\alpha(p)$ defines a vector subspace $E_p \subset T_p(C\mathbf{R}^2)$:

$$E_p = \ker \alpha(p) = \left\{ \mathbf{v}_p \in T_p(C\mathbf{R}^2) \mid \alpha(p)(\mathbf{v}_p) = 0 \right\};$$

see Fig. 6.4.

In this terminology, we can rephrase the condition for γ to be the lift of a curve in $\mathbf{R}^2$ as follows:

Proposition 6.1.3. *Let $\gamma : I \to C\mathbf{R}^2$ be a curve in $C\mathbf{R}^2$, with $\gamma(t) = (x(t), y(t), m(t))$ and $\dot{x}(t) \neq 0$ for all $t \in I$. Consider the one-form $\alpha = dy - mdx$ on $C\mathbf{R}^2$, whose kernel we denote by $E_p = \ker \alpha(p)$. Then γ is the lift of a curve c in $\mathbf{R}^2$ if and only if $\dot{\gamma}(t) \in E_{\gamma(t)}$ for all $t \in I$. Equivalently, γ is the lift of a curve c if and only if $\gamma^* \alpha = 0$.*

For this reason, we now consider a wavefront in $C\mathbf{R}^2$ to be a curve $\gamma : I \to C\mathbf{R}^2$ that is the lift of a curve $c : I \to \mathbf{R}^2$.

Returning to our effort to express Huygens' principle in terms of a diffeomorphism, we see that the proper setting for the wavefront map ϕ_r above is as a map

$$\phi_r : C\mathbf{R}^2 \to C\mathbf{R}^2.$$

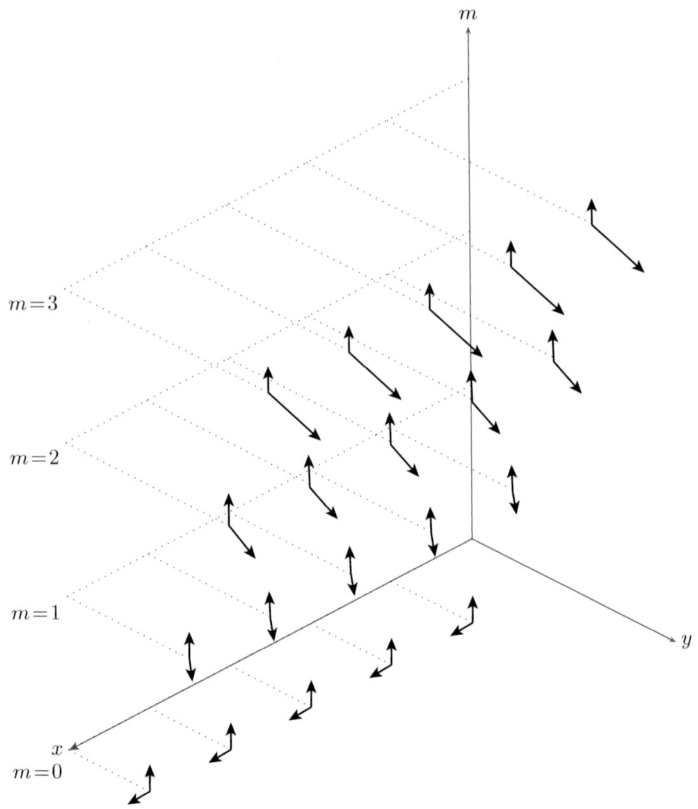

Fig. 6.4 Bases for the subspaces $E_p = \ker \alpha_p$.

Proposition 6.1.4. *The map* $\phi_r : C\mathbf{R}^2 \to C\mathbf{R}^2$ *given by*

$$\phi_r(x, y, m) = \left(x - \frac{rm}{\sqrt{1+m^2}}, y + \frac{r}{\sqrt{1+m^2}}, m \right)$$

is a diffeomorphism.

Proof. See Exercise 3.22. □

In the framework we have been developing, the most important property of ϕ_r is expressed in the following proposition.

Proposition 6.1.5. *Let* $\phi_r : C\mathbf{R}^2 \to C\mathbf{R}^2$ *be the diffeomorphism of Proposition 6.1.4, and let* $\alpha = dy - m\,dx$ *be the special one-form from Proposition 6.1.3. Then*

$$\phi_r^* \alpha = \alpha.$$

Proof.

$$\begin{aligned}
\phi_r^*\alpha &= d\left(y + \frac{r}{\sqrt{1+m^2}}\right) - md\left(x - \frac{rm}{\sqrt{1+m^2}}\right) \\
&= \left(dy - \frac{rm}{(1+m^2)^{3/2}}dm\right) - m\left(dx - \frac{r}{(1+m^2)^{3/2}}dm\right) \\
&= dy - mdx \\
&= \alpha. \qquad \square
\end{aligned}$$

An immediate consequence of Proposition 6.1.5 is the following statement.

Corollary 6.1.6. *Suppose $\gamma_0 : I \to C\mathbf{R}^2$ is a wavefront. Then $\gamma_r = \phi_r \circ \gamma_0$ is also a wavefront. In other words, ϕ_r maps wavefronts to wavefronts.*

Proof. We assume that γ_0 is the lift of a curve in $\mathbf{R}^2$, so by Proposition 6.1.3 we have $\gamma_0^*\alpha = 0$ for the one-form $\alpha = dy - mdx$. But then

$$\begin{aligned}
(\phi_r \circ \gamma_0)^*\alpha &= \gamma_0^*\phi_r^*\alpha \\
&= \gamma_0^*\alpha \qquad \text{by Proposition 6.1.5} \\
&= 0.
\end{aligned}$$

Hence, again by Proposition 6.1.3, γ_r is the lift of a curve on $\mathbf{R}^2$. □

We conclude this section by highlighting the role of the special differential form α in the formulation of Huygens' principle in the setting of contact elements. First, the one-form α allows us to distinguish which curves in $C\mathbf{R}^2$ represent wavefronts. The fact that the diffeomorphism ϕ_r constructed according to the requirements of Huygens' principle has the property that ϕ_r preserves α thus translates into the fact that ϕ_r transforms wavefronts into wavefronts, and so describes a model for the propagation of light.

6.2 Motivation II: Differential Equations and Contact Elements

In the previous section, we saw how the set of contact elements provides a setting for a geometric description of Huygens' principle for the propagation of light. Huygens did not use the mathematical framework of contact elements. The first place that the set of contact elements ("lineal elements") was used explicitly was in Sophus Lie's work in the late nineteenth century. Much of Lie's work can be seen as an attempt to use geometric and algebraic techniques for analyzing problems in differential equations. In this section, we present some of these ideas in the context of solving

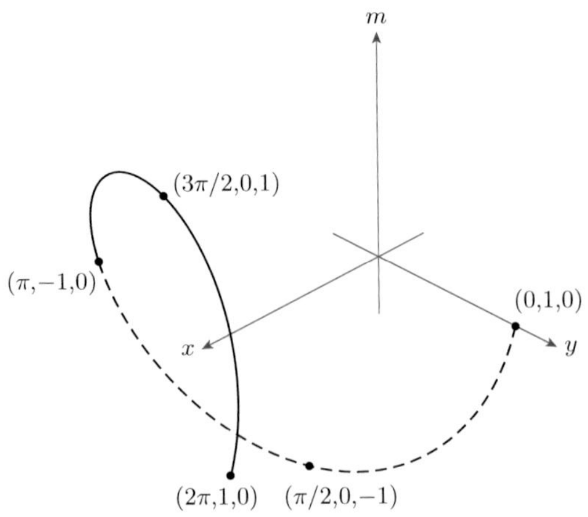

Fig. 6.5 The image of the one-jet of the function $u : \mathbf{R} \to \mathbf{R}$ given by $u(t) = \cos t$.

first-order ordinary differential equations, although our treatment really derives from methods designed for first-order *partial* differential equations. We will return to this more general setting at the end of the chapter.

Consider for the sake of illustration the initial value problem of finding solutions $u : \mathbf{R} \to \mathbf{R}$ to the ordinary differential equation

$$(2u - t)u' = u - t, \quad u(0) = 1, \tag{6.5}$$

where u is considered a function of the independent variable t. The reader is invited to review which methods encountered in an elementary course in ordinary differential equations might be applied to solve this equation. Our goal is to formulate a different approach to this ordinary differential equation by recasting it in the setting of contact elements.

Recall that a curve $c : I \to \mathbf{R}^2$ gives rise to a lift $\tilde{c} : I \to C\mathbf{R}^2$ defined by $\tilde{c}(t) = (x(t), y(t), m(t))$, where $c(t) = (x(t), y(t))$ and

$$m(t) = \frac{\dot{y}(t)}{\dot{x}(t)}$$

(provided, of course, that $\dot{x}(t) \neq 0$ for $t \in I$). In the same way, a smooth real-valued function $u : \mathbf{R} \to \mathbf{R}$ gives rise to a curve $\tilde{u} : \mathbf{R} \to C\mathbf{R}^2$ as follows:

$$\tilde{u}(t) = (t, u(t), u'(t)) .$$

The parameterized curve $\tilde{u}$ is called the *one-jet* of the function u. The one-jet can be thought of as the graph of u in the set of contact elements. See Fig. 6.5.

Note that $\tilde{u}^*\alpha = 0$ for the special one-form $\alpha = dy - mdx$. The partial converse of this statement is slightly more involved than what was encountered in Proposition 6.1.3.

Proposition 6.2.1. *Let $c : I \to C\mathbf{R}^2$ be a smooth curve in $C\mathbf{R}^2$,*

$$c(s) = (x(s), y(s), m(s)),$$

satisfying $c^*\alpha = 0$, where $\alpha = dy - mdx$. Suppose further that there is $s_0 \in I$ such that $\dot{x}(s_0) \neq 0$. Then there exist an interval I_1 containing $x(s_0)$ and a function $g : I_1 \to \mathbf{R}$ such that the image of the one-jet $\tilde{g} : I_1 \to C\mathbf{R}^2$ coincides with the image of $c : I_0 \to C\mathbf{R}^2$ on some subinterval I_0 of I.

Proof. Let $I_0 \subset I$ be an interval containing s_0 such that $\dot{x}(s) \neq 0$ for all $s \in I_0$, so that the function x is monotone on I_0. By the one-dimensional version of the inverse function theorem, x is invertible on I_0: There exist an interval I_1 containing $x(s_0)$ and a one-to-one, onto function $\chi : I_1 \to I_0$ such that $(\chi \circ x)(s) = s$ for all $s \in I_0$ and $(x \circ \chi)(t) = t$ for all $t \in I_1$.

Define $g : I_1 \to \mathbf{R}$ by $g(t) = (y \circ \chi)(t)$. Note that for each $t \in I_1$, there is $s = \chi(t) \in I_0$ such that $x(s) = t$. Further,

$$\begin{aligned} g'(t) &= (y \circ \chi)'(t) \\ &= y'(\chi(t)) \cdot \chi'(t) \\ &= y'(s) \cdot \frac{1}{x'(s)} \\ &= m(s) \qquad \text{since } c^*\alpha = 0. \end{aligned}$$

All this together shows that the image of the one-jet $\tilde{g} : I_1 \to C\mathbf{R}^2$ with component functions $\tilde{g}(t) = (t, g(t), g'(t))$ coincides with the image of $c : I_0 \to C\mathbf{R}^2$ given by $c(s) = (x(s), y(s), m(s))$. In fact, $c \circ \chi = \tilde{g}$. □

Consider now a first-order ordinary differential equation of the form

$$F(t, u, u') = 0, \tag{6.6}$$

where we consider $F : C\mathbf{R}^2 \to \mathbf{R}$ to be a smooth real-valued function. Equation (6.5) at the outset of this section is an example of such an ordinary differential equation, with

$$F(x, y, m) = (2y - x)m + x - y.$$

We will say that a function $u : I \to \mathbf{R}$ is a solution to the ordinary differential equation Eq. (6.6) if the graph of the one-jet $\tilde{u}$ of u lies on the level set

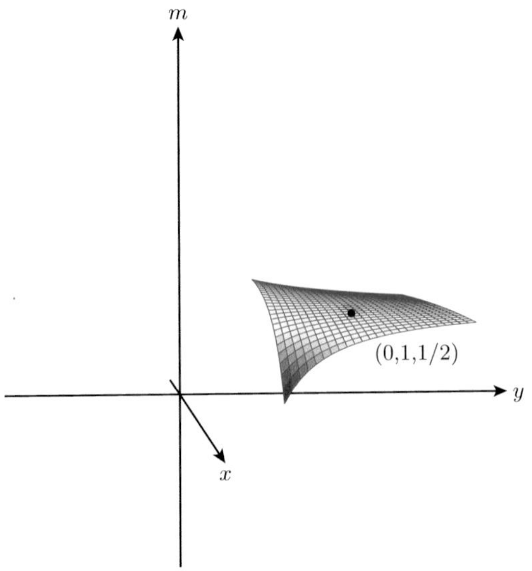

Fig. 6.6 The surface S_F in $C\mathbf{R}^2 = \mathbf{R}^3$ when $F(x,y,m) = (2y - x)m + x - y$.

$$S_F = \left\{(x,y,m) \in C\mathbf{R}^2 \mid F(x,y,m) = 0\right\},$$

or equivalently, if $F \circ \tilde{u} = 0$. In other words, u is a solution to Eq. (6.6) if $\tilde{u}$ is a parameterized curve on the geometric set S_F.

An initial condition of the form $u(t_0) = u_0$ corresponds to the statement that the graph of the one-jet $\tilde{u}$ also intersects the line in $C\mathbf{R}^2$ given by $\{(x,y,m) \mid x = t_0, y = u_0\}$. In our model example, the initial condition $u(0) = 1$, combined with the condition that $F(x,y,m) = 0$, together amount to the requirement that $(0,1,1/2) \in \text{Image}(\tilde{u})$. See Fig. 6.6.

Having described a first-order ordinary differential equation as a two-dimensional level set in the three-dimensional space of contact elements $C\mathbf{R}^2$, we will now outline a geometric technique to solve the ordinary differential equation. The solution, which as we have seen above will be viewed as a curve in $C\mathbf{R}^2$, will be obtained by means of the existence theorem for systems of ordinary differential equations in the guise of integrating a particular vector field, called the characteristic vector field. After outlining the method, known as the method of characteristics, we will then adjust the method according to a technique developed by Lie.

In order to employ this method in the setting described above, we must assume that the one-forms $\alpha = dy - m dx$ and dF are linearly independent at all points of the level set S_F. The reader can verify that for $(x,y,m) \in S_F$, this condition is satisfied when either $F_m \neq 0$ or $F_x \neq -mF_y$, where F_x, F_y, and F_m are the partial derivatives with respect to the indicated variables. This condition ensures that at

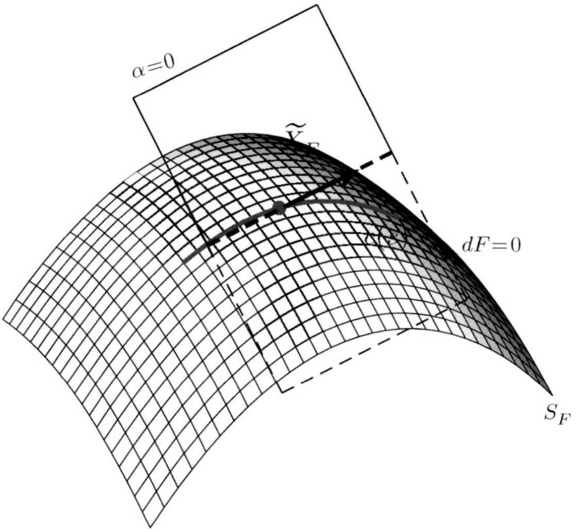

Fig. 6.7 The characteristic vector field $\tilde{X}_F$.

each point $p = (x, y, m)$ of S_F, the tangent plane $T_pS_F \subset T_pC\mathbf{R}^2$ intersects the plane $E_p \subset T_pC\mathbf{R}^2$ defined by

$$E_p = \ker \alpha(p) = \left\{ Y_p \in T_pC\mathbf{R}^2 \mid \alpha(p)(Y_p) = 0 \right\}$$

in a one-dimensional subspace of $T_pC\mathbf{R}^2$ for all $p \in S_F$.

Definition 6.2.2. Let $F : C\mathbf{R}^2 \to \mathbf{R}$ be a smooth function and let S_F be the level set $\{F = 0\}$ subject to the assumption that α and dF are linearly independent for all $p \in S_F$. Define $\tilde{X}_F = \left\langle \tilde{X}_F^1, \tilde{X}_F^2, \tilde{X}_F^3 \right\rangle$ to be the vector field uniquely determined by the following three conditions:

- $\alpha(\tilde{X}_F) = 0$;
- $dF(\tilde{X}_F) = 0$;
- $\tilde{X}_F^1 = -F_m$.

In coordinates,

$$\tilde{X}_F = \langle -F_m, -mF_m, F_x + mF_y \rangle .$$

The field $\tilde{X}_F$ is called the *characteristic vector field* associated with the function F. See Fig. 6.7.

The assumption that dF and α are independent at each $p \in S_F$ ensures that the vector field $\tilde{X}_F$ is nowhere zero, and hence spans the vector subspace $T_pS_F \cap E_p$. The third condition in Definition 6.2.2 is a normalization condition that ensures that $\tilde{X}_F$ is unique.

Theorem 6.2.3. *With F, $\tilde{X}_F$, and S_F as in Definition 6.2.2, let $c : I \to C\mathbf{R}^2$ described as $c(s) = (x(s), y(s), m(s))$ be the integral curve of $\tilde{X}_F$ through a point $c(0) = (x_0, y_0, m_0) \in S_F$. Then there is an interval I_1 such that the function $u : I_1 \to \mathbf{R}$ defined by $u(t) = (y \circ \chi)(t)$ is a solution of the initial value problem described by $F(t, u, u') = 0$ and $u(x_0) = y_0$. Here, as in the proof of Proposition 6.2.1, $x \circ \chi = \mathrm{Id}$.*

Proof. Most of the work has already been expended in the constructions. By the chain rule,

$$\begin{aligned}\frac{d}{ds}\left(F(c(s))\right) &= \frac{\partial F}{\partial x}\frac{dx}{ds} + \frac{\partial F}{\partial y}\frac{dy}{ds} + \frac{\partial F}{\partial m}\frac{dm}{ds} \\ &= dF(\tilde{X}_F) \qquad \text{since } c \text{ is an integral curve of } \tilde{X}_F \\ &= 0, \qquad \text{by the definition of } \tilde{X}_F.\end{aligned}$$

Hence F is constant along c, and so

$$F(c(s)) = F(c(0)) = F(x_0, y_0, m_0) = 0,$$

since $c(0) \in S_F$.

Further, the condition that $\alpha(\tilde{X}_F) = 0$ is equivalent to $c^*\alpha = 0$, which by Proposition 6.2.1 implies that c coincides, up to reparameterization, with the lift of a function $u = y \circ \chi$. Hence u represents a solution to the ordinary differential equation Eq. (6.6) satisfying the initial condition $u(x_0) = y_0$. □

Theorem 6.2.3 summarizes what is called the *method of characteristics*: In order to find solutions of a differential equation, find the integral curves of the corresponding characteristic vector field. Lie adapted the method by exploiting the fact that S_F is not just any level set, but the zero level set.

Proposition 6.2.4. *Let $\tilde{X}_F$ be the characteristic vector field associated to the smooth function F as above. Define a new vector field X_F as follows: $X_F = \tilde{X}_F + F\partial_y$, or in coordinates, $X_F = \langle -F_m, F - mF_m, F_x + mF_y\rangle$. Then the integral curve of X_F through a point $p \in S_F$ agrees with the integral curve of $\tilde{X}_F$ through p.*

Proof. According to the proof of Theorem 6.2.3, $F = 0$ along the integral curves $\tilde{c}$ of $\tilde{X}_F$. However, at all points $p \in S_F$, we have $X_F(p) = \tilde{X}_F(p)$. So by the uniqueness of flows, the integral curves c of X_F and $\tilde{c}$ of $\tilde{X}_F$ through a point $p \in S_F$ must coincide. □

The vector field X_F defined in Proposition 6.2.4 will be called the *Lie characteristic vector field* of F.

One of the goals of this chapter will be to recast the results of this section into the language of contact geometry, including the geometric distinction between the characteristic vector field $\tilde{X}_F$ and the Lie characteristic field X_F. We only point out here that for the purposes of solving ordinary differential equations, it is often easier to integrate to obtain the flow lines of X_F than to obtain those of $\tilde{X}_F$.

Example 6.2.5. We employ the method of characteristics to solve the initial value problem (6.5). Recall that the corresponding smooth function $F : C\mathbf{R}^2 \to \mathbf{R}$ is

$$F(x, y, m) = (2y - x)m + x - y.$$

Hence the characteristic vector field is given by

$$\tilde{X}_F = \left\langle x - 2y, (x - 2y)m, 2m^2 - 2m + 1 \right\rangle,$$

while the Lie characteristic field for F is given by

$$X_F = \left\langle x - 2y, x - y, 2m^2 - 2m + 1 \right\rangle.$$

Note that the system of ordinary differential equations associated to finding the integral curves of X_F decouples into a linear system of first-order equations in x and y, along with a separable ordinary differential equation involving only the dependent variable m:

$$\begin{cases} \frac{dx}{ds} &= x - 2y, \\ \frac{dy}{ds} &= x - y, \\ \frac{dm}{ds} &= 2m^2 - 2m + 1. \end{cases}$$

Using the standard techniques for solving first-order linear systems of ordinary differential equations, the reader can verify that

$$(x(s), y(s)) = (-2\sin s, \cos s - \sin s)$$

satisfies the first two equations with the initial conditions

$$(x(0), y(0)) = (0, 1).$$

We could now proceed to solve the third equation for m. However, we already have enough information to solve the differential equation (6.5). After all, writing

$$t = x(s) = -2\sin s,$$

we obtain $s(t) = \sin^{-1}(-t/2)$ on the interval $I_1 = (-2, 2)$ ($s(t)$ is what we called $\chi(t)$ in the discussion above), so by Proposition 6.2.1,

$$u(t) = y(s(t)) = \frac{\sqrt{4 - t^2}}{2} + \frac{t}{2}$$

is a solution to the differential equation

$$(2u - t)u' = u - t, \quad u(0) = 1.$$

As mentioned, the third equation in this case was superfluous to solving the differential equation. We leave it as an exercise to verify that $m(s) = \dfrac{y'(s)}{x'(s)}$ satisfies both the third differential equation arising from the characteristic vector field and $m'(s) = u'(t)$, where $t = -2 \sin s$.

As we have mentioned, the importance of the method of characteristics cannot be fully appreciated in the setting of ordinary differential equations. We will later show its full power in the setting of first-order partial differential equations. We have presented the method in this context only to show how the set of contact elements provides a natural setting for first-order differential equations—Lie's main interest in initially taking the first formal steps in contact geometry. As in Sect. 6.1, the reader will take note of the role played by a special one-form α. The special one-form is just one of several basic objects of contact geometry that are already in play in the method of characteristics.

6.3 Basic Concepts

Having spent the prior two sections showcasing what might be called the first example of a "contact space," the set of contact elements, we turn to the general setting. In the course of doing so, we will have a chance to look back and see several of the objects wc have encountered so far during the motivations in a more general setting.

In this section, as mentioned in the introduction to this chapter, we will work only in $\mathbf{R}^3$, although we may be picturing it differently—for example as $C\mathbf{R}^2$, the set of contact elements. All of the definitions and theorems generalize to higher dimensions, which will be the subject of the final section.

The primary object of study in contact geometry is a plane field $E \subset T\mathbf{R}^3$ with certain properties that we will detail below. The reader can keep in mind the role that $E = \ker \alpha$ played in the description of wavefronts in our formulation of Huygens' principle and again in the method of characteristics.

A plane field, which we briefly encountered in Chap. 6 in our discussion of sectional curvature, is defined to be a smoothly varying family of two-dimensional subspaces $E_p \subset T_p\mathbf{R}^3$, smooth in the sense that there are smooth vector fields X, Y such that for all $p \in \mathbf{R}^3$, $\{X(p), Y(p)\}$ is a basis for E_p. A plane field on $\mathbf{R}^3$ can also be described, however, as the kernel of a nowhere-zero one-form. In fact, in this chapter we will consider only plane fields of this type. So given a plane field $E \subset T\mathbf{R}^3$, we will assume that there is a smooth, nowhere-zero one-form α such that for all $p \in \mathbf{R}^3$,

$$E_p = \left\{X_p \in T_p\mathbf{R}^3 \mid \alpha(p)(X_p) = 0\right\}.$$

Note that α does not *uniquely* describe the plane field E. In fact, for any nowhere-zero function $f : \mathbf{R}^3 \to \mathbf{R}$, $\ker \alpha = \ker(f\alpha)$.

Definition 6.3.1. A *contact form* on $\mathbf{R}^3$ is a one-form α on $\mathbf{R}^3$ that is nondegenerate in the following sense: For all $p \in \mathbf{R}^3$,

$$\alpha_p \wedge d\alpha_p \neq 0,$$

i.e., for any basis $\{X_p, Y_p, Z_p\} \subset T_p\mathbf{R}^3$, we have $(\alpha_p \wedge d\alpha_p)(X_p, Y_p, Z_p) \neq 0$. The *contact distribution* E_α associated to α is the plane field defined by $E_\alpha = \ker \alpha$.

We will illustrate the significance of the nondegeneracy condition repeatedly throughout the chapter.

In analogy with Definition 5.1.2, we say that a domain U with a contact form α is a *contact space*, which we write (U, α). In this chapter, we will almost exclusively consider $U = \mathbf{R}^3$ or, in the last section, $U = \mathbf{R}^{2n+1}$.

We will now turn to some examples.

Example 6.3.2. Consider $\mathbf{R}^3$ with the usual coordinates (x, y, z). Let

$$\alpha_0 = x dy + dz.$$

We will call α_0 the *standard contact form* on $\mathbf{R}^3$. Note that $d\alpha_0 = dx \wedge dy$, and so

$$\alpha_0 \wedge d\alpha_0 = dx \wedge dy \wedge dz,$$

the standard volume three-form on $\mathbf{R}^3$.

The contact distribution is seen to be

$$E_0 = \left\{\left\langle X^1, X^2, X^3\right\rangle \mid X^3 = -xX^2\right\},$$

which has as a basis, for example, the vector fields

$$\{\langle 1, 0, 0\rangle, \langle 0, 1, -x\rangle\}.$$

See Fig. 6.8.

Example 6.3.3. The set of contact elements $C\mathbf{R}^2$ can be considered the set $\mathbf{R}^3$ with coordinates (x, y, m). The special one-form that we encountered in the first two sections,

$$\alpha_1 = dy - m dx,$$

is a contact form, since $\alpha_1 \wedge d\alpha_1 = -dx \wedge dy \wedge dm$. The corresponding contact distribution is given by

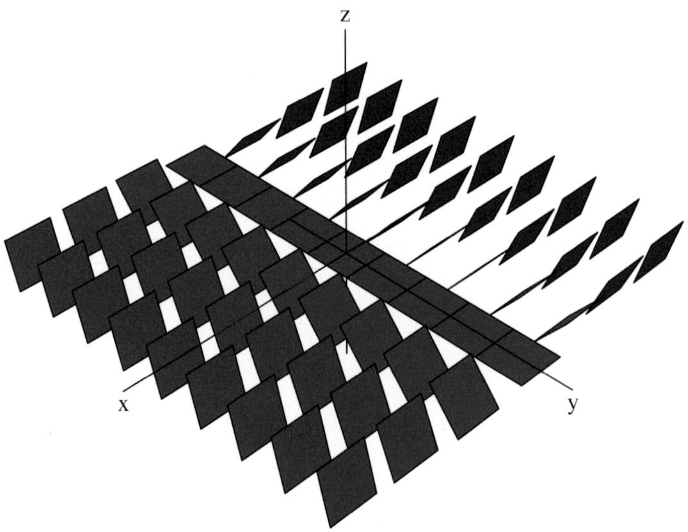

Fig. 6.8 The contact distribution for the standard contact structure induced by α_0.

$$E_1 = \{\langle X^1, X^2, X^3 \rangle \mid X^2 = mX^1\}$$

with basis $\{\langle 1, m, 0\rangle, \langle 0, 0, 1\rangle\}$.

In this setting, we can rephrase Proposition 6.1.3 as saying that a parameterized curve $c : I \to \mathbf{R}^3$ is the lift of a curve in $\mathbf{R}^2$ if and only if $\dot{c}(t)$ is in the contact distribution for all $t \in I$. We also note that the characteristic vector field $\tilde{X}_F$ of a smooth function $F : \mathbf{R}^3 \to \mathbf{R}$ (see Definition 6.2.2) satisfies $\tilde{X}_F(p) \in (E_1)_p$ for all $p \in \mathbf{R}^3$.

Example 6.3.4. Define

$$\alpha_2 = dz + xdy - ydx.$$

Since $\alpha_2 \wedge d\alpha_2 = 2dx \wedge dy \wedge dz$, α_2 is a contact form with contact distribution E_2 spanned by the vector fields $V_1 = \langle 1, 0, y\rangle$ and $V_2 = \langle 0, 1, -x\rangle$.

Example 6.3.5. Let $\alpha_3 = (\cos z)dx + (\sin z)dy$. We have

$$d\alpha_3 = (\sin z)dx \wedge dz - (\cos z)dy \wedge dz,$$

so

$$\alpha_3 \wedge d\alpha_3 = -dx \wedge dy \wedge dz,$$

and hence α_3 is a contact form. The contact distribution E_3 is spanned by the vector fields

$$\{\langle 0, 0, 1\rangle, \langle -\sin z, \cos z, 0\rangle\}.$$

Example 6.3.6. Consider the one-form on $\mathbf{R}^3$ expressed in cylindrical coordinates as

$$\alpha_4 = \cos r\, dz + r \sin r\, d\theta,$$

where as usual, $r^2 = x^2+y^2$ and $\tan\theta = y/x$. There are issues with the smoothness of α_4 that are masked somewhat by the cylindrical coordinates, in particular along the z-axis (when $r=0$). Explicitly, we have

$$d\theta = \frac{-y\,dx + x\,dy}{r^2},$$

and so, written in rectangular coordinates,

$$\alpha_4 = \cos r\, dz + \left(\frac{\sin r}{r}\right)(-y\,dx + x\,dy),$$

$$d\alpha_4 = \left(\frac{-x\sin r}{r}\right) dx \wedge dz - \left(\frac{y\sin r}{r}\right) dy \wedge dz + \left(\cos r + \frac{\sin r}{r}\right) dx \wedge dy.$$

In order to address the differentiability issues when $r=0$, we can replace $\dfrac{\sin r}{r}$ with

$$f(r) = \begin{cases} \frac{\sin r}{r} & \text{if } r > 0, \\ 1 & \text{if } r = 0. \end{cases}$$

It is an exercise in calculus to show that f is not only continuous (from the right) at $r=0$, but also differentiable (from the right) at $r=0$. Although we will continue to write α_4 in terms of $\frac{\sin r}{r}$, we will in fact mean f, and so α_4 is differentiable.

The calculations above show that

$$\alpha_4 \wedge d\alpha_4 = \left(1 + \cos r \cdot \frac{\sin r}{r}\right) dx \wedge dy \wedge dz,$$

which is nowhere zero, and so α_4 is a contact form. The contact distribution E_4 is spanned by the vector fields

$$\left\{(\cos r)\partial_x + y\left(\frac{\sin r}{r}\right)\partial_z, (\cos r)\partial_y - x\left(\frac{\sin r}{r}\right)\partial_z\right\},$$

or in cylindrical coordinates,

$$\{\partial_r, (\cos r)\partial_\theta - (r\sin r)\partial_z\}.$$

We will have many occasions to return to these examples throughout this chapter and in the exercises. For the rest of the section, though, we will investigate the meaning and some of the implications of the nondegeneracy condition for a contact form.

The differential form $\alpha \wedge d\alpha$, as a three-form on $\mathbf{R}^3$, must be of the form $f dV$, where dV is a basis three-form for the one-dimensional vector space $\Lambda(\mathbf{R}^3)$, for example $dx \wedge dy \wedge dz$. Hence the nondegeneracy condition $\alpha \wedge d\alpha \neq 0$ implies that $\alpha \wedge d\alpha$ differs from dV by a factor of a nowhere-zero function f. For this reason, $\alpha \wedge d\alpha$ may be regarded as a "volume form" on $\mathbf{R}^3$ when α is nondegenerate.

Referring to a fixed volume form dV, it is possible to classify α as "positive" or "negative" according to whether the corresponding f is everywhere positive or everywhere negative. For example, relative to the standard volume form $dx \wedge dy \wedge dz$ on $\mathbf{R}^3$, the contact forms in Examples 6.3.3 and 6.3.5 are negative, while those in Examples 6.3.4 and 6.3.6 are positive. In fact, in $\mathbf{R}^3$, being positive or negative depends only on the contact distribution E: For any nowhere-zero function g, the reader can check that the forms α and $g\alpha$ are either both positive or both negative.

These preliminary comments on the nondegeneracy condition have had an algebraic character. We now turn to the more geometric side of the condition. In particular, we will discuss the meaning of the nondegeneracy condition of α from the perspective of the contact distribution E_α. To do so, we turn to the question of "integrating" a plane field.

Consider first the example of a smooth function $F : \mathbf{R}^3 \to \mathbf{R}$. Assuming that $dF(p) \neq 0$ for all $p \in \mathbf{R}^3$, the function F defines a plane field E_F on $\mathbf{R}^3$ in a natural way: For any $p \in \mathbf{R}^3$, let S_p be the level surface of F through p, and let $E_F(p) = T_p S_p \subset T_p \mathbf{R}^3$. In fact, $E_F = \ker dF$.

This example raises the following question: Starting with an arbitrary smooth plane field $E \subset T\mathbf{R}^3$, is there a smooth function $F : \mathbf{R}^3 \to \mathbf{R}$ such that $E = \ker dF$ at each point $p \in \mathbf{R}^3$? If so, E can be considered to be tangent to the level sets of F, and E is said to be *integrable* (also called *completely integrable*). The terminology comes from the analogy of "integrating" a one-form dF to obtain the function F. Indeed, the question is directly analogous to "integrating" a vector field (a "line field") to obtain integral curves.

The answer to the question of the preceding paragraph is given by Frobenius's theorem. For the sake of the statement of the theorem we define a *k-dimensional distribution* $E \subset T\mathbf{R}^n$ ($k \leq n$) to be a smoothly varying k-dimensional subspace of the tangent space. We can think of a k-dimensional distribution as the subspace spanned by k smoothly varying vector fields $X_1, \ldots, X_k$ that are linearly independent at each point.

Theorem 6.3.7 (Frobenius). *Let E be a k-dimensional distribution in $\mathbf{R}^n$ with the property that for all $p \in \mathbf{R}^n$ and $X_p, Y_p \in E_p$, we have $[X_p, Y_p] \in E_p$. Then E is locally integrable in the following sense: For each $p \in \mathbf{R}^n$, there exist a domain $U_p \subset \mathbf{R}^k$ and a smooth regular parameterization $\phi : U_p \to \mathbf{R}^n$ such that $p \in S = \phi(U_p)$ and such that for all $q \in S$, $E_q = T_q S$. Moreover, there exist a domain $V_p \subset \mathbf{R}^n$ containing p and a smooth function $F : V_p \to \mathbf{R}^{n-k}$ such that $S \cap V_p = F^{-1}(\mathbf{0})$.*

Conversely, if E is integrable, then for all $p \in \mathbf{R}^n$ and all vector fields X, Y such that $X_p, Y_p \in E_p$ for all $p \in \mathbf{R}^n$, $[X_p, Y_p] \in E_p$.

A plane field E with the property that $[X, Y] \in E$ for all $X, Y \in E$ is called *involutive*. As a consequence of Frobenius's theorem, note that if E is an involutive distribution of dimension $k = n - 1$, then there exist a domain V containing p and a smooth function $F : V \to \mathbf{R}$ such that for each $q \in V$, $E_q = \ker dF(q)$.

We will not present the proof. The interested reader may find a detailed discussion of the theorem and proof in [40]; see also the more intuitive presentation in [3, Appendix 4].

We will, however, prove an equivalent version of Frobenius's theorem assuming Theorem 6.3.7. The case that we will be most concerned with in this chapter will be that in which $k = 2$ and $n = 3$, i.e., the case that E is a plane field in $\mathbf{R}^3$.

Theorem 6.3.8. *Let E be a plane field in $\mathbf{R}^3$ defined as the kernel of a one-form α, so that $E = \ker \alpha$. Then E is integrable if and only if $\alpha \wedge d\alpha = 0$.*

Proof. Suppose first that E is integrable. Then there is a function $F : \mathbf{R}^3 \to \mathbf{R}$ such that $\ker \alpha = \ker dF$. This implies that there is a smooth, nowhere-zero function $a : \mathbf{R}^3 \to \mathbf{R}$ such that

$$\alpha = adF.$$

So $d\alpha = da \wedge dF$, and

$$\alpha \wedge d\alpha = (a\, dF) \wedge (da \wedge dF) = 0,$$

a consequence of Proposition 4.2.7.

On the other hand, suppose that $\alpha \wedge d\alpha = 0$. For $p \in \mathbf{R}^3$, let U, V be vector fields such that $\{U(q), V(q)\}$ is a basis for E_q at each point q near p. Let W be a vector field also defined near p such that $\{U, V, W\}$ is a basis for $T\mathbf{R}^3$, and in particular, $W(q) \notin E_q$. Then

$$\begin{aligned} 0 &= (\alpha \wedge d\alpha)(U, V, W) \\ &= \alpha(U)d\alpha(V, W) - \alpha(V)d\alpha(U, W) + \alpha(W)d\alpha(U, V) \\ &= \alpha(W)d\alpha(U, V) \quad \text{since } U, V \in \ker \alpha. \end{aligned}$$

Since $W \notin \ker \alpha$, this implies that $d\alpha(U, V) = 0$. But according to Proposition 4.7.22,

$$\begin{aligned} 0 &= d\alpha(U, V) \\ &= U\left[\alpha(V)\right] - V\left[\alpha(U)\right] - \alpha([U, V]) \\ &= -\alpha([U, V]), \end{aligned}$$

and so $[U, V] \in \ker \alpha = E$.

Since $\{U, V\}$ is (locally) a basis for E, for any two vector fields $X, Y \in E$, we have $[X, Y] \in E$ as a consequence of Proposition 4.7.6. Hence, by Theorem 6.3.7, E is integrable. □

This short detour in discussing the integrability of plane fields has been for the purpose of reformulating the nondegeneracy condition for contact forms:

The contact distribution $E = \ker\alpha$ is not integrable.

In fact, it is often said that the nondegeneracy condition for a contact form means that the contact distribution is "maximally nonintegrable," since not only is $\alpha \wedge d\alpha$ not identically zero, it is in fact nowhere zero.

Before concluding this section, we introduce one other fundamental object in contact geometry. The nondegeneracy of the contact form implies the existence of a distinguished vector field that is transverse to the contact distribution. That fact, in turn, is a consequence of the following analogue to Proposition 5.1.9.

Proposition 6.3.9. *Let Ω be a nowhere-zero three-form on $\mathbf{R}^3$, in the sense that for all $p \in \mathbf{R}^3$, $\Omega_p \neq 0_p$. For $p \in \mathbf{R}^3$, let $(\Lambda_2)_p$ be the vector space of two-forms on $\mathbf{R}^3$ at p. Then for all $p \in \mathbf{R}^3$, Ω induces a vector space isomorphism*

$$\tilde{\Omega}_p : T_p\mathbf{R}^3 \to (\Lambda_2)_p$$

defined by

$$\tilde{\Omega}_p(X_p) = i(X_p)\Omega_p.$$

Moreover, the association $p \mapsto \Omega_p$ is smooth in the sense that if X is a smooth vector field on $\mathbf{R}^3$, then $\tilde{\Omega}(X)$ is a smooth two-form on $\mathbf{R}^3$.

Proof. Exercise. □

The isomorphism of Proposition 6.3.9 will appear more than once in the following sections. The first consequence is the existence of the distinguished vector field mentioned above.

Theorem 6.3.10. *Let α be a contact form on $\mathbf{R}^3$. There is a unique vector field ξ with the properties that*

- *$\alpha(\xi) = 1$, and*
- *$i(\xi)d\alpha = 0$.*

The vector field ξ is called the Reeb field *of α.*

Proof. The fact that the contact form α is nondegenerate means exactly that the three-form $\Omega = \alpha \wedge d\alpha$ is nowhere zero and so satisfies the hypotheses of Proposition 6.3.9. So define

$$\xi = \tilde{\Omega}^{-1}(d\alpha).$$

Note that for any vector field X, we have

$$\begin{aligned}(i(\xi)d\alpha)(X) &= (i(\xi)(i(\xi)\Omega))(X) \\ &= \Omega(\xi, \xi, X) \\ &= 0,\end{aligned}$$

and so $i(\xi)d\alpha = 0$.

Further, if $E = \ker \alpha$ is the contact distribution of α, then $\xi \notin E$. Otherwise, if we had $\xi_p \in E_p$ for some $p \in \mathbf{R}^3$, we could find a basis $\{\xi_p, B_p^2, B_p^3\}$ of $T_p\mathbf{R}^3$ such that $\{\xi_p, B_p^2\}$ was a basis for the two-dimensional subspace $E_p \subset T_p\mathbf{R}^3$. But this, along with the fact that $i(\xi)d\alpha = 0$, would contradict the nondegeneracy of α:

$$\begin{aligned}0 &\neq (\alpha \wedge d\alpha)(\xi_p, B_p^2, B_p^3) \\ &= \alpha(\xi_p)d\alpha(B_p^2, B_p^3) - \alpha(B_p^2)d\alpha(\xi_p, B_p^3) + \alpha(B_p^3)d\alpha(\xi_p, B_p^2).\end{aligned}$$

Hence $\xi \notin E$ for all p.

To verify the second property of ξ, for each $p \in \mathbf{R}^3$, let X_p, Y_p be any two linearly independent vectors in E_p. Note that $\{\xi_p, X_p, Y_p\}$ then forms a basis for $T_p\mathbf{R}^3$. By a calculation exactly like the one in the preceding paragraph,

$$d\alpha(X_p, Y_p) \neq 0;$$

we may in fact assume that

$$d\alpha(X_p, Y_p) = 1$$

by replacing X_p with $\frac{1}{a}X_p$, where $a = d\alpha(X_p, Y_p)$ if necessary. Then

$$\begin{aligned}(\Omega_p)(\xi_p, X_p, Y_p) &= (i(\xi_p)\Omega_p)(X_p, Y_p) \\ &= d\alpha_p(X_p, Y_p) \\ &= 1,\end{aligned}$$

and so

$$\begin{aligned}1 &= (\alpha \wedge d\alpha)(\xi_p, X_p, Y_p) \\ &= \alpha(\xi_p)d\alpha(X_p, Y_p) - \alpha(X_p)d\alpha(\xi_p, Y_p) + \alpha(Y_p)d\alpha(\xi_p, X_p) \\ &= \alpha(\xi_p) \qquad \text{since } X_p, Y_p \in E_p \text{ and } d\alpha(X_p, Y_p) = 1.\end{aligned}$$

Finally, to show that ξ so defined is unique, assume that ξ' is a vector field such that $\alpha(\xi') = 1$ and $i(\xi')d\alpha = 0$. Then

$$\begin{aligned} i(\xi')(\Omega) &= i(\xi')\,(\alpha \wedge d\alpha) \\ &= \alpha(\xi')d\alpha - \alpha \wedge i(\xi')d\alpha \\ &= d\alpha, \end{aligned}$$

so $\tilde{\Omega}(\xi') = \tilde{\Omega}(\xi)$. But since $\tilde{\Omega}$ is one-to-one, we have $\xi' = \xi$. □

A number of times in the discussion and proofs above, we have noted that the Reeb field ξ of a contact form α is transverse to the contact distribution E_α, in the sense that ξ together with a basis for E_α yields a basis for $T\mathbf{R}^3$. For example, the Reeb field for the standard contact form $\alpha_0 = xdy + dz$ on $\mathbf{R}^3$ is $\xi_0 = \dfrac{\partial}{\partial z}$. In the exercises, the reader is invited to find the Reeb fields for the other contact forms in the examples above.

An immediate consequence of the transversality of the Reeb field to the contact distribution is the fact that every vector field can be decomposed into a "vertical" and a "horizontal" part.

Proposition 6.3.11. *Let α be a contact form on $\mathbf{R}^3$ with corresponding Reeb field ξ and contact distribution E. Then every smooth vector field X on $\mathbf{R}^3$ can be written uniquely as*

$$X = f\xi + H(X),$$

where $f : \mathbf{R}^3 \to \mathbf{R}$ is a smooth function and $H(X)$ is a smooth vector field such that for all $p \in \mathbf{R}^3$, $H_p(X) \in E_p$.

Proof. For any $p \in \mathbf{R}^3$, let

$$f(p) = \alpha(p)(X_p).$$

So defined, f is smooth by virtue of the smoothness of α and X.

Define $H(X) = X - f\xi$. Then $H(X) \in E$ by construction. Uniqueness follows from the definition of f. □

One calls $H(X)$ the *horizontal component of* X. A vector field X satisfying $H(X) = 0$, and hence parallel to the Reeb field, is called *vertical*. Likewise, a vector field for which $X = H(X)$ is called *horizontal*.

It is important to notice that the Reeb field is defined essentially in terms of the contact form α and not in terms of the contact distribution, which was identified at the beginning of this section as the essential object of study of contact geometry. In particular, for a nowhere-zero function $f : \mathbf{R}^3 \to \mathbf{R}$, the contact forms α and $f\alpha$ have the same contact distribution but different Reeb fields. In this sense, the Reeb field is not really an object of contact geometry per se, but rather an object of "contact dynamics." We will return later in the chapter to discuss the relationship between the Reeb field for a contact form α and that of $f\alpha$.

6.4 Contact Diffeomorphisms

Our emphasis in this text has been geometry as the study of a structure defined by a tensor field on the tangent bundle. In the previous chapter, we singled out from the set of all diffeomorphisms certain special ones—the isometries—that "preserve the Riemannian metric structure." Isometries were defined in terms of the metric tensor g as those diffeomorphisms ϕ for which $\phi^* g = g$.

The situation for contact geometry is complicated somewhat by the fact that the object we have identified as central to contact geometry is not the contact form α, but rather the contact distribution associated to the contact form. For that reason, we will distinguish between two special types of diffeomorphisms—those that preserve the distribution and those that preserve the contact form itself. One of the goals of this section is to highlight the distinctions between these.

Definition 6.4.1. Let (U_1, α_1) and (U_2, α_2) be contact spaces, where $U_1, U_2 \subset \mathbf{R}^3$ are domains, and let $\phi : U_1 \to U_2$ be a diffeomorphism. We say that ϕ is a *contact diffeomorphism* if there is a nowhere-zero function $f : U_1 \to \mathbf{R}$ such that

$$\phi^* \alpha_2 = f \alpha_1.$$

We write $\phi : (U_1, \alpha_1) \to (U_2, \alpha_2)$.

Before illustrating the definition, we prove a proposition that captures the essence of the condition $\phi^* \alpha_2 = f \alpha_1$.

Proposition 6.4.2. *For domains $U_1, U_2 \subset \mathbf{R}^3$, let (U_1, α_1) and (U_2, α_2) be contact spaces with associated contact distributions E_1 and E_2. Let $\phi : U_1 \to U_2$ be a diffeomorphism. Then $\phi : (U_1, \alpha_1) \to (U_2, \alpha_2)$ is a contact diffeomorphism if and only if*

$$\phi_*(E_1) = E_2.$$

Proof. First, assume that $\phi : (U_1, \alpha_1) \to (U_2, \alpha_2)$ is a contact diffeomorphism, so there is a nowhere-zero function $f : U_1 \to \mathbf{R}$ such that $\phi^* \alpha_2 = f \alpha_1$. For $p \in U_1$, let $Y_{\phi(p)} \in (E_2)_{\phi(p)}$ and let $X_p \in (E_1)_p$ be such that $(\phi_*)_p(X_p) = Y_{\phi(p)}$, since $(\phi_*)_p : T_p U_1 \to T_{\phi(p)} U_2$ is a vector space isomorphism. We have

$$\begin{aligned} 0 &= \alpha_2(\phi(p))(Y_{\phi(p)}) \\ &- \alpha_2(\phi(p))((\phi_*)_p(X_p)) \\ &= (\phi^* \alpha_2)(p)(X_p) \\ &= f(p) \alpha_1(p)(X_p). \end{aligned}$$

In other words, $X_p \in (E_1)_p$, since f is nowhere zero, and so $Y_{\phi(p)} \in (\phi_*)_p (E_1)_p$. This shows that $E_2 \subset \phi_*(E_1)$.

Similarly, let $X_p \in (E_1)_p$. Then

$$\begin{aligned}\alpha_2(\phi(p))\left((\phi_*)_p(X_p)\right) &= (\phi^*\alpha_2)(p)(X_p)\\ &= f(p)\alpha_1(p)(X_p)\\ &= 0,\end{aligned}$$

and so $(\phi_*)_p(X_p) \in (E_2)_{\phi(p)}$. Hence $\phi_*(E_1) \subset E_2$, which together with the previous paragraph shows that

$$\phi_*(E_1) = E_2.$$

Conversely, assume that ϕ is a diffeomorphism satisfying $\phi_*(E_1) = E_2$. Let ξ_1 be the Reeb field associated to α_1. Define $f : U_1 \to \mathbf{R}$ by

$$f(p) = \alpha_2(\phi(p))((\phi_*)_p((\xi_1)_p)) = (\phi^*\alpha_2)(p)((\xi_1)_p).$$

Note first that f is nowhere zero: If $f(p) = 0$, then $(\phi_*)_p(\xi_1)_p \in (E_2)_{\phi(p)}$, and so $(\xi_1)_p \in (E_1)_p$, since ϕ_* is an isomorphism and since $E_2 \subset \phi_*(E_1)$ by assumption. This contradicts the fact that $\alpha_1(\xi_1) = 1$.

We will now show that for f so constructed, $\phi^*\alpha_2 = f\alpha_1$, i.e., ϕ is a contact diffeomorphism. For $p \in U_1$, choose $X_p \in T_pU_1$. We write, following the decomposition given in Proposition 6.3.11,

$$X_p = V_p + H_p,$$

where $V_p = (\alpha_1(p)(X_p))\,(\xi_1)_p$ and $H_p = X_p - V_p \in (E_1)_p$. We have, using the linearity of ϕ_* and α,

$$\begin{aligned}(\phi^*\alpha_2)\,(p)(X_p) &= \alpha_2(\phi(p))((\phi_*)_p(X_p))\\ &= \alpha_2(\phi(p))((\phi_*)_p(V_p) + (\phi_*)_p(H_p))\\ &= \alpha_2(\phi(p))((\phi_*)_p(V_p)) + \alpha_2(\phi(p))((\phi_*)_p(H_p))\\ &= \alpha_2(\phi(p))((\phi_*)_p(V_p)) \quad \text{since } \phi_*(E_1) \subset E_2\\ &= (\alpha_1(p)(X_p)) \cdot \alpha_2(\phi(p))((\phi_*)_p(\xi_1)_p)\\ &= f(p) \cdot (\alpha_1(p)(X_p))\,.\end{aligned}$$

Hence $\phi^*\alpha_2 = f\alpha_1$, and the proposition is proved. □

The following definition is a more restrictive "special" type of diffeomorphism, which is closer in spirit to the other structure-preserving diffeomorphisms we see in this text (although further in spirit from the essence of contact geometry considered as the study of the contact distribution independent of the contact form).

Definition 6.4.3. Let (U_1, α_1) and (U_2, α_2) be contact spaces, where $U_1, U_2 \subset \mathbf{R}^3$ are domains, and let $\phi : U_1 \to U_2$ be a diffeomorphism. Then ϕ is a *strictly contact diffeomorphism* if

$$\phi^* \alpha_2 = \alpha_1.$$

Every strictly contact diffeomorphism is a contact diffeomorphism according to Definition 6.4.1, with $f = 1$, and so preserves the contact distribution. In addition, the following proposition shows that strictly contact diffeomorphisms preserve the Reeb field.

Proposition 6.4.4. *Let (U_1, α_1) and (U_2, α_2) be two contact spaces with corresponding Reeb fields ξ_1 and ξ_2. A contact diffeomorphism $\phi : (U_1, \alpha_1) \to (U_2, \alpha_2)$ is strictly contact if and only if $\phi_* \xi_1 = \xi_2$.*

Proof. Assume first that ϕ is strictly contact, so that $\phi^* \alpha_2 = \alpha_1$. Our goal will be to show that $\alpha_2(\phi_* \xi_1) = 1$ and $i(\phi_* \xi_1) d\alpha_2 = 0$, which by the uniqueness of the Reeb field proved in Theorem 6.3.10 will show that $\xi_2 = \phi_* \xi_1$.

The first of these conditions is straightforward:

$$\begin{aligned}
\alpha_2(\phi_* \xi_1) &= (\phi^* \alpha_2)(\xi_1) \\
&= \alpha_1(\xi_1) \qquad \text{by Definition 6.4.3} \\
&= 1 \qquad \text{by Theorem 6.3.10.}
\end{aligned}$$

To verify the second condition, for $p \in U_1$, choose any $W_{\phi(p)} \in T_{\phi(p)} U_2$. Since ϕ is a diffeomorphism, there is a tangent vector $V_p \in T_p U_1$ such that $(\phi_*)_p(V_p) = W_{\phi(p)}$. We have

$$\begin{aligned}
\left(i((\phi_*)_p(\xi_1)_p) d\alpha_2(\phi(p))\right)(W_{\phi(p)}) &= d\alpha_2(\phi(p)) \left((\phi_*)_p(\xi_1)_p, W_{\phi(p)}\right) \\
&= d\alpha_2(\phi(p)) \left((\phi_*)_p(\xi_1)_p, (\phi_*)_p(V_p)\right) \\
&= (\phi^* d\alpha_2)(p)((\xi_1)_p, V_p) \\
&= d(\phi^* \alpha_2)(p)((\xi_1)_p, V_p) \\
&= d\alpha_1(p)((\xi_1)_p, V_p) \\
&= 0 \quad \text{by Theorem 6.3.10.}
\end{aligned}$$

We have relied on the assumption that ϕ is strictly contact for the second-to-last equality. Together, these show that $\phi_* \xi_1 = \xi_2$, where ξ_2 is the Reeb field for α_2.

Now assume that ϕ is a contact diffeomorphism, so that $\phi^* \alpha_2 = f \alpha_1$ for some nowhere-zero function $f : U_1 \to \mathbf{R}$. We need to show that under the additional assumption that $\phi_* \xi_1 = \xi_2$, then $f(p) = 1$ for all $p \in U_1$. To see this, note that for all $p \in U_1$,

$$\begin{aligned}
f(p) &= i((\xi_1)_p)(f(p) \cdot \alpha(p)) \\
&= i((\xi_1)_p)(\phi^* \alpha_2)(p)
\end{aligned}$$

$$= \phi^*\left(i(\phi_*(\xi_1)_p)\alpha_2\right)(p) \quad \text{by Theorem 4.6.19(5)}$$
$$= \phi^*\left(i(\xi_2)\alpha_2\right)(p) \quad \text{by assumption}$$
$$= \phi^*(1)(p)$$
$$= 1. \qquad \Box$$

As we have mentioned, the central object of study in contact geometry is to be the contact distribution, not the contact form. For that reason, strictly contact diffeomorphisms that preserve the form itself are not the most important structure-preserving diffeomorphisms. However, because of the important role of the contact form in our presentation, and in particular the importance of the Reeb field, the more restrictive strictly contact diffeomorphisms will find a place throughout the chapter.

We now turn to examples of contact diffeomorphisms.

Example 6.4.5. Let $\alpha_0 = xdy + dz$ and $\alpha_1 = dy - mdx$ be the contact forms described in Examples 6.3.2 and 6.3.3. Then $\phi_1 : (\mathbf{R}^3, \alpha_0) \to (\mathbf{R}^3, \alpha_1)$ given by

$$\phi_1(x, y, z) = (y, z, -x)$$

satisfies $\phi_1^* \alpha_1 = \alpha_0$. Hence ϕ_1 is a strictly contact diffeomorphism. Note that ϕ_1 is linear.

Example 6.4.6. Let $\alpha_0 = xdy + dz$ and $\alpha_2 = dz + xdy - ydx$ be the contact forms described in Examples 6.3.2 and 6.3.4. Let $\phi_2 : (\mathbf{R}^3, \alpha_0) \to (\mathbf{R}^3, \alpha_2)$ be given by

$$\phi_2(x, y, z) = \left(x, \frac{y}{2}, z + \frac{xy}{2}\right).$$

As in the previous example, $\phi_2^* \alpha_2 = \alpha_0$, so ϕ_2 is a strictly contact diffeomorphism.

Example 6.4.7. Consider $(\mathbf{R}^3, \alpha_0)$, where $\alpha_0 = xdy + dz$ is the standard contact form on $\mathbf{R}^3$. Let $\mathbf{a} = (a_1, a_2, a_3)$ be an ordered triple of constants, and define the diffeomorphism $T_{\mathbf{a}} : (\mathbf{R}^3, \alpha_0) \to (\mathbf{R}^3, \alpha_0)$ by

$$T_{\mathbf{a}}(x, y, z) = (x + a_1, y + a_2, z + a_3 - a_1 y),$$

Then $T_{\mathbf{a}}$ is a strictly contact diffeomorphism. We call such diffeomorphisms "contact translations."

Strictly contact diffeomorphisms are the formal analogy in the contact setting to isometries in the Riemannian setting: They are diffeomorphisms that preserve the structure tensor. When there is a strictly contact transformation between two contact spaces, not only are their contact structures the "same" in the sense that they have corresponding contact distributions, but they also have corresponding Reeb fields. The examples above (and Exercise 6.9) show that the contact forms in Examples 6.3.2 through 6.3.5 in fact describe isomorphic contact structures with corresponding Reeb fields.

As in the case of isometries, the strictly contact condition amounts to saying that the component functions of the diffeomorphism satisfy a system of nonlinear partial differential equations. For example, let $(\mathbf{R}^3, \alpha_0)$ be the standard contact space, i.e., $\alpha_0 = xdy + dz$. Let $\phi : (\mathbf{R}^3, \alpha_0) \to (\mathbf{R}^3, \alpha_0)$ be a strictly contact diffeomorphism with component functions $\phi = (\phi^1, \phi^2, \phi^3)$. Then the functions $\phi^i : \mathbf{R}^3 \to \mathbf{R}$ satisfy the system

$$\begin{cases} \phi^1 \dfrac{\partial \phi^2}{\partial x} + \dfrac{\partial \phi^3}{\partial x} & = 0, \\ \phi^1 \dfrac{\partial \phi^2}{\partial y} + \dfrac{\partial \phi^3}{\partial y} & = x, \\ \phi^1 \dfrac{\partial \phi^2}{\partial z} + \dfrac{\partial \phi^3}{\partial z} & = 1. \end{cases} \tag{6.7}$$

Part of the subtlety of contact geometry arises from the fact that the Reeb field is considered an "extra" structure, related to the contact form but incidental to the contact distribution. Contact diffeomorphisms—which preserve the contact distribution but not necessarily the Reeb field—are thus considered the proper structure-preserving diffeomorphisms in contact geometry.

Example 6.4.8. Let α_0 be the standard contact form on $\mathbf{R}^3$ and let λ be a nonzero real number. Let $D_\lambda : (\mathbf{R}^3, \alpha_0) \to (\mathbf{R}^3, \alpha_0)$ be defined by

$$D_\lambda(x, y, z) = (\lambda x, \lambda y, \lambda^2 z).$$

Then D_λ is a contact diffeomorphism, since $D_\lambda^* \alpha_0 = \lambda^2 \alpha_0$. Note that D_λ is not strictly contact when $\lambda \neq 1$.

There is an algebraic structure on the set of all contact diffeomorphisms given by the following theorem.

Theorem 6.4.9. *For a domain $U \subset \mathbf{R}^3$, let (U, α) be a contact space with contact distribution $E_\alpha = \ker \alpha$, and let Diff (U, E_α) be the set of all contact diffeomorphisms from (U, α) to itself. Then:*

1. *If $\phi_1, \phi_2 \in$ Diff $(\mathbf{R}^3, E_\alpha)$, then $\phi_1 \circ \phi_2 \in$ Diff $(\mathbf{R}^3, E_\alpha)$.*
2. *The identity map $\mathrm{Id} \in$ Diff $(\mathbf{R}^3, E_\alpha)$.*
3. *If $\phi \in$ Diff $(\mathbf{R}^3, E_\alpha)$, then $\phi^{-1} \in$ Diff $(\mathbf{R}^3, E_\alpha)$.*

In other words, Diff $(\mathbf{R}^3, E_\alpha)$ is a group.

Proof. Statements (1) and (3) follow from the fact that for a diffeomorphism $\phi : U \to U$, a function $f : U \to \mathbf{R}$, and a one-form α on U,

$$\phi^*(f\alpha) = (f \circ \phi)(\phi^* \alpha).$$

In particular, since $(\phi \circ \phi^{-1})^* \alpha = \alpha$, we have

$$(\phi^{-1})^* \alpha = \left(\frac{1}{f \circ \phi^{-1}} \right) \alpha.$$

Statement (2) is just the fact that $(\mathrm{Id})^* \alpha = \alpha$. □

The following question arises naturally in the context of contact diffeomorphisms: Given two contact spaces, is there a contact diffeomorphism between them? In other words, how can we determine when two contact spaces are the same? For example, the contact structure defined by the contact form of Example 6.3.6 is *not* the same as the standard contact structure defined by $\alpha_0 = xdy + dz$. That fact, noted in 1983 by Bennequin, became a landmark in contact geometry and gave rich texture to the field, including the distinction between "tight" and "overtwisted" contact structures. Techniques developed by Eliashberg to classify contact structures are topological in nature and are hence beyond the scope of this text. We will return to this topic when we discuss Darboux's theorem.

6.5 Contact Vector Fields

We now investigate the "infinitesimal contact diffeomorphisms" in analogy with our treatment of Killing vector fields as the "infinitesimal isometries" in the Riemannian setting (see the remark following Definition 5.6.6). In so doing, we will again see the importance of the nondegeneracy condition for the contact form. We will also develop a methodology that in principle, if not always effectively in practice, yields interesting examples of contact diffeomorphisms for a given contact space (U, α).

Definition 6.5.1. Let (U, α) be a contact space, and let X be a smooth vector field on $U \subset \mathbf{R}^3$. Then X is a *contact vector field* if $\mathcal{L}_X \alpha = g\alpha$ for some smooth function $g : U \to \mathbf{R}$. We say that X is a *strictly contact vector field* if $\mathcal{L}_X \alpha = 0$.

We will shortly show how to characterize contact vector fields in such a way that generating examples will be easy. One example, however, is immediate.

Example 6.5.2. Let (U, α) be a contact space with Reeb field ξ. Then ξ is a strictly contact vector field:

$$\begin{aligned}\mathcal{L}_\xi \alpha &= i(\xi)d\alpha + d\left(i(\xi)\alpha\right)\\ &= 0 + d(1)\\ &= 0.\end{aligned}$$

The terminology in Definition 6.5.1 is due to the following result.

Theorem 6.5.3. *Let X be a contact vector field on the contact space (U, α), so that $\mathcal{L}_X \alpha = g\alpha$ for some smooth $g : U \to \mathbf{R}$. Let ϕ_t be the flow generated by X. Then for all t such that ϕ_t is defined, ϕ_t is a contact diffeomorphism. If X is a strictly contact vector field, then ϕ_t is a strictly contact diffeomorphism.*

Proof. The second statement is a direct result of Theorem 4.7.23. In fact, the first statement follows by a slight modification of the same argument. Using Proposition 4.7.2, we have

$$\begin{aligned}\frac{d}{dt}(\phi_t^*\alpha) &= \phi_t^*(\mathcal{L}_X\alpha)\\ &= \phi_t^*(g\alpha)\\ &= (g\circ\phi_t)(\phi_t^*\alpha).\end{aligned}$$

Expressing this in coordinates reveals a system of separable first-order ordinary differential equations that together yield $\phi_t^*\alpha = f\alpha$, where

$$f(t,p) = \exp\left(\int_0^t (g\circ\phi_s)(p)ds\right). \qquad \square$$

Theorem 6.5.3 also justifies referring to contact vector fields as "infinitesimal contact diffeomorphisms."

A fundamental fact about contact vector fields is the following consequence of the nondegeneracy condition for α. The reader should verify that the set $\mathcal{X}(\mathbf{R}^3,\alpha)$ of contact vector fields, with the operations of pointwise tangent vector addition and scalar multiplication by smooth real-valued functions, is a vector space.

Theorem 6.5.4. *Let (U,α) be a contact space, where $U \subset \mathbf{R}^3$ is a domain. Let $\mathcal{X}(U,\alpha)$ be the vector space of contact vector fields on (U,α), and let $C^\infty(U)$ be the vector space of smooth real-valued functions on U. Then the map $\Phi : \mathcal{X}(U,\alpha) \to C^\infty(U)$ given by*

$$\Phi(X) = \alpha(X)$$

is a vector space isomorphism.

Proof. The fact that Φ is a linear map is a consequence of the linearity of α.

To show that Φ is one-to-one, let X be a contact vector field and suppose $\Phi(X) = 0$. Then there exists a smooth function $g : U \to \mathbf{R}$ such that $\mathcal{L}_X\alpha = g\alpha$; we will first show that $g \equiv 0$. Let ξ be the Reeb field associated to α. By supposition,

$$\begin{aligned}\mathcal{L}_X\alpha &= i(X)d\alpha + d\left(i(X)\alpha\right)\\ &= i(X)d\alpha + d(0)\\ &= i(X)d\alpha,\end{aligned}$$

and so we have

$$\begin{aligned}g &= g\cdot i(\xi)\alpha \quad \text{since } \alpha(\xi) = 1\\ &= i(\xi)(g\alpha)\\ &= i(\xi)(\mathcal{L}_X\alpha)\\ &= i(\xi)(i(X)d\alpha)\\ &= -i(X)(i(\xi)d\alpha)\\ &= 0.\end{aligned}$$

The two prior calculations in fact show that if $\Phi(X) = 0$, then $i(X)d\alpha = 0$. But then

$$\begin{aligned} i(X)(\alpha \wedge d\alpha) &= (\alpha(X))d\alpha - \alpha \wedge i(X)d\alpha \\ &= 0, \end{aligned}$$

and so by the nondegeneracy of α, $X = 0$. Hence Φ is one-to-one.

To show that Φ is onto, we will rely on the following lemma, which will occasionally be of interest on its own.

Lemma 6.5.5. *Let (U, α) be a contact space with associated Reeb field ξ and contact distribution $E = \ker \alpha$. Let $\mathcal{B}$ be the set of one-forms β with the property that $\beta(\xi) = 0$ (such one-forms are called* semibasic*). Then the map $A : E \to \mathcal{B}$ given by*

$$A(H) = i(H)d\alpha$$

induces a vector space isomorphism $E_p \to \mathcal{B}_p$ for all $p \in U$.

Proof (of Lemma 6.5.5). We leave to the reader to show that the set $\mathcal{B}_p$ of semibasic one-forms at p is a vector subspace of $\Lambda_1(T_pU)$ for all $p \in U$.

The map A is linear due to properties of the interior product. Further, for every vector field X, we have

$$\begin{aligned} (i(X)d\alpha)(\xi) &= -i(X)(i(\xi)d\alpha) \\ &= 0 \quad \text{by the definition of the Reeb field,} \end{aligned}$$

and so in particular, if $H \in E$, then $A(H) \in \mathcal{B}$.

To show that A is one-to-one, suppose that for $H \in E$ we have $A(H) = 0$. In the proof above that Φ is one-to-one, we showed that the conditions $\alpha(H) = 0$ and $i(H)d\alpha = 0$, along with the nondegeneracy of α, together imply that $H = 0$. But H is horizontal, so $\alpha(H) = 0$ by definition, and $A(H) = 0$ by assumption, so $i(H)d\alpha = 0$. Hence $H = 0$ and A is one-to-one.

To show that A is onto, suppose $\beta \in \mathcal{B}$, so that $\beta(\xi) = 0$. It is an exercise to show that the nondegeneracy of α guarantees that there are vector fields $\{B_1, B_2, B_3\}$ such that for each $p \in U$, $\{B_1(p), B_2(p), B_3(p)\}$ is a basis for $T_p(U)$ and such that $B_1 = \xi$ and $B_2, B_3 \in E$ with $d\alpha(B_2, B_3) = 1$.

So for a basis $\{B_1, B_2, B_3\}$ such that $B_1 = \xi$ and $B_2, B_3 \in E$ with $d\alpha(B_2, B_3) = 1$, define $H_\beta = b_2 B_2 + b_3 B_3$, where $b_2 = \beta(B_3)$ and $b_3 = -\beta(B_2)$. Note that by construction, $H_\beta \in E$. Moreover, for any $V = v_1 B_1 + v_2 B_2 + v_3 B_3$, we have

$$\begin{aligned} (i(H_\beta)d\alpha)(V) &= d\alpha(H_\beta, V) \\ &= (b_2 v_3 - b_3 v_2) d\alpha(B_2, B_3) \end{aligned}$$

$$= v_2\beta(B_2) + v_3\beta(B_3)$$
$$= \beta(V) \qquad \text{since } \beta(B_1) = 0 \text{ by supposition.}$$

Hence $A(H_\beta) = \beta$, and so A is onto. □

Returning now to the proof of Theorem 6.5.4, let $f \in C^\infty(U)$, and consider the one-form $\beta_f = \xi\,[f]\,\alpha - df$. Note that the definition of β_f ensures that $\beta_f(\xi) = 0$. Hence we can define

$$X_f = f\xi + H_f,$$

where $H_f = A^{-1}(\beta_f)$ in the notation of Lemma 6.5.5, i.e., $i(H_f)d\alpha = \beta_f$.

We have, then, that

$$\begin{aligned} i(X_f)\alpha &= \alpha(f\xi) + \alpha(H_f) \\ &= f\alpha(\xi) \quad \text{since } H_f \in E \\ &= f. \end{aligned}$$

Further, we can verify that X_f is a contact vector field:

$$\begin{aligned} \mathcal{L}_{X_f}\alpha &= i(X_f)d\alpha + d\,(i(X_f)\alpha) \\ &= i(f\xi)d\alpha + i(H_f)d\alpha + d(f) \\ &= \xi\,[f]\,\alpha - df + df \\ &= \xi\,[f]\,\alpha, \end{aligned}$$

and so $\mathcal{L}_{X_f}\alpha = g\alpha$ for $g = \xi\,[f]$. □

The contact vector field $X_f = \Phi^{-1}(f)$ is called the *contact gradient* of f, in analogy with Definition 5.1.10. Note the essential role of the nondegeneracy condition in both constructions. For reasons that will be clear in the next chapter, X_f is also referred to as the *contact Hamiltonian* of f.

The following result follows immediately from the proof of Theorem 6.5.4.

Corollary 6.5.6. *Let (U, α) be a contact space with Reeb field ξ. Suppose that X_f is the contact gradient of a smooth function $f : U \to \mathbf{R}$. Then X_f is a* strictly *contact vector field if and only if*

$$\xi\,[f] = 0.$$

We will conclude with several examples to illustrate the main ideas of this section.

Example 6.5.7. Let $(\mathbf{R}^3, \alpha)$ be a contact space with Reeb field ξ and let $f : \mathbf{R}^3 \to \mathbf{R}$ be a constant function, $f(p) = c$ for all $p \in \mathbf{R}^3$. Then the contact gradient X_c

is given by $X_c = c\xi$. By Corollary 6.5.6, X_c is strictly contact. In particular, when $f \equiv 1$, $X_f = X_1 = \xi$. Also note that when $f \equiv 0$, $X_f = X_0 = 0$; the flow of X_0 is the identity diffeomorphism.

Example 6.5.8. Let $(\mathbf{R}^3, \alpha_0)$ be the standard contact space with $\alpha_0 = xdy + dz$, so that the Reeb field ξ is equal to $\frac{\partial}{\partial z}$ and, writing $X = \langle X^1, X^2, X^3 \rangle$,

$$\begin{aligned} A(X) &= i(X)d\alpha_0 \\ &= i(X)\,(dx \wedge dy) \\ &= -X^2 dx + X^1 dy. \end{aligned}$$

Hence for a semibasic one-form $\beta = b_1 dx + b_2 dy$, we have

$$A^{-1}(\beta) = \langle b_2, -b_1, xb_1 \rangle .$$

In particular, for a smooth function $f : \mathbf{R}^3 \to \mathbf{R}$, the semibasic form β_f (in the notation of Theorem 6.5.4) is given by

$$\begin{aligned} \beta_f &= \xi\,[f]\,\alpha - df \\ &= f_z(xdy + dz) - (f_x dx + f_y dy + f_z dz) \\ &= -f_x dx + (xf_z - f_y)dy, \end{aligned}$$

so that

$$H_f = A^{-1}(\xi\,[f]\,\alpha - df) = \langle -f_y + xf_z, f_x, -xf_x \rangle .$$

These calculations show that for the standard contact form α_0, the contact gradient is given by

$$X_f = f\xi + H_f = \langle xf_z - f_y, f_x, f - xf_x \rangle . \tag{6.8}$$

For example, consider the function $f(x, y, z) = yz$. The contact gradient is then given, according to Eq. (6.8), by

$$X_f = \langle xy - z, 0, yz \rangle .$$

Integrating X_f to find the integral curve through an initial point (x_0, y_0, z_0) is possible, since the corresponding system

$$\begin{cases} \dot{x} &= xy - z, \\ \dot{y} &= 0, \\ \dot{z} &= yz, \end{cases}$$

is partially decoupled. We leave to the reader to carry out the integration, which yields the integral curve $c : I \to \mathbf{R}^3$ given by

$$\begin{aligned} c(t) &= (x(t), y(t), z(t)) \\ &= \left(e^{y_0 t}(x_0 + z_0 t), y_0, z_0 e^{y_0 t}\right). \end{aligned}$$

Note that these integral curves are defined for all t ($I = \mathbf{R}$), and so in particular, it is possible to consider the time-one flow $\phi_f : \mathbf{R}^3 \to \mathbf{R}^3$ generated by X_f, given by

$$\phi_f(x, y, z) = (e^y(x + z), y, ze^y).$$

The reader may verify that ϕ_f is a contact diffeomorphism with $\phi_f^* \alpha_0 = e^y \alpha_0$.

There are obvious obstructions to constructing globally defined contact diffeomorphisms in this way, namely the problems encountered in Chap. 3 of integrating vector fields. To illustrate this, consider the function $g(x, y, z) = xz$. The contact gradient X_g is given by

$$X_g = \left\langle x^2, z, 0 \right\rangle,$$

so the integral curve $c : I \to \mathbf{R}^3$ through $c(0) = (x_0, y_0, z_0)$ satisfies

$$\begin{cases} \dot{x} &= x^2, \\ \dot{y} &= z, \\ \dot{z} &= 0. \end{cases}$$

Again, this system is partially decoupled and can be integrated to obtain

$$c(t) = \left(\frac{x_0}{1 - x_0 t}, y_0 + tz_0, z_0\right),$$

which is defined on the interval $(-\infty, 1/x_0)$ if $x_0 > 0$, or $(1/x_0, +\infty)$ if $x_0 < 0$.

In general, it will be impossible to globally define the time-one flow for X_g. However, for each point $p_0 = (x_0, y_0, z_0)$ with $x_0 > 0$, it is possible to choose an $a > 0$ and a domain U_a containing p_0 such that the map $\phi_a : U_a \to \phi_a(U_a)$ (the "time-a flow") given by

$$\phi_a(x, y, z) = \left(\frac{x}{1 - ax}, y + az, z\right)$$

is a diffeomorphism onto its image. For example, if $0 < x_0 < 1$, choose a to satisfy $1 < a < 1/x_0$ and choose

$$U_a = \{(x, y, z) \mid 0 < x < 1/a\}.$$

In fact, $\phi_a : U_a \to \phi_a(U_a)$ is by construction a contact diffeomorphism (not defined on all of $\mathbf{R}^3$), with $\phi_a^* \alpha_0 = \dfrac{1}{1-ax}\alpha_0$.

We close this section by returning to one of our motivating problems for contact geometry, the method of characteristics.

Example 6.5.9. Let us consider the set of contact elements $C\mathbf{R}^2$ to be $\mathbf{R}^3$ with coordinates (x, y, m) and the contact form $\alpha_1 = dy - mdx$, so that the Reeb field ξ_1 is given by $\xi_1 = \frac{\partial}{\partial y}$ and $A(X) = -X^3 dx + X^1 dm$ for $X = \langle X^1, X^2, X^3 \rangle$. By almost identical calculations as in the previous example, we have for a smooth function $F : \mathbf{R}^3 \to \mathbf{R}$ that

$$H_F = \langle -F_m, -mF_m, F_x + mF_y \rangle ,$$

and so the contact gradient of F for the contact form α_1 is given by

$$X_F = F\xi_1 + H_F = \langle -F_m, F - mF_m, F_x + mF_y \rangle .$$

Note that we encountered both of these vector fields in Sect. 6.2 in the context of using the method of characteristics to solve the first-order ordinary differential equation represented by the equation $F(t, u, u') = 0$. In fact, H_F was what we called the characteristic vector field (and denoted by $\tilde{X}_F$) of the equation $F = 0$, while X_F was what we called the Lie characteristic vector field. In hindsight, we can see that X_F is a contact vector field, and so its flow determines a contact diffeomorphism of $C\mathbf{R}^2$, whereas H_F is a horizontal vector field.

6.6 Darboux's Theorem

We now turn to a fundamental result in contact geometry. This entire section will be devoted to understanding its statement and proof. We will take advantage of this context to illustrate several other basic theorems and techniques of differential geometry that are of interest in their own right.

Darboux's theorem essentially says that from a local point of view, all contact structures are "the same." More precisely, near any point, a form α on $\mathbf{R}^3$ can be expressed as the standard contact form $\alpha_0 = xdy + dz$ by means of a change of coordinates, i.e., a local diffeomorphism. In this sense, Darboux's theorem is a kind of "normal form" theorem that is common, for example, in the field of linear algebra.

More important from a geometric perspective, Darboux's theorem states that there are no local properties to distinguish one contact structure from another. This is in marked contrast to Riemannian geometry, where tensor quantities like curvature have a distinctly local character.

Differences between contact structures, if they arise, must then be found at the global level, like topological invariants or the global behavior of integral curves.

Indeed, this fact shapes the contours of the subject at a more advanced level. It has the distinct disadvantage that many of the most interesting questions in contact geometry will fall outside the bounds of this introductory text.

We begin this section with a basic theorem of differential geometry, a kind of "straightening out" theorem that says that locally, the flow of a vector field can be represented as a line. While it properly belongs in the setting of Chap. 3, on advanced calculus, we present it here for two reasons. First, we will use the result (and its corollary) only in the proof of Darboux's theorem. More importantly, though, it can be considered a first example in a family of "normal form" theorems to which Darboux's theorem belongs.

Theorem 6.6.1. *Let X be a smooth vector field on $\mathbf{R}^n$ and let $p \in \mathbf{R}^n$ be such that $X(p) \neq \mathbf{0}_p$. Then there exist a domain U containing p and a diffeomorphism $\phi : U \to \phi(U) \subset \mathbf{R}^n$ such that $\phi(p) = p$ and*

$$\phi_*(X(x)) = \left.\frac{\partial}{\partial x_1}\right|_{\phi(x)}$$

for all $x \in U$, where $(x_1, \ldots, x_n)$ are understood to be the coordinates of $\mathbf{R}^n$.

Proof. First, for $p = (p_1, \ldots, p_n) \in \mathbf{R}^n$, define $t_p : \mathbf{R}^n \to \mathbf{R}^n$ by

$$t_p(x_1, \ldots, x_n) = (x_1 - p_1, \ldots, x_n - p_n).$$

Note that $t_p(p) = 0$ and $(t_p)_*(X) = X \circ t_p$. In particular, $(t_p)_*(X(p)) = X(0)$.

Second, suppose that Y is a vector field on $\mathbf{R}^n$ satisfying $Y(0) \neq \mathbf{0}$. Let $\{\mathbf{e}_1, \ldots, \mathbf{e}_n\}$ be a basis for $T_0\mathbf{R}^n$ such that $\mathbf{e}_1 = Y(0)$. Define a linear isomorphism (and hence diffeomorphism) $A : \mathbf{R}^n \to \mathbf{R}^n$ in such a way that $A_*\mathbf{e}_i = \left.\frac{\partial}{\partial x_i}\right|_0$ for all $i = 1, \ldots, n$ (we rely on the fact that $A(0) = 0$). In particular, $A_*Y(0) = \left.\frac{\partial}{\partial x_1}\right|_0$.

Now suppose that Z is a vector field on $\mathbf{R}^n$ with coordinates $(z_1, \ldots, z_n)$ such that $Z(0) = \frac{\partial}{\partial z_1}(0)$. Let $\psi : V \times I \to \mathbf{R}^n$ be the flow generated by Z, where $V \subset \mathbf{R}^n$ is a domain containing 0 and $I \subset \mathbf{R}$ is an interval about 0. Here we are writing

$$\psi(x_1, \cdots, x_n, s) = \psi_s(x_1, \ldots, x_n) = (z_1, \ldots, z_n),$$

where ψ has the properties (from Theorem 3.9.2)

$$\psi_*\left(\frac{\partial}{\partial s}(x, s)\right) = Z(\psi_s(x))$$

and

$$\psi(x, 0) = x \quad \text{for all } x \in V.$$

Consider the function $\sigma : V_1 \to \mathbf{R}^n$, where $V_1 \subset \mathbf{R} \times \mathbf{R}^{n-1}$, defined by

$$\sigma(x_1, x_2, \ldots, x_n) = \psi_{x_1}(0, x_2, \ldots, x_n) = \psi(0, x_2, \ldots, x_n, x_1).$$

More precisely, $V_1 = I_1 \times U_1$, where $I_1 \subset I$ is a subinterval containing 0 and $U_1 \subset \mathbf{R}^{n-1}$ is a domain containing $(0, \ldots, 0)$ such that $\sigma(V_1) \subset V$. Note that $\sigma(0) = 0$. As a result of the properties of the flow, the function σ satisfies $(\sigma)_*(0) = \mathrm{Id}$ and $\sigma_*(\frac{\partial}{\partial x_1}) = Z \circ \sigma$ on V_1. In particular, $(\sigma_*)(0)$ is an isomorphism, so by the inverse function theorem 3.6.12, there is a domain W containing 0 such that $\sigma : W \to \sigma(W)$ is a diffeomorphism. The inverse $\sigma^{-1} : \sigma(W) \to W$ is differentiable and has the property that $(\sigma^{-1})_*(Z) = \frac{\partial}{\partial x_1} \circ \sigma^{-1}$.

For $p \in \mathbf{R}^n$, we now construct the desired diffeomorphism in an appropriate domain containing p. If X is a vector field such that $X(p) \neq 0$, then the vector field Y defined by $Y(y) = (t_p)_*(X(x))$, where $y = t_p(x)$, has the property that $Y(0) \neq 0$. Hence there is a linear diffeomorphism $A : \mathbf{R}^n \to \mathbf{R}^n$ such that the vector field Z defined by $Z(z) = A_*(Y(y))$, where $z = A(y)$, has the property that $Z(0) = \frac{\partial}{\partial z_1}(0)$. So there are a domain W around 0 and a diffeomorphism $\sigma : W \to \sigma(W)$ such that $\sigma_*(\frac{\partial}{\partial x_1}(w)) = Z(u)$, for all $w \in W$, where $u = \sigma(w)$.

Define the domain U by $U = (t_p^{-1} \circ A^{-1} \circ \sigma)(W)$; we have $x \in U$ exactly when there is an element $w \in W$ such that

$$\sigma(w) = A(t_p(x)).$$

In particular, $p \in U$, since $0 \in W$ and

$$0 = \sigma(0) = A(t_p(p)).$$

Define the diffeomorphism $\phi : U \to \phi(U)$ by

$$\phi = t_p^{-1} \circ \sigma^{-1} \circ A \circ t_p.$$

The reader can verify that $\phi(p) = p$.

Then, for $x \in U$, we have

$$\begin{aligned}
\phi_* X(x) &= (t_p^{-1} \circ \sigma^{-1} \circ A \circ t_p)_* X(x) \\
&= (t_p^{-1} \circ \sigma^{-1} \circ A)_* Y(y) \quad \text{where } y = t_p(x) \\
&= (t_p^{-1} \circ \sigma^{-1})_* Z(z) \quad \text{where } z = A(t_p(x)) \\
&= (t_p^{-1} \circ \sigma^{-1})_* Z(\sigma(w)) \quad \text{where } w \in W \text{ is such that } \sigma(w) = z \\
&= (t_p^{-1})_* \frac{\partial}{\partial w_1}(w) \\
&= \frac{\partial}{\partial x_1}(\phi(x)) \quad \text{by an explicit calculation of } (t_p^{-1})_*. \qquad \square
\end{aligned}$$

We will need the following corollary, applying Theorem 6.6.1 to the context of the problem of (locally) first-order partial differential equations of a particular form.

Corollary 6.6.2. *Let X be a smooth vector field on $\mathbf{R}^n$ and let $g : \mathbf{R}^n \to \mathbf{R}$ be a smooth function. Then for every $p \in \mathbf{R}^n$ such that $X(p) \neq 0$, there exist a domain U containing p and a smooth function $f : U \to \mathbf{R}$ such that*

$$X[f] = g\big|_U.$$

In other words, writing $X = \sum_{i=1}^{n} X^i \frac{\partial}{\partial x_i}$ for smooth functions $X^i : \mathbf{R}^n \to \mathbf{R}$, the first-order partial differential equation

$$X^1 \frac{\partial f}{\partial x_1} + \cdots + X^n \frac{\partial f}{\partial x_n} = g$$

has a smooth solution f defined in a domain about each point p for which $X(p) \neq 0$.

Proof. For a given $p = (p_1, \ldots, p_n) \in \mathbf{R}^n$, let U be a domain containing p and let $\phi : U \to \phi(U)$ be a diffeomorphism satisfying the conclusions of Theorem 6.6.1, in particular $\phi_* X = \frac{\partial}{\partial x_1}$ on $\phi(U)$ and $\phi(p) = p$. We can assume (by choosing a smaller domain if necessary) that U has the property that for $x = (x_1, x_2, \ldots, x_n) \in \phi(U)$, we have $(s, x_2, \ldots, x_n) \in \phi(U)$ for all $s \in [p_1, x_1]$ if $p_1 \leq x_1$ and for all $s \in [x_1, p_1]$ if $x_1 < p_1$.

Then for all $x = (x_1, x_2, \ldots, x_n) \in \phi(U)$, define the smooth function $F : \phi(U) \to \mathbf{R}$ by

$$F(x_1, x_2, \ldots, x_n) = \int_{p_1}^{x_1} g(s, x_2, \ldots, x_n) ds;$$

by construction, $\frac{\partial F}{\partial x_1} = g$.

Now define $f = F \circ \phi$ on U. Then

$$\begin{aligned} X[f] &= df(X) \\ &= d(F \circ \phi)(X) \\ &= (\phi^* dF)(X) \\ &= dF(\phi_* X) \\ &= dF(\frac{\partial}{\partial x_1}) \\ &= \frac{\partial F}{\partial x_1} = g. \qquad \square \end{aligned}$$

The second result we will need before turning to Darboux's theorem concerns *time-dependent* vector fields and forms. These are properly thought of as smoothly varying one-parameter families of vector fields or differential forms in the same sense as any vector field generates a time-dependent family of diffeomorphisms via the existence and uniqueness theorem (Theorem 3.9.2). As a matter of fact, without going into all the technical details, we shall assume that Theorem 3.9.2 extends to time-dependent vector fields. In particular, given a time-dependent vector field X_t, then for each $t = t_0$ for which X_t is defined and around each point p there are a domain U and a time-dependent family of diffeomorphisms $\phi_t : U \to \phi_t(U)$ defined for t in an interval around t_0 having the property that $\frac{d}{dt}\phi_t(x) = X_t(\phi_t(x))$.

In fact, one basic way in which a time-dependent vector field might arise is by differentiating a smoothly varying family of diffeomorphisms. Given such a family $\phi_t : \mathbf{R}^n \to \mathbf{R}^n$, define

$$X_t(\phi_t(x)) = \frac{d\phi_t}{dt}(x) = \phi_*(x,t)\left(\frac{\partial}{\partial s}\right),$$

where $\phi : \mathbf{R}^n \times \mathbf{R} \to \mathbf{R}^n$ is defined by $\phi(x,s) = \phi_s(x)$.

Some of the details will emerge in the proof of the following theorem of advanced calculus. The reader should compare this to Proposition 4.7.2.

Theorem 6.6.3. *Let α_t be a smoothly varying, time-dependent family of differential k-forms on $\mathbf{R}^n$. Let $\phi_t : \mathbf{R}^n \to \mathbf{R}^n$ be a smoothly varying time-dependent family of diffeomorphisms, and let X_t be the associated time-dependent vector field satisfying $X_t(\phi_t(x)) = \frac{d\phi_t}{dt}(x)$. Then*

$$\frac{d}{dt}\left(\phi_t^*\alpha_t\right) = \phi_t^*\left(\frac{d\alpha_t}{dt} + \mathcal{L}_{X_t}\alpha_t\right). \tag{6.9}$$

Proof. We will present the proof only in the special case we will encounter here, namely $n = 3$ and $k = 1$. In other words, α_t is a family of one-forms on $\mathbf{R}^3$. The general case is more complicated only by virtue of notation.

We first prove that Eq. (6.9) holds in a special, adapted case, namely for the family of diffeomorphisms

$$\psi_t : \mathbf{R}^3 \times I \to \mathbf{R}^3 \times I,$$

where I is an open interval containing 0 and

$$\psi_t(x,s) = (x, s+t)$$

is defined when $s \in I$, $s+t \in I$, and $x = (x_1, x_2, x_3) \in \mathbf{R}^3$. Note that the vector field X_t associated to ψ_t is

$$X_t(x,s) = \frac{d}{dt}\psi_t(x,s) \circ \psi_t^{-1} = \langle 0,1\rangle_{(x,s)} = \left.\frac{\partial}{\partial s}\right|_{(x,s)}.$$

Now let σ_t be a time-dependent family of one-forms on $\mathbf{R}^3 \times I$, which can be written in components as follows:

$$\sigma_t(x,s) = a_1(x,s,t)dx_1 + a_2(x,s,t)dx_2 + a_3(x,s,t)dx_3 + b(x,s,t)ds$$

for smooth functions $a_1, a_2, a_3, b : \mathbf{R}^3 \times I \times I_1 \to \mathbf{R}$ (here I_1 is the interval on which the family σ_t is defined).

We will prove that Eq. (6.9) holds in this special case, for the family of one-forms σ_t and the family of diffeomorphisms ψ_t, by performing three coordinate calculations.

First, note that

$$\begin{aligned}\psi_t^*\sigma_t(x,s,t) &= (a_1 \circ \psi_t)dx_1 + (a_2 \circ \psi_t)dx_2 + (a_3 \circ \psi_t)dx_3 + (b \circ \psi_t)d(s+t) \\ &= a_1(x,s+t,t)dx_1 + a_2(x,s+t,t)dx_2 \\ &\quad + a_3(x,s+t,t)dx_3 + b(x,s+t,t)ds,\end{aligned}$$

so that the chain rule yields

$$\begin{aligned}\frac{d}{dt}(\psi_t^*\sigma_t)(x,s,t) = &\left[\frac{\partial a_1}{\partial s}(x,s+t,t) + \frac{\partial a_1}{\partial t}(x,s+t,t)\right] dx_1 \\ &+ \left[\frac{\partial a_2}{\partial s}(x,s+t,t) + \frac{\partial a_2}{\partial t}(x,s+t,t)\right] dx_2 \\ &+ \left[\frac{\partial a_3}{\partial s}(x,s+t,t) + \frac{\partial a_3}{\partial t}(x,s+t,t)\right] dx_3 \\ &+ \left[\frac{\partial b}{\partial s}(x,s+t,t) + \frac{\partial b}{\partial t}(x,s+t,t)\right] ds.\end{aligned} \tag{6.10}$$

Second, since

$$\begin{aligned}\frac{d}{dt}(\sigma_t)(x,s,t) = &\frac{\partial a_1}{\partial t}(x,s,t)dx_1 + \frac{\partial a_2}{\partial t}(x,s,t)dx_2 \\ &+ \frac{\partial a_3}{\partial t}(x,s,t)dx_3 + \frac{\partial b}{\partial t}(x,s,t)ds,\end{aligned}$$

we have

$$\begin{aligned}\psi_t^*\left[\frac{d}{dt}(\sigma_t)\right](x,s,t) = &\frac{\partial a_1}{\partial t}(x,s+t,t)dx_1 + \frac{\partial a_2}{\partial t}(x,s+t,t)dx_2 \\ &+ \frac{\partial a_3}{\partial t}(x,s+t,t)dx_3 + \frac{\partial b}{\partial t}(x,s+t,t)d(s+t)\end{aligned}$$

$$= \frac{\partial a_1}{\partial t}(x, s+t, t)dx_1 + \frac{\partial a_2}{\partial t}(x, s+t, t)dx_2 \qquad (6.11)$$

$$+ \frac{\partial a_3}{\partial t}(x, s+t, t)dx_3 + \frac{\partial b}{\partial t}(x, s+t, t)ds.$$

Note that t is constant with respect to the d operator on $\mathbf{R}^3 \times I$.

Finally, applying Proposition 4.7.15 to X_t, with $X_t^1 = X_t^2 = X_t^3 = 0$ and $X_t^4 = 1$ as above, we have

$$\mathcal{L}_{X_t}\sigma_t(x, s, t) = \frac{\partial a_1}{\partial s}(x, s, t)dx_1 + \frac{\partial a_2}{\partial s}(x, s, t)dx_2$$

$$+ \frac{\partial a_3}{\partial s}(x, s, t)dx_3 + \frac{\partial b}{\partial s}(x, s, t)ds.$$

Hence

$$\psi_t^* \left(\mathcal{L}_{X_t}\sigma_t\right)(x, s, t) = \left(\frac{\partial a_1}{\partial s} \circ \psi_t\right) dx_1 + \left(\frac{\partial a_2}{\partial s} \circ \psi_t\right) dx_2$$

$$+ \left(\frac{\partial a_3}{\partial s} \circ \psi_t\right) dx_3 + \left(\frac{\partial b}{\partial s} \circ \psi_t\right) d(s+t)$$

$$= \frac{\partial a_1}{\partial s}(x, s+t, t)dx_1 + \frac{\partial a_2}{\partial s}(x, s+t, t)dx_2 \qquad (6.12)$$

$$+ \frac{\partial a_3}{\partial s}(x, s+t, t)dx_3 + \frac{\partial b}{\partial s}(x, s+t, t)ds.$$

Comparing Eq. (6.10) with the sum of Eqs. (6.11) and (6.12) yields Eq. (6.9) in the specific case of ψ_t.

We now turn to the general case. Let $\phi_t : \mathbf{R}^3 \to \mathbf{R}^3$ be a smooth family of diffeomorphisms that generates the time-dependent vector field X_t. Consider the smooth function $\phi : \mathbf{R}^3 \times I \to \mathbf{R}^3$ given by

$$\phi(x, s) = \phi_s(x),$$

where $x \in \mathbf{R}^3$ and $s \in I$. Here I is the open interval on which the family ϕ_t is defined. Note that by construction,

$$(\phi_*)(x, t) \left(\left.\frac{\partial}{\partial s}\right|_{(x,t)} \right) = X_t(\phi_t(x)).$$

Define the smooth map $i : \mathbf{R}^3 \to \mathbf{R}^3 \times I$ by $i(x) = (x, 0)$ and note that

$$\phi_t = \phi \circ \psi_t \circ i,$$

where $\psi_t : \mathbf{R}^3 \times I \to \mathbf{R}^3 \times I$ is the special function considered at the outset. Then for a smooth family α_t of one-forms, we have

$$\begin{aligned}
\frac{d}{dt}(\phi_t^*\alpha_t) &= \frac{d}{dt}(i^*\psi_t^*\phi^*\alpha_t) \\
&= i^*\left(\frac{d}{dt}(\psi_t^*\phi^*\alpha_t)\right) \\
&= i^*\left(\psi_t^*\left[\frac{d}{dt}(\phi^*\alpha_t) + \mathcal{L}_{\frac{\partial}{\partial s}}\phi^*\alpha_t\right]\right) \quad \text{(the case proved above)} \\
&= i^*\left(\psi_t^*\left[\phi^*\left(\frac{d}{dt}\alpha_t\right) + \phi^*(\mathcal{L}_{X_t}\alpha_t)\right]\right) \quad \text{by Proposition 4.7.13} \\
&= i^*\psi_t^*\phi^*\left(\frac{d}{dt}\alpha_t + \mathcal{L}_{X_t}\alpha_t\right) \\
&= \phi_t^*\left(\frac{d}{dt}\alpha_t + \mathcal{L}_{X_t}\alpha_t\right),
\end{aligned}$$

proving Eq. (6.9) in general. □

We are now in a position to state and prove the contact version of Darboux's theorem.

Theorem 6.6.4 (Darboux's theorem for contact geometry). *Let $(\mathbf{R}^3, \alpha)$ be a contact space and let $\alpha_0 = x\,dy + dz$ be the standard contact form on $\mathbf{R}^3$. For all $p \in \mathbf{R}^3$, there are a domain U containing p and a diffeomorphism $\phi : U \to \phi(U) \subset \mathbf{R}^3$ such that $\phi(p) = p$ and such that $\phi^*\alpha = \alpha_0$ on U.*

Proof. In broad outline, we will follow the strategy of Theorem 6.6.1. That is, we will first translate the problem to one "centered" at the origin. We then use a linear map to give the contact one-form a standard form *at the origin*. The third step is where the details differ. We will construct a "path" of one-forms from a given form to the standard contact form, which will in turn allow us to construct a one-parameter family of diffeomorphisms that ultimately gives the appropriate change of coordinates. Finally, we translate the problem back to the base point p, being careful to ensure that the work we have done in the third step is not distorted in the process of translation.

Let $p = (p_1, p_2, p_3) \in \mathbf{R}^3$. Define the diffeomorphism $t_p : \mathbf{R}^3 \to \mathbf{R}^3$ by

$$t_p(x, y, z) = (x + p_1, y + p_2, z + p_3).$$

Define $\beta_0 = t_p^*\alpha$ and note that $\beta_0(0) = \alpha(p)$. Since t_p is a diffeomorphism, the one-form β_0 is nondegenerate.

Next, let $\xi_0 = \frac{\partial}{\partial z}$ and ξ_1 be the Reeb fields for $\alpha_0 = x dy + dz$ and β_0 respectively, and let $\{\zeta_1, \zeta_2\}$ be a basis for $E_1 = \ker \beta_0$. Define the linear isomorphism $A_* : T_0\mathbf{R}^3 \to T_0\mathbf{R}^3$ by

$$A_*(\frac{\partial}{\partial x}) = \zeta_1, \quad A_*(\frac{\partial}{\partial y}) = \zeta_2, \quad A_*(\frac{\partial}{\partial z}) = \xi_1.$$

Extend A_* to a linear map $A : \mathbf{R}^3 \to \mathbf{R}^3$; A is a diffeomorphism. Further, $(A^*\beta_0)(0) = \alpha_0(0)$ by construction, since $A(0) = 0$ and the maps agree on the basis vectors $\frac{\partial}{\partial x}\big|_0, \frac{\partial}{\partial y}\big|_0, \frac{\partial}{\partial z}\big|_0$. Define $\beta_1 = A^*\beta_0$.

We now consider the one-parameter family of smooth one-forms

$$\alpha_t = (1-t)\alpha_0 + t\beta_1,$$

where $t \in [0,1]$. It can be considered a "line segment" of one-forms with initial "point" α_0 and terminal "point" $\alpha_1 = \beta_1$. Note that for all t, $\alpha_t(0) = \alpha_0(0) = \beta_1(0)$; in particular, $\alpha_t(0)$ is nondegenerate for all t. By a topological argument involving the continuity of $\alpha_t \wedge d\alpha_t$ and the compactness of $[0,1]$, there is a domain V_1 containing $0 \in \mathbf{R}^3$ such that α_t is nondegenerate, and so α_t is a contact form, on V_1 for all t. Let ξ_t denote the Reeb field of the contact form α_t

The core of the proof is the construction of a smooth family of diffeomorphisms $\phi_t : V \to \phi_t(V)$ for some domain $V \subset V_1$ containing $0 \in \mathbf{R}^3$ having the property that $\phi_t^*\alpha_t = \alpha_0$ for all $t \in [0,1]$. We will do this by constructing a time-dependent vector field X_t with properties that, together with Theorem 6.6.3, will ensure that its flow will give the desired family of diffeomorphisms.

We first construct the vertical component of X_t in the sense of the definition following Proposition 6.3.11. This amounts to specifying a smoothly varying family of smooth functions f_t such that $\alpha_t(X_t) = f_t$. To do this, define the function

$$g_t = -\dot{\alpha}_t(\xi_t),$$

where $\dot{\alpha}_t = \dfrac{d}{dt}\alpha_t$. By Corollary 6.6.2, there exist a domain $V_2 \subset V_1 \subset \mathbf{R}^3$ containing 0 and a family of functions $f_t : V_2 \to \mathbf{R}$ such that

$$\xi_t[f_t] = g_t.$$

(The existence of a *single* domain V_2 that exists for all $t \in [0,1]$ depends on a topological argument again involving the compactness of $[0,1]$.)

In fact, we may make several additional assumptions on the functions f_t. First, we may assume that $f_t(0) = 0$ for $t \in [0,1]$. If not, replace f_t with $\tilde{f}_t = f_t - f_t(0)$, which still has the property that $\xi_t\left[\tilde{f}_t\right] = g_t$.

Second, we may assume that $df_t(0) = 0$. To see this, note first that for $t \in [0,1]$, $df_t(0)$ is semibasic relative to the Reeb field ξ_t:

$$\begin{aligned}
df_t(0)(\xi_t(0)) &= \xi_t(0)\,[f_t(0)] \\
&= g_t(0) \\
&= -\dot{\alpha}_t(0)(\xi_t(0)) \\
&= (\alpha_0(0) - \beta_1(0))\,(\xi_t(0)) \\
&= 0.
\end{aligned}$$

Then by Theorem 6.6.1, there exist a domain $V_3 \subset V_2$ containing 0 and coordinates (s_1, s_2, s_3) such that $\xi_t\big|_{V_3} = \frac{\partial}{\partial s_3}$, in which case $df_t(0) = a_1 ds_1 + a_2 ds_2$ for some constants a_1, a_2. So if $df_t(0) \neq 0$, replace f_t with $\tilde{f}_t = f_t - a_1 s_1 - a_2 s_2$; then $\xi_t\left[\tilde{f}_t\right] = \xi_t[f_t]$, $\tilde{f}_t(0) = f_t(0)$, and $d\tilde{f}(0) = 0$.

We now construct the horizontal component of X_t. We observe that the one-form $-\dot{\alpha}_t - df_t$ is semibasic relative to ξ_t:

$$\begin{aligned}
(-\dot{\alpha}_t - df_t)\,(\xi_t) &= -\dot{\alpha}_t(\xi_t) - df_t(\xi_t) \\
&= -\dot{\alpha}_t(\xi_t) - g_t \\
&= 0 \quad \text{by the definition of } g_t.
\end{aligned}$$

Hence, applying Lemma 6.5.5, there is a family of horizontal vector fields Y_t satisfying

$$i(Y_t)d\alpha_t = -\dot{\alpha}_t - df_t.$$

Note that since $df_t(0) = 0$ by assumption and since

$$\dot{\alpha}_t(0) = \beta_1(0) - \alpha_0(0) = 0,$$

we have $Y_t(0) = 0$ for all t.

Define

$$X_t = f_t\xi_t + Y_t.$$

We will show that the flow ϕ_t of X_t, defined on a domain $V \subset V_2$ with the property that $\phi_t(V) \subset V_2$ for all $t \in [0,1]$, satisfies $\phi_t^*\alpha_t = \alpha_0$ for all t. By the usual properties of flows, $\phi_0 = \mathrm{Id}$. In addition, $\phi_t(0) = 0$ for all t, since $X_t(0) = 0$ for all t. By Theorem 6.6.3,

$$\begin{aligned}
\frac{d}{dt}(\phi_t^*\alpha_t) &= \phi_t^*\left(\dot{\alpha}_t + \mathcal{L}_{X_t}\alpha_t\right) \\
&= \phi_t^*\left(\dot{\alpha}_t + i(X_t)d\alpha_t + d(\alpha_t(X_t))\right) \\
&= \phi_t^*\left(\dot{\alpha}_t + i(Y_t)d\alpha_t + df_t\right) \\
&= \phi_t^*\left(\dot{\alpha}_t - \dot{\alpha}_t - df_t + df_t\right) \\
&= 0.
\end{aligned}$$

Hence $\phi_t^*\alpha_t$ is constant with respect to t, and so

$$\phi_t^*\alpha_t = \phi_0^*\alpha_0 = \alpha_0 \text{ for all } t;$$

in particular,

$$\phi_1^*\alpha_1 = \phi_1^*\beta_1 = \alpha_0.$$

Finally, let $\tilde{t}_p$ be the contact translation by (p_1, p_2, p_3); see Example 6.4.7. As demonstrated in that example, this is a strictly contact diffeomorphism of α_0, i.e., $\tilde{t}_p^*\alpha_0 = \alpha_0$.

We now define

$$\phi = t_p \circ A \circ \phi_1 \circ \tilde{t}_p^{-1}$$

on the domain $U = \tilde{t}_p(V)$, which contains p since $p = \tilde{t}_p(0)$ and $0 \in V$.

Then

$$\begin{aligned}
\phi(p) &= t_p(A(\phi_1(\tilde{t}_p^{-1}(p)))) \\
&= t_p(A(\phi_1(0))) \\
&= t_p(A(0)) \\
&= t_p(0) \\
&= p,
\end{aligned}$$

and

$$\begin{aligned}
\phi^*\alpha &= (t_p \circ A \circ \phi_1 \circ \tilde{t}_p^{-1})^*\alpha \\
&= (\tilde{t}_p^{-1})^*\phi_1^* A^* t_p^*\alpha \\
&= (\tilde{t}_p^{-1})^*\phi_1^* A^*\beta_0 \\
&= (\tilde{t}_p^{-1})^*\phi_1^*\beta_1 \\
&= (\tilde{t}_p^{-1})^*\alpha_0 \\
&= \alpha_0. \qquad \square
\end{aligned}$$

We state the following corollary, which makes precise the statement, "All contact structures are locally the same."

Corollary 6.6.5. *Let* $(\mathbf{R}^3, \alpha_1)$ *and* $(\mathbf{R}^3, \alpha_2)$ *be two contact spaces. Then for each* $p \in \mathbf{R}^3$, *there exist a domain* U *containing* p *and a diffeomorphism* $\phi : U \to V$ *such that* $\phi^*\alpha_2 = \alpha_1$ *and* $\phi(p) = p$.

Proof. By Darboux's theorem, given $p \in \mathbf{R}^3$, for $j = 1, 2$, there are domains U_j with $p \in U_j$, along with diffeomorphisms $\phi_j : U_j \to \phi_j(U_j)$ with $\phi_j(p) = p$, such that $\phi_j^*\alpha_j = \alpha_0$, where α_0 is the standard contact form on $\mathbf{R}^3$. Let $U = \phi_1(U_1) \cap U_2$. Note that $\phi = \phi_2 \circ \phi_1^{-1} : U \to \phi(U)$ satisfies $\phi(p) = p$ and

$$\begin{aligned}\phi^*\alpha_2 &= (\phi_2 \circ \phi_1^{-1})^*\alpha_2 \\ &= (\phi_1^{-1})^*\phi_2^*\alpha_2 \\ &= (\phi_1^{-1})^*\alpha_0 \\ &= \alpha_1. \qquad \square\end{aligned}$$

In closing, we cite an example from Bennequin's groundbreaking 1983 paper [8] to show that Darboux's theorem is really a *local* result.

Example 6.6.6. Consider the contact form $\alpha_2 = xdy - ydx + dz$ on $\mathbf{R}^3$, which can be written in cylindrical coordinates (r, θ, z) as $\alpha_2 = dz + r^2 d\theta$. Let $f : \mathbf{R}^3 \to \mathbf{R}$ be the function given by

$$f(r, \theta, z) = \frac{1}{(1 + r^2 + z^2)^2}$$

and let $\beta = f\alpha_2$. Since f is nowhere zero, β is also a contact form.

We claim that there is no diffeomorphism $\phi : \mathbf{R}^3 \to \mathbf{R}^3$ such that $\phi^*\alpha_2 = \beta$ (and hence, in light of Example 6.4.6, no diffeomorphism that pulls back the standard contact form to β). Suppose, to the contrary, that there were a diffeomorphism ϕ such that $\phi^*\alpha_2 = \beta$. Then, by Proposition 6.4.4, we would have $\phi_*\xi_b = \xi_2$, where ξ_2 (resp. ξ_b) is the Reeb field of α_2 (resp. β). Now on the one hand, we ask the reader to show in Exercise 6.6 that $\xi_2 = \frac{\partial}{\partial z}$. On the other hand, the reader should verify (Exercise 6.10) that $\xi_b = X_{1/f}$, the contact gradient of $1/f$ relative to the contact form α_2. After the calculations of Exercise 6.13, the reader may show that

$$\xi_b = X_{1/f} = 2(1 + r^2 + z^2)\left(\frac{\partial}{\partial \theta} + zr\frac{\partial}{\partial r} + \left(\frac{1 + z^2 - r^2}{2}\right)\frac{\partial}{\partial z}\right).$$

Now consider the integral curve c of ξ_b through the point $(1, 0, 0)$. Since $\xi_b(1, 0, 0) = 4\frac{\partial}{\partial \theta}$, the image of c is a circle of radius 1 in the plane $z = 0$ in $\mathbf{R}^3$. But according to Proposition 3.8.8, ϕ must exchange the integral curves of ξ_b with those of $\phi_*\xi_b = \xi_2$, which are all lines parallel to the z-axis in $\mathbf{R}^3$. A basic fact of topology is that the image of a circle by a diffeomorphism cannot be a line, a contradiction. Hence there can be no diffeomorphism $\phi : \mathbf{R}^3 \to \mathbf{R}^3$ satisfying $\phi^*\alpha_2 = \beta$.

6.7 Higher Dimensions

We opened this chapter by noting the potential advantages in viewing geometry not as the study of "points in space," but rather as the study of structures on more abstract sets of objects—such as, for instance, the set of contact elements $C\mathbf{R}^2$. As a set, $C\mathbf{R}^2$ is the same as $\mathbf{R}^3$, and we often treat them as the same. But viewing the "points" of $\mathbf{R}^3$ as contact elements in $C\mathbf{R}^2$ gives a natural setting for such seemingly unrelated problems as Huygens' principle in geometric optics and the method of characteristics for solving differential equations.

A natural generalization would be to define contact structures in higher dimensions. We noted at the outset that we would carry out our main presentation of contact geometry in the context of $\mathbf{R}^3$ exactly to exploit the naive familiarity with $\mathbf{R}^3$ that the student develops, for example, in a first course in multivariable calculus. The goal of this section is to extend the basic concepts of this chapter to higher dimensions. In so doing, some of the fundamental features of contact geometry will stand out more clearly than they do in the specific setting of $\mathbf{R}^3$. In fact, we will see in this section certain geometric objects that might not attract as much attention when viewed solely from the three-dimensional perspective.

Many of the propositions of this section are just higher-dimensional versions of earlier results in this chapter. We will leave the proofs of such statements to the reader. The emphasis in this section will be on results that are more closely tied to the dimension, an aspect of the subject that until now has been in the background.

We will then return to the space of contact elements, now in higher dimensions, in order to formulate the method of characteristics to solve first-order *partial* differential equations, which, as mentioned in Sect. 6.2, was the initial motivation for the method.

The first task in generalizing contact geometry to higher dimensions is to properly understand the contact condition of the nondegeneracy of one-forms. Let α be a nowhere-zero one-form on $\mathbf{R}^m$. At each point $p \in \mathbf{R}^m$, the vector subspace $E_p \subset T_p\mathbf{R}^m$ given by $E_p = \ker \alpha(p)$ has dimension $(m-1)$.

Now consider the two-form $d\alpha(p)$ restricted to E_p. Recall that $d\alpha(p)$ is said to be nondegenerate on E_p if for all $\mathbf{v}_p \in E_p$ with $\mathbf{v}_p \neq \mathbf{0}_p$, there is a $\mathbf{w}_p \in E_p$ such that $d\alpha(p)(\mathbf{v}_p, \mathbf{w}_p) \neq 0$. In the terminology of Definition 2.10.1, if $d\alpha(p)$ is nondegenerate, then $d\alpha(p)$ is a linear symplectic form on E_p.

It is a consequence of Theorem 2.10.4 that the nondegeneracy condition implies that there is a basis of E_p consisting of pairs of tangent vectors $\{\mathbf{e}_i, \mathbf{f}_i\}$ $(i = 1.\dots, n)$ such that

$$(d\alpha(p))^n(\mathbf{e}_1, \mathbf{f}_1, \dots, \mathbf{e}_n, \mathbf{f}_n) = (d\alpha(p) \wedge \cdots \wedge d\alpha(p))(\mathbf{e}_1, \mathbf{f}_1, \dots, \mathbf{e}_n, \mathbf{f}_n) = 1.$$

In particular, $\dim(E_p) = 2n$.

Finally, for each $p \in \mathbf{R}^m$, the fact that $\dim E_p = m - 1$ implies that there is a tangent vector $\mathbf{g}_p \notin E_p$ such that

$$\{\mathbf{e}_1, \mathbf{f}_1, \dots, \mathbf{e}_n, \mathbf{f}_n, \mathbf{g}_p\}$$

is a basis for $T_p\mathbf{R}^m$, and so $m = 2n + 1$. Note then that

$$\begin{aligned}(\alpha(p) \wedge (d\alpha(p))^n)(\mathbf{e}_1, \mathbf{f}_1, \ldots, \mathbf{e}_n, \mathbf{f}_n, \mathbf{g}_p) \\ = \alpha(p)(\mathbf{g}_p) \cdot (d\alpha(p))^n(\mathbf{e}_1, \mathbf{f}_1, \ldots, \mathbf{e}_n, \mathbf{f}_n)) \\ \neq 0,\end{aligned}$$

since $\mathbf{g} \notin \ker \alpha(p)$ but $\mathbf{e}_i, \mathbf{f}_i \in \ker \alpha(p)$ for all $i = 1, \ldots, n$.

This discussion prompts the following definition.

Definition 6.7.1. A *contact form α on a domain $U \subset \mathbf{R}^{2n+1}$* is a nowhere-zero differential one-form that is nondegenerate in the sense that at each point $p \in U$, $d\alpha(p)$ is nondegenerate on $E_p = \ker \alpha(p)$. Equivalently, α is a contact form if at each point p we have the following inequality of $(2n+1)$-forms:

$$\alpha(p) \wedge (d\alpha(p))^n \neq 0.$$

We will call the pair (U, α) a *contact space* and $E = \ker \alpha$ the *contact distribution* associated to α.

Note that as a consequence of the discussion above, *a contact space must be odd-dimensional.* This has been incorporated into the definition by the odd dimension $(2n+1)$.

As in the three-dimensional case, the central object of study is the contact distribution E, which, by virtue of having dimension one less than that of the contact space, is often called the *contact hyperplane*.

We note that the existence of a Reeb field associated with a contact form α follows from an argument similar to one given in the proof of Theorem 6.3.10.

Theorem 6.7.2. *Let (U, α) be a contact space, where $U \subset \mathbf{R}^{2n+1}$ is a domain. Then there exists a unique vector field ξ on U, called the* Reeb field *for α, having the properties that $\alpha(\xi) = 1$ and $i(\xi)d\alpha = 0$.*

Example 6.7.3. In analogy with Example 6.3.2, the *standard contact form* on $\mathbf{R}^{2n+1}$ with coordinates $(x_1, y_1, \ldots, x_n, y_n, z)$ is defined to be

$$\alpha_0 = dz + \sum_{i=1}^{n} x_i dy_i.$$

The reader will note that in keeping with the discussion at the beginning of the section, the coordinates are "paired," with the exception of the distinguished z-coordinate. In fact, the Reeb field ξ_0 of α_0 is just

$$\xi_0 = \frac{\partial}{\partial z}.$$

For all $p = (x_1, y_1, \ldots, x_n, y_n, z) \in \mathbf{R}^{2n+1}$, the vector field ξ_0 is transverse to the contact distribution $(E_0)_p = \ker \alpha_0(p) \subset T_p\mathbf{R}^{2n+1}$,

$$(E_0)_p = \left\{ \langle X^1, Y^1, \ldots, X^n, Y^n, Z \rangle_p \mid Z = -x_1 Y^1 - \cdots - x_n Y^n \right\},$$

in the sense that $(\xi_0)_p \notin (E_0)_p$.

Example 6.7.4. A higher-dimensional version of Example 6.3.4 is given by the one-form α_2 on $\mathbf{R}^{2n+1}$ with coordinates $(x_1, y_1, \ldots, x_n, y_n, z)$ as follows:

$$\alpha_2 = dz + \sum_{i=1}^{n} (x_i dy_i - y_i dx_i).$$

Again, the reader will note the "paired" coordinates (x_i, y_i), which in this example appear in a more symmetric form than in Example 6.7.3. Again there is the distinguished z-coordinate associated to the Reeb field

$$\xi_2 = \frac{\partial}{\partial z}$$

that is transverse to the contact distribution

$$(E_2)_p = \left\{ \langle X^1, Y^1, \ldots, X^n, Y^n, Z \rangle_p \mid Z = \sum_{i=1}^{n} (y_i X^i - x_i Y^i) \right\}.$$

A number of the other definitions and statements from the earlier sections translate directly to the higher-dimensional setting. We list the most important of them here for reference, with the understanding that the proofs follow by adapting the methods directly from the three-dimensional setting.

Definition 6.7.5. Suppose (U_1, α_1) and (U_2, α_2) are contact spaces, where $U_1, U_2 \subset \mathbf{R}^{2n+1}$ are domains. Let $\phi : U_1 \to U_2$ be a diffeomorphism. Then $\phi : (U_1, \alpha_1) \to (U_2, \alpha_2)$ is a *contact diffeomorphism* if there is a nowhere-zero function $f : U_1 \to \mathbf{R}$ such that $\phi^* \alpha_2 = f \alpha_1$. In the case that $\phi^* \alpha_2 = \alpha_1$, ϕ is said to be a *strictly contact diffeomorphism*.

Theorem 6.7.6. *For domains $U_1, U_2 \subset \mathbf{R}^{2n+1}$, suppose $\phi : (U_1, \alpha_1) \to (U_2, \alpha_2)$ is a diffeomorphism between contact spaces with contact distributions $E_1 = \ker \alpha_1$ and $E_2 = \ker \alpha_2$ and Reeb fields ξ_1 and ξ_2. Then ϕ is a contact diffeomorphism if and only if $\phi_*(E_1) = E_2$. Further, ϕ is a strictly contact diffeomorphism if both $\phi_*(E_1) = E_2$ and $\phi_*(\xi_1) = \xi_2$.*

Definition 6.7.7. Let $U \subset \mathbf{R}^{2n+1}$ be a domain. A vector field X on the contact space (U, α) is said to be a *contact vector field* if there is a function $g : U \to \mathbf{R}$ such that $\mathcal{L}_X \alpha = g\alpha$. Furthermore, X is a *strictly contact vector field* if $\mathcal{L}_X \alpha = 0$.

The flow ϕ_t of a contact vector field preserves the contact distribution of α (and so is a contact diffeomorphism for each t), while the flow of a strictly contact vector field preserves the contact form itself (and so is a strictly contact diffeomorphism for each t).

Theorem 6.7.8. *For a domain $U \subset \mathbf{R}^{2n+1}$, let $\mathcal{X}(U, \alpha)$ be the vector space of contact vector fields on a contact space (U, α), and let $C^\infty(U)$ be the vector space of smooth real-valued functions on U. Consider the map $\Phi : \mathcal{X}(U, \alpha) \to C^\infty(U)$ given by*

$$\Phi(X) = \alpha(X).$$

Then Φ is a vector space isomorphism whose inverse is given by

$$\Phi^{-1}(f) = f\xi + H,$$

where ξ is the Reeb field of α and H is the unique horizontal vector field satisfying $i(H)d\alpha = (\xi\,[f])\alpha - df$.

As before, we call $X_f = \Phi^{-1}(f)$ the *contact gradient* of f.

We state the contact version of Darboux's theorem in higher dimensions.

Theorem 6.7.9. *Let (U, α) be a contact space, where $U \subset \mathbf{R}^{2n+1}$, and let α_0 be the standard contact form on $\mathbf{R}^{2n+1}$. Then for each $p \in U$, there exist a domain $V \subset U$ containing p and a diffeomorphism $\phi : V \to \phi(V) \subset \mathbf{R}^{2n+1}$ such that $\phi(p) = p$ and $\phi^*\alpha = \alpha_0$ on V.*

Having indicated the concepts that have direct generalizations to higher dimensions, we now turn to those concepts that are best seen in the general setting as opposed to that of $\mathbf{R}^3$.

As in the three-dimensional case, Frobenius's theorem (Theorem 6.3.7) guarantees that the contact distribution $E = \ker \alpha$ is not integrable due to the nondegeneracy condition $\alpha \wedge (d\alpha)^n \neq 0$. The natural question, then, is whether there are geometric sets of lesser dimension that are tangent to the contact distribution at each point. The following theorem answers the question in the affirmative—up to a point.

Theorem 6.7.10. *Let $S = \phi(U) \subset \mathbf{R}^{2n+1}$ be a parameterized set in the contact space $(\mathbf{R}^{2n+1}, \alpha)$, where $U \subset \mathbf{R}^k$ is a domain and $\phi : U \to \mathbf{R}^{2n+1}$ is a regular parameterization. Suppose that for all $p \in S$, we have $T_pS \subset E_p$, where $E = \ker \alpha$ is the contact distribution. Then $k \leq n$.*

Proof. Suppose to the contrary that for some $p \in S$, T_pS contains $(n+1)$ linearly independent tangent vectors $\mathbf{e}_1, \ldots, \mathbf{e}_{n+1}$, where by hypothesis, $\mathbf{e}_i \in E_p$ for all i. We can complete the basis of E_p with tangent vectors $\mathbf{f}_1, \ldots, \mathbf{f}_{n-1} \in E_p$; in this way, the set

$$\{\mathbf{e}_1, \ldots, \mathbf{e}_{n+1}, \mathbf{f}_1, \ldots, \mathbf{f}_{n-1}, \xi\}$$

is a basis for $T_p\mathbf{R}^{2n+1}$, where ξ is the Reeb field for α. We have

$$(\alpha \wedge (d\alpha)^n)(\mathbf{e}_1, \ldots, \mathbf{e}_{n+1}, \mathbf{f}_1, \ldots, \mathbf{f}_{n-1}, \xi) = (d\alpha)^n(\mathbf{e}_1, \ldots, \mathbf{e}_{n+1}, \mathbf{f}_1, \ldots, \mathbf{f}_{n-1}),$$

since $\alpha(\xi) = 1$ and $\alpha(\mathbf{e}_i) = \alpha(\mathbf{f}_j) = 0$. The exterior product on the right-hand side is a sum of terms each of which has n factors of the form $d\alpha(\mathbf{e}_i, \mathbf{e}_j)$, $d\alpha(\mathbf{e}_i, \mathbf{f}_j)$, or

$d\alpha(\mathbf{f}_i, \mathbf{f}_j)$. In fact, a counting argument shows that each term must have at least one factor of the form $d\alpha(\mathbf{e}_i, \mathbf{e}_j)$. Since S is integrable by assumption as a parameterized set in $\mathbf{R}^{2n+1}$, Frobenius's theorem (Theorem 6.3.7) and an argument exactly like that in the proof of Theorem 6.3.8 together imply $d\alpha(\mathbf{e}_i, \mathbf{e}_j) = 0$. Hence

$$(\alpha \wedge (d\alpha)^n)(\mathbf{e}_1, \ldots, \mathbf{e}_{n+1}, \mathbf{f}_1, \ldots, \mathbf{f}_{n-1}, \xi) = 0,$$

contradicting the nondegeneracy of α. This shows that T_pS cannot contain $(n+1)$ linearly independent tangent vectors. □

The integrable geometric sets of a contact space $(\mathbf{R}^{2n+1}, \alpha)$ having maximum dimension n will have a special role in contact geometry.

Definition 6.7.11. Let $S = \phi(U)$ be a parameterized geometric set, where $U \subset \mathbf{R}^n$ is a domain and $\phi : U \to \mathbf{R}^{2n+1}$ is a regular parameterization. If S is such that $T_pS \subset \ker \alpha(p)$ for all $p \in S$, then S is called a *Legendre set*. Equivalently, S is a Legendre set if $\phi^*\alpha = 0$.

Example 6.7.12 (Legendre curves). The case of $\mathbf{R}^3$, which has been the context for most of this chapter, is the case $n = 1$ for Theorem 6.7.10. The Legendre sets in this case will be called *Legendre curves*. The integrable sets of maximal dimension are curves $c : I \to \mathbf{R}^3$, where $I \subset \mathbf{R}$ is an open interval, satisfying $\dot{c}(t) \in \ker \alpha(c(t))$ for all $t \in I$. We have already seen the special role these curves play in important examples, namely as the wavefronts of Sect. 7.1 and as the curves in $C\mathbf{R}^2$ in Sect. 7.2 that are lifts of smooth functions.

Example 6.7.13. Consider $\mathbf{R}^5$ with coordinates (x_1, y_1, x_2, y_2, z) and standard contact form $\alpha_0 = dz + x_1 dy_1 + x_2 dy_2$. Consider $U = \mathbf{R}^2$ with coordinates (s, t) and let $\phi_1 : \mathbf{R}^2 \to \mathbf{R}^5$ be given by

$$\phi_1(s, t) = (s, 0, t, 0, 0).$$

Then $S_1 = \phi_1(\mathbf{R}^2)$ is a Legendre set of dimension two, which we will call a *Legendre surface*. In fact, $E = \text{Span}\,\{e_1, e_2\}$, where $e_1 = (\phi_1)_*(\frac{\partial}{\partial s}) = \frac{\partial}{\partial x_1}$ and $e_2 = (\phi_1)_*(\frac{\partial}{\partial t}) = \frac{\partial}{\partial x_2}$.

Example 6.7.14. Let $\phi_2 : U = \mathbf{R}^2 \to (\mathbf{R}^5, \alpha_0)$ be the map given by

$$\phi_2(s, t) = \left(s, -s, t, -t, \frac{1}{2}(s^2 + t^2) \right).$$

Then the reader may verify that $S_2 = \phi_2(\mathbf{R}^2)$ is again a Legendre surface in $(\mathbf{R}^5, \alpha_0)$. We can write $E = \text{Span}\{e_1, e_2\}$, where

$$e_1 = (\phi_2)_* \left(\frac{\partial}{\partial s} \right) = \frac{\partial}{\partial x_1} - \frac{\partial}{\partial y_1} + x_1 \frac{\partial}{\partial z}$$

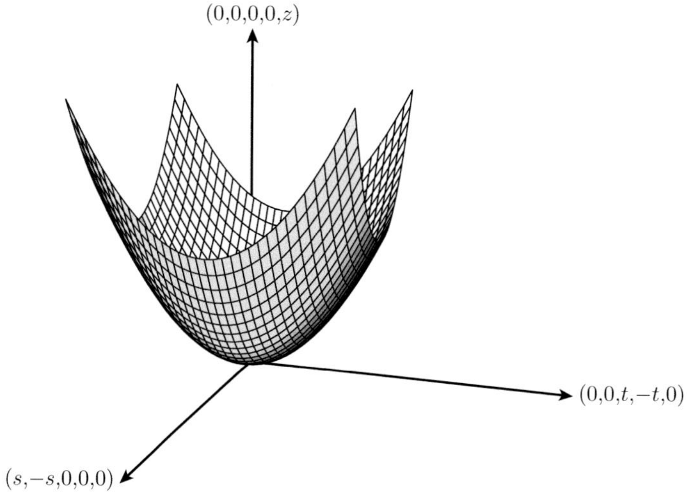

Fig. 6.9 A paraboloid embedded as a Legendre set in $\mathbf{R}^5$.

and

$$e_2 = (\phi_2)_* \left(\frac{\partial}{\partial t}\right) = \frac{\partial}{\partial x_2} - \frac{\partial}{\partial y_2} + x_2 \frac{\partial}{\partial z}.$$

Note that S_2 can be visualized in $\mathbf{R}^5$ as the graph of the paraboloid described by

$$z = \frac{1}{2}(u^2 + v^2)$$

over the uv-plane, where the u-axis is along the vector $(1, -1, 0, 0, 0) \in \mathbf{R}^5$ and the v-axis is along $(0, 0, 1, -1, 0)$. See Fig. 6.9.

Example 6.7.15 (The Whitney sphere). Consider the standard contact space $(\mathbf{R}^{2n+1}, \alpha_0)$ with coordinates $(x_1, y_1, \ldots, x_n, y_n, z)$ and $\alpha_0 = dz + \sum x_i dy_i$. Let $U \subset \mathbf{R}^n$ be the domain defined by

$$U = \left\{(u_1, \ldots, u_n) \mid u_1^2 + \cdots + u_n^2 < 1\right\}$$

and define $\phi : U \to (\mathbf{R}^{2n+1}, \alpha_0)$ by

$$\phi(u_1, \ldots, u_n) = \left(u_1, -2u_0u_1, \ldots, u_n, -2u_0u_n, 2u_0 - (4/3)(u_0)^3\right),$$

where $u_0 = \sqrt{1 - \sum u_i^2}$. Since $u_0^2 + \sum u_i^2 = 1$, we have

$$u_0 du_0 + \sum u_i du_i = 0,$$

which, in turn, along with the usual properties of the exterior derivative, shows that $\phi^*\alpha_0 = 0$. Hence $S = \phi(U)$ is a Legendre set in $\mathbf{R}^{2n+1}$, known as the *Whitney sphere*. The fact that ϕ is regular and one-to-one onto its image S means that S can be considered as a part of an n-dimensional sphere "embedded" in the contact space $(\mathbf{R}^{2n+1}, \alpha_0)$.

We now return to the method of characteristics described in Sect. 6.2. As mentioned there, the method of characteristics was originally developed as a technique to solve not ordinary differential equations, but first-order partial differential equations. We will outline the general approach here, then illustrate the method with an example in the two-dimensional setting.

A first-order partial differential equation in n variables is an equation relating the n independent variables $x_1, \ldots, x_n$, a function $u : \mathbf{R}^n \to \mathbf{R}$ of these n variables, and the n first partial derivatives of u, $u_{x_1} = \dfrac{\partial u}{\partial x_1}, \ldots, u_{x_n} = \dfrac{\partial u}{\partial x_n}$. Following our treatment in Sect. 6.2 (when $n = 1$), we represent the first-order partial differential equation as the zero set of a smooth, real-valued function $F : \mathbf{R}^{2n+1} \to \mathbf{R}$,

$$S_F = \{(x_1, \ldots, x_n, y, m_1, \ldots, m_n) \mid F(x_1, \ldots, x_n, y, m_1, \ldots, m_n) = 0\}.$$

Here we consider the set of contact elements $C\mathbf{R}^{n+1}$ to be the same as $\mathbf{R}^{2n+1}$.

In analogy with Sect. 6.2, we consider the special one-form

$$\alpha_1 = dy - m_1 dx_1 - \cdots - m_n dx_n$$

on $\mathbf{R}^{2n+1}$. Since α_1 is nondegenerate, we can consider the contact space $(\mathbf{R}^{2n+1}, \alpha_1)$.

For a domain $U \subset \mathbf{R}^n$ and any smooth function $u : U \to \mathbf{R}$, define the *one-jet* $\tilde{u} : U \to \mathbf{R}^{2n+1}$ to be the function

$$\tilde{u}(p) = (p_1, \ldots, p_n, u(p), u_{x_1}(p), \ldots, u_{x_n}(p)),$$

where $p = (p_1, \ldots, p_n) \in \mathbf{R}^n$. In this way, we can say that a function $u : U \to \mathbf{R}$ is a solution to the partial differential equation described by $F = 0$ if $F \circ \tilde{u} = 0$ on U. Note that $\tilde{u}^*\alpha_1 = 0$, and so $\tilde{u}(U)$ is a Legendre set in $\mathbf{R}^{2n+1}$, provided that $\tilde{u}$ is regular.

Suppose $c : \mathbf{R}^n \to \mathbf{R}^{2n+1}$ is a smooth function satisfying $c^*\alpha_1 = 0$. In general, there may be no smooth function $u : \mathbf{R}^n \to \mathbf{R}$ such that $\tilde{u} = c$, for example in the case of the map

$$c(x_1, \ldots, x_n) = (0, \ldots, 0, 0, x_1, \ldots, x_n).$$

However, under special circumstances that generalize the condition that $\dot{x}(s_0) \neq 0$ in Proposition 6.2.1, we can be assured that c can be viewed as the one-jet of some smooth function (up to reparameterization).

Proposition 6.7.16. *Let $c : \mathbf{R}^n \to \mathbf{R}^{2n+1}$ be a smooth function, where for $\bar{t} = (t_1, \ldots, t_n) \in \mathbf{R}^n$,*

$$c(\bar{t}) = (x(\bar{t}), y(\bar{t}), m(\bar{t})).$$

Here $x : \mathbf{R}^n \to \mathbf{R}^n$, $y : \mathbf{R}^n \to \mathbf{R}$, and $m : \mathbf{R}^n \to \mathbf{R}^n$ are written with component functions $x(\bar{t}) = (x_1(\bar{t}), \ldots, x_n(\bar{t})) \in \mathbf{R}^n$, $y(\bar{t}) \in \mathbf{R}$, $m(\bar{t}) = (m_1(\bar{t}), \ldots, m_n(\bar{t})) \in \mathbf{R}^n$. Suppose that

$$c^*\alpha_1 = 0.$$

Suppose further that for some $\tau \in \mathbf{R}^n$, $(x_*)(\tau)$ is an isomorphism. Then there are a domain $U \subset \mathbf{R}^n$ containing $x(\tau)$ and a smooth function $u : U \to \mathbf{R}$ such that $\tilde{u} \circ x = c$.

Proof. Since $(x_*)(\tau)$ is an isomorphism, the inverse function theorem (Theorem 3.6.12) guarantees the existence of a domain $V \subset \mathbf{R}^n$ containing τ on which $x : V \to x(V)$ is a diffeomorphism. Hence on the domain $U = x(V)$, we have a differentiable inverse $x^{-1} : U \to V$.

Define then $u : U \to \mathbf{R}$ by

$$u = y \circ x^{-1}.$$

We claim that $\tilde{u} = c \circ x^{-1}$. To see this, let $\bar{t} \in V$ and let $p = x(\bar{t}) \in U$. We have $u(p) = u(x(\bar{t})) = y(\bar{t})$. Further, applying the chain rule, we have

$$\begin{aligned}
u_{x_i}(p) &= \left.\frac{\partial}{\partial x_i}\right|_p \left[y \circ x^{-1}\right] \\
&= (y \circ x^{-1})_*(p) \left[\left.\frac{\partial}{\partial x_i}\right|_p\right] \\
&= ((y_*)(x^{-1}(p)) \circ (x^{-1})_*(p)) \left[\left.\frac{\partial}{\partial x_i}\right|_p\right] \\
&= (y_*)(\bar{t}) \left[(x_*)(\bar{t})\right]^{-1} \left[\left.\frac{\partial}{\partial x_i}\right|_p\right] \\
&= (y_*)(\bar{t}) \left[\left.\frac{\partial}{\partial t_i}\right|_{\bar{t}}\right] \\
&= \frac{\partial y}{\partial t_i}(\bar{t}) \\
&= m_i(\bar{t}) \quad \text{since } c^*\alpha_1 = 0.
\end{aligned}$$

Hence, for all $p \in U$,

$$\tilde{u}(p) = (p, u(p), u_{x_i}(p)) = c(x(\bar{t}), y(\bar{t}), m(\bar{t})) = (c \circ x^{-1})(p). \qquad \square$$

In order to expect results for existence and uniqueness in the setting of partial differential equations, special care must be taken in describing the initial (or boundary) conditions in a way that generalizes the standard initial conditions for ordinary differential equations. We will consider here the *Cauchy problem* for first-order partial differential equations: Find a function $u : \mathbf{R}^n \to \mathbf{R}$ that satisfies $F \circ \tilde{u} = 0$, when u additionally has specified values on some $(n-1)$-dimensional geometric set of $\mathbf{R}^n$. This indeed generalizes the ordinary differential equation case of $n = 1$: a connected, 0-dimensional geometric set is a point in $\mathbf{R}$, so specifying a value of the function there amounts to specifying $u(x_0) = y_0$.

For this exposition, we will adopt the following conventions.

Definition 6.7.17. For a first-order partial differential equation in n variables $x_1, \ldots, x_n$, the *Cauchy data* consists of an $(n-1)$-dimensional geometric set $S = \phi(U)$ described by a regular parameterization $\phi : U \to \mathbf{R}^n$, where $U \subset \mathbf{R}^{n-1}$ is a domain, along with a smooth function $g : U \to \mathbf{R}$. We write the Cauchy data in the form (U, ϕ, g).

Using these conventions, we have the following formulation of the Cauchy problem for first-order partial differential equations:

> Cauchy Problem: Let $F : \mathbf{R}^{2n+1} \to \mathbf{R}$ be a smooth function on the set of contact elements $\mathbf{R}^{2n+1}$ with contact form α_1 and Cauchy data (U, ϕ, g). Find all smooth functions u : $\mathbf{R}^n \to \mathbf{R}$ that satisfy
>
> - $F \circ \tilde{u} = 0$;
> - For all $s \in U$, $u(\phi(s)) = g(s)$.

As in the one-dimensional case, a central role in the construction of solutions will be played by a special vector field, known again as the characteristic vector field. Note that for every smooth function $f : \mathbf{R}^{2n+1} \to \mathbf{R}$, we can consider the contact gradient X_f of f relative to the contact form α_1 (see Theorem 6.7.8). Let $\tilde{X}_f$ be the horizontal part of the contact gradient, i.e., $\tilde{X}_f = X_f - f\xi_1$, where ξ_1 is the Reeb field for α_1.

Definition 6.7.18. The *characteristic vector field* for the partial differential equation represented by the smooth function $F : (\mathbf{R}^{2n+1}, \alpha_1) \to \mathbf{R}$ is $\tilde{X}_F$, the horizontal part of the contact gradient of F relative to α_1. In components, writing

$$\tilde{X}_F = \left\langle X^1, \ldots, X^n, Y, M^1, \ldots, M^n \right\rangle,$$

we have

$$\begin{cases} X^i &= -F_{m_i} \quad \text{for } i = 1, \ldots, n, \\ Y &= -\sum_{i=1}^n m_i F_{m_i}, \\ M^i &= F_{x_i} + m_i F_y \quad \text{for } i = 1, \ldots, n. \end{cases}$$

The reader can verify that just as in the case $n = 1$, the integral curves of $\tilde{X}_F$ have important properties relative to the Cauchy problem.

Proposition 6.7.19. *Let* $F : (\mathbf{R}^{2n+1}, \alpha_1) \to \mathbf{R}$ *be a smooth function with associated characteristic vector field* $\tilde{X}_F$. *Let* $c : I \to \mathbf{R}^{2n+1}$ *be an integral curve of* $\tilde{X}_F$ *through a point* $p = c(0) \in \mathbf{R}^{2n+1}$ *such that* $F(p) = 0$. *Then for all* $t \in I$, $\dot{c}(t)$ *is tangent to both the zero set* S_F *and the contact distribution* $E = \ker \alpha_1$. *In other words, for all* $t \in I$,

- $\alpha_1(\dot{c}(t)) = 0$, *and*
- $dF(\dot{c}(t)) = 0$.

We are now in a position to state an existence and uniqueness theorem for the Cauchy problem.

Theorem 6.7.20. *Let* $U \subset \mathbf{R}^{n-1}$ *be a domain with coordinates* $(s_1, \ldots, s_{n-1})$ *and let* $\phi : U \to \mathbf{R}^n$ *be a smooth, regular parameterization with component functions* $\phi = (\phi^1, \ldots, \phi^n)$. *For the Cauchy problem associated to the smooth function* $F : \mathbf{R}^{2n+1} \to \mathbf{R}$ *(with zero set* S_F*) and the Cauchy data* (U, ϕ, g), *suppose that:*

(A) There exist an element $\bar{s} \in U$ *and an element* $\sigma = (x_0, y_0, m_0) \in S_F \subset \mathbf{R}^{2n+1}$ *with* $x_0 = \phi(\bar{s}) \in \mathbf{R}^n$, $y_0 = g(\bar{s}) \in \mathbf{R}$, *and* $m_0 = (\mu_1, \ldots, \mu_n) \in \mathbf{R}^n$ *satisfying*

$$\sum_{i=1}^{n} \left(\mu_i \frac{\partial \phi^i}{\partial s_j}(\bar{s}) \right) = \frac{\partial g}{\partial s_j}(\bar{s}) \quad \text{for } j = 1, \ldots, n-1.$$

(B) The tangent vector $V_0 \in T_{x_0}\mathbf{R}^n$ *given by*

$$V_0 = \sum_{i=1}^{n} F_{m_i}(\sigma) \cdot \left. \frac{\partial}{\partial x_i} \right|_{x_0}$$

is not in $T_{x_0}S$, *where* $S = \phi(U) \subset \mathbf{R}^n$.

Then there are a domain $W \subset \mathbf{R}^n$ *containing* x_0 *and a unique smooth function* $u : W \to \mathbf{R}$ *that is a solution to the Cauchy problem* $F \circ \tilde{u} = 0$ *and* $u(\phi(s)) = g(s)$ *for all* $s \in U$ *such that* $\phi(s) \in W$.

Proof. The proof will proceed in three steps. In the first, we will extend the parameterization of the geometric set $S = \phi(U) \subset \mathbf{R}^n$ given as part of the Cauchy data to an $(n-1)$-dimensional geometric set $S' \subset \mathbf{R}^{2n+1}$ in such a way that $T_pS' \subset E_p = \ker \alpha_1(p)$ for all $p \in S'$. Then, we "thicken" S' using the characteristic field of F to obtain a new n-dimensional geometric set in $\mathbf{R}^{2n+1}$, which is in fact a Legendre set. Finally, we show that the resulting parameterization of the Legendre set in fact satisfies the hypotheses of Proposition 6.7.16, and so defines a function $u : \mathbf{R}^n \to \mathbf{R}$ with the desired properties.

Our goal in the first step is to define a function $c : U \to \mathbf{R}^{2n+1}$, where U is the domain in $\mathbf{R}^{n-1}$ given as part of the Cauchy data, such that $c(\bar{s}) = \sigma$ and $c^*\alpha_1 = 0$. In addition, if we write

$$c(s) = (x(s), y(s), m(s)),$$

we will require that $x(s) = \phi(s)$ and $y(s) = g(s)$ for all $s \in U$. In this way, we can say that c will extend the Cauchy data to a geometric set in $\mathbf{R}^{2n+1}$.

With prescribed functions for x and y, constructing c amounts to finding a function $m : U \to \mathbf{R}^n$ that is consistent with the desired requirements for c. To do this, we construct a smooth function

$$\Phi : U \times \mathbf{R}^n \to \mathbf{R}^n$$

with component functions $\Phi = (\Phi^1, \dots, \Phi^n)$ as follows: For $j = 1, \dots, n-1$, define

$$\Phi^j(s_1, \dots, s_{n-1}, m_1, \dots, m_n) = \left(\sum_{i=1}^{n} m_i \frac{\partial \phi^i}{\partial s_j}(s_1, \dots, s_{n-1}) \right) - \frac{\partial g}{\partial s_j}(s_1, \dots, s_{n-1})$$

and

$$\begin{aligned} &\Phi^n(s_1, \dots, s_{n-1}, m_1, \dots, m_n) \\ &\quad = F(\phi(s_1, \dots, s_{n-1}), g(s_1, \dots, s_{n-1}), m_1, \dots, m_n). \end{aligned}$$

Note that assumption (A) in the hypotheses of the theorem implies that

$$\Phi(\bar{s}, m_0) = (0, \dots, 0).$$

Computing the matrix of partial derivatives of Φ with respect to the variables m_j gives

$$\left[\frac{\partial \Phi^i}{\partial m_j} \right] = \begin{bmatrix} \frac{\partial \phi^1}{\partial s_1} & \cdots & \frac{\partial \phi^n}{\partial s_1} \\ \vdots & & \vdots \\ \frac{\partial \phi^1}{\partial s_{n-1}} & \cdots & \frac{\partial \phi^n}{\partial s_{n-1}} \\ \frac{\partial F}{\partial m_1} & \cdots & \frac{\partial F}{\partial m_n} \end{bmatrix}.$$

Assumption (B) of the hypotheses guarantees that this matrix has rank n at the point $(\bar{s}, m_0)$. Hence by the implicit function theorem, there exist a domain $U_1 \subset U \subset \mathbf{R}^{n-1}$ containing $\bar{s}$ and a smooth function $m : U_1 \to \mathbf{R}^n$ such that $m(\bar{s}) = m_0$ and for all $s \in U_1$, $\Phi(s, m(s)) = 0$. In particular, since $\Phi^i(s, m(s)) = 0$ for all $i = 1, \dots, n-1$, we have $c^*\alpha_1 = 0$. Also, since $\Phi^n(s, m(s)) = 0$, we have $F \circ c = 0$, where $c : U \to \mathbf{R}^{2n+1}$ is given by $c(s) = (\phi(s), g(s), m(s))$. This concludes the first step of the proof.

Proceeding to the second step, we suppose that we are given a smooth function $c : U \to \mathbf{R}^{2n+1}$, where U is a domain in $\mathbf{R}^{n-1}$, satisfying $c^*\alpha_1 = 0$ and $F \circ c = 0$ on U. Our goal in this step is to construct a domain $V \subset \mathbf{R}^n$ along with a smooth function $\tilde{c} : V \to \mathbf{R}^{2n+1}$ such that $\tilde{c}^*\alpha_1 = 0$ and $F \circ \tilde{c} = 0$. The domain will have

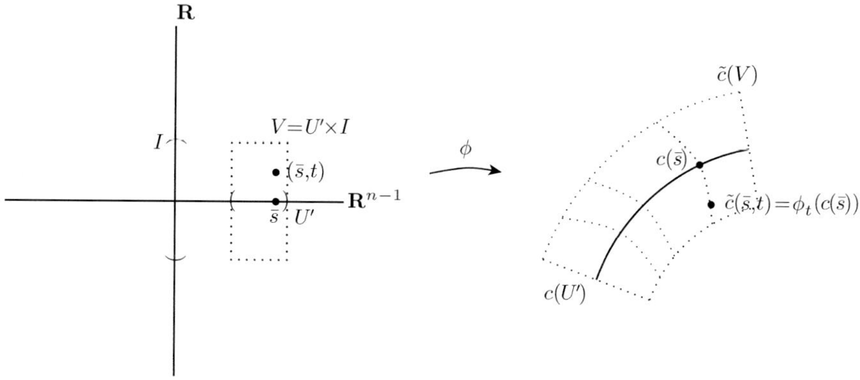

Fig. 6.10 The construction of $\tilde{c}$.

the form $V = U' \times I$, where $U' \subset U \subset \mathbf{R}^{n-1}$ is a domain and $I \subset \mathbf{R}$ is an open interval containing 0. We would like $\tilde{c}$ to extend c on U' with $\bar{s} \in U'$, in the sense that for $s \in U'$, $\tilde{c}(s,0) = c(s)$.

To accomplish this, we rely heavily on the definition of the characteristic field $\tilde{X}_F$. By Theorem 3.9.2, there exist a domain $U' \subset U$ containing $\bar{s}$, an interval $I \subset \mathbf{R}$ containing 0, and a differentiable map $\phi : U' \times I \to \mathbf{R}^{2n+1}$ such that

- $\phi(s,0) = c(s)$;
- $\dfrac{d}{dt}(\phi(c(s),t)) = \tilde{X}_F(\phi(c(s),t))$; and
- $\phi(s,t_1+t_2) = \phi(\phi(s,t_1),t_2)$.

As usual, we will write $\phi_t : \mathbf{R}^n \to \mathbf{R}^n$ to mean $\phi_t(s) = \phi(s,t)$.

We then define the domain $V = U' \times I$ and the map $\tilde{c} : V \to \mathbf{R}^{2n+1}$ to be $\tilde{c}(s,t) = \phi_t(c(s))$. See Fig. 6.10. We claim that V and $\tilde{c}$ have the desired properties. First,

$$\frac{d}{dt}(F \circ \tilde{c}) = dF(\dot{\tilde{c}}) = dF(\tilde{X}_F) = 0 \quad \text{by Proposition 6.7.19,}$$

and so

$$(F \circ \tilde{c})(s,t) = (F \circ \tilde{c})(s,0) = F(c(s)) = 0$$

by the assumptions on c.

Likewise, consider the t-derivative of $\tilde{c}^*\alpha_1$:

$$\begin{aligned}
\frac{d}{dt}(\tilde{c}^*\alpha_1) &= \frac{d}{dt}(\phi_t^*\alpha_1) \\
&= \phi_t^*\left(\mathcal{L}_{\tilde{X}_F}\alpha_1\right) \quad \text{by Proposition 4.7.2} \\
&= \tilde{c}^*\left(i(\tilde{X}_F)d\alpha_1 + di(\tilde{X}_F)\alpha_1\right) \quad \text{by Theorem 4.7.18} \\
&= \tilde{c}^*\left((i(\xi_1)dF)\alpha_1 - dF\right) \quad \tilde{X}_F \text{ is horizontal and Definition 6.7.18}
\end{aligned}$$

$$= \tilde{c}^*(G\alpha_1) - \tilde{c}^*(dF) \quad \text{where } G = \xi_1[F]$$
$$= (G \circ \tilde{c})(\tilde{c}^*\alpha_1) \quad \text{by the calculation above, since } \tilde{c}^* dF = d\tilde{c}^* F = 0.$$

Note that in the coordinates (s, t) on V, the one-form $\tilde{c}^*\alpha_1$ is independent of dt, and so is a linear combination of the forms ds_i. This follows from the fact that

$$\begin{aligned}(\tilde{c}^*\alpha_1)\left(\frac{\partial}{\partial t}\right) &= \alpha_1\left(\tilde{c}_*\frac{\partial}{\partial t}\right) \\ &= \alpha_1(\tilde{X}_F) \\ &= 0.\end{aligned}$$

Hence the equality of differential forms

$$\frac{d}{dt}(\tilde{c}^*\alpha_1) = (G \circ \tilde{c})\tilde{c}^*\alpha_1$$

amounts to a system of first-order linear ordinary differential equations whose solution yields

$$\tilde{c}^*\alpha_1 = \sum_{i=1}^{n} b_i \exp(H_i) ds_i,$$

where b_i are constants and $H_i : \mathbf{R}^n \to \mathbf{R}$ are smooth functions obtained by integrating $\int (G \circ \tilde{c}) dt$ and using initial conditions. However, the initial condition

$$c^*\alpha_1 = \tilde{c}^*\alpha_1\Big|_{t=0} = 0$$

implies that the constants b_i are zero for all $i = 1, \ldots, n$. Hence

$$\tilde{c}^*\alpha_1 = 0,$$

as desired.

The final step is to show that the function $\tilde{c} : \mathbf{R}^n \to \mathbf{R}^{2n+1}$ constructed according to the second step satisfies the hypotheses of Proposition 6.7.16. This is a direct consequence of assumptions (A) and (B). By assumption (A), there is $\bar{s} \in U \subset \mathbf{R}^{n-1}$ such that $\sigma = c(\bar{s}) = (x(\bar{s}), y(\bar{s}), m(\bar{s}))$ satisfies $F(\sigma) = 0$ and $c^*\alpha_1(\bar{s}) = 0$. Then, according to the construction of step 2, $(\bar{s}, 0) \in V$ is such that $\tilde{c}(\bar{s}, 0) = \sigma$, and so $F(\tilde{c}(\bar{s}, 0)) = 0$ and $\tilde{c}^*\alpha_1(\bar{s}, 0) = 0$. Writing

$$\tilde{c}(s, t) = (x(s, t), y(s, t), m(s, t)),$$

we will show that $(x_*)(\bar{s}, 0)$ is an isomorphism. This is a consequence of

$$x_* = \begin{bmatrix} \frac{\partial x_1}{\partial s_1} & \cdots & \frac{\partial x_1}{\partial s_{n-1}} & \frac{dx_1}{dt} \\ \vdots & & \vdots & \vdots \\ \frac{\partial x_n}{\partial s_1} & \cdots & \frac{\partial x_n}{\partial s_{n-1}} & \frac{dx_n}{dt} \end{bmatrix} = \begin{bmatrix} \frac{\partial x_1}{\partial s_1} & \cdots & \frac{\partial x_1}{\partial s_{n-1}} & -\frac{\partial F}{\partial m_1} \\ \vdots & & \vdots & \vdots \\ \frac{\partial x_n}{\partial s_1} & \cdots & \frac{\partial x_n}{\partial s_{n-1}} & -\frac{\partial F}{\partial m_n} \end{bmatrix},$$

the second equality due to Definition 6.7.18. Assumption (B), then, is precisely that $(x_*)(s, 0)$ has rank n, and so is invertible. Hence the hypotheses of Proposition 6.7.16 are satisfied.

We summarize the above steps to prove Theorem 6.7.20: For a first-order partial differential equation represented by the equation $F(x, y, m) = 0$ along with the Cauchy data (U, ϕ, g), there is a smooth function $c : U' \to \mathbf{R}^{2n+1}$ given by

$$c(s) = (\phi(s), g(s), m(s))$$

for some function $m : U' \to \mathbf{R}^n$, where $U' \subset U$ is a domain containing $\bar{s}$. This function c satisfies the conditions $c^*\alpha_1 = 0$ and $F \circ c = 0$. Further, there exist a domain $V \subset \mathbf{R}^n$ and a function $\tilde{c} : V \to \mathbf{R}^{2n+1}$ such that $V = U'' \times I$, where $U'' \subset U' \subset \mathbf{R}^{n-1}$, is a domain containing $\bar{s}$, $I \subset \mathbf{R}$ is an interval containing 0, $\tilde{c}(s, 0) = c(s)$, $\tilde{c}^*\alpha_1 = 0$, and $F \circ \tilde{c} = 0$. Then, by Proposition 6.7.16, there exist a domain $W \subset V$ containing $\phi(s_0)$ and a function $u : W \to \mathbf{R}$ satisfying $F \circ \tilde{u} = 0$ and for all $x = \phi(s) \in \phi(U) \cap W$, $u(x) = g(s)$.

The uniqueness of the solution u follows from the uniqueness of $\tilde{c}$ from Theorem 3.9.2. □

In practice, the following analogue of Proposition 6.2.4 is often useful.

Proposition 6.7.21. *Let $F : (\mathbf{R}^{2n+1}, \alpha_1) \to \mathbf{R}$ be a smooth function with zero set S_F. Then the integral curves of the contact gradient X_F of F and those of its horizontal part $\tilde{X}_F = H(X_F)$ through a point $p \in S_F$ coincide.*

This allows Lie's adaptation of the method of characteristics, integrating the contact vector field X_F as the crucial step in finding solutions to a first-order partial differential equation. We illustrate this method with an example; several more can be found in the exercises.

Example 6.7.22. Consider the first-order partial differential equation

$$\frac{\partial u}{\partial x} + \frac{\partial u}{\partial y} + u = 0$$

with initial condition

$$u(x, y) = 1 \quad \text{on } x^2 + y^2 = 1.$$

We can represent this as the zero set S_F, where $F : \mathbf{R}^5 \to \mathbf{R}$ is the function on the set of contact elements $C\mathbf{R}^3 = \mathbf{R}^5$ with coordinates (x_1, x_2, w, m_1, m_2) given by

$$F(x_1, x_2, w, m_1, m_2) = m_1 + m_2 + w.$$

For the Cauchy data, we will consider the domain $U = (-\pi/4, 3\pi/4)$ along with the parameterization $\phi : U \to \mathbf{R}^2$ given by

$$\phi(s) = (\cos(s), \sin(s)).$$

We have chosen the domain in such a way that the vector field on $\mathbf{R}^2$ given by

$$V_F = F_{m_1}\frac{\partial}{\partial x} + F_{m_2}\frac{\partial}{\partial y} = \frac{\partial}{\partial x} + \frac{\partial}{\partial y}$$

is nowhere tangent to $\phi(U)$. Finally, we define $g : U \to \mathbf{R}$ by $g(s) = 1$.

Lie's characteristic vector field X_F is given by

$$X_F = -\frac{\partial}{\partial x_1} - \frac{\partial}{\partial x_2} + w\frac{\partial}{\partial w} + m_1\frac{\partial}{\partial m_1} + m_2\frac{\partial}{\partial m_2}.$$

Hence we will look for the integral curves $c : I \to \mathbf{R}^5$ given by $c(t) = (x_1(t), x_2(t), w(t), m_1(t), m_2(t))$ satisfying

$$\begin{cases} \dot{x}_1 = -1, \\ \dot{x}_2 = -1, \\ \dot{w} = w, \\ \dot{m}_1 = m_1, \\ \dot{m}_2 = m_2, \end{cases} \tag{6.13}$$

with $(x_1(0), x_2(0), w(0), m_1(0), m_2(0)) = (\cos(s), \sin(s), 1, a(s), -1 - a(s))$. (Here $a : U \to \mathbf{R}$ is the function obtained by solving the system that comes from condition (A) of Theorem 6.7.20:

$$\Phi^1 = -m_1 \sin s + m_2 \cos s = 0,$$

$$\Phi^2 = m_1 + m_2 + 1 = 0,$$

for m_1 and m_2 in terms of s.)

Solving the first three equations of the system (6.13) yields

$$(x_1(t), x_2(t), w(t)) = (-t + \cos(s), -t + \sin(s), e^t).$$

Eliminating the parameter s by writing

$$1 = \cos^2 s + \sin^2 s = (x_1 + t)^2 + (x_2 + t)^2$$

and solving the quadratic equation (in t)

$$2t^2 + 2(x_1 + x_2)t + (x_1^2 + x_2^2 - 1) = 0$$

yields

$$t = \frac{-(x_1 + x_2)}{2} + \frac{\sqrt{2 - (x_1 - x_2)^2}}{2},$$

with the sign chosen to ensure that $t = 0$ when $x_1^2 + x_2^2 = 1$. Note that on $\phi(U)$, we have $x_1 + x_2 > 0$.

Hence the unique solution to the partial differential equation

$$\frac{\partial u}{\partial x} + \frac{\partial u}{\partial y} + u = 0$$

with the given Cauchy data is

$$u(x, y) = \exp\left(\frac{\sqrt{2 - (x - y)^2}}{2} - \frac{(x + y)}{2}\right).$$

This function is defined on a region in the xy-plane between the lines $x - y = \sqrt{2}$ and $x - y = -\sqrt{2}$, which in particular contains the circle $x^2 + y^2 = 1$. Recall that in our choice of t we also relied on the fact that $x + y > 0$, further restricting the domain on which the solution u is defined.

6.8 For Further Reading

Since contact geometry made its first explicit appearance in Lie's 1888 three-volume series [29] and then in his 1896 *Geometry of Contact Transformations* [28], the subject for decades was relegated to appendices and chapters in books whose main theme was symplectic geometry. See, for example, the classic texts of Kobayashi and Nomizu [24, vol. 2, Note 28] and Arnold [3, Appendix 4]. Libermann and Marle [27] devote the last chapter of their text to contact structures. None of these treatments are elementary. The 1985 text *Applied Differential Geometry* [11], by Burke, does emphasize the importance—even primacy—of contact structures in physics, but again the treatment is at the level of a graduate student in mathematics or physics.

Geiges deserves special mention for not only what is sure to be the standard reference [19] for graduate students studying contact structures for some years to come, but also for several survey articles [17, 18], which inspired the presentation here of Huygens' principle.

The method of characteristics is standard in a first course in partial differential equations, although it is not always presented in the context of the set of contact elements; see, for example, Folland's introductory text [16]. Our presentation here follows closely that of Giaquinta and Hildebrandt [20].

Finally, the unpublished manuscript "Introduction to Contact Geometry" by A. Banyaga exerted a strong influence on the author's intuition at a time when expositions were sparse and largely inadequate.

6.9 Exercises

6.1. For each of the curves $\tilde{c} : \mathbf{R} \to C\mathbf{R}^2$ below, decide whether it is a lift of a curve $c : \mathbf{R} \to \mathbf{R}^2$:

(a) $\tilde{c}(t) = (t, t^2, t^3)$;
(b) $\tilde{c}(t) = (e^t, e^t, t)$;
(c) $\tilde{c}(t) = (t, e^t, e^t)$;
(d) $\tilde{c}(t) = (e^t, e^{2t}, 2e^t)$.

6.2. Find the image of the lifts of the following plane curves under the wavefront diffeomorphism ϕ_r of Proposition 6.1.4. Sketch the graph of the projection of $\phi_r(I)$ onto the first two coordinates.

(a) A nonvertical line through (x_0, y_0) with unit direction vector (a, b).
(b) The upper half of a circle with center (x_0, y_0) and radius R.
(c) The parabola $y = x^2$.

6.3. Supply the details to the proof of Proposition 6.1.4.

6.4. Use the method of characteristics to solve the first-order ordinary differential equations below:

(a) $u' - 2u = t^2$, $u(0) = 1$.
(b) $t^2 u' + tu = 1$, $u(1) = 2$.
(c) $uu' - 4t = 4$, $u(0) = 1$.

6.5. Let u be the solution to the ordinary differential equation

$$(2u - t)u' = u - t, \quad u(0) = 1;$$

see Example 6.2.5. Verify that $m(s) = u'(-2 \sin s)$, where $m(s) = \dfrac{y'(s)}{x'(s)}$, and x and y are defined as in the example. Moreover, verify that

$$m'(s) = 2m^2 - 2m + 1.$$

6.6. Write the components of the Reeb field for the following contact forms on $\mathbf{R}^3$:

(a) $\alpha_1 = dy - m dx$ from Example 6.3.3.
(b) $\alpha_2 = dz + x dy - y dx$ from Example 6.3.4.
(c) $\alpha_3 = (\cos z) dx + (\sin z) dy$ from Example 6.3.5.
(d) $\alpha_4 = (\cos r) dz + (r \sin r) d\theta$; see the discussion from Example 6.3.6.

6.7. Let $CL(\mathbf{R}^3, \alpha_0)$ be the set of all *linear* contact diffeomorphisms of the standard contact space $(\mathbf{R}^3, \alpha_0)$, so that $T \in CL(\mathbf{R}^3, \alpha_0)$ if and only if there is a 3×3 matrix A such that $[T] = A$ and $T^* \alpha_0 = a\alpha_0$ for some nonzero constant $a \in \mathbf{R}$.

(a) Show that $CL(\mathbf{R}^3, \alpha_0)$ is a subgroup of $\text{Diff}(\mathbf{R}^3, \alpha_0)$, i.e., that $\text{Id} \in CL(\mathbf{R}^3, \alpha_0)$, that if $T_1, T_2 \in CL(\mathbf{R}^3, \alpha_0)$, then $T_1 \circ T_2 \in CL(\mathbf{R}^3, \alpha_0)$ and that if $T \in CL(\mathbf{R}^3, \alpha_0)$, then $T^{-1} \in CL(\mathbf{R}^3, \alpha_0)$.
(b) Find conditions on the entries of a matrix A such that $A = [T]$ with $T \in CL(\mathbf{R}^3, \alpha_0)$.
(c) What are the conditions for a matrix A to correspond to a linear *strictly* contact diffeomorphism?

6.8. Use the partial differential equations (6.7) to show that the only *affine* strictly contact diffeomorphisms of the standard contact space $(\mathbf{R}^3, \alpha_0)$ are the those $\phi : \mathbf{R}^3 \to \mathbf{R}^3$ having the form $\phi(x, y, z) = (x/a, ay + b, z + c)$ for some constants $a, b, c \in \mathbf{R}$, where $a \neq 0$. (Recall that an affine map is one that can be written using matrix/vector notation as $\phi(\mathbf{x}) = A\mathbf{x} + \mathbf{b}$ for some matrix A and constant vector $\mathbf{b}$.)

6.9. Let $\alpha_0 = xdy + dz$ and $\alpha_3 = (\cos z)dx + (\sin z)dy$ be the contact forms of Examples 6.3.2 and 6.3.5. Define $\phi_3 : (\mathbf{R}^3, \alpha_0) \to (\mathbf{R}^3, \alpha_3)$ by

$$\phi_3(x, y, z) = (z \cos y + x \sin y, z \sin y - x \cos y, y).$$

Compute ϕ_3^{-1} to verify that ϕ_3 is a diffeomorphism. Then show that $\phi_3^* \alpha_3 = \alpha_0$.

6.10. Let α be a contact form on $\mathbf{R}^3$ with corresponding Reeb field ξ. Let $f : \mathbf{R}^3 \to \mathbf{R}$ be a nowhere-zero function. Show that the Reeb field of $f\alpha$ is given by $X_{1/f}$, the contact gradient of the function $1/f$ relative to α.

6.11. Let $\text{Diff}(\mathbf{R}^3, \alpha)$ be the set of all *strictly* contact diffeomorphisms of the contact space $(\mathbf{R}^3, \alpha)$. Show that:

(a) If $\phi_1, \phi_2 \in \text{Diff}(\mathbf{R}^3, \alpha)$, then $\phi_1 \circ \phi_2 \in \text{Diff}(\mathbf{R}^3, \alpha)$.
(b) If $\phi \in \text{Diff}(\mathbf{R}^3, \alpha)$, then $\phi^{-1} \in \text{Diff}(\mathbf{R}^3, \alpha)$.

6.12. Let (U, α) be a contact space, where $U \subset \mathbf{R}^3$ is a domain. Show that there are vector fields $\{B_1, B_2, B_3\}$ such that for each $p \in U$, $\{B_1(p), B_2(p), B_3(p)\}$ is a basis for $T_p(U)$ and such that $B_1 = \xi$ and $B_2, B_3 \in E$ with $d\alpha(B_2, B_3) = 1$.

6.13. Write the components of the contact gradient of a function $f : \mathbf{R}^3 \to \mathbf{R}$ relative to the following contact forms on $\mathbf{R}^3$:

(a) $\alpha_2 = dz + xdy - ydx$ from Example 6.3.4.
(b) $\alpha_3 = (\cos z)dx + (\sin z)dy$ from Example 6.3.5 .
(c) $\alpha_4 = (\cos r)dz + (r \sin r)d\theta$; see the discussion from Example 6.3.6.

6.14. Let $\mathbf{X}$ be a vector field on $\mathbf{R}^3$ with flow ϕ_t. Prove the three-dimensional *transport theorem*:

Let $f_t : \mathbf{R}^3 \to \mathbf{R}$ be a time-dependent family of smooth functions, and let $W \subset \mathbf{R}^3$ be a region of integration. Then

$$\frac{d}{dt}\left[\iiint_{W_t} f_t dV\right] = \iiint_{W_t} \left[\frac{Df_t}{dt} + f_t \text{div}\,\mathbf{X}\right] dV,$$

where $W_t = \phi_t(W)$.

In this statement, we are using terminology from physics. The expression

$$\frac{Df_t}{dt} = \frac{\partial f_t}{\partial t} + \nabla f_t \cdot \mathbf{X}$$

is the *material derivative* of f_t, which involves both the "time" t derivative of f_t and the standard vector calculus gradient of f_t in the direction of (dot product with) the vector field $\mathbf{X}$. In this context, note that div $\mathbf{X} : \mathbf{R}^3 \to \mathbf{R}$ is the smooth function defined by $d\,(i(\mathbf{X})dV) = (\text{div}\,\mathbf{X})\,dV$.

Hint: Use the change of variables theorem (Theorem 4.5.4) and apply Theorem 6.6.3 to the n-form $\mu_t = f_t dV$.

6.15. For each of the contact spaces below, give three examples of Legendre curves:

(a) $(\mathbf{R}^3, \alpha_0)$, where $\alpha_0 = xdy + dz$;
(b) $(\mathbf{R}^3, \alpha_1)$, where $\alpha_1 = dy - mdx$;
(c) $(\mathbf{R}^3, \alpha_2)$, where $\alpha_2 = dz + xdy - ydx$;
(d) $(\mathbf{R}^3, \alpha_3)$, where $\alpha_3 = (\cos z)dx + (\sin z)dy$.

6.16. (This exercise uses concepts from Chap. 6.) Let g be the metric tensor on $\mathbf{R}^3$ defined by the matrix

$$[g_{ij}] = \begin{bmatrix} 1 & 0 & 0 \\ 0 & 1+x^2 & x \\ 0 & x & 1 \end{bmatrix}.$$

We will consider the interaction of g with the standard contact form $\alpha_0 = xdy + dz$.

(a) Show that g considered as a metric on the contact space $(\mathbf{R}^3, \alpha_0)$ has the property that the Reeb field ξ_0 is orthogonal to any vector $X \in \ker \alpha_0$.
(b) Let $c : I \to (\mathbf{R}^3, \alpha_0)$ be a Legendre curve. Show that the g-length of c is the same as the standard Euclidean length of the projection of c into the xy-plane in $\mathbf{R}^2$.

6.17. Consider Example 6.7.22.

(a) What is special about the points $(\frac{1}{\sqrt{2}}, -\frac{1}{\sqrt{2}})$ and $(-\frac{1}{\sqrt{2}}, \frac{1}{\sqrt{2}})$, corresponding to $s = -\pi/4$ and $s = 3\pi/4$ respectively? Why does the method of characteristics fail for intervals containing these two values of s?
(b) Redo Example 6.7.22 for the parameterization given by the interval $U_1 = (-\pi/4, 3\pi/4)$ and $\phi : U_1 \to \mathbf{R}^2$ given by $\phi(s) = (-\cos s, \sin s)$, again with $g(s) = 1$ for all $s \in U_1$.

(c) Redo Example 6.7.22 with different Cauchy data: $g(s) = \cos s$ on the line parameterized by $\phi(s) = (s, 0)$.

6.18. Use the method of characteristics to solve the first-order partial differential equations below. Be sure to state an appropriate domain on which the hypotheses of Theorem 6.7.20 apply.

(a) $uu_x + u_y = 1$, with $u = 0$ on the line $y = x$. What happens when the initial conditions are changed to $u = 1$ on $y = x$?

(b) $u_x + u_y + u_z = u$, with $u(x, y, z) = 1$ on the sphere

$$x^2 + y^2 + z^2 = 1.$$

(c) $u_x + u_y + u_z = u$, with $u(x, y, z) = 1$ on the plane $x + y + z = 1$.

Chapter 7
Symplectic Geometry

After the previous two chapters, the reader may have begun to develop a sense in which a structure tensor determines a geometry on $\mathbf{R}^n$. The structure tensor—the metric tensor in the case of Riemannian geometry or the contact form in the case of contact geometry—determines concepts or objects that are singled out by or defined in terms of the tensor. The set of those diffeomorphisms that "preserve the structure" in an appropriate sense gives a way of discussing when two objects are "the same" from a geometric point of view, in the same sense that two shapes in the plane are "the same," or congruent, if one can be transformed onto the other by means of translations, rotations, and reflections.

The present chapter on symplectic geometry provides another example of geometry as a set equipped with a structure tensor. The word "symplectic" was invented by the German mathematician Hermann Weyl in 1939 in the context of studying certain sets of transformations and their invariants. It was formed by replacing the Latin-based word "complex" (com + plex, or "woven together") with the corresponding Greek roots sym + plectic, in recognition of the many similarities between symplectic structures and structures based on the complex numbers.

In this text we will not be able to explore the many connections between symplectic geometry and complex geometry, although the reader will notice similarities throughout. The main similarity to complex numbers will be that the symplectic structure "weaves together" pairs of coordinates (x, y) in a manner analogous to the way that a complex number z combines a pair of real numbers (x, y) as $z = x + iy$.

Symplectic geometry shares fundamental similarities with Riemannian geometry. First, as in Riemannian geometry, the structure tensor is a $(0, 2)$-tensor. The important difference between the two structures is that the condition that the metric tensor be symmetric is replaced with the condition that the symplectic tensor be skew-symmetric, and in particular a differential form. In Chap. 2, we already saw some of the far-reaching consequences of this "minor" change in the context

A. McInerney, *First Steps in Differential Geometry: Riemannian, Contact, Symplectic*, Undergraduate Texts in Mathematics, DOI 10.1007/978-1-4614-7732-7_7,

of linear algebra. Second, as in Riemannian geometry but in contrast to contact geometry, the fundamental object of study will be the structure tensor itself, and not some other structure like the contact distribution defined by the structure tensor.

Despite these similarities, one of the fundamental features of symplectic geometry will be its closer connection to contact geometry. A major theme of this chapter will be to highlight the relationships between these two geometric structures.

In fact, the close relationship between symplectic and contact geometry might be one of the reasons for what could be called a historical underappreciation of the latter. Key ideas of symplectic geometry began to emerge distinctly as early as 1868 in Lagrange's studies of celestial mechanics. Hamilton's work some 20 years later solidified this foundation, giving the analytic techniques described by symplectic geometry a central role in mechanics. While these results were not given a geometric formulation until near the turn of the twentieth century with the works of Lie, Poincaré and É. Cartan, the fact that the ideas had so permeated treatments of classical mechanics ensured that by the 1960s, symplectic geometry had come to be seen as the mathematical framework for classical mechanics. This interest ensured that symplectic geometry would receive more attention than contact geometry, despite their close relationship.

We will outline in broad strokes the way that symplectic geometry arose from problems in mechanics in the first section of this chapter. This section will also motivate the basic definitions and concepts that follow. The basic example here will be the "phase space," which plays the same motivating role in symplectic geometry that the space of contact elements plays in contact geometry.

After discussing the diffeomorphisms that preserve the symplectic structure and the associated distinguished vector fields, we will turn to describing special geometric sets in a symplectic space. We will pay particular attention to the so-called Lagrange sets, which are the symplectic analogues of the Legendre sets in contact geometry. We also consider the special case of particular geometric sets in a symplectic space, the hypersurfaces of contact type, to illustrate the close connection between symplectic and contact geometry.

As in contact geometry, there is a kind of local "normal form" Darboux's theorem for symplectic structures. This normal form involves the close pairing of coordinates mentioned earlier in analogy with complex structures. This implies a kind of symmetry that distinguishes symplectic geometry from both contact and Riemannian geometry. As with contact geometry, the symplectic Darboux's theorem shows that there are no local ways of distinguishing one symplectic structure from another, again in sharp contrast to Riemannian geometry.

In the last section, we indicate the sorts of invariants that allow one to distinguish symplectic structures. This is necessarily a "global" question, and as we have indicated in Chap. 7, that fact places severe limitations on how far one can go in this direction at an introductory level. In fact, the search for global invariants has been one of the main stimuli for research in symplectic geometry, and had its biggest breakthroughs only in the 1980s.

7.1 Motivation: Hamiltonian Mechanics and Phase Space

In this section, we will give an overview of Hamiltonian mechanics, which was historically the main motivation for symplectic geometry. Our exposition will necessarily be missing some details, because the transition from Newtonian mechanics to Hamiltonian mechanics involves an intermediate stage of Lagrangian mechanics, where the mathematical techniques presuppose the calculus of variations—a topic outside the scope of this text. At that point in our exposition, we will refer the reader to appropriate sources.

The main goal of the section will be to describe the natural symplectic structure on what physicists call the "phase space," known to mathematicians as the cotangent bundle $T^*\mathbf{R}^n$, i.e., the dual of the tangent bundle $T\mathbf{R}^n$. The phase space gives a natural setting for Hamiltonian mechanics in the same sense as the set of contact elements provides a setting for Huygens' principle or the method of characteristics. In those cases, we have seen how the contact form distinguishes certain special curves (or, more generally, special geometric sets), namely wavefronts or solutions to a given differential equation. In the case of the phase space, we will again see how a special differential form distinguishes those curves that represent motions of a given physical system.

Newton's lasting contribution to human knowledge has a twofold character. On the one hand, he developed a mathematical model that was able to describe a wide variety of physical phenomena with a striking degree of accuracy. This was the beginning of the modern era in science. In addition, however, Newton also developed the mathematical techniques—calculus—through which the model could be exploited. The first of these contributions belongs to physics, the second to mathematics.

The Newtonian description of mechanics is premised on describing the position $\mathbf{x}$ of a particle in three-dimensional space as a smooth function of time t. The function $\mathbf{x} : I \to \mathbf{R}^3$ is known as the *motion* of the particle. In modern terminology, Newton's second law states that the motion of a particle in space can be described as a smooth solution to a second-order system of ordinary differential equations, written in vector notation as

$$\ddot{\mathbf{x}}(t) = \mathbf{F}(\mathbf{x}, \dot{\mathbf{x}}, t).$$

In particular, the existence and uniqueness theorem of ordinary differential equations states that for given forces, the motion is completely described by specifying the initial position and velocity. It is in this sense that Newtonian physics is said to be "deterministic." Note that we have suppressed the scalar quantity representing mass that is familiar from the formulation $\mathbf{F} = m\mathbf{a}$. It is understood in this model that describing a physical system amounts to describing the "force function" $\mathbf{F}$ appropriate to the system.

For our presentation, we will make two assumptions to simplify the description. First, we assume that the function $\mathbf{F}$ describing the forces is independent of time.

In physics, this is a kind of classical relativity: "The equations of motion are invariant with respect to Galilean translations, including time translations." Second, we will assume that the forces involved are conservative, i.e., that there is a scalar function U of position such that $\mathbf{F} = -\nabla U$ (where, as in vector calculus, ∇ represents the gradient). This is a nontrivial assumption that does not accurately describe all physical systems. In the case of a conservative system, the function U is called the *potential energy*, and the negative sign in the definition of $\mathbf{F}$ implies that particles move in a conservative system in such a way that the potential energy decreases with time.

There is a second scalar quantity associated to a physical system known as the *kinetic energy*, or energy of motion. For example, for a point whose motion is described by the vector function $\mathbf{x} : I \to \mathbf{R}^3$, the kinetic energy at a given time t is given by the expression $T(t) = \frac{1}{2}(\dot{\mathbf{x}} \cdot \dot{\mathbf{x}})$. Unlike potential energy, which depends on position, kinetic energy is independent of position but dependent on velocity. The *total energy* E of a system is defined to be the sum of the potential energy and kinetic energy,

$$E = U + T.$$

Using this notation, we have the following formulation of the law of conservation of energy:

Proposition 7.1.1. *For a particle whose motion is described by a function* $\mathbf{x} : I \to \mathbf{R}^3$ *satisfying* $\ddot{\mathbf{x}} = -\nabla U$*, the total energy* E *is constant with respect to time.*

Proof. We have

$$\begin{aligned}
\frac{d}{dt}(E) &= \frac{dU}{dt} + \frac{dT}{dt} \\
&= \nabla U \cdot \dot{\mathbf{x}} + 2\left(\frac{1}{2}\dot{\mathbf{x}} \cdot \ddot{\mathbf{x}}\right) \\
&= -\ddot{\mathbf{x}} \cdot \dot{\mathbf{x}} + \dot{\mathbf{x}} \cdot \ddot{\mathbf{x}} \\
&= 0.
\end{aligned}$$

The second equality is an application of the chain rule and the product rule. □

The main advantage of the Newtonian description of mechanics is that a number of basic but important physical systems can be described completely using the tools of vector calculus. First-year physics students are still taught the basic laws of mechanics from the seventeenth- and eighteenth-century Newtonian vantage point.

It quickly became apparent, however, that for more complicated physical systems, the Newtonian framework had severe practical problems. One significant problem, of course, is the difficulty in integration that arises in producing explicit solutions to the second-order ordinary differential equations involved.

Several mathematicians, most notably Euler, Lagrange, and Hamilton, recast Newton's model and at the same time developed powerful new mathematical machinery. The essential shift, first in the Lagrangian model and then in the Hamiltonian model, involved taking as the basic objects not the forces describing the system, but rather the energy of the system as a whole.

According to the Lagrangian model, suppose we are given a domain $V \subset \mathbf{R}^n$ with coordinates $(q_1, \dots, q_n)$. These correspond to the parameters of the positions of a finite number of particles whose motion is constrained to lie on some geometric set. Associated to any curve $\mathbf{x} : I \to V$ given by $\mathbf{x}(t) = (q_1(t), \dots, q_n(t))$ representing the evolution of the system with time, we can associate two scalar functions $T, U : I \to \mathbf{R}$, where

$$T(t) = \frac{1}{2}\sum_{i=1}^{n} \dot{q}_i^2(t) \quad \text{and} \quad U(t) = U(q_1(t), \dots, q_n(t)).$$

Note that the function U depends only on the position coordinates q_i.

For a given physical system, consider the function $L = T - U$ as a function of the $2n$ variables $q_1, \dots, q_n$ (representing position) and $v_1, \dots, v_n$ (representing velocity). The function L is called the *Lagrangian function* for the system. The fundamental result of Lagrangian mechanics is the following:

Theorem 7.1.2. *A smooth function* $\mathbf{x} : I \to V$ *described as* $\mathbf{x}(t) = (q_1(t), \dots, q_n(t))$ *is a solution to the equations of motion described by the system* $\ddot{\mathbf{x}} = -\nabla U$ *with potential energy* U*, kinetic energy* T*, and Lagrangian function* $L(\mathbf{x}, \mathbf{v}) = T(\mathbf{v}) - U(\mathbf{x})$*, if and only if*

$$\frac{d}{dt}\left(\frac{\partial L}{\partial v_i}\right) - \frac{\partial L}{\partial q_i} = 0, \quad i = 1, \dots, n,$$

where $L(t) = L(\mathbf{x}(t), \dot{\mathbf{x}}(t))$.

The n equations of Theorem 7.1.2 are known as the *Euler–Lagrange equations.*

The proof of this theorem is outside the scope of this text. It involves finding the minimum (or extreme) value of the so-called "action functional," defined as $\int_I L dt$, over the set of all possible paths $\mathbf{x} : I \to V$ with fixed endpoints. This is a problem in the calculus of variations. The reader may consult, for example, [3], [15], or [26].

From the point of view of differential geometry, the proper way to think of the Lagrangian function L is as a function whose domain is the tangent bundle $T\mathbf{R}^n$, where the coordinates

$$(q_1, \dots, q_n, v_1, \dots, v_n)$$

represent a tangent vector $\langle v_1, \dots, v_n \rangle_q \in T_q\mathbf{R}^n$ at the point $q = (q_1, \dots, q_n)$. Note that the fundamental Newtonian objects of interest, the total force $\mathbf{F}$ and the total momentum $\mathbf{p}$ of the system, are recovered as

$$F_i = \frac{\partial L}{\partial q_i} \quad \text{and} \quad p_i = \frac{\partial L}{\partial v_i},$$

where F_i and p_i are the components of $\mathbf{F}$ and $\mathbf{p}$ respectively. In this setting, physicists sometimes call the tangent bundle $T\mathbf{R}^n$ the *state space* of the system.

Example 7.1.3. In order to compare the Newtonian and Lagrangian approaches to finding the motions of a physical system, we consider one of the simplest examples, uniform circular motion in a plane. The reader can imagine, for example, the case of a weight being swung in a circle with constant radius r and constant speed ν.

In the Newtonian case, we consider rectangular coordinates (x, y) centered at the origin representing the center of the motion. The fact that the object's velocity changes direction but not magnitude implies the existence of a constraint centripetal force $\mathbf{F} = \langle F^1, F^2 \rangle = \langle -b^2 x, -b^2 y \rangle$ directed toward the center; here $b = \nu/r$ is a constant representing the angular speed. The Newtonian equations of motion then are

$$\begin{cases} \ddot{x} = -b^2 x, \\ \ddot{y} = -b^2 y, \end{cases}$$

with initial position $(r\cos\theta_0, r\sin\theta_0)$ and initial velocity $(-br\sin\theta_0, br\cos\theta_0)$. This is a decoupled system, which yields, by the standard techniques of solving ordinary differential equations,

$$(x(t), y(t)) = (r\cos(bt + \theta_0), r\sin(bt + \theta_0)).$$

By contrast, the motion of the system can be described in terms of a single parameter $q = \theta(t)$ via the map $\mathbf{x}(q) = (r\cos q, r\sin q)$. Since the system has no external forces acting on it, there is no potential energy, i.e., $U = 0$. It does have kinetic energy, however:

$$T = \frac{1}{2}\left(\dot{x}^2 + \dot{y}^2\right) = \frac{1}{2} r^2 \dot{q}^2.$$

Hence, the Lagrangian function for this system is given by

$$L(q, v) = \frac{1}{2} r^2 v^2.$$

Since $\frac{\partial L}{\partial q} = 0$ and $\frac{\partial L}{\partial v} = r^2 v$, the Euler–Lagrange equation according to Theorem 7.1.2 then is

$$\frac{d}{dt}(r^2 v) = 0,$$

so $r^2 v = r^2(\nu/r)$ (using the fact that the angular speed at 0 is given to be ν/r). Hence $v(t) = b$ for all $t \in \mathbf{R}$. Now since $v = \dot{q}$, we have

$$\frac{d}{dt}(q) = b,$$

and so $q(t) = bt + \theta_0$, the same result we obtained through the Newtonian approach.

This example gives a sense of the way in which the Lagrange method simplifies the differential equations describing a motion. In the end, the technique involves solving two consecutive first-order ordinary differential equations instead of second-order ones—similar, in fact, to the standard techniques of solving second-order ordinary differential equations by introducing an extra variable to represent the first derivative. The key to the technique's power, though, involves the way that the Lagrangian function is adapted to the given system.

We present only the Lagrangian setting, however, as a segue to the setting of most interest to us. Following Hamilton, we adapt the Lagrange method by means of a change of coordinates. Recalling that the momentum coordinates can be written as $p_i = \frac{\partial L}{\partial v_i}(\mathbf{q}, \mathbf{v})$, we attempt to solve those n equations for v_i in terms of p_i and q_i. This is essentially a problem of the type addressed by the implicit function theorem, and can be done as long as the matrix

$$\left[\frac{\partial^2 L}{\partial v_i \partial v_j} \right]$$

is invertible. (This condition is satisfied for most physical systems. In fact, for many physical systems, the functions $\frac{\partial L}{\partial v_i}$ are linear in the v-coordinates.)

Assume, then, that we can define the n functions $v_i = v_i(\mathbf{q}, \mathbf{p})$ $(i = 1, \dots, n)$ such that

$$p_i = \frac{\partial L}{\partial v_i}\left(q_1, \dots, q_n, v_1, \dots, v_n\right).$$

Using these functions, we can define a new function $H : \mathbf{R}^n \times \mathbf{R}^n \to \mathbf{R}$ as follows:

$$H(\mathbf{q}, \mathbf{p}) = \left(\sum_{i=1}^{n} p_i v_i \right) - L(\mathbf{q}, \mathbf{v}(\mathbf{q}, \mathbf{p})).$$

The new function H is called the *Hamiltonian function* for the system.

Theorem 7.1.4. *Let $\mathbf{x} : I \to V \subset \mathbf{R}^n$ be a smooth curve and let L be a given Lagrangian function with associated Hamiltonian function H. Then $(\mathbf{q} = \mathbf{x}, \dot{\mathbf{q}} = \dot{\mathbf{x}})$ is a solution to the Euler–Lagrange equations of Theorem 7.1.2 if and only if $(\mathbf{q} = \mathbf{x}, \mathbf{p} = \frac{\partial L}{\partial v}(\mathbf{x}, \dot{\mathbf{x}}))$ is a solution to the system given by the $2n$ equations*

$$\dot{q}_i = \frac{\partial H}{\partial p_i}, \quad \dot{p}_i = -\frac{\partial H}{\partial q_i} \quad (i = 1, \dots, n).$$

In particular, if either of the equivalent conditions is satisfied, then $\mathbf{x}$ is a solution to the Newtonian equations of motion $\ddot{\mathbf{x}} = -\nabla U$.

The equations of Theorem 7.1.4 are called *Hamilton's equations* for the system.

Proof. On the one hand, we have

$$dH = \sum_{i=1}^{n} \left(\frac{\partial H}{\partial q_i} dq_i + \frac{\partial H}{\partial p_i} dp_i \right).$$

On the other hand, by the definition of the Hamiltonian function H, we have

$$\begin{aligned}
dH &= d\left(\sum p_i v_i - L\right) \\
&= \sum(p_i dv_i) + \sum(v_i dp_i) - \sum(\frac{\partial L}{\partial q_i} dq_i) - \sum(\frac{\partial L}{\partial v_i} dv_i) \\
&= \sum(v_i dp_i) - \sum(\frac{\partial L}{\partial q_i} dq_i) \quad \text{since } p_i = \tfrac{\partial L}{\partial v_i}.
\end{aligned}$$

Comparing the components of dH in the two expressions, we have

$$\frac{\partial H}{\partial p_i} = v_i \quad \text{and} \quad \frac{\partial H}{\partial q_i} = -\frac{\partial L}{\partial q_i}.$$

Since $p_i = \frac{\partial L}{\partial v_i}$, the second equation implies the equivalence of the Euler–Lagrange equations to the equations $\dot{p}_i = -\frac{\partial H}{\partial q_i}$, and the first equation amounts to $\dot{q}_i = \frac{\partial H}{\partial p_i}$, since we are considering motions for which $v_i(t) = \dot{q}_i(t)$. □

Example 7.1.5. Returning to Example 7.1.3, we had $L(q, v) = \frac{1}{2}r^2 v^2$. The momentum coordinate is given by $p = \frac{\partial L}{\partial v} = r^2 v$, and so we can solve for v to obtain $v(p, q) = p/r^2$. Hence

$$H(q, p) = pv - L = p(p/r^2) - \frac{1}{2}r^2(p/r^2)^2 = \frac{p^2}{2r^2}.$$

Hamilton's equations for this system, then, are

$$\begin{cases} \dot{q} = \frac{\partial H}{\partial p} = \frac{p}{r^2}, \\ \dot{p} = -\frac{\partial H}{\partial q} = 0. \end{cases}$$

This is a partially decoupled system. The second equation implies that $p(t) = r^2(\nu/r) = r\nu$ is constant with respect to t. The first equation then implies that $q = (\nu/r)t + \theta_0 = bt + \theta_0$, as was previously established.

We now step back from the physics and its concern for the equations of motion in order to discuss the proper mathematical setting for the Hamiltonian formalism. The transition from the Lagrangian function L to the Hamiltonian function H is a particular case of a general mathematical construction known as a *Legendre transformation*. We will explore the Legendre transformation more in the exercises. From a more advanced perspective, the Legendre transformation gives a natural way of associating to each real-valued function on the tangent bundle $T\mathbf{R}^n$, a new real-valued function on the cotangent bundle $T^*\mathbf{R}^n$. For this and other reasons, momentum is properly viewed as a covector, i.e., an element of $T_q^*\mathbf{R}^n$, and the Hamiltonian function then is a map $H : T^*\mathbf{R}^n \to \mathbf{R}$. In the context of Hamiltonian mechanics, the cotangent bundle $T^*\mathbf{R}^n$ is called the *phase space* of the system, as opposed to the state space $T\mathbf{R}^n$.

There is a natural one-form on the phase space $T^*\mathbf{R}^n$, which we describe as follows. Consider the coordinates $(q_1, \dots, q_n, p_1, \dots, p_n)$ on $T^*\mathbf{R}^n$ corresponding to the cotangent vector

$$p = p_1 dx_1 + \cdots + p_n dx_n \in T_q^*\mathbf{R}^n,$$

where $q = (q_1, \dots, q_n)$. There is the standard projection map

$$\pi : T^*\mathbf{R}^n \to \mathbf{R}^n$$

given by $\pi(p) = q$, where $p \in T_q^*\mathbf{R}^n$. In coordinates,

$$\pi(q_1, \dots, q_n, p_1, \dots, p_n) = (q_1, \dots, q_n).$$

Abusing notation somewhat, we have

$$\pi_* \frac{\partial}{\partial q_i} = \frac{\partial}{\partial x_i} \quad \text{and} \quad \pi_* \frac{\partial}{\partial p_i} = 0,$$

where $\pi_* : T(T^*\mathbf{R}^n) \to T\mathbf{R}^n$ is the tangent map of π. Now a one-form on $T^*\mathbf{R}^n$ is a smoothly varying field of maps $T_{(q,p)}(T^*\mathbf{R}^n) \to \mathbf{R}$. So for a tangent vector

$$\mathbf{A}_{(q,p)} = a_1 \frac{\partial}{\partial q_1} + \cdots + a_n \frac{\partial}{\partial q_n} + b_1 \frac{\partial}{\partial p_1} + \cdots + b_n \frac{\partial}{\partial p_n} \in T_{(q,p)}T^*\mathbf{R}^n,$$

define the one-form α_0 on $T^*\mathbf{R}^n$ as

$$\begin{aligned}
(\alpha_0(q,p))(\mathbf{A}_{(q,p)}) &= (p \circ \pi_*)(\mathbf{A}_{(q,p)}) \\
&= p(\pi_*(\mathbf{A}_{(q,p)})) \\
&= p\left(a_1 \frac{\partial}{\partial x_1} + \cdots + a_n \frac{\partial}{\partial x_n}\right) \\
&= p_1 a_1 + \cdots + p_n a_n,
\end{aligned}$$

where $p = p_1 dx_1 + \cdots + p_n dx_n \in T_q^*\mathbf{R}^n$. In coordinates, α_0 can be expressed as

$$\alpha_0(q,p) = \sum_{i=1}^n p_i dq_i.$$

This special one-form on $T^*\mathbf{R}^n$ is called the *Liouville form*.

Now define a two-form ω_0 on $T^*\mathbf{R}^n$ by $\omega_0 = d\alpha_0$. In coordinates,

$$\omega_0 = \sum_{i=1}^{n} dp_i \wedge dq_i.$$

Note that on each tangent space $T_{(q,p)}(T^*\mathbf{R}^n)$, the two-form $\omega_0(q,p)$ is nondegenerate, so in the language of Sect. 2.10, $\omega_0(q,p)$ is a linear symplectic form.

Now for any smooth function $H : T^*\mathbf{R}^n \to \mathbf{R}$, define the vector field

$$X_H = \sum_{i=1}^{n} \left(\frac{\partial H}{\partial p_i} \frac{\partial}{\partial q_i} - \frac{\partial H}{\partial q_i} \frac{\partial}{\partial p_i} \right).$$

The reader may verify that

$$i(X_H)\omega_0 = -dH;$$

in fact, it is a consequence of Theorem 2.10.5 that X_H is uniquely defined by this property.

We may now restate Theorem 7.1.4 in the language of the vector field X_H, defined in turn in terms of the two-form ω_0.

Theorem 7.1.6. *Let H be a smooth Hamiltonian function for a physical system defined on the phase space $T^*\mathbf{R}^n$. Then the motions of the system are precisely the integral curves of the vector field X_H.*

Proof. In coordinates, the differential equations defining the integral curves of X_H are exactly Hamilton's equations. □

We summarize this discussion by saying that symplectic geometry provides a particularly elegant formulation for Hamiltonian mechanics. In fact, we will continue to see its relevance for rephrasing physical phenomena in a geometric way. We will now turn to the extent to which some of these concepts occur in a more general setting.

7.2 Basic Concepts

If the central object of Riemannian geometry is the metric $(0,2)$-tensor and the central object of contact geometry is the contact one-form (more precisely, its kernel), then the central object of symplectic geometry is the symplectic two-form.

Definition 7.2.1. A *symplectic form* on a domain $U \subset \mathbf{R}^m$ is a smooth differential two-form ω satisfying the following properties:

- ω is nondegenerate: For all $p \in U$, if $\mathbf{v}_p \in T_pU$ is such that $\omega(p)(\mathbf{v}_p, \mathbf{w}_p) = 0$ for all $\mathbf{w}_p \in T_pU$, then $\mathbf{v}_p = \mathbf{0}_p$.
- ω is closed: $d\omega = 0$.

We call the pair (U, ω) a *symplectic space*.

The nondegeneracy condition says that for all $p \in U$, $\omega(p)$ is a linear symplectic form on T_pU.

We will review some of the consequences of the nondegeneracy condition, which because of its algebraic character will replicate many of the results of Sect. 2.10. The first immediate consequence is a restriction on the dimension of a symplectic space. The following is an adaptation of Theorem 2.10.4 into the smooth setting, and the proof is identical.

Proposition 7.2.2. *Let (U,ω) be a symplectic space, where $U \subset \mathbf{R}^m$ is a domain. Then $m = 2n$ for some positive integer n. In other words, the dimension of a symplectic space is even.*

For this reason, we will normally write the domain $U \subset \mathbf{R}^{2n}$.

Proposition 7.2.3. *Let ω be a differential two-form on $U \subset \mathbf{R}^{2n}$. Then ω is nondegenerate if and only if*

$$\omega^n = \omega \wedge \cdots \wedge \omega \neq 0 \quad \textit{on } U,$$

in other words, if ω^n is nowhere equal to the zero $2n$-form on U.

Proof. Apply the argument preceding Definition 6.3.1 to show that ω nondegenerate implies $\omega^n \neq 0$. We leave the converse as an exercise in linear algebra. □

For a symplectic space (U,ω) with $U \subset \mathbf{R}^{2n}$, the top-dimensional $(2n)$-form ω^n on U is called the *volume form associated to* ω. In the case $n = 1$, a symplectic form, as a two-form on $\mathbf{R}^2$, is already a top-dimensional form. Hence the symplectic structure can be thought of as a sort of generalized oriented area form.

A key consequence of the nondegeneracy condition is the following smooth analogue of Theorem 2.10.5, which was mentioned at the end of the last section.

Proposition 7.2.4. *Let $\mathcal{X}(U)$ be the vector space of smooth vector fields on a symplectic space (U,ω), where $U \subset \mathbf{R}^{2n}$ is a domain, and let $\Lambda_1(U)$ be the vector space of one-forms on U. Then the map $\Phi : \mathcal{X}(U) \to \Lambda_1(U)$ given by $\Phi(X) = i(X)\omega$ is a vector space isomorphism.*

Before turning to examples, we note that in comparison to the consequences of the nondegeneracy condition, the condition that the symplectic form must be closed is of a more geometric nature. We mention here two immediate consequences; more will be seen in subsequent sections.

The following is a direct application of the Gauss–Stokes theorem (Theorem 4.5.9).

Proposition 7.2.5. *Suppose S is an oriented three-dimensional region of integration in a symplectic space (U,ω), where $U \subset \mathbf{R}^{2n}$, and let R be the boundary of S. Then*

$$\int_R \omega = 0.$$

In particular, if S is the image of the unit ball in $\mathbf{R}^3$, and so R is the image of the unit sphere, under a regular parameterization, then $\int_R \omega = 0$. This can be seen as a kind of symmetry condition, where oriented areas "cancel out."

Proposition 7.2.6. *Let ω be a symplectic form defined on a domain $U \subset \mathbf{R}^{2n}$. Then for each point $p \in U$, there exist a domain V containing p and a one-form α defined on V such that $\omega = d\alpha$ on V.*

Proof. Apply Poincaré's lemma (Theorem 4.4.11). □

We now turn to examples.

Example 7.2.7. Starting with the case of $n = 1$, a symplectic form on $\mathbf{R}^2$ with coordinates (x, y) is just a nowhere-zero multiple of the top-dimensional area form: $\omega = f dx \wedge dy$ for some nowhere-zero function $f : \mathbf{R}^2 \to \mathbf{R}$. The closed condition is automatically fulfilled for dimensional reasons, $d\omega$ being a three-form on $\mathbf{R}^2$.

Example 7.2.8. Considering $T^*\mathbf{R}^n$ to be the same as $\mathbf{R}^{2n}$ with coordinates $(q_1, \ldots, q_n, p_1, \ldots, p_n)$, then the two-form described in the previous section,

$$\omega_0 = \sum_{i=1}^{n} (dp_i \wedge dq_i),$$

is a symplectic form. It is closed, since it is exact: $\omega_0 = d\alpha_0$, where α_0 is the Liouville form of the prior section. It is nondegenerate, since

$$\omega_0^n = (n!) dp_1 \wedge \cdots \wedge dp_n \wedge dq_1 \wedge \cdots \wedge dq_n.$$

In fact, we will adjust this example slightly to define the *standard symplectic structure* on $\mathbf{R}^{2n}$. Let $(x_1, y_1, \ldots, x_n, y_n)$ be coordinates on $\mathbf{R}^{2n}$. Then the standard symplectic form on $\mathbf{R}^{2n}$ will be

$$\omega_0 = \sum_{i=1}^{n} (dx_i \wedge dy_i).$$

It is closed, since $\omega_0 = d\alpha_0$, where $\alpha_0 = \sum x_i dy_i$, and it is nondegenerate by the same calculation as the symplectic form on $T^*\mathbf{R}^n$ above. The pair $(\mathbf{R}^{2n}, \omega_0)$ will be called the *standard symplectic space*.

As we have noted from the outset, even in the linear setting of Chap. 2, a symplectic structure implies a clear pairing of coordinates, which is explicit in the standard symplectic structure with pairs of coordinates (x_i, y_i). The notation accentuates the pairing. In the case of the standard symplectic form ω_0, the integral $\int_S \omega_0$ is literally a "sum of oriented areas" of a region projected into the n coordinate planes defined by the coordinates (x_i, y_i). See Fig. 7.1.

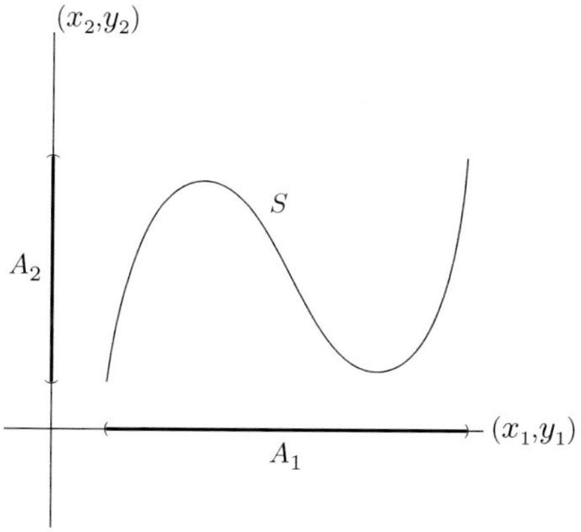

Fig. 7.1 In $(\mathbf{R}^4, \omega_0)$, $\int_S \omega_0 = \text{Area}(A_1) + \text{Area}(A_2)$.

For the standard symplectic space $(\mathbf{R}^{2n}, \omega_0)$, consider a vector field X written in coordinates as

$$X = \langle X^1, Y^1, \ldots, X^n, Y^n \rangle = \sum_{i=1}^{n} \left(X^i \frac{\partial}{\partial x_i} + Y^i \frac{\partial}{\partial y_i} \right).$$

In these coordinates, the isomorphism Φ described in Proposition 7.2.4 is given by

$$\Phi(X) = \sum_i (X^i dy_i - Y^i dx_i)$$

and

$$\Phi^{-1}(\sum a_i dx_i + b_i dy_i) = \langle b_1, -a_1, \ldots, b_n, -a_n \rangle .$$

We will return often to the standard symplectic space $(\mathbf{R}^{2n}, \omega_0)$. When we do so, we will describe $\mathbf{R}^{2n}$ by the *coordinate pairs* (x_i, y_i) rather than the explicit listing $(x_1, y_1, \ldots, x_n, y_n)$.

Example 7.2.9. More generally, consider $\mathbf{R}^{2n}$ with coordinate pairs (x_i, y_i), and consider the two-form

$$\omega_a = \sum_i a_i dx_i \wedge dy_i,$$

where the $a_i : \mathbf{R}^{2n} \to \mathbf{R}$ are smooth functions.

Note that

$$\omega_a^n = (n!) a_1 \cdots a_n \cdot dx_1 \wedge dy_1 \wedge \cdots \wedge dx_n \wedge dy_n;$$

hence for ω_a to be nondegenerate, we must have that the functions a_i are all nowhere zero.

Further, we have

$$d\omega_a = \sum_{i,j} \left(\frac{\partial a_i}{\partial x_j} dx_j \wedge dx_i \wedge dy_i + \frac{\partial a_i}{\partial y_j} dy_j \wedge dx_i \wedge dy_i \right).$$

So in order for ω_a to be closed, we must have

$$\frac{\partial a_i}{\partial x_j} = \frac{\partial a_i}{\partial y_j} = 0 \quad \text{for } i \neq j.$$

In other words, the functions a_i may depend only on the corresponding paired coordinates: $a_i = a_i(x_i, y_i)$. In this case, the closed condition complements the pairing of coordinates that arises essentially from the nondegeneracy condition.

Symplectic forms on $\mathbf{R}^{2n}$ of the form ω_a are called *warped* symplectic forms.

Example 7.2.10. Let (U_1, ω_1) and (U_2, ω_2) be two symplectic spaces, where $U_1 \subset \mathbf{R}^{2m}$ and $U_2 \subset \mathbf{R}^{2n}$ are domains. In this example, we show how to define a symplectic form Ω on the product $U_1 \times U_2 \subset \mathbf{R}^{2m} \times \mathbf{R}^{2n}$ in a natural way.

We first define the two projection maps $p_1 : U_1 \times U_2 \to U_1$ and $p_2 : U_1 \times U_2 \to U_2$ by $p_1(a, b) = a$ and $p_2(a, b) = b$. Define

$$\Omega = p_1^* \omega_1 + p_2^* \omega_2.$$

On the one hand, we show that Ω is closed. We have

$$\begin{aligned} d\Omega &= d(p_1^* \omega_1 + p_2^* \omega_2) \\ &= p_1^* d\omega_1 + p_2^* d\omega_2 \\ &= 0 \quad \text{since } \omega_1 \text{ and } \omega_2 \text{ are closed.} \end{aligned}$$

To show that Ω is nondegenerate, consider the $(m+n)$-fold exterior product

$$(p_1^* \omega_1 + p_2^* \omega_2)^{(m+n)}.$$

It is a sum of terms having the form

$$(p_1^* \omega_1)^k \wedge (p_2^* \omega_2)^l,$$

where $k + l = m + n$. Note that $(p_1^* \omega_1)^k = 0$ when $k > m$ and $(p_2^* \omega_2)^l = 0$ when $l > n$ for dimensional reasons. These two statements together imply that

$$(p_1^* \omega_1 + p_2^* \omega_2)^{(m+n)} = \binom{m+n}{n} (p_1^* \omega_1)^m \wedge (p_2^* \omega_2)^n.$$

For $(v, w) \in U_1 \times U_2$, let $\{\mathbf{e}_1, \ldots, \mathbf{e}_{2m}, \mathbf{f}_1, \ldots, \mathbf{f}_{2n}\}$ be the standard basis for $T_{(v,w)}(U_1 \times U_2)$. Note that for all j, $i(\mathbf{e}_j) p_2^* \omega_2 = 0$ and $i(\mathbf{f}_j) p_1^* \omega_1 = 0$, so that

$$(p_1^*\omega_1 + p_2^*\omega_2)^{(m+n)}(\mathbf{e}_1, \ldots, \mathbf{e}_{2m}, \mathbf{f}_1, \ldots, \mathbf{f}_{2n}) =$$

$$= \binom{m+n}{n} \left[(p_1^*\omega_1)^m \wedge (p_2^*\omega_2)^n\right](\mathbf{e}_1, \ldots, \mathbf{e}_{2m}, \mathbf{f}_1, \ldots, \mathbf{f}_{2n})$$

$$= \binom{m+n}{n} (p_1^*\omega_1)^m(\mathbf{e}_1, \ldots, \mathbf{e}_{2m}) \cdot (p_2^*\omega_2)(\mathbf{f}_1, \ldots, \mathbf{f}_{2n})$$

$$\neq 0 \quad \text{since } \omega_1 \text{ and } \omega_2 \text{ are both nondegenerate.}$$

Hence, by Proposition 7.2.3, Ω is nondegenerate and Ω is a symplectic form on $U_1 \times U_2$.

We will sometimes use the notation

$$\Omega = (\omega_1) \oplus (\omega_2)$$

for the symplectic form $p_1^*\omega_1 + p_2^*\omega_2$ on the product space $U_1 \times U_2$. Note that Example 7.2.9 is a special case of Example 7.2.10 applied to n copies of $(\mathbf{R}^2, a_i dx_i \wedge dy_i)$.

The following example hints at the close relationship between symplectic and contact forms.

Example 7.2.11. Let (U, α) be a contact space, where $U \subset \mathbf{R}^{2n+1}$ is a domain. We will construct a symplectic form ω on $U \times \mathbf{R}$ as follows: Let (x, t) be coordinates on $U \times \mathbf{R} \subset \mathbf{R}^{2n+2}$, and define the projection

$$p_1 : U \times \mathbf{R} \to U$$

by $p_1(x, t) = x$.

Define $\tilde{\alpha} = p_1^*\alpha$ and $\omega = d(e^t\tilde{\alpha})$. Then ω is closed, since it is exact. To verify that ω is nondegenerate, note that

$$\omega = e^t(dt \wedge \tilde{\alpha} + d\tilde{\alpha}),$$

and so

$$\omega^{n+1} = (n+1)e^{(n+1)t} dt \wedge \tilde{\alpha} \wedge (d\tilde{\alpha})^n,$$

since $d\tilde{\alpha}^{(n+1)} = 0$ and $\tilde{\alpha} \wedge \tilde{\alpha} = 0$, since $\tilde{\alpha}$ is a one-form. Because α is a contact form, this shows that $\omega^{n+1} \neq 0$.

The symplectic space $(U \times \mathbf{R}, \omega)$ constructed in this way from a contact space (U, α) is called the *symplectization* of (U, α). The symplectic form ω in this case is another example of an *exact* symplectic form, in that there is a one-form α such that $\omega = d\alpha$. The standard symplectic form is another example of an exact symplectic form.

7.3 Symplectic Diffeomorphisms

Having defined a geometric structure on the tangent space by means of a closed, nondegenerate two-form, we will now turn our attention to those diffeomorphisms that preserve this structure.

Definition 7.3.1. Let (U_1, ω_1) and (U_2, ω_2) be two symplectic spaces, for domains $U_1, U_2 \subset \mathbf{R}^{2n}$, and let $\phi : U_1 \to U_2$ be a diffeomorphism. Then ϕ is called a *symplectic diffeomorphism* if

$$\phi^* \omega_2 = \omega_1.$$

We write $\phi : (U_1, \omega_1) \to (U_2, \omega_2)$.

The symplectic diffeomorphisms are sometimes referred to as *symplectomorphisms*. We reserve this term only for the *linear* symplectic maps described in Chap. 2.

The reader will notice the parallels between symplectic diffeomorphisms on the one hand and isometries (in the setting of Riemannian geometry) and *strictly* contact diffeomorphisms (in the contact setting) on the other. In these cases, structure-preserving diffeomorphisms are those that preserve the structure tensor.

Before proceeding to examples, we prove an immediate geometric consequence of Definition 7.3.1. Recall from Proposition 7.2.3 that to every symplectic space $(\mathbf{R}^{2n}, \omega)$ there is an associated volume form $\Omega = \omega^n$.

Proposition 7.3.2. *For domains $U_1, U_2 \subset \mathbf{R}^{2n}$, let (U_1, ω_1) and (U_2, ω_2) be symplectic spaces with associated volume forms Ω_1 and Ω_2. Let $\phi : (U_1, \omega_1) \to (U_2, \omega_2)$ be a symplectic diffeomorphism. Then $\phi^* \Omega_2 = \Omega_1$, i.e., ϕ preserves the volume form associated to ω.*

Proof. Applying Proposition 4.4.16, we obtain

$$\phi^* \Omega_2 = \phi^* (\omega_2^n) = (\phi^* \omega_2)^n = \omega_1^n = \Omega_1. \qquad \square$$

As we have seen, in dimension two, a symplectic form is a top-dimensional form and hence is an oriented area ("two-dimensional volume") form. In this case, a symplectic diffeomorphism is nothing more than an area-preserving diffeomorphism. One of the interesting questions in symplectic geometry is understanding the extent to which the set of symplectic diffeomorphisms is different from the set of volume-preserving diffeomorphisms in higher dimensions. We will return to this question in Sect. 7.7.

Example 7.3.3. Consider the standard symplectic space $(\mathbf{R}^{2n}, \omega_0)$ with coordinate pairs (x_i, y_i), so that $\omega_0 = \sum dx_i \wedge dy_i$. For any constant

$$\mathbf{a} = (a_1, b_1, \ldots, a_n, b_n),$$

the translation $T_{\mathbf{a}} : (\mathbf{R}^{2n}, \omega_0) \to (\mathbf{R}^{2n}, \omega_0)$ given by

$$T_{\mathbf{a}}(x_1, y_1, \ldots, x_n, y_n) = (x_1 + a_1, y_1 + b_1, \ldots, x_n + a_n, y_n + b_n)$$

is a symplectic diffeomorphism.

It is useful to contrast this first simple example of a symplectic diffeomorphism with the comparable contact translation of the standard contact structure in Example 6.4.7. In that case, there is a distinguished direction (the z-direction) corresponding to the direction of the Reeb field and transverse to the contact distribution. In addition, there is an asymmetry to the paired (x_i, y_i)-coordinates, where translations in the y-directions preserves the contact structure, while translations in the x-directions do not, at least without modification. In the case of the standard symplectic structure, the paired coordinates (x_i, y_i) are indistinguishable.

Example 7.3.4. The linear symplectomorphisms of Sect. 2.10 are symplectic diffeomorphisms of the standard symplectic space $(\mathbf{R}^{2n}, \omega_0)$. Recall from Theorem 2.10.20 that if $T_A : \mathbf{R}^{2n} \to \mathbf{R}^{2n}$ is a linear symplectic diffeomorphism with matrix representation A, then the condition $T_A^* \omega_0 = \omega_0$ is equivalent to the matrix condition $A \in \mathrm{Sp}(2n)$, i.e.,

$$A^T J A = J.$$

Here J is the special matrix written in block matrix notation as

$$J = \begin{bmatrix} J_0 & 0 & \cdots & 0 \\ 0 & J_0 & \cdots & 0 \\ & & & \\ 0 & 0 & \ddots & 0 \\ 0 & 0 & \cdots & J_0 \end{bmatrix} \quad \text{and} \quad J_0 = \begin{bmatrix} 0 & -1 \\ 1 & 0 \end{bmatrix}.$$

For example, given nonzero constants $c_1, \ldots, c_n$, the map

$$\phi(x_1, y_1, \ldots, x_n, y_n) = \left(c_1 x_1, \frac{1}{c_1} y_1, \ldots, c_n x_n, \frac{1}{c_n} y_n \right)$$

is a (linear) symplectic diffeomorphism.

At this point we can also give a simple proof of Theorem 2.10.22 regarding the determinant of a linear symplectomorphism.

Corollary 7.3.5. *Let $A \in \mathrm{Sp}(2n)$ be a symplectic matrix. Then* $\det A = 1$.

Proof. Since A is a symplectic matrix, the corresponding linear transformation $T_A : \mathbf{R}^{2n} \to \mathbf{R}^{2n}$ given by $T_A(\mathbf{x}) = A\mathbf{x}$ satisfies $T_A^* \omega_0 = \omega_0$. By Proposition 7.3.2, $T_A^* \Omega_0 = \Omega_0$, where

$$\Omega_0 = \omega_0^n = dx_1 \wedge \cdots \wedge dx_n \wedge dy_1 \wedge \cdots \wedge dy_n$$

is the standard volume form on $\mathbf{R}^{2n}$. But by Proposition 4.4.20,

$$T_A^* \Omega_0 = (\det(T_A)_*)\Omega_0$$

and $(T_A)_* = A$, since T_A is linear. Hence $\det A = 1$. □

In fact, the linear condition $A^T J A = J$ gives the system of nonlinear partial differential equations that the component functions of a symplectic diffeomorphism $\phi : (\mathbf{R}^{2n}, \omega_0) \to (\mathbf{R}^{2n}, \omega_0)$ must satisfy. We illustrate this in the following example.

Example 7.3.6. For the case $n = 1$, a symplectic diffeomorphism

$$\phi : (\mathbf{R}^2, \omega_0) \to (\mathbf{R}^2, \omega_0)$$

with component functions $\phi = (\phi^x, \phi^y)$ corresponding to coordinates (x, y) must satisfy

$$\begin{aligned}
\omega_0 &= \phi^* \omega_0 \\
&= d\phi^x \wedge d\phi^y \\
&= (\frac{\partial \phi^x}{\partial x} dx + \frac{\partial \phi^x}{\partial y} dy) \wedge (\frac{\partial \phi^y}{\partial x} dx + \frac{\partial \phi^y}{\partial y} dy) \\
&= \left(\frac{\partial \phi^x}{\partial x} \frac{\partial \phi^y}{\partial y} - \frac{\partial \phi^x}{\partial y} \frac{\partial \phi^y}{\partial x} \right) dx \wedge dy,
\end{aligned}$$

so the symplectic condition in this case is the area-preserving condition

$$det(\phi_*) = \frac{\partial \phi^x}{\partial x} \frac{\partial \phi^y}{\partial y} - \frac{\partial \phi^x}{\partial y} \frac{\partial \phi^y}{\partial x} = 1.$$

In this case, the matrix form

$$[\phi_*]^T J_0 [\phi_*] = J_0$$

appears as

$$\begin{bmatrix} \frac{\partial \phi^x}{\partial x} & \frac{\partial \phi^x}{\partial y} \\ \frac{\partial \phi^y}{\partial x} & \frac{\partial \phi^y}{\partial y} \end{bmatrix}^T \begin{bmatrix} 0 & -1 \\ 1 & 0 \end{bmatrix} \begin{bmatrix} \frac{\partial \phi^x}{\partial x} & \frac{\partial \phi^x}{\partial y} \\ \frac{\partial \phi^y}{\partial x} & \frac{\partial \phi^y}{\partial y} \end{bmatrix} = \begin{bmatrix} 0 & -1 \\ 1 & 0 \end{bmatrix},$$

which on its face yields four equations. The reader may check, though, that three of the four equations are redundant.

In contrast to the previous case $n = 1$, the case $n = 2$ shows that the partial differential equations represented by the symplectic condition $\phi^* \omega_0 = \omega_0$ are in fact quite distinct from the volume-preserving condition $\det(\phi_*) = 1$. In this case, the condition $\phi^* \omega_0 = \omega_0$ (or equivalently, $[\phi_*]^T J [\phi_*] = J$), yields six independent partial differential equations.

As with contact spaces, symplectic spaces are all locally "the same." This is the content of Darboux's theorem for symplectic spaces. We shall see that this result is a consequence of the contact Darboux's theorem (Theorem 6.7.9) and a "contactification" lemma.

The following lemma can be thought of as parallel to the symplectization procedure of Example 7.2.11.

Lemma 7.3.7. *Let* $(\mathbf{R}^{2n}, \omega)$ *be an* exact *symplectic space, so that there is a one-form* β *such that* $\omega = d\beta$*. Then* $\alpha = \beta + dt$ *is a contact form on* $\mathbf{R}^{2n+1}$ *with Reeb field* $\xi = \frac{\partial}{\partial t}$*. Here we consider* $\mathbf{R}^{2n+1} = \mathbf{R}^{2n} \times \mathbf{R}$ *with coordinates* $(x_1, \ldots, x_{2n}, t)$.

Proof. We have $d\alpha = d(\beta + dt) = d\beta$, and so $(d\alpha)^n = (d\beta)^n = \omega^n$. Hence

$$\begin{aligned}\alpha \wedge (d\alpha)^n &= (\beta + dt) \wedge \omega^n \\ &= \beta \wedge \omega^n + dt \wedge \omega^n.\end{aligned}$$

But $\beta \wedge \omega^n$ is a $(2n+1)$-form involving only the $2n$ coordinates $x_1, \ldots, x_{2n}$, and so $\beta \wedge \omega^n = 0$. Moreover, since ω^n is nowhere zero on $\mathbf{R}^{2n}$, the $(2n+1)$-form $dt \wedge \omega^n$ is nowhere zero on $\mathbf{R}^{2n+1}$, and so α is nondegenerate.

To show that $\xi = \frac{\partial}{\partial t}$, note that $i(\frac{\partial}{\partial t})d\alpha = i(\frac{\partial}{\partial t})d\beta = 0$ and $\beta(\frac{\partial}{\partial t}) = 0$, again since β involves only the coordinates $x_1, \ldots, x_{2n}$. Also, $i(\frac{\partial}{\partial t})\alpha = \beta(\frac{\partial}{\partial t}) + dt(\frac{\partial}{\partial t}) = 1$. Hence, by Theorem 6.3.10, $\frac{\partial}{\partial t}$ is the Reeb field for α. □

Theorem 7.3.8 (Darboux's theorem for symplectic geometry). *Let* $(\mathbf{R}^{2n}, \omega)$ *be a symplectic space. For each point* $p \in \mathbf{R}^{2n}$*, there exist a domain* U *containing* p *and a diffeomorphism* $\Phi : U \to \Phi(U) \subset \mathbf{R}^{2n}$ *such that* $\Phi(p) = p$ *and on* U*,* $\Phi^*\omega = \omega_0$*, where* $\omega_0 = \sum(dx_i \wedge dy_i)$ *is the standard symplectic form on* $\mathbf{R}^{2n}$ *with coordinate pairs* (x_i, y_i).

Proof. Since ω is closed, Poincaré's lemma (Theorem 4.4.11) guarantees that for every point $p \in \mathbf{R}^{2n}$, there exist a domain $V_1 \subset \mathbf{R}^{2n}$ containing p and a one-form β defined on V_1 such that $\omega = d\beta$ for all points in V_1.

Now consider the one-form $\alpha = \beta + dt$ described in Lemma 7.3.7, so that $(V_1 \times \mathbf{R}, \alpha)$ is a contact space. By the contact Darboux's theorem (Theorem 6.6.4), there are a domain $V_2 \times I$ containing $(p, 0)$ and a diffeomorphism $\phi : (V_2 \times I, \alpha_0) \to (W, \alpha)$ such that $\phi(p, 0) = (p, 0)$ and $\phi^*\alpha = \alpha_0$, where $\alpha_0 = dz + \sum x_i dy_i$ is the standard contact form on $\mathbf{R}^{2n+1}$. Here $W = \phi(V_2 \times I) \subset \mathbf{R}^{2n+1}$ is a domain containing $(p, 0)$, which we can write $W = W' \times I'$, with $W' \subset \mathbf{R}^{2n}$ a domain containing p and I' an interval containing 0. We may assume that $V_2 \subset V_1$.

By Proposition 6.4.4, ϕ must exchange the Reeb fields: $\phi_*\xi_0 = \xi$. Using column vector notation with coordinates $(x_1, y_1, \ldots, x_n, y_n, z)$ and component functions

$$\phi = (\phi^1, \ldots, \phi^{2n+1}),$$

this condition says, in light of Lemma 7.3.7, that

$$\begin{bmatrix} \frac{\partial \phi^1}{\partial x_1} & \cdots & \frac{\partial \phi^1}{\partial y_n} & \frac{\partial \phi^1}{\partial z} \\ \vdots & & & \vdots \\ \frac{\partial \phi^{2n+1}}{\partial x_1} & \cdots & \frac{\partial \phi^{2n+1}}{\partial y_n} & \frac{\partial \phi^{2n+1}}{\partial z} \end{bmatrix} \begin{bmatrix} 0 \\ \vdots \\ 0 \\ 1 \end{bmatrix} = \begin{bmatrix} 0 \\ \vdots \\ 0 \\ 1 \end{bmatrix}.$$

This shows that for $i = 1, \dots, 2n$, $\phi^i = \phi^i(x_1, y_1, \dots, x_n, y_n)$ are independent of z and that $\phi^{2n+1}(x_1, y_1, \dots, x_n, y_n, z) = z + \psi(x_1, y_1, \dots, x_n, y_n)$ for some smooth function $\psi : \mathbf{R}^{2n} \to \mathbf{R}$.

Define the function $\Phi : V_2 \to W'$ by means of the component functions ϕ^i:

$$\Phi = (\phi^1, \dots, \phi^{2n}).$$

The preceding paragraph shows that writing

$$D\psi = \begin{bmatrix} \frac{\partial \psi}{\partial x_1} & \frac{\partial \psi}{\partial y_1} & \cdots & \frac{\partial \psi}{\partial x_n} & \frac{\partial \psi}{\partial y_n} \end{bmatrix},$$

we have

$$\phi_* = \begin{bmatrix} \Phi_* & 0 \\ D\psi & 1 \end{bmatrix},$$

and so

$$0 \neq \det(\phi_*) = \det(\Phi_*).$$

By the inverse function theorem (Theorem 3.6.12), there exist a domain $U \subset V_2$ containing p and a domain $U' \subset W'$, also containing p, on which $\Phi : U \to U'$ is a diffeomorphism.

Define the map $i : \mathbf{R}^{2n} \to \mathbf{R}^{2n+1}$ by $i(x) = (x, 0)$ and the map $\pi : \mathbf{R}^{2n+1} = \mathbf{R}^{2n} \times \mathbf{R} \to \mathbf{R}^{2n}$ by $\pi(x, t) = x$. Note that $\Phi = \pi \circ \phi \circ i$ and that $\Phi(p) = \pi(\phi(p, 0)) = \pi(p, 0) = p$. Further,

$$\begin{aligned} \Phi^* \beta &= i^* \phi^* \pi^* \beta \\ &= i^* \phi^* (\alpha - dz) \quad \text{since } \alpha = \beta + dz \\ &= i^* (\phi^* \alpha - \phi^* dz) \\ &= i^* (\alpha_0 - d(z + \psi)) \\ &= i^* (\sum x_i dy_i + dz - dz - d\psi) \\ &= \sum x_i dy_i - d\psi, \end{aligned}$$

and so

$$\begin{aligned} \Phi^* \omega &= \Phi^* (d\beta) \\ &= d\Phi^* \beta \end{aligned}$$

$$= d(\sum x_i dy_i - d\psi)$$
$$= \sum (dx_i \wedge dy_i)$$
$$= \omega_0.$$

Hence the domain U and the diffeomorphism Φ are the ones desired in the statement of the theorem. □

As in the contact case, Darboux's theorem shows that differences between symplectic structures occur at the global (topological) level, and hence outside the scope of this text. In fact, it was only in 1985 that Gromov's landmark paper [22] demonstrated the existence of a symplectic structure on $\mathbf{R}^4$ that is not equivalent to the standard one. In 1990, Bates and Peschke [7] published an explicit example of an "exotic" (i.e., not equivalent to the standard) symplectic structure on $\mathbf{R}^4$.

Like the sets of Riemannian and contact structure-preserving diffeomorphisms, the set of all symplectic diffeomorphisms has an algebraic structure. The following is the symplectic analogue to Theorems 5.6.10 and 6.4.9.

Theorem 7.3.9. *Let $(\mathbf{R}^{2n}, \omega)$ be a symplectic space and let Diff $(\mathbf{R}^{2n}, \omega)$ denote the set of all symplectic diffeomorphisms $\phi : \mathbf{R}^{2n} \to \mathbf{R}^{2n}$. Then Diff$(\mathbf{R}^{2n}, \omega)$ is a group under the operation of function composition. In other words,*

- *If $\phi_1, \phi_2 \in$ Diff $(\mathbf{R}^{2n}, \omega)$, then $\phi_1 \circ \phi_2 \in$ Diff $(\mathbf{R}^{2n}, \omega)$;*
- $\mathrm{Id} \in$ *Diff $(\mathbf{R}^{2n}, \omega)$, where* $\mathrm{Id}(x) = x$ *for all $x \in \mathbf{R}^{2n}$; and*
- *If $\phi \in$ Diff $(\mathbf{R}^{2n}, \omega)$, then $\phi^{-1} \in$ Diff $(\mathbf{R}^{2n}, \omega)$.*

Proof. Exercise. □

We close this section by noting that the algebraic structure described by Theorem 7.3.9, as well as the corresponding theorem in the contact setting, has an impact on the geometry of the space. In 1872, Felix Klein set out the task of classifying the new geometric structures that were appearing in the wake of the non-Euclidean revolution. In his Erlangen program, Klein identified geometry as the study of objects preserved by a given group of transformations. Hence the following question arises: If two geometric spaces have the same transformation group (up to group isomorphism), are the geometric spaces the same (up to geometric isomorphism)? More precisely, given two geometric spaces, suppose there is an isomorphism between their groups of structure-preserving, one-to-one and onto maps. Does this imply the existence of a structure-preserving map between the two geometric spaces? An affirmative answer would imply that the algebraic structure of the transformation group would "determine the geometric structure."

The case of Riemannian geometry shows that this program is not true in general. Consider the case of two spheres with different radii endowed with the metric tensors obtained from the standard Euclidean metric tensor; see Definition 5.2.6. These two geometric sets have the same isometry groups. However, since they have different scalar curvatures, they can not be isometric.

In 1988, Banyaga was able to show that under certain fairly general assumptions, the group of symplectic diffeomorphisms does in fact determine the geometry of a symplectic space in the sense of Klein's Erlangen program. In 1995, this result was extended to the contact setting independently in [6, 35].

We have already seen ways in which Riemannian geometry is fundamentally different from contact or symplectic geometry. For example, there is (and can be) no version of Darboux's theorem for Riemannian geometry, since there are objects that are local in character, such as the curvature tensor, that can distinguish one Riemannian metric from another.

The fact that the group of isometries does not determine a given Riemannian structure is yet another way in which Riemannian geometry differs from contact or symplectic geometry. Without elaborating, we may say that there are "not enough" isometries in general compared to the "many" contact and symplectic diffeomorphisms that exist for a given space.

7.4 Symplectic and Hamiltonian Vector Fields

Considering vector fields as "infinitesimal automorphisms," we have the following analogue to Killing vector fields and (strictly) contact vector fields.

Definition 7.4.1. Let X be a vector field on a symplectic space (U, ω), where $U \subset \mathbf{R}^{2n}$. Then X is a *symplectic vector field* if

$$\mathcal{L}_X \omega = 0.$$

In light of Theorem 4.7.23, the symplectic vector fields are those whose one-parameter group of diffeomorphisms consists entirely of symplectic diffeomorphisms.

The requirement that a symplectic form be closed has an important consequence for symplectic vector fields.

Proposition 7.4.2. *Let X be a vector field on the symplectic space $(\mathbf{R}^{2n}, \omega)$. Then X is a symplectic vector field if and only if for each point $p \in \mathbf{R}^{2n}$, there exist a domain $U \subset \mathbf{R}^{2n}$ containing p and a smooth function $f : U \to \mathbf{R}$ such that*

$$i(X)\omega = -df \quad \text{on } U.$$

Proof. We rely on the Cartan formula (Theorem 4.7.18):

$$\begin{aligned} \mathcal{L}_X \omega &= di(X)\omega + i(X)d\omega \\ &= di(X)\omega \quad \text{since } \omega \text{ is closed.} \end{aligned}$$

So $\mathcal{L}_X \omega = 0$ implies that $i(X)\omega$ is closed. In that case, by Poincaré lemma (Theorem 4.4.11), $i(X)\omega$ is locally exact: At each point $p \in \mathbf{R}^{2n}$, there are a domain

U and a function f such that $i(X)\omega = -df$ on U (the sign chosen arbitrarily for reasons discussed below).

Conversely, given the existence of a domain U containing p and a smooth function $f : U \to \mathbf{R}$ satisfying $i(X)\omega = -df$, then

$$\mathcal{L}_X\omega = d(i(X)\omega) = d(-df) = 0. \qquad \square$$

Comparing the condition $i(X)\omega = -df$ to the discussion preceding Theorem 7.1.6, we can summarize Proposition 7.4.2 by saying that the symplectic vector fields are *locally Hamiltonian vector fields*.

If X is a symplectic vector field defined on a domain $U \subset \mathbf{R}^{2n}$, it may not be the case that there is a single function $f : U \to \mathbf{R}$ for which $i(X)\omega = -df$. This is a global question, and so lies outside the scope of this text. However, the following shows that every function does give rise to a distinguished vector field.

Theorem 7.4.3. *Let $f : U \to \mathbf{R}$ be a smooth function on the symplectic space (U, ω), where $U \subset \mathbf{R}^{2n}$. There is a unique symplectic vector field X_f on U such that $i(X_f)\omega = -df$. The field X_f is called the* Hamiltonian vector field *with* Hamiltonian function *f.*

Proof. For the smooth function f, let X_f be the vector field defined uniquely, according to Proposition 7.2.4, by

$$X_f = \Phi^{-1}(-df),$$

where Φ is the vector space isomorphism between vector fields and one-forms induced by the nondegenerate form ω. It is a symplectic vector field almost by construction:

$$\mathcal{L}_{X_f}\omega = d(i(X_f)\omega) = d(-df) = 0. \qquad \square$$

The reader should compare this theorem with Theorem 6.5.4, guaranteeing the existence of a contact gradient vector field for a smooth function $f : U \to \mathbf{R}$ defined on a contact space (U, α) with $U \subset \mathbf{R}^{2n+1}$. For that reason, the contact gradient is also sometimes called the contact Hamiltonian vector field. In both cases, as well as in the case of the gradient in Riemannian geometry, the key condition guaranteeing the existence of such a vector field associated to a function is the nondegeneracy of the structure tensor. In contrast to the contact case, however, in which there is an isomorphism between smooth functions and contact vector fields, there may be symplectic vector fields that do not arise as the (global) Hamiltonian vector field of a function. Furthermore, different functions may give rise to the same Hamiltonian vector field: $X_f = X_g$ when $f - g = c$ for some constant c.

Example 7.4.4. Consider the standard symplectic space $(\mathbf{R}^{2n}, \omega_0)$ with coordinate pairs (x_i, y_i), so that $\omega = \sum dx_i \wedge dy_i$. Then the condition that $i(X_f)\omega = -df$ can be expressed using coordinates $X_f = \left\langle X^1, Y^1, \ldots, X^n, Y^n \right\rangle$ as

$$\sum \left(X^i dy_i - Y^i dx_i\right) = \sum \left(-\frac{\partial f}{\partial x_i} dx_i - \frac{\partial f}{\partial y_i} dy_i\right),$$

and so

$$X_f = \sum -\frac{\partial f}{\partial y_i}\frac{\partial}{\partial x_i} + \frac{\partial f}{\partial x_i}\frac{\partial}{\partial y_i}.$$

Note that the differential equations defining the integral curves of X_f are

$$\begin{cases} \frac{dx_i}{dt} = -\frac{\partial f}{\partial y_i}, \\ \frac{dy_i}{dt} = \frac{\partial f}{\partial x_i}, \end{cases}$$

which are exactly Hamilton's equations for the system defined by the Hamiltonian function f from Sect. 7.1 (although the momentum p variables here correspond to the x-coordinates while the position q variables correspond to the y-coordinates).

The following algebraic proposition follows from the fact that the map Φ of Proposition 7.2.4 is a vector space isomorphism.

Proposition 7.4.5. *Let $f, g : U \to \mathbf{R}$ be smooth functions on the symplectic space (U, ω) and let c be any constant. Then $X_{f+g} = X_f + X_g$ and $X_{cf} = cX_f$.*

Proof. Exercise. □

Built into the definition of the Hamiltonian vector field is a mathematical expression for conservation of energy, when the function $f : \mathbf{R}^{2n} \to \mathbf{R}$ is interpreted as the energy of a physical system. Compare this to Proposition 7.1.1.

Proposition 7.4.6. *Let $f : U \to \mathbf{R}$ be a smooth function on a symplectic space (U, ω). Then the integral curves of X_f lie on level sets of f.*

Proof. Let $c : I \to U$ be an integral curve of X_f. Then

$$\begin{aligned} \frac{d}{dt}\left(f(c(t))\right) &= df(X_f) \quad \text{by the chain rule} \\ &= \left(-i(X_f)\omega\right)(X_f) \quad \text{by the definition of } X_f \\ &= -\omega(X_f, X_f) \\ &= 0 \quad \text{by the skew-symmetry of } \omega. \end{aligned}$$

Hence, for all $t \in I$, $f(c(t)) = k$ for some constant k, and so $c(t)$ lies on the level set $f = k$. See Fig. 7.2. □

Proposition 7.4.6 also gives an example of geometric methods in solving differential equations, of which the method of characteristics in Chap. 7 is another. With the inherent difficulties in explicitly solving a system of differential equations of the form $\frac{dx_i}{dt} = X^i$, where X^i are some functions of the variables $(x_1, \ldots, x_n)$, it is often more practical to find level sets on which the solutions (considered as curves

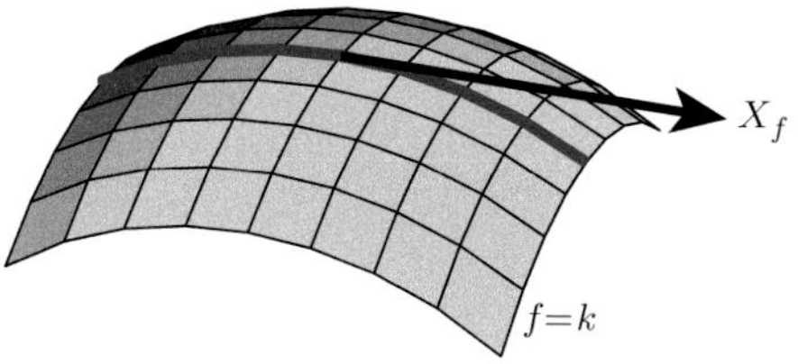

Fig. 7.2 The integral curves of the Hamiltonian vector field of f lie on level sets of f.

in $\mathbf{R}^n$) must lie. This potentially allows for a qualitative analysis of the system of ordinary differential equations, even when an explicit solution cannot be produced.

Example 7.4.7. Consider the system

$$\begin{cases} \frac{dx}{dt} &= 2y - 2, \\ \frac{dy}{dt} &= -2x - 4. \end{cases}$$

Solving this system explicitly is possible using standard techniques of solving ordinary differential equations, for example by making an appropriate change of variables to convert this system to a linear one. However, observing that the system describes the integral curves of the Hamiltonian vector field X_f, where $f(x, y) = (x+2)^2 + (y-1)^2$ in the standard symplectic space $(\mathbf{R}^2, dx \wedge dy)$, it is immediate that solution curves lie on circles about the center $(-2, 1)$, the radius depending on the initial conditions.

This line of reasoning prompts the following definition, which does not depend on the symplectic structure.

Definition 7.4.8. Let X be a smooth vector field on a domain $U \subset \mathbf{R}^n$. A *first integral* of X is a smooth function $f : U \to \mathbf{R}$ such that $X[f] = df(X) = 0$.

In the setting of a symplectic space (U, ω), Proposition 7.4.6 says simply that f is a first integral of X_f.

We will now develop a new operation that will give a way of discussing the problem of finding first integrals of a Hamiltonian vector field. This operation on the set of smooth functions has independent interest as well.

Definition 7.4.9. Let $f, g : U \to \mathbf{R}$ be smooth, real-valued functions on the symplectic space (U, ω). The *Poisson bracket* $\{f, g\} : U \to \mathbf{R}$ is defined to be the smooth function

$$\{f, g\} = \left.\frac{d}{dt}\right|_{t=0} (g \circ \phi_t),$$

where ϕ_t is the flow generated by the Hamiltonian vector field X_f.

An immediate consequence of the definition is the following alternative characterization of the Poisson bracket, which in fact could have been used as the definition.

Proposition 7.4.10. *Let $f, g : U \to \mathbf{R}$ be smooth functions on the symplectic space (U, ω) with Hamiltonian vector fields X_f and X_g. Then*

$$\{f, g\} = \omega(X_f, X_g).$$

Proof. Let $\phi_t : \mathbf{R}^{2n} \to \mathbf{R}^{2n}$ be the flow generated by the vector field X_f. Then

$$\begin{aligned}
\{f, g\} &= \left.\frac{d}{dt}\right|_{t=0} (g \circ \phi_t) \\
&= dg(X_f) \quad \text{by the chain rule} \\
&= (-i(X_g)\omega)(X_f) \quad \text{by the definition of } X_g \\
&= -\omega(X_g, X_f) \\
&= \omega(X_f, X_g).
\end{aligned}$$

□

The advantage of Proposition 7.4.10 is that it highlights the symmetry (more accurately, the skew-symmetry) in the roles played by the functions f and g. In Definition 7.4.9, it appears that the functions are playing different roles, one being differentiated while the other integrated.

Corollary 7.4.11. *Let $f, g : U \to \mathbf{R}$ be smooth, real-valued functions defined on the symplectic space (U, ω). Then:*

- *The map $(f, g) \mapsto \{f, g\}$ is bilinear:*

  $$\{c_1 f_1 + c_2 f_2, g\} = c_1\{f_1, g\} + c_2\{f_2, g\}$$

  and

  $$\{f, c_1 g_1 + c_2 g_2\} = c_1\{f, g_1\} + c_2\{f, g_2\}$$

  for all smooth functions $f, f_1, f_2, g, g_1, g_2 : U \to \mathbf{R}$ and all constants c_1, c_2.
- *The map $(f, g) \mapsto \{f, g\}$ is skew-symmetric:*

  $$\{g, f\} = -\{f, g\}.$$

- *The map $(f, g) \mapsto \{f, g\}$ obeys a product rule in each component:*

  $$\{f_1 f_2, g\} = f_1\{f_2, g\} + f_2\{f_1, g\}$$

  and

  $$\{f, g_1 g_2\} = g_1\{f, g_2\} + g_2\{f, g_1\}$$

  for all smooth functions $f, f_1, f_2, g, g_1, g_2 : U \to \mathbf{R}$.

Proof. The first two statements follow from the bilinearity and skew-symmetry of ω as well as the linearity of the map Φ used to define X_f.

We illustrate the proof of the product rule in the second component, the other verification being identical.

$$\begin{aligned}\{f, g_1 g_2\} &= d(g_1 g_2)(X_f) \quad \text{as in the proof of Proposition 7.4.10}\\ &= (g_1 dg_2 + g_2 dg_1)(X_f)\\ &= g_1 dg_2(X_f) + g_2 dg_1(X_f)\\ &= g_1\{f, g_2\} + g_2\{f, g_1\}.\end{aligned}$$ □

Corollary 7.4.12. *A function g is a first integral for a Hamiltonian vector field X_f if and only if $\{f, g\} = 0$.*

Proof. The proof of Proposition 7.4.10 shows that

$$X_f\,[g] = dg(X_f) = \{f, g\}.$$

Hence $X_f\,[g] = 0$ exactly when $\{f, g\} = 0$. □

The next example provides a coordinate expression for the Poisson bracket in the standard symplectic space $(\mathbf{R}^{2n}, \omega_0)$.

Example 7.4.13. We showed in Example 7.4.4 that for the standard symplectic form $\omega_0 = \sum dx_i \wedge dy_i$ on $\mathbf{R}^{2n}$ with coordinate pairs (x_i, y_i), the Hamiltonian vector field for the smooth Hamiltonian function f is given by

$$X_f = \sum_{i=1}^{n} \left(-\frac{\partial f}{\partial y_i}\frac{\partial}{\partial x_i} + \frac{\partial f}{\partial x_i}\frac{\partial}{\partial y_i} \right).$$

Hence by Proposition 7.4.10, we have

$$\begin{aligned}\{f, g\} &= \omega_0(X_f, X_g)\\ &= \sum_{i=1}^{n} \left(\left(-\frac{\partial f}{\partial y_i}\right)\left(\frac{\partial g}{\partial x_i}\right) - \left(\frac{\partial f}{\partial x_i}\right)\left(-\frac{\partial g}{\partial y_i}\right) \right)\\ &= \sum_{i=1}^{n} \left(\left(\frac{\partial f}{\partial x_i}\right)\left(\frac{\partial g}{\partial y_i}\right) - \left(\frac{\partial f}{\partial y_i}\right)\left(\frac{\partial g}{\partial x_i}\right) \right).\end{aligned}$$

Consider, for instance, the function $H : \mathbf{R}^4 \to \mathbf{R}$ given by

$$H(x_1, y_1, x_2, y_2) = \frac{1}{2}\left(x_1^2 + y_1^2 + x_2^2 + y_2^2\right).$$

The Hamiltonian vector field corresponding to H is given by

$$X_H = -y_1\frac{\partial}{\partial x_1} + x_1\frac{\partial}{\partial y_1} - y_2\frac{\partial}{\partial x_2} + x_2\frac{\partial}{\partial y_2},$$

so a first integral $f : \mathbf{R}^4 \to \mathbf{R}$ of X_H must satisfy

$$0 = \{H, f\} = X_H[f] = -y_1 \frac{\partial f}{\partial x_1} + x_1 \frac{\partial f}{\partial y_1} - y_2 \frac{\partial f}{\partial x_2} + x_2 \frac{\partial f}{\partial y_2}.$$

For example,

$$f(x_1, y_1, x_2, y_2) = x_1 x_2 + y_1 y_2$$

is a first integral of X_H.

The bracket notation invites comparison to the Lie bracket of vector fields. In fact, the similarities are by no means coincidental, as the following propositions demonstrate.

Proposition 7.4.14. *Suppose that $f, g : U \to \mathbf{R}$ are smooth, real-valued functions on the symplectic space (U, ω). Then the Hamiltonian vector field of the Poisson bracket $\{f, g\}$ is the Lie bracket of the vector fields X_f, X_g:*

$$X_{\{f,g\}} = [X_f, X_g].$$

Proof. We apply Proposition 4.7.22, which, in the case $k = 2$, states that

$$\begin{aligned} d\omega(Z, X_1, X_2) &= Z[\omega(X_1, X_2)] - X_1[\omega(Z, X_2)] + X_2[\omega(Z, X_1)] \\ &\quad - \omega([Z, X_1], X_2) + \omega([Z, X_2], X_1) - \omega([X_1, X_2], Z). \end{aligned}$$

Let Z be any vector field on U, and let X_f, X_g be the Hamiltonian vector fields for f and g. By the previous equality and using the fact that ω is closed, we have

$$\begin{aligned} (i([X_f, X_g])\omega)(Z) &= \omega([X_f, X_g], Z) \\ &= Z[\omega(X_f, X_g)] - X_f[\omega(Z, X_g)] + X_g[\omega(Z, X_f)] \\ &\quad - \omega([Z, X_f], X_g) + \omega([Z, X_g], X_f) - d\omega(Z, X_f, X_g) \\ &= Z[\{f, g\}] - X_f[dg(Z)] + X_g[df(Z)] \\ &\quad - dg([Z, X_f]) + df([Z, X_g]) \quad \text{since } \omega \text{ is closed} \\ &= Z[\{f, g\}] - X_f[Z[g]] + X_g[Z[f]] \\ &\quad - ([Z, X_f])[g] + ([Z, X_g])[f] \\ &= Z[\{f, g\}] - X_f[Z[g]] + X_g[Z[f]] \\ &\quad - (ZX_f - X_f Z)[g] + (ZX_g - X_g Z)[f] \\ &= Z[\{f, g\}] - Z[X_f[g]] + Z[X_g[f]] \\ &= Z[\{f, g\}] - Z[\{f, g\}] - Z[\{f, g\}] \quad \text{by Corollary 7.4.11} \\ &= -Z[\{f, g\}] \\ &= -d\{f, g\}(Z), \end{aligned}$$

all of which shows that

$$i([X_f, X_g])\omega = -d\{f, g\}.$$

Hence by the uniqueness of the Hamiltonian vector field,

$$X_{\{f,g\}} = [X_f, X_g]. \qquad \square$$

Proposition 7.4.15. *The Poisson bracket satisfies the Jacobi identity: For all smooth functions* $f, g, h : U \to \mathbf{R}$ *on the symplectic space* (U, ω),

$$\{\{f, g\}, h\} + \{\{g, h\}, f\} + \{\{h, f\}, g\} = 0.$$

Proof. We rely repeatedly on the fact, first noted in the proof of Proposition 7.4.10 and used again several times in the preceding proof, that $\{f, g\} = X_f\,[g]$.

In particular,

$$\begin{aligned}
\{\{f, g\}, h\} &= X_{\{f,g\}}\,[h] = ([X_f, X_g])[h], \quad \text{by Proposition 7.4.14;}\\
\{\{g, h\}, f\} &= -\{f, \{g, h\}\} = -X_f[\{g, h\}] = -X_f[X_g[h]]; \quad \text{and}\\
\{\{h, f\}, g\} &= -\{g, \{h, f\}\} = \{g, \{f, h\}\} = X_g[\{f, h\}] = X_g[X_f[h]].
\end{aligned}$$

Adding the three equations and relying on the fact that $[X, Y] = XY - YX$ proves the Jacobi identity for the Poisson bracket. $\square$

Corollary 7.4.16. *Suppose that the smooth functions* $f, g : U \to \mathbf{R}$ *are first integrals of a Hamiltonian vector field* X_H *with Hamiltonian function* $H : U \to \mathbf{R}$. *Then* $\{f, g\}$ *is also a first integral of* X_H.

Proof. Suppose that $\{H, f\} = \{H, g\} = 0$. Relying on the Jacobi identity, we have

$$\begin{aligned}
X_H[\{f, g\}] &= \{H, \{f, g\}\}\\
&= \{\{g, H\}, f\} + \{\{H, f\}, g\}\\
&= \{0, f\} + \{0, g\} = 0.
\end{aligned} \qquad \square$$

This shows that given two first integrals of a Hamiltonian vector field, we can generate "new" first integrals.

Example 7.4.17. On the standard symplectic space $(\mathbf{R}^4, \omega_0)$, consider the function $H : \mathbf{R}^4 \to \mathbf{R}$ given by

$$H(x_1, y_1, x_2, y_2) = \frac{1}{2}(x_1^2 + y_1^2 + x_2^2 + y_2^2).$$

In Example 7.4.13, we saw that the function $f : \mathbf{R}^4 \to \mathbf{R}$ given by

$$f(x_1, y_1, x_2, y_2) = x_1x_2 + y_1y_2$$

is a first integral for the Hamiltonian vector field X_H.

The reader may verify that

$$g_1(x_1, y_1, x_2, y_2) = x_1 y_2 - y_1 x_2$$

is also a first integral for X_H. Hence, by Corollary 7.4.16,

$$g_2 = \{f, g_1\} = x_1^2 + y_1^2 - x_2^2 - y_2^2$$

is another first integral of X_H.

It is worth noting that first integrals generated as the Poisson bracket of other first integrals may not be independent of the original ones. We will return to the topic of first integrals in the next section.

While there are "more" symplectic vector fields than Hamiltonian vector fields, it is always easy to construct Hamiltonian vector fields from symplectic ones.

Proposition 7.4.18. *Let X and Y be symplectic vector fields on a symplectic space (U, ω). Then the Lie bracket $[X, Y]$ is a Hamiltonian vector field (and hence also symplectic) with Hamiltonian function $f = \omega(X, Y)$.*

Proof. Assume that X and Y are symplectic vector fields, so that

$$d(i(X)\omega) = d(i(Y)\omega) = 0.$$

We will again rely on Proposition 4.7.22. On the one hand, as in Proposition 7.4.14, we have, for any vector field Z on U,

$$\begin{aligned}(i([X,Y])\omega)(Z) &= \omega([X,Y], Z) \\ &= X[\omega(Y,Z)] - Y[\omega(X,Z)] + Z[\omega(X,Y)] \\ &\quad + \omega([X,Z], Y) - \omega([Y,Z], X).\end{aligned}$$

On the other hand, applying Proposition 4.7.22 to the one-forms $i(X)\omega$ and $i(Y)\omega$ yields

$$\begin{aligned}\omega([X,Z], Y) &= -(i(Y)\omega)([X,Z]) \\ &= -(X[(i(Y)\omega)(Z)] - Z[(i(Y)\omega)(X)] - (di(Y)\omega)(X,Z)) \\ &= -X[\omega(Y,Z)] + Z[\omega(Y,X)] \quad \text{since } di(Y)\omega = 0 \text{ by assumption},\end{aligned}$$

and likewise

$$\begin{aligned}\omega([Y,Z], X) &= -(i(X)\omega)([Y,Z]) \\ &= -Y[\omega(X,Z)] + Z[\omega(X,Y)].\end{aligned}$$

Substituting the last two expressions into the first yields

$$\begin{aligned}\omega([X,Y],Z) &= X\left[\omega(Y,Z)\right] - Y\left[\omega(X,Z)\right] + Z\left[\omega(X,Y)\right] \\ &\qquad - X\left[\omega(Y,Z)\right] + Z\left[\omega(Y,X)\right] \\ &\qquad\quad + Y\left[\omega(X,Z)\right] - Z\left[\omega(X,Y)\right] \\ &= -Z\left[\omega(X,Y)\right] \\ &= -df(Z), \quad \text{where } f = \omega(X,Y).\end{aligned}$$

Hence $[X,Y] = X_f$ is a Hamiltonian vector field. □

We conclude this section by examining the effect of symplectic diffeomorphisms on Hamiltonian vector fields and Poisson brackets.

Proposition 7.4.19. *Let $\phi : (U,\omega) \to (U,\omega)$ be a symplectic diffeomorphism and let $f : U \to \mathbf{R}$ be a smooth function. Then*

$$\phi^* X_f = X_{f\circ\phi},$$

where the pullback of a vector field is defined, as usual, by $\phi^* X = (\phi^{-1})_* X$.

Proof. We will show that $i(\phi^* X_f)\omega = -d(f \circ \phi)$. To do this, suppose Y is any vector field on U. We have

$$\begin{aligned}(i(\phi^* X_f)\omega)(Y) &= (i(\phi^* X_f)(\phi^*\omega))(Y) \quad \text{since } \phi \text{ is symplectic} \\ &= (\phi^*\omega)(\phi^* X_f, Y) \\ &= \omega(\phi_*\phi^* X_f, \phi_* Y) \\ &= \omega(X_f, \phi_* Y) \\ &= -df(\phi_* Y) \\ &= -(\phi^* df)(Y) \\ &= -(d\phi^* f)(Y) \\ &= -d(f \circ \phi)(Y),\end{aligned}$$

and so $i(\phi^* X_f)\omega = -d(f \circ \phi)$, as desired. □

We have introduced the Poisson bracket as a secondary object defined in terms of Hamiltonian vector fields. We now show that it is in fact an object that belongs to the realm of symplectic geometry, in the sense that it is preserved by symplectic diffeomorphisms. What is more, symplectic diffeomorphisms can actually be characterized as those diffeomorphisms that preserve the Poisson bracket.

Proposition 7.4.20. *Let (U,ω) be a symplectic space and let $\phi : U \to U$ be a diffeomorphism. Then ϕ is a symplectic diffeomorphism if and only if for all smooth functions $f, g : U \to \mathbf{R}$,*

$$\{f \circ \phi, g \circ \phi\} = \{f, g\} \circ \phi.$$

Proof. Assume first that ϕ is symplectic, i.e., $\phi^*\omega = \omega$. Then

$$\begin{aligned}
\{f \circ \phi, g \circ \phi\} &= \omega(X_{f\circ\phi}, X_{g\circ\phi}) \\
&= \omega(\phi^* X_f, \phi^* X_g) \quad \text{by Proposition 7.4.19} \\
&= (\phi^*\omega)(\phi^* X_f, \phi^* X_g) \quad \text{since } \phi \text{ is symplectic} \\
&= (\omega(\phi_*\phi^* X_f, \phi_*\phi^* X_g)) \circ \phi \\
&= (\omega(X_f, X_g)) \circ \phi \\
&= \{f, g\} \circ \phi.
\end{aligned}$$

Suppose now that for all real-valued functions $f, g : U \to \mathbf{R}$ we have

$$\{f \circ \phi, g \circ \phi\} = \{f, g\} \circ \phi.$$

The key observation will be that the Poisson bracket uniquely determines the associated symplectic form, in the sense that if ω_1 and ω_2 are two symplectic forms on $\mathbf{R}^{2n}$ such that the associated Poisson brackets agree for all f, g, i.e., $\{f, g\}_1 = \{f, g\}_2$, then $\omega_1 = \omega_2$. We leave this as an exercise for the reader.

Now given a symplectic form $\omega_1 = \omega$, the pullback $\omega_2 = \phi^*\omega$ is also a symplectic form. The goal is to show that $\omega_1 = \omega_2$. Denote the Hamiltonian vector field of a function f relative to the symplectic form ω_1 by X_f, and the Poisson bracket relative to ω_1 by $\{\ ,\ \}_1$; the corresponding objects relative to the symplectic form ω_2 will be written $\tilde{X}_f$ and $\{\ ,\ \}_2$.

Note that $\tilde{X}_f = X_{f\circ\phi^{-1}}$. After all, we have

$$\begin{aligned}
-df &= i(\tilde{X}_f)\omega_2 \\
&= i(\tilde{X}_f)(\phi^*\omega) \\
&= \phi^*\left(i(\phi_*\tilde{X}_f)\omega\right),
\end{aligned}$$

and so

$$i(\phi_*\tilde{X}_f)\omega = -(\phi^{-1})^* df = -d(f \circ \phi^{-1}).$$

Hence, for all $x \in U$,

$$\begin{aligned}
\{f, g\}_2(x) &= (\phi^*\omega)(x)(\tilde{X}_f(x), \tilde{X}_g(x)) \\
&= \omega(\phi(x))(\phi_*\tilde{X}_f(\phi(x)), \phi_*\tilde{X}_g(\phi(x))) \\
&= \omega(\phi(x))(X_{f\circ\phi^{-1}}(\phi(x)), X_{g\circ\phi^{-1}}(\phi(x)))
\end{aligned}$$

$$
\begin{aligned}
&= (\{f\circ\phi^{-1}, g\circ\phi^{-1}\}_1\circ\phi)(x)\\
&= \{(f\circ\phi^{-1})\circ\phi, (g\circ\phi^{-1})\circ\phi\}_1(x)\\
&= \{f,g\}_1(x).
\end{aligned}
$$

We have used the assumption that ϕ preserves the Poisson bracket in the second-to-last equality.

It follows that $\omega_2 = \omega_1$, and so ϕ is a symplectic diffeomorphism. □

7.5 Geometric Sets in Symplectic Spaces

In this section, we consider special geometric sets in a symplectic space $(\mathbf{R}^{2n}, \omega)$. Geometric sets, introduced in general in Chap. 3, will be in this section either parameterized sets $S = \phi(U)$, where $U \subset \mathbf{R}^k$, $1 \leq k \leq 2n$, is a domain and $\phi : U \to \mathbf{R}^{2n}$ is a regular parameterization, or level sets

$$S = S_F = \left\{x \in \mathbf{R}^{2n} \mid F(x) = a\right\},$$

where $F : \mathbf{R}^{2n} \to \mathbf{R}^l$, $1 \leq l < 2n$, is a smooth map and $a \in \mathbf{R}^l$ is a regular value of F. The motivating concept in this section will be the notion of ω-orthogonality introduced in Sect. 2.10, now in the setting of the tangent space.

Recall that for every point $p \in \mathbf{R}^{2n}$, tangent vectors $\mathbf{v}_p, \mathbf{w}_p \in T_p\mathbf{R}^{2n}$ are said to be *ω-orthogonal* if $\omega(\mathbf{v}_p, \mathbf{w}_p) = 0$. For every subset $W_p \subset T_p\mathbf{R}^{2n}$, the *$\omega$-orthogonal complement of W_p* is the vector subspace

$$W_p^\omega = \left\{\mathbf{v}_p \in T_p\mathbf{R}^{2n} \mid \omega(\mathbf{v}_p, \mathbf{w}_p) = 0 \text{ for all } \mathbf{w}_p \in W_p\right\}.$$

The following definitions are the nonlinear analogues of those in Definition 2.10.13.

Definition 7.5.1. Let S be a geometric set in a symplectic space $(\mathbf{R}^{2n}, \omega)$. Then S is called:

- *isotropic* if for all $p \in S$, $T_pS \subset (T_pS)^\omega$;
- *coisotropic* if for all $p \in S$, $(T_pS)^\omega \subset T_pS$;
- *Lagrangian* if for all $p \in S$, $(T_pS)^\omega = T_pS$;
- *symplectic* if for all $p \in S$, $(T_pS)^\omega \cap T_pS = \{\mathbf{0}_p\}$.

We leave the proof of the following alternative characterizations of ω-orthogonality relations to the reader as a review of the symplectic linear algebra of Sect. 2.10.

Proposition 7.5.2. *Let S be a geometric set in a symplectic space $(\mathbf{R}^{2n}, \omega)$. Then:*

- *S is isotropic if and only if for all $\mathbf{v}_p, \mathbf{w}_p \in T_pS$, $\omega(\mathbf{v}_p, \mathbf{w}_p) = 0$.*
- *If S is a parameterized set, i.e., $S = \phi(U)$ for some domain $U \subset \mathbf{R}^k$ and regular parameterization $\phi : U \to \mathbf{R}^{2n}$, then S is isotropic if and only if $\phi^*\omega = 0$.*
- *S is coisotropic if and only if for all $\mathbf{v}_p \in T_p\mathbf{R}^{2n}$ with $\mathbf{v}_p \notin T_pS$, there is a tangent vector $\mathbf{w}_p \in T_pS$ such that $\omega(\mathbf{v}_p, \mathbf{w}_p) \neq 0$.*
- *S is Lagrangian if and only if S is isotropic (or coisotropic) and*

$$\dim(T_pS) = n.$$

- *S is symplectic if and only if $\omega\big|_S$ is nondegenerate.*

In all of these special symplectic geometric sets, any parameterization of the set must be well adapted to the "pairing" of the symplectic coordinates established by the symplectic form. We illustrate this with the following linear examples in the standard symplectic space $(\mathbf{R}^{2n}, \omega_0)$.

Example 7.5.3. Let $(\mathbf{R}^{2n}, \omega_0)$ be the standard symplectic space with coordinate pairs (x_i, y_i). Let $S_1 = \phi_1(U_1)$, where $U_1 = \mathbf{R}^k$ for $1 \leq k \leq n$ and

$$\phi_1(u_1, \ldots, u_k) = (u_1, 0, \ldots, u_k, 0, 0, \ldots, 0).$$

In other words,

$$S_1 = \{(x_1, y_1, \ldots, x_n, y_n) \mid y_i = 0 \text{ for all } i \text{ and } x_j = 0 \text{ for all } j > k\}.$$

Then S_1 is isotropic.

Example 7.5.4. Let $(\mathbf{R}^{2n}, \omega_0)$ be the standard symplectic space with coordinate pairs (x_i, y_i). Let $S_2 = \phi_2(U_2)$, where $U_2 = \mathbf{R}^k$ for $k = n + j$ with $1 \leq j \leq n$ and

$$\phi_2(u_1, \ldots, u_n, u_{n+1}, \ldots, u_{n+j}) = (u_1, u_{n+1}, \ldots, u_j, u_{n+j}, u_{j+1}, 0, \ldots, u_n, 0).$$

In other words,

$$S_2 = \{(x_1, y_1, \ldots, x_n, y_n) \mid y_i = 0 \text{ when } i > j\}$$

and

$$T_pS_2 = \left\{ \langle a_1, b_1, \ldots, a_n, b_n \rangle_p \mid b_i = 0 \text{ for } i > j \right\}.$$

Then S_2 is coisotropic. Indeed, for $\mathbf{w}_p \in (T_pS_2)^\omega$, the condition that

$$\omega_0(\mathbf{w}_p, \mathbf{v}_p) = 0 \text{ for all } \mathbf{v}_p \in T_pS_2$$

means that for

$$\mathbf{w}_p = \langle a_1, b_1, \ldots, a_n, b_n \rangle_p \quad \text{and} \quad \mathbf{v}_p = \langle s_1, t_1, \ldots, s_j, t_j, s_{j+1}, 0, \ldots, s_n, 0 \rangle_p,$$

we have

$$0 = \sum_{i=1}^{j} (a_i t_i - b_i s_i) + \sum_{i=j+1}^{n} (-b_i s_i)$$

This in turn shows, by choosing $\mathbf{v}_p$ to be the $n + j$ standard basis vectors for T_pS, that $b_i = 0$ for all $i = 1, \dots, n$ and $a_i = 0$ for $i = 1, \dots, j$. In particular, $b_i = 0$ for $i = j + 1, \dots, n$, and so $\mathbf{w}_p \in T_pS_2$. Hence $(T_pS_2)^\omega \subset T_pS_2$.

Example 7.5.5. For the standard symplectic space with coordinate pairs (x_i, y_i), let $S_3 = \phi_3(U_3)$, where $U_3 = \mathbf{R}^n$ and

$$\phi_3(u_1, \dots, u_n) = (u_1, 0, \dots, u_n, 0).$$

In other words,

$$S_3 = \{(x_1, y_1, \dots, x_n, y_n) \mid y_i = 0 \text{ for } i = 1, \dots, n\}.$$

Then S_3 is Lagrangian. This follows from Example 7.5.3 and Proposition 7.5.2, since $\dim T_pS_3 = n$.

Example 7.5.6. Still considering the standard symplectic space $(\mathbf{R}^{2n}, \omega_0)$, now define $S_4 = \phi_4(U_4)$, where $U_4 = \mathbf{R}^{2k}$, $1 \leq k \leq n$, and $\phi_4 : U_4 \to \mathbf{R}^{2n}$ is given by

$$\phi_4(u_1, \dots, u_{2k}) = (u_1, \dots, u_{2k}, 0, \dots, 0).$$

Then S_4 is symplectic.

We have listed the above linear examples in detail to give a sense of the most basic special (in terms of ω-orthogonality) geometric sets. More interesting nonlinear examples, however, are also easy to construct.

Example 7.5.7. Every one-dimensional parametric set (i.e., a parameterized curve) is an isotropic set. Indeed, let $I \subset \mathbf{R}$ be an interval and $c : I \to (\mathbf{R}^{2n}, \omega)$ a smooth regular map. Then $c^*\omega$ is a two-form on the one-dimensional set I, and so $c^*\omega = 0$ for purely dimensional reasons.

Example 7.5.8. Let $S_F \subset (\mathbf{R}^{2n}, \omega)$ be a hypersurface defined as the zero set of a smooth function $F : \mathbf{R}^{2n} \to \mathbf{R}$, i.e.,

$$S_F = \left\{x \in \mathbf{R}^{2n} \mid F(x) = 0\right\},$$

where 0 is a regular value of F. We have $T_pS_F = \ker dF(p)$. Then S_F is coisotropic. Supposing that $\mathbf{w}_p \in (T_pS_F)^\omega$, we have $\omega(\mathbf{w}_p, \mathbf{v}_p) = 0$ for all $\mathbf{v}_p \in T_pS_F$. Let $\mathbf{u}_p \in T_p\mathbf{R}^{2n}$ be the tangent vector such that $i(\mathbf{u}_p)\omega = dF(p)$, which exists by virtue of Proposition 7.2.4. In fact, $\mathbf{u}_p \in T_pS_F$, since

$$dF(p)(\mathbf{u}_p) = (i(\mathbf{u}_p)\omega)(\mathbf{u}_p) = \omega(\mathbf{u}_p, \mathbf{u}_p) = 0.$$

Then since $\omega(\mathbf{w}_p, \mathbf{u}_p) = 0$, we have $\mathbf{w}_p \in T_pS_F$:

$$dF(p)(\mathbf{w}_p) = (i(\mathbf{u}_p)\omega)(\mathbf{w}_p) = \omega(\mathbf{u}_p, \mathbf{w}_p) = 0.$$

This shows that $(T_pS_F)^\omega \subset T_pS_F$.

Example 7.5.9. Consider the cotangent bundle $T^*\mathbf{R}^n$, introduced in Sect. 7.1, which we consider to be $\mathbf{R}^{2n}$ with coordinates $(q_1, \dots, q_n, p_1, \dots, p_n)$ describing the one-form

$$p_1 dx_1 + \cdots + p_n dx_n \in T^*_{(q_1,\dots,q_n)}\mathbf{R}^n.$$

Recall that there is a distinguished one-form, called the Liouville form,

$$\alpha_0 = \sum p_i dq_i$$

along with a corresponding symplectic two-form $\omega_0 = d\alpha_0$.

Every one-form $\beta = \beta_1 dx_1 + \cdots + \beta_n dx_n$ on $\mathbf{R}^n$ defines a map

$$\tilde{\beta} : \mathbf{R}^n \to T^*\mathbf{R}^n$$

by

$$\tilde{\beta}(x_1, \dots, x_n) = (x_1, \dots, x_n, \beta_1(x), \dots, \beta_n(x)).$$

Note that $\tilde{\beta}^*\alpha_0 = \beta$.

Now suppose that β is a *closed* one-form. Then $S = \tilde{\beta}(\mathbf{R}^n)$ is a Lagrangian subset of $T^*\mathbf{R}^n$ with symplectic form ω_0. The geometric set S has dimension n, since $\tilde{\beta}_*$ has rank n everywhere. Further,

$$\begin{aligned}
\tilde{\beta}^*\omega_0 &= \tilde{\beta}^*(d\alpha_0) \\
&= d(\tilde{\beta}^*\alpha_0) \\
&= d\beta \\
&= 0 \quad \text{since } \beta \text{ is closed.}
\end{aligned}$$

In the special case that β is *exact*, there is a smooth function $f : \mathbf{R}^n \to \mathbf{R}$ such that $\beta = df$. In this case, f is called the *generating function* of the Lagrangian set $S = \widetilde{df}(\mathbf{R}^n)$.

One way to generate examples of special geometric sets in a symplectic space is to look at subsets of the product of two symplectic spaces (U_1, ω_1) and (U_2, ω_2), where $U_1, U_2 \subset \mathbf{R}^{2n}$ are domains. Consider the symplectic space formed as the product $U_1 \times U_2$ with symplectic form $\Omega = (-\omega_1) \oplus (\omega_2)$, using the notation following Example 7.2.10. Note that the verifications in that example still hold, although we have changed the sign of the first component.

Example 7.5.10. Let $\phi : (U_1, \omega_1) \to (U_2, \omega_2)$ be a symplectic diffeomorphism, i.e., $\phi^*\omega_2 = \omega_1$. Let $\Omega = (-\omega_1) \oplus (\omega_2)$ be the symplectic form on the product space $U_1 \times U_2$ of Example 7.2.10. Define the geometric set $S = \Phi(U_1)$, where $\Phi : U_1 \to U_1 \times U_2$ is defined by

$$\Phi(x) = (x, \phi(x)).$$

Then S is a Lagrangian set in $(U_1 \times U_2, \Omega)$. The map Φ has maximal rank $2n$, so $\dim T_pS = 2n$ for all $p \in S$. Further,

$$\Phi^*((-\omega_1) \oplus (\omega_2)) = \phi^*\omega_2 - \omega_1 = \omega_1 - \omega_1 = 0.$$

We mention here one fact, which is a kind of converse to the result of the previous example. It is in fact the tip of the iceberg of deeper theorems relating the study of symplectic diffeomorphisms to the study of Lagrangian sets.

Proposition 7.5.11. *Let (U_1, ω_1) and (U_2, ω_2) be symplectic spaces, and let $\phi : U_1 \to U_2$ be a diffeomorphism. Then ϕ is a symplectic diffeomorphism if and only if the set*

$$\Gamma_\phi = \{(x, \phi(x)) \mid x \in U_1\} \subset (U_1 \times U_2, \Omega)$$

is a Lagrangian set, where $\Omega = (-\omega_1) \oplus (\omega_2)$.

Proof. We have shown in Example 7.5.10 that if ϕ is a symplectic diffeomorphism, then Γ_ϕ is a Lagrangian set.

Suppose now that ϕ is a diffeomorphism such that Γ_ϕ is a Lagrangian set. Using the notation of the previous example along with Proposition 7.5.2, this means that $\Phi^*((-\omega_1) \oplus (\omega_2)) = 0$. We have, for any tangent vector $\mathbf{v}_p \in T_pU_1$, that $\Phi_*\mathbf{v}_p = (\mathbf{v}_p, \phi_*\mathbf{v}_p)$, and so for any two tangent vectors $\mathbf{v}_p$, $\mathbf{w}_p \in T_pU_1$,

$$\begin{aligned}
0 &= (\Phi^*\Omega)(\mathbf{v}_p, \mathbf{w}_p) \\
&= \Omega(\Phi_*\mathbf{v}_p, \Phi_*\mathbf{w}_p) \\
&= \Omega((\mathbf{v}_p, \phi_*\mathbf{v}_p), (\mathbf{w}_p, \phi_*\mathbf{w}_p)) \\
&= \omega_2(\phi_*\mathbf{v}_p, \phi_*\mathbf{w}_p) - \omega_1(\mathbf{v}_p, \mathbf{w}_p) \\
&= (\phi^*\omega_2)(\mathbf{v}_p, \mathbf{w}_p) - \omega_1(\mathbf{v}_p, \mathbf{w}_p).
\end{aligned}$$

Hence

$$\phi^*\omega_2 = \omega_1,$$

and so ϕ is a symplectic diffeomorphism. □

We next see that the condition for a geometric set to be coisotropic has a natural formulation in terms of the Poisson bracket. First, though, we note that it is possible to determine whether a Hamiltonian vector field with Hamiltonian function f is ω-orthogonal to a geometric set by considering the action of tangent vectors on f.

Proposition 7.5.12. *Let S be a geometric set in a symplectic space $(\mathbf{R}^{2n}, \omega)$. Let $f : \mathbf{R}^{2n} \to \mathbf{R}$ be a smooth function with Hamiltonian vector field X_f. Then $X_f \in (TS)^\omega$ if and only if $V[f] = 0$ for all $V \in TS$.*

Proof. Since for any $V \in TS$, $df(V) = (-i(X_f)\omega)(V) = -\omega(X_f, V)$, the condition that $df(V) = 0$ for all $V \in TS$ is equivalent to $X_f \in (TS)^\omega$. □

The preceding proposition allows the following alternative characterization of coisotropic geometric sets.

Proposition 7.5.13. *Let S be a geometric set in a symplectic space $(\mathbf{R}^{2n}, \omega)$. Then S is coisotropic if and only if for every point $p \in S$, there is a domain $U \subset \mathbf{R}^{2n}$ containing p such that if f and g are any two smooth functions on U that are constant on $S \cap U$, then $\{f, g\} = 0$ on U.*

Proof. We may suppose without loss of generality that S is a parameterized set. By Corollary 3.6.14, there exist a domain W containing p and k smooth functions $h_1, \ldots, h_k$ defined on W such that

$$S \cap W = \{x \in W \mid h_1(x) = \cdots = h_k(x) = 0\}.$$

They are functionally independent as a consequence of being the component functions of a diffeomorphism.

Suppose first that S is coisotropic, so that $(TS)^\omega \subset TS$, and let f and g be smooth functions defined on U such that for all $x \in S \cap U$, $f(x) = c_1$ and $g(x) = c_2$ for some constants c_1 and c_2. Note that the proof of Theorem 3.4.13 shows that $df(\mathbf{v}_p) = dg(\mathbf{v}_p) = 0$ for all $\mathbf{v}_p \in T_pS$. In particular, since $(T_pS)^\omega \subset T_pS$, if $\mathbf{v}_p \in (T_pS)^\omega$, then $df(\mathbf{v}_p) = dg(\mathbf{v}_p) = 0$. By Proposition 7.5.12, the Hamiltonian vector field X_g satisfies $X_g(p) \in (T_pS)^\omega$, and so

$$\{f, g\} = \omega(X_f, X_g) = -df(X_g) = 0.$$

Conversely, suppose that for all smooth functions f, g such that f and g are constant on S, $\{f, g\} = 0$. In particular, for the k functions $h_1, \ldots, h_k$, $\{h_i, h_j\} = 0$.

Note that a tangent vector $\mathbf{v}_p$ is in T_pS if and only if $dh_i(\mathbf{v}_p) = 0$ for all $i = 1, \ldots, k$. This implies that for every $\mathbf{v}_p \in T_pS$,

$$\omega(X_{h_i}(p), \mathbf{v}_p) = -dh_i(\mathbf{v}_p) = 0,$$

and so $X_{h_i}(p) \in (T_pS)^\omega$ for all i. In fact, since the h_i are functionally independent, the tangent vectors $X_{h_1}(p), \ldots, X_{h_k}(p)$ are linearly independent and so form a basis for $(T_pS)^\omega$, since $\dim(T_pS)^\omega = k$ by Theorem 2.10.12. But since $dh_i(X_{h_j}) = -\{h_i, h_j\} = 0$ for all $i = 1, \ldots, k$, this shows that $X_{h_j}(p) \in T_pS$ for all $j = 1, \ldots, k$. Hence $(T_pS)^\omega \subset T_pS$, and so S is coisotropic. □

Example 7.5.14. Consider the standard symplectic space $(\mathbf{R}^4, \omega_0)$ with coordinates (x_1, y_1, x_2, y_2), so that $\omega_0 = dx_1 \wedge dy_1 + dx_2 \wedge dy_2$. Let $S \subset \mathbf{R}^4$ be the geometric set described as the level set

$$S = \left\{(x_1, y_1, x_2, y_2) \mid x_1^2 + y_1^2 + x_2^2 + y_2^2 = 1,\ x_1x_2 + y_1y_2 = 1\right\};$$

The set S can be visualized as the surface of intersection of the three-dimensional slanted cylinder $u_1^2 + u_2^2 = 3$, where $u_1 = x_1 + x_2$ and $u_2 = y_1 + y_2$, with the three-dimensional unit sphere in $\mathbf{R}^4$.

The set S satisfies the hypotheses of Proposition 7.5.13, where

$$f_1(x_1, y_1, x_2, y_2) = x_1^2 + y_1^2 + x_2^2 + y_2^2 - 1,\ f_2(x_1, y_1, x_2, y_2) = x_1x_2 + y_1y_2 - 1.$$

Indeed, using the coordinate expression for the Poisson bracket of Example 7.4.13,

$$\{f_1, f_2\} = (2x_1)(y_2) - (2y_1)(x_2) + (2x_2)(y_1) - (2y_2)(x_1) = 0.$$

Hence S is coisotropic and, since $\dim T_pS = 2$, Lagrangian.

We invite the reader to verify directly that S is coisotropic without relying on Proposition 7.5.13.

7.6 Hypersurfaces of Contact Type

In this section, we explore some of the more direct relationships between symplectic and contact geometry. We begin by returning to the symplectization of a contact space first introduced in Example 7.2.11. Recall that the symplectization of a contact space $(\mathbf{R}^{2n-1}, \alpha)$ is the symplectic space $(\mathbf{R}^{2n}, \omega)$, where

$$\omega = d(e^t\alpha) = e^t(dt \wedge \alpha + d\alpha).$$

Here the t variable is the coordinate of the second component of the product $\mathbf{R}^{2n-1} \times \mathbf{R} = \mathbf{R}^{2n}$. (Note that we have changed our convention on the dimension of the contact space to be $2n - 1$ as opposed to $2n + 1$ in Chap. 7. This has the explicit advantage that the dimension of the symplectization is $2n$; implicitly, the main object of interest in this section is the symplectic space.)

Proposition 7.6.1. *Let $S = \phi(U) \subset \mathbf{R}^{2n-1}$ be a parameterized set in the contact space $(\mathbf{R}^{2n-1}, \alpha)$, where $U \subset \mathbf{R}^{n-1}$ is a domain and $\phi : U \to \mathbf{R}^{2n-1}$ is a regular parameterization. Then S is a Legendre set in $(\mathbf{R}^{2n-1}, \alpha)$ if and only if $S \times \mathbf{R}$ is a Lagrangian set in the symplectization $(\mathbf{R}^{2n}, \omega)$, where $S \times \mathbf{R}$ is parameterized as $S \times \mathbf{R} = \tilde{\phi}(U \times \mathbf{R})$ for $\tilde{\phi} : U \times \mathbf{R} \to \mathbf{R}^{2n-1} \times \mathbf{R}$ given by $\tilde{\phi}(x, t) = (\phi(x), t)$.*

Proof. Suppose that S is a Legendre set in $(\mathbf{R}^{2n-1}, \alpha)$, so that $\phi^*\alpha = 0$. We have

$$\begin{aligned}\tilde{\phi}^*\omega &= \tilde{\phi}^*(d(e^t\alpha)) \\ &= d(\tilde{\phi}^*(e^t\alpha)) \\ &= d(e^t(\phi^*\alpha)) \\ &= 0.\end{aligned}$$

Hence, since $\dim T_p(S \times \mathbf{R}) = n$ for all $p \in S \times \mathbf{R}$, we have that $S \times \mathbf{R}$ is a Lagrangian set in $(\mathbf{R}^{2n}, \omega)$, as desired.

Now suppose $S \times \mathbf{R}$ is a Lagrangian set in $(\mathbf{R}^{2n}, \omega)$, and so $\tilde{\phi}^*\omega = 0$. In particular,

$$\begin{aligned}0 &= i\left(\frac{\partial}{\partial t}\right)(\tilde{\phi}^*\omega) \\ &= i\left(\frac{\partial}{\partial t}\right)(d(e^t(\phi^*\alpha))) \\ &= i\left(\frac{\partial}{\partial t}\right)(e^t(dt \wedge \phi^*\alpha + \phi^* d\alpha)) \\ &= e^t\phi^*\alpha \quad \text{since } \phi^* d\alpha \text{ does not involve } dt,\end{aligned}$$

and so $\phi^*\alpha = 0$. □

Proposition 7.6.2. *Let $(\mathbf{R}^{2n-1}, \alpha)$ be a contact space with symplectization $(\mathbf{R}^{2n}, \omega)$. For any diffeomorphism $\phi : \mathbf{R}^{2n-1} \to \mathbf{R}^{2n-1}$, ϕ is a contact diffeomorphism of $(\mathbf{R}^{2n-1}, \alpha)$ with $\phi^*\alpha = f\alpha$ for a positive function $f : \mathbf{R}^{2n-1} \to \mathbf{R}$ if and only if the diffeomorphism $\tilde{\phi} : \mathbf{R}^{2n} \to \mathbf{R}^{2n}$ defined by*

$$\tilde{\phi}(x,t) = (\phi(x), t - \ln f(x))$$

is a symplectic diffeomorphism of $(\mathbf{R}^{2n}, \omega)$.

Proof. We leave it as an exercise for the reader to show that the map $\tilde{\phi}$ defined in the proposition is in fact a diffeomorphism.

If $\phi^*\alpha = f\alpha$ for some positive function f, then we have

$$\begin{aligned}\tilde{\phi}^*\omega &= \tilde{\phi}^*(e^t(dt \wedge \alpha + d\alpha)) \\ &= e^{t-\ln f}\left(d(t - \ln f) \wedge \phi^*\alpha + d(\phi^*\alpha)\right) \\ &= e^t \cdot e^{-\ln f}\left(dt \wedge (f\alpha) - d(\ln f) \wedge (f\alpha) + d(f\alpha)\right) \\ &= e^t(1/f)\Big(f dt \wedge \alpha - (1/f) df \wedge (f\alpha) + df \wedge \alpha + f d\alpha\Big)\end{aligned}$$

$$
\begin{aligned}
&= e^t(1/f)\Big(f dt \wedge \alpha - df \wedge \alpha + df \wedge \alpha + f d\alpha\Big) \\
&= e^t \left(dt \wedge \alpha + d\alpha\right) \\
&= \omega.
\end{aligned}
$$

Here we have used the fact that $d(\ln f) = (1/f)df$.

On the other hand, assuming that $\tilde{\phi}^*\omega = \omega$, we have

$$
\begin{aligned}
e^t\alpha &= i\left(\frac{\partial}{\partial t}\right)\omega \\
&= i\left(\frac{\partial}{\partial t}\right)(\tilde{\phi}^*\omega) \\
&= i\left(\frac{\partial}{\partial t}\right)\left[e^{t-\ln f}(dt \wedge (\phi^*\alpha) - d(\ln f) \wedge (\phi^*\alpha) + d(\phi^*\alpha))\right] \\
&= e^{t-\ln f}\phi^*\alpha.
\end{aligned}
$$

This implies that $\phi^*\alpha = f\alpha$. □

The following proposition illustrates the relationship between the contact gradient and the Hamiltonian vector field of a function f.

Proposition 7.6.3. *Let $(\mathbf{R}^{2n-1}, \alpha)$ be a contact space with symplectization $(\mathbf{R}^{2n}, \omega)$. For any smooth function $f : \mathbf{R}^{2n-1} \to \mathbf{R}$, define the function $F : \mathbf{R}^{2n-1} \times \mathbf{R} \to \mathbf{R}$ by $F(x,t) = e^t f(x)$. Then the Hamiltonian vector field $\tilde{X}_F$ of F in $(\mathbf{R}^{2n}, \omega)$ is given by*

$$
\tilde{X}_F = X_f + g\frac{\partial}{\partial t},
$$

where X_f is the contact gradient of f in $(\mathbf{R}^{2n-1}, \alpha)$ (with no $\frac{\partial}{\partial t}$ component) and $g = -df(\xi)$, where ξ is the Reeb field corresponding to α.

Proof. Let $\tilde{X}_F$ be the Hamiltonian vector field for the function F. We write $\tilde{X}_F = X + g\frac{\partial}{\partial t}$, where X is a vector field on $\mathbf{R}^{2n-1}$ (although from the outset, the component functions might involve all $(2n)$ variables, including t) and g is some function of all variables (x,t).

Note first that

$$
dF = e^t(f dt + df),
$$

so the condition $i(\tilde{X}_F)\omega = -dF$ can be expressed as

$$
\begin{aligned}
-e^t(f\,dt + df) &= i\left(X + g\frac{\partial}{\partial t}\right)\omega \\
&= i\left(X + g\frac{\partial}{\partial t}\right)\left(e^t(dt\wedge\alpha + d\alpha)\right) \\
&= e^t\Big(g\alpha - \alpha(X)dt + i(X)d\alpha\Big),
\end{aligned}
$$

i.e., $-e^t f = e^t(-\alpha(X))$ and $-e^t df = e^t(g\alpha + i(X)d\alpha)$. This shows that $i(X)\alpha = f$ and that $i(X)d\alpha = -g\alpha - df$. Contracting the second equation with the Reeb field on both sides shows that

$$0 = -g - df(\xi),$$

and so $g = -df(\xi)$, as desired.

In that case, then,

$$i(X)d\alpha = df(\xi)\alpha - df,$$

which together with $i(X)\alpha = f$ shows that X is the contact gradient of f. □

Having described some of the ways that objects in contact geometry correspond to objects in the symplectic setting, our goal now is to show that the particular way that we have "symplectized" a contact space $(\mathbf{R}^{2n-1}, \alpha)$ inside a symplectic space $(\mathbf{R}^{2n}, \omega)$ of higher dimension has a natural generalization. This, in turn, gives a glimpse into the broader setting where contact geometry takes place: contact *manifolds*.

Recall that a *hypersurface* in a symplectic space $(\mathbf{R}^{2n}, \omega)$ is a geometric set of the form $S = f^{-1}(a)$, where $f : \mathbf{R}^{2n} \to \mathbf{R}$ is a smooth function with regular value $a \in \mathbf{R}$, so that if $f(x) = a$, then $df(x) \neq 0$.

Definition 7.6.4. A hypersurface S in a symplectic space $(\mathbf{R}^{2n}, \omega)$ is said to be *of contact type* if there is a smooth one-form α defined on a domain $U \subset \mathbf{R}^{2n}$, where $S \subset U$, such that

- α is nondegenerate on S; and
- $d\alpha\big|_S = \omega\big|_S$.

This definition has some important differences from the way the term is generally defined in the more general setting of a symplectic manifold. First, there is generally the topological assumption of the compactness of S, which implies, for example, that S has finite $(2n-1)$-volume and also ensures that vector fields on S are complete. Second, in the setting of a manifold, the one-form α is defined initially only on S and not necessarily on a domain $U \subset \mathbf{R}^{2n}$.

Our model for a hypersurface of contact type will be the hyperplane $t = 0$ in the symplectization of a contact space.

Example 7.6.5. Let $(\mathbf{R}^{2n-1}, \alpha)$ be a contact space and let $(\mathbf{R}^{2n}, \omega)$ be the corresponding symplectization, with (x,t)-coordinates on $\mathbf{R}^{2n}$ and symplectic form

$\omega = d(e^t\alpha)$. Then the set $S = \mathbf{R}^{2n-1} \times \{0\}$ is a hypersurface of contact type, with the contact form α playing the role of the distinguished one-form in the definition. In this case, the function f is given by $f(x,t) = t$, and so $S = f^{-1}(0)$.

The most important nontrivial example of a hypersurface of contact type is the sphere in $(\mathbf{R}^{2n}, \omega_0)$.

Example 7.6.6. Let $(\mathbf{R}^{2n}, \omega_0)$ be the standard symplectic space with coordinate pairs (x_i, y_i), so that $\omega_0 = \sum dx_i \wedge dy_i$. Let $f : \mathbf{R}^{2n} \to \mathbf{R}$ be defined as

$$f(x_1, y_1, \ldots, x_n, y_n) = -1 + \sum (x_i^2 + y_i^2),$$

and define $S = f^{-1}(0)$, the unit sphere in $\mathbf{R}^{2n}$.

Note that by Theorem 3.4.13, $X \in TS$ if and only if $df(X) = 0$. Writing X using coordinate pairs $\langle X^i, Y^i \rangle$, this means that $X \in TS$ if and only if

$$\sum (x_i X^i + y_i Y^i) = 0.$$

Let

$$\alpha = \frac{1}{2} \sum (x_i dy_i - y_i dx_i),$$

so that $d\alpha = \omega_0$. For all $p \in S$, $\alpha(p)$ is nondegenerate on T_pS. To see this, it is enough to show that $\alpha \wedge (d\alpha)^{n-1}$ is nonzero on some (and hence every) basis of T_pS. We have

$$(d\alpha)^{n-1} = (n-1)! \sum_{i=1}^{n} dx_1 \wedge dy_1 \wedge \cdots \wedge \widehat{dx_i} \wedge \widehat{dy_i} \wedge \cdots dx_n \wedge dy_n,$$

and so

$$\alpha \wedge (d\alpha)^{n-1} = \frac{1}{2}(n-1)! \sum_{i=1}^{n} \Big(x_i dx_1 \wedge dy_1 \wedge \cdots \wedge \widehat{dx_i} \wedge dy_i \wedge \cdots dx_n \wedge dy_n$$
$$- y_i dx_1 \wedge dy_1 \wedge \cdots \wedge dx_i \wedge \widehat{dy_i} \wedge \cdots dx_n \wedge dy_n \Big),$$

where the "hat" notation $\widehat{dx_i}$ indicates that the factor dx_i is omitted.

To complete the verification that $\alpha(p)$ is nondegenerate on T_pS, we will specify a basis of T_pS. Since $p = (p_1, q_1, \ldots, p_n, q_n) \in S$, at least one of the p_i or q_i must be nonzero. Assume for the sake of discussion that $q_n \neq 0$. Then for every $X(p) \in T_pS$ we have

$$Y^n(p) = -\frac{p_1}{q_n} X^1(p) - \frac{q_1}{q_n} Y^1(p) - \cdots - \frac{p_n}{q_n} X^n(p),$$

and so we can use as a basis for T_pS the $(2n-1)$ tangent vectors

$$\left\{\mathbf{e}_1 = \left\langle 1,0,0,\ldots,0,-\frac{p_1}{q_n}\right\rangle_p, \mathbf{f}_1 = \left\langle 0,1,0,\ldots,0,-\frac{q_1}{q_n}\right\rangle_p, \ldots, \right.$$
$$\left. \mathbf{e}_n = \left\langle 0,\ldots,0,1,-\frac{p_n}{q_n}\right\rangle_p \right\}.$$

Then

$$\begin{aligned}\frac{2}{(n-1)!}&\left(\alpha \wedge (d\alpha)^{n-1}\right)(p)(\mathbf{e}_1,\mathbf{f}_1,\ldots,\mathbf{e}_{n-1},\mathbf{f}_{n-1},\mathbf{e}_n)\\ &= p_1\left(-\frac{p_1}{q_n}\right) - q_1\left(\frac{q_1}{q_n}\right) + \\ &\quad \ddots \\ &\quad + p_n\left(-\frac{p_n}{q_n}\right) - q_n(1) \\ &= \frac{-p_1^2 - q_1^2 - \cdots - p_n^2 - q_n^2}{q_n} = \frac{-1}{q_n} \neq 0,\end{aligned}$$

and so $\alpha(p)$ is nondegenerate on T_pS.

These calculations show that S is a hypersurface of contact type in $(\mathbf{R}^{2n}, \omega_0)$.

We performed our calculations in the previous example under the assumption that $y_n \neq 0$, an assumption that really saved us performing the same calculation $2n$ times (once for $x_1 \neq 0$, once for $y_1 \neq 0$, etc.). This is in fact an admission that it is not possible to describe the sphere completely using a single coordinate system, and so the sphere is topologically different from a domain in $\mathbf{R}^{2n}$.

In order to give an alternative description of a hypersurface of contact type, we return to the example of the symplectization of a contact space $(\mathbf{R}^{2n-1}, \alpha)$. The vector field $X_t = \dfrac{\partial}{\partial t}$ in the symplectization $(\mathbf{R}^{2n}, \omega)$, with coordinates (x,t) for $\mathbf{R}^{2n} = \mathbf{R}^{2n-1} \times \mathbf{R}$, has several notable properties. First, at each point on the hypersurface $S = \mathbf{R}^{2n-1} \times \{0\}$, X_t is transverse to S. Second, we have seen that $i(X_t)\omega = e^t\alpha$. This in turn implies both that $i(X_t)\omega\big|_S = \alpha$ and that

$$\mathcal{L}_{X_t}\omega = di(X_t)\omega = d(e^t\alpha) = \omega.$$

This prompts the following definition.

Definition 7.6.7. Let (U,ω) be a symplectic space. A vector field X on U is called a *Liouville vector field* if

$$\mathcal{L}_X\omega = \omega.$$

The relevance of Liouville vector fields to the present discussion is given in the following result.

Proposition 7.6.8. *Let* $(\mathbf{R}^{2n}, \omega)$ *be a symplectic space and let* S *be a hypersurface described as the level set of a smooth function* $f : \mathbf{R}^{2n} \to \mathbf{R}$ *with regular value* $a \in \mathbf{R}$*, i.e.,* $S = \left\{x \in \mathbf{R}^{2n} \mid f(x) = a\right\}$*, and* $df(x) \neq 0$ *for all* $x \in S$*. Then* S *is a hypersurface of contact type if and only if there is a Liouville vector field defined on a domain* $U \subset \mathbf{R}^{2n}$ *containing* S *that is transverse to* S *at all points* $p \in S$*.*

Proof. Suppose first that there is a one-form α defined on a domain $U \subset \mathbf{R}^{2n}$ containing S such that $d\alpha|_S = \omega|_S$ and such that $\alpha|_S$ is nondegenerate. Since ω is nondegenerate, Proposition 7.2.4 guarantees that there is a unique vector field X on U such that $i(X)\omega = \alpha$. This vector field by definition has the property that

$$\mathcal{L}_X\omega = di(X)\omega = d\alpha = \omega \quad \text{on } S,$$

so X is a Liouville vector field defined on U.

To show that X is transverse to S, suppose to the contrary that there is $p \in S$ such that $X(p) \in T_pS$. Since $\alpha|_S$ is nondegenerate, there is a tangent vector $Y \in T_pS$ such that $\alpha(Y) = 1$ and $i(Y)d\alpha = 0$; this follows by mimicking the proof of the existence of the Reeb field; see Theorem 6.3.10. Let $E_p(S) = \ker \alpha|_S$. We have $X(p) \in E_p(S)$, since

$$\alpha(p)(X(p)) = \omega(p)(X(p), X(p)) = 0.$$

Complete a basis $\{\mathbf{e}_1 = X(p), \mathbf{e}_2, \dots, \mathbf{e}_{2n-2}\}$ of $E_p(S)$ in such a way that

$$\{Y, \mathbf{e}_1, \dots, \mathbf{e}_{2n-2}\}$$

is a basis for T_pS. Since $\alpha(p)$ is nondegenerate on S by supposition,

$$\begin{aligned} 0 &\neq \left(\alpha \wedge (d\alpha)^{n-1}\right)(p)(Y(p), \mathbf{e}_1, \dots, \mathbf{e}_{2n-2}) \\ &= \left(\alpha \wedge \omega^{n-1}\right)(p)(Y(p), \mathbf{e}_1, \dots, \mathbf{e}_{2n-2}). \end{aligned}$$

But every term of the latter exterior product has a factor either of the form

$$\begin{aligned} \alpha(Y)\omega(\mathbf{e}_1, \mathbf{e}_j) &= \alpha(\mathbf{e}_j) \\ &= 0, \end{aligned}$$

since $\mathbf{e}_j \in E_p(S)$, or of the form

$$\alpha(\mathbf{e}_i)\omega(Y, \mathbf{e}_j) = 0, \quad \text{since } i(Y)d\alpha = 0.$$

This contradiction shows that $X(p) \notin T_pS$ for all $p \in S$, and $X(p)$ is transverse to S as desired.

Now suppose that there is a Liouville vector field X defined on a domain $U \subset \mathbf{R}^{2n}$ containing S that is transverse to S for all points $p \in S$. Let $\alpha = i(X)\omega$. The condition $\mathcal{L}_X\omega = \omega$ implies that

$$d\alpha = d(i(X)\omega) = \mathcal{L}_X\omega = \omega.$$

To show that α is nondegenerate on S, let Y be the vector field such that $i(Y)\omega = df$, again relying on Proposition 7.2.4. Since $df(Y) = \omega(Y,Y) = 0$, we have $Y(p) \in T_pS$ for all $p \in S$. Note that on S,

$$\alpha(Y) = \omega(X,Y) = -\omega(Y,X) = -df(X) \neq 0,$$

since X is transverse to S. Now, in the manner of Theorem 2.10.4, construct a basis for $T_p\mathbf{R}^{2n}$,

$$\{\mathbf{e}_1 = Y(p), \mathbf{f}_1 = X(p), \dots, \mathbf{e}_n, \mathbf{f}_n\},$$

with the properties that $\omega(\mathbf{e}_i, \mathbf{f}_i) \neq 0$ but

$$\omega(\mathbf{e}_i, \mathbf{e}_j) = \omega(\mathbf{f}_i, \mathbf{f}_j) = \omega(\mathbf{e}_i, \mathbf{f}_j) = 0 \quad \text{when } i \neq j.$$

In particular, for $j = 2, \dots, n$, we have

$$0 = \omega(\mathbf{e}_1, \mathbf{e}_j) = \omega(Y, \mathbf{e}_j) = df(\mathbf{e}_j),$$

and so $\mathbf{e}_j \in T_pS$; likewise, $\mathbf{f}_j \in T_pS$. This means that since $\dim T_pS = 2n-1$,

$$\{\mathbf{e}_1, \mathbf{e}_2, \mathbf{f}_2, \dots, \mathbf{e}_n, \mathbf{f}_n\}$$

is a basis for T_pS. Calculating on S, we obtain

$$\begin{aligned}\left(\alpha \wedge (d\alpha)^{n-1}\right)(\mathbf{e}_1, \mathbf{e}_2, \mathbf{f}_2, \dots, \mathbf{e}_n, \mathbf{f}_n) &= \left(\alpha \wedge \omega^{n-1}\right)(\mathbf{e}_1, \mathbf{e}_2, \mathbf{f}_2, \dots, \mathbf{e}_n, \mathbf{f}_n) \\ &= \alpha(\mathbf{e}_1)\omega(\mathbf{e}_2, \mathbf{f}_2)\cdots\omega(\mathbf{e}_n, \mathbf{f}_n) \\ &\neq 0.\end{aligned}$$

Hence α is nondegenerate on S. □

We again remind the reader that Definition 7.6.4 is a simplification of the standard definition used in the context of symplectic manifolds, where α is defined only on S. In that case, the statement corresponding to Proposition 7.6.8 requires a slightly more complicated proof. Methods similar to those presented here give a Liouville vector field *on* S, but that vector field must be extended to a vector field defined *in a domain containing* S. We refer the more advanced reader to [31, pp. 113–114].

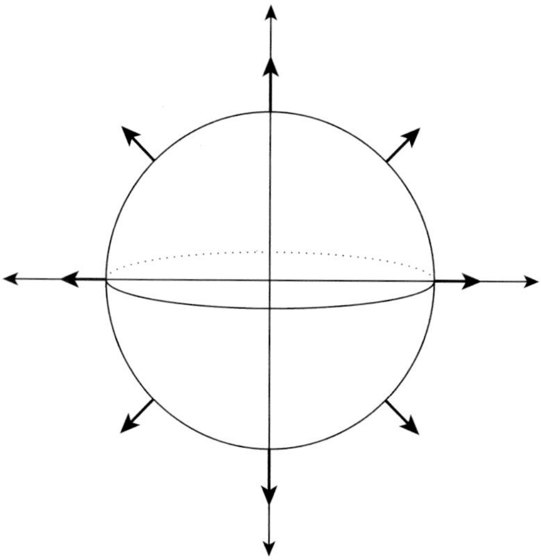

Fig. 7.3 A Liouville vector field on the sphere.

Example 7.6.9. We showed in Example 7.6.6 that the unit sphere in $(\mathbf{R}^{2n}, \omega_0)$ is a hypersurface of contact type. A Liouville vector field is given by

$$X = \frac{1}{2} \sum \left(x_i \frac{\partial}{\partial x_i} + y_i \frac{\partial}{\partial y_i} \right).$$

Indeed, note that

$$i(X)\omega_0 = \frac{1}{2} \sum (x_i dy_i - y_i dx_i),$$

and so

$$\mathcal{L}_X \omega_0 = di(X)\omega_0 = \sum dx_i \wedge dy_i = \omega_0.$$

Further, for the function $f : \mathbf{R}^{2n} \to \mathbf{R}$ given by

$$f(x_1, y_1, \ldots, x_n, y_n) = -1 + \sum (x_i^2 + y_i^2),$$

we have

$$df(X) = \sum (x_i^2 + y_i^2) = 1$$

on $S = f^{-1}(0)$, so X is transverse to S. In fact, this is an alternative way of proving that S is a hypersurface of contact type (Fig. 7.3).

Example 7.6.10. In the standard symplectic space $(\mathbf{R}^4, \omega_0)$ with coordinates (x_1, y_1, x_2, y_2), let $f : \mathbf{R}^4 \to \mathbf{R}$ be defined by

$$f(x_1, y_1, x_2, y_2) = x_2 y_1 - 1,$$

and let S be the geometric set described by the level set

$$f^{-1}(0) = \{(x_1, y_1, x_2, y_2) \mid x_2 y_1 = 1\}.$$

It is a three-dimensional version of a hyperbolic cylinder. Consider the vector field

$$X = x_2 \frac{\partial}{\partial x_1} + y_1 \frac{\partial}{\partial y_1} + x_2 \frac{\partial}{\partial x_2} - y_1 \frac{\partial}{\partial y_2}.$$

We have

$$i(X)\omega_0 = x_2 dy_1 - y_1 dx_1 + x_2 dy_2 + y_1 dx_2,$$

and so

$$\mathcal{L}_X \omega_0 = di(X)\omega_0 = dx_2 \wedge dy_1 - dy_1 \wedge dx_1 + dx_2 \wedge dy_2 + dy_1 \wedge dx_2 = \omega_0;$$

X is a Liouville vector field of ω_0.

Also,

$$\begin{aligned} df(X) &= (x_2 dy_1 + y_1 dx_2)(X) \\ &= (x_2)(y_1) + (y_1)(x_2) \\ &= 2 \neq 0 \quad \text{on } S. \end{aligned}$$

So X is transverse to S, which shows that S is a hypersurface of contact type.

7.7 Symplectic Invariants

A theme in this text has been the presentation of geometry as the study of a set with a given structure defined by a tensor. This tensor in turn singles out diffeomorphisms that preserve the structure. The proper objects of study for a given geometric structure are those sets, quantities, objects, or relations that are preserved by structure-preserving diffeomorphisms.

Most of the objects that we have seen in this chapter that are preserved by symplectic diffeomorphisms have been "dynamical" in nature, i.e., related to vector fields and their integral curves. For example, Hamiltonian vector fields are preserved by symplectic diffeomorphisms in the sense of Proposition 7.4.19. The Poisson bracket, intimately associated with Hamiltonian vector fields, is preserved by symplectic diffeomorphisms according to Proposition 7.4.20. Hence both are properly considered objects of study in symplectic geometry.

In terms of the effect of symplectic diffeomorphisms on subsets of $S \subset \mathbf{R}^{2n}$, the only invariant we have encountered so far is the volume.

Proposition 7.7.1. *In a symplectic space* $(\mathbf{R}^{2n},\omega)$*, let* $S \subset \mathbf{R}^{2n}$ *be a* $2n$*-dimensional region of integration. Let* $\Omega = \omega^n$ *be the volume form associated to* ω*, and define*

$$vol(S,\omega) = \int_S \Omega.$$

Let $\phi : (\mathbf{R}^{2n},\omega) \to (\mathbf{R}^{2n},\omega)$ *be a symplectic diffeomorphism. Then*

$$vol(S,\omega) = vol(\phi(S),\omega).$$

Proof. This is a consequence of applying Proposition 7.3.2 to Theorem 4.5.4. □

However, when $n > 1$, there are many diffeomorphisms that preserve volume but are not symplectic diffeomorphisms. For example, the reader can verify that the (linear) diffeomorphism $\phi : (\mathbf{R}^4,\omega_0) \to (\mathbf{R}^4,\omega_0)$ given by

$$\phi(x_1,y_1,x_2,y_2) = \left(2x_1, y_1, \frac{x_2}{2}, y_2\right)$$

satisfies $\phi^*\Omega = \Omega$, but not $\phi^*\omega_0 = \omega_0$. For this reason, volume might be considered a "crude" invariant of symplectic geometry in the sense that it is a quantity that is actually preserved by a larger class of diffeomorphisms, the volume-preserving ones.

One of the main challenges in the pioneering days of symplectic geometry was the search for "finer" invariants that are preserved by symplectic diffeomorphisms but not by nonsymplectic volume-preserving diffeomorphisms. The aim of this section is to introduce this line of thinking in a way that indicates some of the major concepts in symplectic geometry and topology.

We start by illustrating the basic ideas in the linear setting. Let ω_0 be the standard linear symplectic form on $\mathbf{R}^{2n}$. Recall that $\omega_0(\mathbf{v},\mathbf{w}) = \mathbf{w}^T J\mathbf{v}$, where, using block matrix notation,

$$J = \begin{bmatrix} J_0 & O & \cdots & O \\ O & J_0 & \cdots & O \\ & & & \\ O & O & \ddots & O \\ O & \cdots & O & J_0 \end{bmatrix}, \qquad J_0 = \begin{bmatrix} 0 & -1 \\ 1 & 0 \end{bmatrix}.$$

A linear symplectomorphism $\Phi : (\mathbf{R}^{2n},\omega_0) \to (\mathbf{R}^{2n},\omega_0)$ is characterized by the property that $\omega_0(\Phi\mathbf{v},\Phi\mathbf{w}) = \omega_0(\mathbf{v},\mathbf{w})$, or, in matrix notation, $\Phi^T J\Phi = J$.

One natural way of describing the geometry of a map $\Phi : \mathbf{R}^{2n} \to \mathbf{R}^{2n}$ is by describing how it transforms some fixed set, such as a cube or a sphere. With this in mind, consider the unit ball $B = B(1) \subset \mathbf{R}^{2n}$, where we use the notation

$$B(r) = \left\{(x_1,y_1,\ldots,x_n,y_n) \mid \sum(x_i^2+y_i^2) \le r^2\right\}.$$

For a linear symplectomorphism Φ, the set $\Phi(B)$ is some distortion of B, which still must contain the origin **0** by virtue of being a linear map. In light of the discussion above, the volumes of B and $\Phi(B)$ must be the same.

Example 7.7.2. Let $\Phi : (\mathbf{R}^4, \omega_0) \to (\mathbf{R}^4, \omega_0)$ be given by

$$\Phi(x_1, y_1, x_2, y_2) = \left(2x_1, \frac{y_1}{2}, 3x_2, \frac{y_2}{3}\right).$$

We leave as a short exercise to confirm that Φ so described is a linear symplectomorphism.

Let

$$E_{2,3} = \left\{(x_1, y_1, x_2, y_2) \;\middle|\; \frac{x_1^2 + y_1^2}{4} + \frac{x_2^2 + y_2^2}{9} \le 1\right\}.$$

Geometrically, $E_{2,3}$ can be thought of as a four-dimensional ellipsoid. Note that $\Phi(B) \subset E_{2,3}$. This is a consequence of the inequalities

$$\begin{aligned} \frac{(2x_1)^2}{4} + \frac{(y_1/2)^2}{4} + \frac{(3x_2)^2}{9} + \frac{(x_2/3)^2}{9} &\le \frac{4x_1^2}{4} + \frac{4y_1^2}{4} + \frac{9x_2^2}{9} + \frac{9y_2^2}{9} \\ &\le 1 \quad \text{for } (x_1, y_1, x_2, y_2) \in B. \end{aligned}$$

However, $\Phi(B) \neq E_{2,3}$, since, for example, $(0, 2, 0, 0) \in E_{2,3}$ but $(0, 2, 0, 0) \notin \Phi(B)$.

Another way of geometrically understanding the map Φ is to consider the effect of Φ on the coordinate pairs (x_1, y_1) and (x_2, y_2). For example, consider the sets

$$Z_1(r) = \left\{(x_1, y_1, x_2, y_2) \mid x_1^2 + y_1^2 \le r^2\right\}$$

and

$$Z_2(r) = \left\{(x_1, y_1, x_2, y_2) \mid x_2^2 + y_2^2 \le r^2\right\},$$

where $r > 0$. These sets can be visualized as cylinders in $\mathbf{R}^4$ whose "axes" are the (x_2, y_2) and (x_1, y_1) coordinate planes, respectively. Note that $B \subset Z_1(1) \cap Z_2(1)$. A slight modification of the inequalities above will confirm that $\Phi(B) \subset Z_1(2)$ and $\Phi(B) \subset Z_2(3)$ (Fig. 7.4).

These sorts of geometric tools to understand a linear symplectomorphism Φ give rise to the following question: What is the smallest r such that $\Phi(B) \subset Z_1(r)$? In the preceding example, the smallest such radius must be 2, since $(1, 0, 0, 0) \in B$ but

$$\Phi(1, 0, 0, 0) = (2, 0, 0, 0) \notin Z_1(r) \quad \text{when } r < 2.$$

The following theorem addresses this question. It states that a linear symplectomorphism cannot "squeeze" the symplectic pair of coordinates (x_1, y_1) too narrowly.

Theorem 7.7.3 (Linear nonsqueezing theorem). *Let $\Phi : (\mathbf{R}^{2n}, \omega_0) \to (\mathbf{R}^{2n}, \omega_0)$ be a linear symplectomorphism. Suppose that $\Phi(B(r)) \subset Z_1(R)$. Then $r \le R$.*

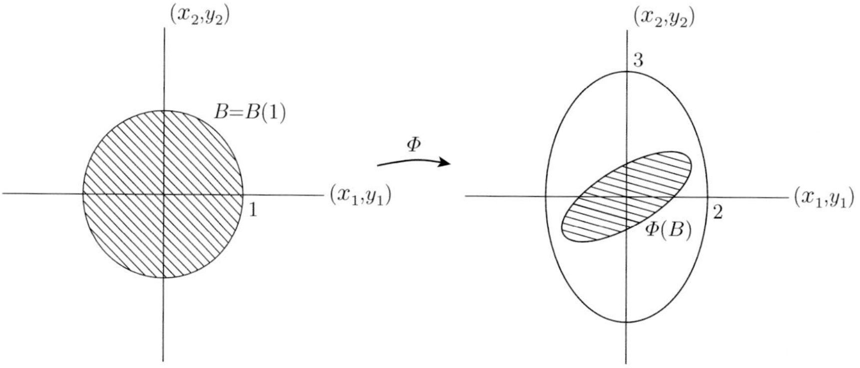

Fig. 7.4 A loose schematic describing the effect of the linear symplectomorphism Φ from Example 7.7.2 on the unit ball B.

Proof. Writing the standard Euclidean inner product as G_0, the reader may verify that $G_0(\mathbf{v}, \mathbf{w}) = \omega_0(\mathbf{v}, J\mathbf{w})$. A topological argument shows that $\Phi(B(r)) \subset Z_1(R)$ if and only if $\Phi\mathbf{x} \in Z_1(R)$ for all $\mathbf{x}$ such that $G_0(\mathbf{x}, \mathbf{x}) = r^2$, i.e., for $\mathbf{x}$ on the boundary of $B(r)$. Writing pairs $(\mathbf{e}_i, \mathbf{f}_i)$ for the standard symplectic basis of $\mathbf{R}^{2n}$, define the vectors $\mathbf{u}_i = \Phi^{-1}\mathbf{e}_i$ and $\mathbf{v}_i = \Phi^{-1}\mathbf{f}_i$. By Theorem 2.10.18, the set

$$\{\mathbf{u}_1, \mathbf{v}_1, \dots, \mathbf{u}_n, \mathbf{v}_n\}$$

is also a symplectic basis, and so in particular $J\mathbf{u}_i = \mathbf{v}_i$ and $J\mathbf{v}_i = -\mathbf{u}_i$. The condition that $\Phi(B(r)) \subset Z_1(R)$ is exactly the condition that

$$[G_0(\Phi\mathbf{x}, \mathbf{e}_1)]^2 + [G_0(\Phi\mathbf{x}, \mathbf{f}_1)]^2 \leq R^2$$

for all $\mathbf{x}$ such that $G_0(\mathbf{x}, \mathbf{x}) = r^2$. But

$$\begin{aligned} G_0(\Phi\mathbf{x}, \mathbf{e}_1) &= \omega_0(\Phi\mathbf{x}, J\mathbf{e}_1) \\ &= \omega_0(\Phi\mathbf{x}, \mathbf{f}_1) \\ &= \omega_0(\Phi\mathbf{x}, \Phi\mathbf{v}_1) \\ &= \omega_0(\mathbf{x}, \mathbf{v}_1) \quad \text{since } \Phi \text{ is a linear symplectomorphism} \\ &= \omega_0(\mathbf{x}, J\mathbf{u}_1) \\ &= G_0(\mathbf{x}, \mathbf{u}_1), \end{aligned}$$

and similarly $G_0(\Phi\mathbf{x}, \mathbf{f}_1) = G_0(\mathbf{x}, \mathbf{v}_1)$. Hence the assumption that $\Phi(B(r)) \subset Z_1(R)$ implies that

$$[G_0(\mathbf{x}, \mathbf{u}_1)]^2 + [G_0(\mathbf{x}, \mathbf{v}_1)]^2 \leq R^2$$

for all $\mathbf{x}$ such that $G_0(\mathbf{x}, \mathbf{x}) = r^2$. In particular, for $\mathbf{x} = r\mathbf{u}_1$,

$$[G_0(r\mathbf{u}_1, \mathbf{u}_1)]^2 + [G_0(r\mathbf{u}_1, \mathbf{v}_1)]^2 = r^2 + 0 \le R^2,$$

which implies that $r \le R$. □

Theorem 7.7.3 can actually be stated and proved in a more general setting. An *affine symplectomorphism* of the standard symplectic vector space $(\mathbf{R}^{2n}, \omega_0)$ is a map $T : \mathbf{R}^{2n} \to \mathbf{R}^{2n}$ of the form

$$T(\mathbf{x}) = A\mathbf{x} + \mathbf{x}_0,$$

where A is a symplectic matrix and $\mathbf{x}_0 \in \mathbf{R}^{2n}$ is a constant vector. Define $A\mathrm{Sp}(\mathbf{R}^{2n})$ to be the set of all affine symplectomorphisms of $\mathbf{R}^{2n}$. We shall refer to the result in this case as the *affine nonsqueezing theorem*; see Exercise 7.23.

In fact, the conclusion of Theorem 7.7.3 characterizes linear symplectomorphisms up to a sign. In order to state and prove this statement more precisely, we introduce the notion of symplectic balls and cylinders.

Definition 7.7.4. A *symplectic ball* of radius r is the image $B_r = \Phi(B(r))$ of the standard ball

$$B(r) = \left\{(x_1, y_1, \ldots, x_n, y_n) \mid \sum (x_i^2 + y_i^2) \le r^2\right\}$$

under a linear symplectomorphism $\Phi \in \mathrm{Sp}(2n)$. Similarly, a *symplectic cylinder* is the image $Z_R = \Psi(Z_1(R))$ of the cylinder

$$Z_1(R) = \left\{(x_1, y_1, \ldots, x_n, y_n) \mid x_1^2 + y_1^2 \le R^2\right\},$$

under a linear symplectomorphism $\Psi \in \mathrm{Sp}(2n)$.

Definition 7.7.5. A linear isomorphism $T : \mathbf{R}^{2n} \to \mathbf{R}^{2n}$ of the standard symplectic space $(\mathbf{R}^{2n}, \omega_0)$ with coordinate pairs (x_i, y_i) has the *nonsqueezing property* if for every symplectic ball B_r and for every symplectic cylinder Z_R, the condition $T(B_r) \subset Z_R$ implies that $r \le R$.

Theorem 7.7.6. *Let $T : \mathbf{R}^{2n} \to \mathbf{R}^{2n}$ be a linear isomorphism. Then both T and T^{-1} have the nonsqueezing property if and only if $T^*\omega_0 = \pm\omega_0$, where ω_0 is the standard symplectic form on $\mathbf{R}^{2n}$.*

Proof. Suppose that T is a linear symplectomorphism, i.e., $T^*\omega_0 = \omega_0$. Suppose further that for all $\Phi, \Psi \in \mathrm{Sp}(2n)$, we have $T(B_r) \subset Z_R$, where $B_r = \Phi(B(r))$ and $Z_R = \Psi(Z_1(R))$. Then $(\Psi^{-1}T\Phi)(B(r)) \subset Z_1(R)$, and so Theorem 7.7.3 guarantees that $r \le R$, since $\Psi^{-1}T\Phi \in \mathrm{Sp}(2n)$ by Proposition 2.10.17. Likewise, also by Proposition 2.10.17, T^{-1} is a linear symplectomorphism and so has the nonsqueezing property.

If now $T^*\omega_0 = -\omega_0$, define the map $\tilde{T} : \mathbf{R}^{2n} \to \mathbf{R}^{2n}$ as $\tilde{T} = T \circ \rho$, where $\rho : \mathbf{R}^{2n} \to \mathbf{R}^{2n}$ is defined by

$$\rho(x_1, y_1, \ldots, x_n, y_n) = (x_1, -y_1, \ldots, x_n, -y_n).$$

Since $\rho^*\omega_0 = -\omega_0$, we have $\tilde{T}^*\omega_0 = \omega_0$, and so as before, both $\tilde{T}$ and $\tilde{T}^{-1}$ have the nonsqueezing property. But since $\rho(B(r)) = B(r)$ and $\rho(Z(R)) = Z(R)$, then T (and T^{-1}) also has the nonsqueezing property.

We prove the converse by contradiction. Assuming that T and T^{-1} have the nonsqueezing property, suppose that there are nonzero vectors $\mathbf{v}, \mathbf{w} \in \mathbf{R}^{2n}$ such that

$$|\omega_0(\mathbf{v}, \mathbf{w})| \neq |\omega_0(T\mathbf{v}, T\mathbf{w})|.$$

We can show that in this case we may in fact assume that such nonzero vectors $\mathbf{v}, \mathbf{w}$ additionally satisfy:

- $\omega_0(\mathbf{v}, \mathbf{w}) = 1$, and
- $|\omega_0(T\mathbf{v}, T\mathbf{w})| < 1$.

To see this, note first that we may assume $|\omega_0(T\mathbf{v}, T\mathbf{w})| < |\omega_0(\mathbf{v}, \mathbf{w})|$. If not, replace T, $\mathbf{v}$, and $\mathbf{w}$ with T^{-1}, $T\mathbf{v}$, and $T\mathbf{w}$ respectively throughout. (It is at this point that we rely on the assumption that both T and T^{-1} have the nonsqueezing property.) Note that this assumption implies that $|\omega_0(\mathbf{v}, \mathbf{w})| > 0$. We can in fact guarantee that $\omega_0(\mathbf{v}, \mathbf{w}) = 1$ by replacing $\mathbf{w}$ with $\tilde{\mathbf{w}} = (1/c)\mathbf{w}$, where $c = \omega_0(\mathbf{v}, \mathbf{w})$.

Further, we can take $|\omega_0(T\mathbf{v}, T\mathbf{w})| > 0$. If not, i.e., if $\omega_0(T\mathbf{v}, T\mathbf{w}) = 0$, let $\mathbf{y}$ be a vector such that $\omega_0(\mathbf{y}, T\mathbf{w}) \neq 0$; such a vector exists, since ω_0 is nondegenerate and $T\mathbf{w} \neq \mathbf{0}$ (since T is one-to-one). Let $\mathbf{x} \in \mathbf{R}^{2n}$ be such that $T\mathbf{x} = \mathbf{y}$ (which exists because T is onto), and consider the vector

$$\tilde{\mathbf{v}} = \mathbf{v} + \epsilon\mathbf{x},$$

where $\epsilon \neq 0$. We have on the one hand that

$$\omega_0(\tilde{\mathbf{v}}, \mathbf{w}) = \omega_0(\mathbf{v}, \mathbf{w}) + \epsilon\omega_0(\mathbf{x}, \mathbf{w})$$

and on the other hand

$$\begin{aligned} \omega_0(T\tilde{\mathbf{v}}, T\mathbf{w}) &= \omega_0(T\mathbf{v}, T\mathbf{w}) + \epsilon\omega_0(T\mathbf{x}, T\mathbf{w}) \\ &= \epsilon\omega_0(\mathbf{y}, T\mathbf{w}) \quad \text{assuming that } \omega_0(T\mathbf{v}, T\mathbf{w}) = 0 \\ &\neq 0. \end{aligned}$$

So by choosing ϵ close enough to 0, we can guarantee that both $\omega_0(\tilde{\mathbf{v}}, \mathbf{w}) \neq 0$ and that $|\omega_0(T\tilde{\mathbf{v}}, T\mathbf{w})| < |\omega_0(\tilde{\mathbf{v}}, \mathbf{w})|$. This completes the justification of the two additional assumptions on $\mathbf{v}, \mathbf{w}$.

Now for nonzero $\mathbf{v}, \mathbf{w}$ satisfying the assumptions above, let

$$\lambda^2 = |\omega_0(T\mathbf{v}, T\mathbf{w})| < 1.$$

We construct two symplectic bases $\mathcal{B}_1$ and $\mathcal{B}_2$ for $(\mathbf{R}^{2n}, \omega_0)$ as follows. First, let

$$\mathcal{B}_1 = \{\mathbf{a}_1, \mathbf{b}_1, \ldots, \mathbf{a}_n, \mathbf{b}_n\}$$

be a symplectic basis constructed in the manner of Theorem 2.10.4, where $\mathbf{a}_1 = \mathbf{v}$ and $\mathbf{b}_1 = \mathbf{w}$. Also, let

$$\mathcal{B}_2 = \{\mathbf{c}_1, \mathbf{d}_1, \ldots, \mathbf{c}_n, \mathbf{d}_n\}$$

be a symplectic basis also constructed according to Theorem 2.10.4, where $\mathbf{c}_1 = (1/\lambda)(T\mathbf{v})$ and $\mathbf{d}_1 = (1/\lambda)(T\mathbf{w})$.

Define two linear isomorphisms $\Phi_1, \Phi_2 : \mathbf{R}^{2n} \to \mathbf{R}^{2n}$ in terms of their actions on the standard symplectic basis $\mathcal{B}_0 = \{\mathbf{e}_1, \mathbf{f}_1, \ldots, \mathbf{e}_n, \mathbf{f}_n\}$:

$$\Phi_1(\mathbf{e}_i) = \mathbf{a}_i, \quad \Phi_1(\mathbf{f}_i) = \mathbf{b}_i,$$

and

$$\Phi_2(\mathbf{e}_i) = \mathbf{c}_i, \quad \Phi_2(\mathbf{f}_i) = \mathbf{d}_i.$$

By Theorem 2.10.18, both Φ_1 and Φ_2 are linear symplectomorphisms by construction.

Define the linear isomorphism $A = \Phi_2^{-1} \circ T \circ \Phi_1$. Since $A(\mathbf{e}_1) = \lambda \mathbf{e}_1$ and $A(\mathbf{f}_1) = \lambda \mathbf{f}_1$, we have $A(B(1)) \subset Z_1(\lambda^2)$, and so $T(\Phi_1(B(1))) \subset \Phi_2(Z_1(\lambda^2))$. The nonsqueezing property of T then implies $1 \leq \lambda^2$, contradicting the assumption that $\lambda^2 < 1$.

Hence $|\omega_0(T\mathbf{v}, T\mathbf{w})| = |\omega_0(\mathbf{v}, \mathbf{w})|$ for all $\mathbf{v}, \mathbf{w} \in \mathbf{R}^{2n}$, i.e., $T^*\omega_0 = \pm\omega_0$. □

The linear nonsqueezing theorem can be thought of as giving the cylinder $Z_1(R)$ of *smallest* radius into which a ball $B(r)$ of fixed radius can be mapped by means of a linear symplectomorphism. There is an alternative viewpoint from which to understand the nonsqueezing theorem, however—one that leads to the notion of a symplectic invariant discussed earlier in this section. Namely, we ask the following question: For an arbitrary set S, what is the ball of *largest* radius that can be mapped into S by means of a linear symplectomorphism? More generally, one could consider the same question but include affine symplectomorphisms to allow translations.

In order to formulate an appropriate definition, we will need terminology from the structure of the set of real numbers. For a set A of real numbers, a real number $s \in \mathbf{R}$ is said to be the *supremum* of A, $s = \sup A$, if $a \leq s$ for all $a \in A$ and if for every real number b such that $a \leq b$ for all $a \in A$, we have $s \leq b$. If for a set A, no such real number exists, we say that $\sup A = +\infty$. The supremum is also known as the least upper bound.

Definition 7.7.7. Let $S \subset \mathbf{R}^{2n}$ be a nonempty set in the standard symplectic space $(\mathbf{R}^{2n}, \omega_0)$. Let

$$R_S = \left\{ r \geq 0 \mid \text{There is } \Phi \in \mathrm{ASp}(\mathbf{R}^{2n}) \text{ such that } \Phi(B(r)) \subset S \right\}.$$

The *linear symplectic width* $w(S)$ of S is the (possibly infinite) supremum of R_S:

$$w(S) = \sup R_S.$$

For every nonempty set S, we have $0 \in R_S$: $B(0)$ consists of simply the origin, which can be mapped by translation (an affine symplectomorphism) into S. Hence R_S is nonempty.

Note that historically, many authors, including McDuff and Salamon in their authoritative *Introduction to Symplectic Topology* [31], define

$$w(S) = \sup \left\{ \pi r^2 \mid r \in R_S \right\}.$$

This has the advantage of agreeing with area in the case $n = 1$ and $S = B(r)$ and suggests, in that way, the view of symplectic width as a kind of higher-dimensional analogue to area. Despite the suggestiveness of that notation, we present the quantity as a radius rather than as an area. This change from what has been standard will imply several other minor dimensional variations from the customary terminology in the discussion below.

Proposition 7.7.8. *The linear symplectic width has the following properties:*

- *Suppose that S and T are subsets of $\mathbf{R}^{2n}$ with the property that there is an affine symplectomorphism $\Psi : \mathbf{R}^{2n} \to \mathbf{R}^{2n}$ such that $\Psi(S) \subset T$. Then $w(S) \leq w(T)$.*
- *Let $S \subset \mathbf{R}^{2n}$ and let $\lambda \in \mathbf{R}$. Define the set $\lambda S = \{\lambda x \mid x \in S\}$. Then $w(\lambda S) = |\lambda| w(S)$.*
- *$w(B(r)) = w(Z_1(r)) = r$.*

Proof. The first two properties follow from the following property of the supremum: If I_1 and I_2 are sets of real numbers with the property that $I_1 \subset I_2$, then $\sup I_1 \leq \sup I_2$.

Addressing the first property, assume that $\Psi(S) \subset T$ and suppose that $r \in R_S$. Then there is an affine symplectomorphism Φ such that $\Phi(B(r)) \subset S$. But then $\Psi(\Phi(B(r))) \subset T$, and so $r \in R_T$, since $\Psi \circ \Phi$ is an affine symplectomorphism. Hence $R_S \subset R_T$, and so $w(S) \leq w(T)$.

Addressing the second property, suppose that $r \in R_{\lambda S}$, so that there is an affine symplectomorphism Φ such that $\Phi(B(r)) \subset \lambda S$. We write $\Phi(\mathbf{x}) = A\mathbf{x} + \mathbf{x}_0$, where A is a linear symplectomorphism and $\mathbf{x}_0 = \Phi(\mathbf{0})$, and define

$$\tilde{\Phi}(\mathbf{x}) = \pm \left(A\mathbf{x} + \frac{\mathbf{x}_0}{\lambda} \right),$$

with the sign chosen to be the same as the sign of λ. The reader may verify that $\tilde{\Phi}(B(r/|\lambda|)) \subset S$, and so $r/|\lambda| \in R_S$ and $r \in |\lambda| R_S$. Hence $R_{\lambda S} \subset |\lambda| R_S$. All these implications may be reversed to show that in fact $R_{\lambda S} = |\lambda| R_S$, and so $w(\lambda S) = |\lambda| w(S)$.

The third property follows from the first along with the affine nonsqueezing theorem. Since the identity map is a linear symplectomorphism, $r \in R_{B(r)}$ and $r \in$

$R_{Z_1(r)}$. Hence $w(B(r)) \geq r$ and $w(Z_1(r)) \geq r$. Further, the linear nonsqueezing theorem implies that $w(Z_1(r)) \leq r$, and so in fact $w(Z_1(r)) = r$. Then by the first property proved here and the fact that $B(r) \subset Z_1(r)$, $w(B(r)) \leq w(Z_1(r)) = r$, and so also $w(B(r)) = r$. □

Corollary 7.7.9. *Let $\Phi : (\mathbf{R}^{2n}, \omega_0) \to (\mathbf{R}^{2n}, \omega_0)$ be a linear symplectomorphism. Then for every set $S \subset \mathbf{R}^{2n}$,*

$$w(\Phi(S)) = w(S).$$

In other words, the linear symplectic width is a symplectic invariant.

Proof. Writing $A = S$ and $B = \Phi(S)$, we have trivially $B \subset \Phi(A)$, so $w(B) \leq w(A)$, and $A \subset \Phi^{-1}(B)$, so $w(A) \leq w(B)$. □

In fact, the property of preserving linear symplectic width characterizes linear symplectomorphisms, up to a sign.

Theorem 7.7.10. *Suppose that $T : \mathbf{R}^{2n} \to \mathbf{R}^{2n}$ is a linear isomorphism. Then T preserves the linear symplectic width of ellipsoids if and only if $T^*\omega_0 = \pm\omega_0$.*

Proof. Assume first that T is a linear symplectomorphism. If E is an ellipsoid, Corollary 7.7.9 guarantees that $w(E) = w(T(E))$. In the case that $T^*\omega_0 = -\omega_0$, apply Corollary 7.7.9 to the map $T \circ \rho$, where

$$\rho(x_1, y_1, \dots, x_n, y_n) = (x_1, -y_1, \dots, x_n, -y_n).$$

Now suppose that T preserves the linear symplectic width of ellipsoids. We will show that T has the linear nonsqueezing property. To that end, suppose $T(B(r_1)) \subset Z_1(r_2)$ for some positive constants r_1, r_2. Then $R_{T(B(r_1))} \subset R_{Z_1(r_2)}$, and so $w(T(B(r_1))) \leq w(Z_1(r_2))$. But since $T(B(r_1))$ and $B(r_1)$ are ellipsoids, $w(T(B(r_1))) = w(B(r_1)) = r_1$, since T preserves the linear symplectic width of ellipsoids. But $w(Z_1(r_2)) = r_2$, and so $r_1 \leq r_2$. Hence T has the nonsqueezing property.

An analogous argument shows that T^{-1} has the nonsqueezing property. Hence, by Theorem 7.7.6,

$$T^*\omega_0 = \pm\omega_0.$$ □

The following theorem relates the linear symplectic width to the notion of the symplectic spectrum of an ellipsoid introduced in Sect. 2.10.

Theorem 7.7.11. *Let $E \subset \mathbf{R}^{2n}$ be an ellipsoid in the standard symplectic space $(\mathbf{R}^{2n}, \omega_0)$ whose symplectic spectrum is $\sigma(E) = (r_1, \dots, r_n)$, with $r_1 \leq \cdots \leq r_n$. Then $w(E) = r_1$.*

Proof. Let E be an ellipsoid in $\mathbf{R}^{2n}$. By Theorem 2.10.25, there is a linear symplectomorphism $\Phi : \mathbf{R}^{2n} \to \mathbf{R}^{2n}$ such that $E = \Phi(E(r_1, \dots, r_n))$. So $w(E) =$

$w(E(r_1, \ldots, r_n))$. We have $B(r_1) \subset E(r_1, \ldots, r_n)$, so $w(E(r_1, \ldots, r_n)) \geq r_1$. But since $E(r_1, \ldots, r_n) \subset Z_1(r_1)$, if there exist $r > 0$ and a linear symplectomorphism Ψ such that $\Psi(B(r)) \subset E(r_1, \ldots, r_n)$, then $r \leq r_1$ by the linear nonsqueezing theorem. Hence $w(E) = r_1$. □

There are a number of other properties of linear symplectic width that can be found, for example, in the exposition in [31]. The purpose of the discussion, though, has been to motivate the following definition in the nonlinear setting.

Definition 7.7.12. Let $(\mathbf{R}^{2n}, \omega)$ be a symplectic space. A *capacity* (or symplectic capacity) c on $(\mathbf{R}^{2n}, \omega)$ is an assignment of a value $c(S)$, where $c(S) \geq 0$ or $c(S) = +\infty$, to each set $S \subset \mathbf{R}^{2n}$ such that the following properties are satisfied:

- Monotonicity: For subsets S and T of $\mathbf{R}^{2n}$, if there exist a domain $U \subset \mathbf{R}^{2n}$ containing S and a symplectic diffeomorphism $\phi : U \to \phi(U)$ such that $\phi(S) \subset T$, then $c(S) \leq c(T)$.
- Conformality: If $S \subset \mathbf{R}^{2n}$ and $\lambda \in \mathbf{R}$, then $c(\lambda S) = |\lambda| c(S)$.
- Nontriviality: $c(B(1)) = c(Z_1(1)) = 1$.

The following proposition is proved exactly in the manner of Corollary 7.7.9.

Proposition 7.7.13. *Suppose that c is a capacity on the symplectic space $(\mathbf{R}^{2n}, \omega)$. If $\phi : (\mathbf{R}^{2n}, \omega) \to (\mathbf{R}^{2n}, \omega)$ is a symplectic diffeomorphism, then for every set $S \subset \mathbf{R}^{2n}$,*

$$c(\phi(S)) = c(S).$$

The nontriviality axiom of a capacity is meant to preclude the "trivial" assignments $c(S) = 0$ for all S or $c(S) = \infty$ for all S. The axiom precludes volume as a capacity. It also happens to be the most difficult ("least trivial") axiom to establish in the construction of a symplectic capacity.

There is more than one way to construct capacities, and there are different capacities that are not equivalent. All of the constructions require advanced techniques, well outside the scope of this text. McDuff and Salamon note that every significant advance in symplectic geometry in the 1980s and 1990s yielded a new way of constructing a capacity. However, we will state the following result, which is fundamental given the centrality of the concept for many of the questions in symplectic geometry.

Theorem 7.7.14. *Let $(\mathbf{R}^{2n}, \omega)$ be a symplectic space. Then there exists a symplectic capacity c on $(\mathbf{R}^{2n}, \omega)$.*

We illustrate some of the consequences of the existence of a symplectic capacity. The first is a nonlinear analogue of the linear nonsqueezing theorem.

Theorem 7.7.15. *Let $(\mathbf{R}^{2n}, \omega_0)$ be the standard symplectic space with coordinate pairs (x_i, y_i). For $r, R \geq 0$, define*

$$B(r) = \left\{ (x_1, y_1, \ldots, x_n, y_n) \;\middle|\; \sum (x_i^2 + y_i^2) \leq r^2 \right\}$$

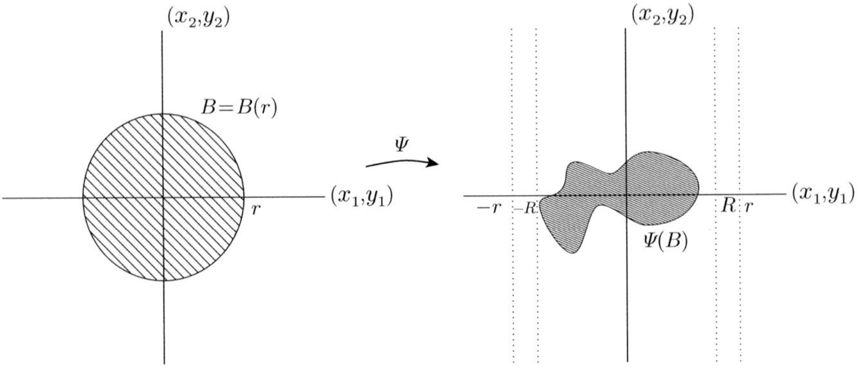

Fig. 7.5 The nonsqueezing theorem prohibits the above map Ψ from being a symplectic diffeomorphism, even if $vol(B) = vol(\Psi(B))$.

and

$$Z_1(R) = \left\{(x_1, y_1, \ldots, x_n, y_n) \mid x_1^2 + y_1^2 \le R^2\right\}.$$

Suppose there is a symplectic diffeomorphism $\phi : U \to \phi(U)$ with U a domain containing $B(r)$ and such that $\phi(B(r)) \subset Z_1(R)$. Then $r \le R$ (Fig. 7.5).

Proof. Let c be a capacity on $(\mathbf{R}^{2n}, \omega_0)$, whose existence is guaranteed by Theorem 7.7.14. Then

$$\begin{aligned} c(B(r)) &= c(rB(1)) \\ &= rc(B(1)) \quad \text{by conformality} \\ &= r(1) \quad \text{by nontriviality} \\ &= r, \end{aligned}$$

and likewise $c(Z(R)) = R$. But since $\phi(B(r)) \subset Z_1(R)$ by assumption, monotonicity gives $c(B(r)) \le c(Z_1(R))$, and so $r \le R$. □

In fact, the conclusion of the nonsqueezing theorem is equivalent to the existence of a capacity; see Exercise 7.24.

Another consequence of the existence of a symplectic capacity is that the limit of a sequence of symplectic diffeomorphisms must be a symplectic diffeomorphism. This shows that the set of all symplectic diffeomorphisms is topologically closed in the set of all diffeomorphisms, and gives further justification to the notion of "symplectic topology."

We state the following topological lemma without proof. See [31, p. 378].

Lemma 7.7.16. *Let* $(\mathbf{R}^{2n}, \omega_0)$ *be the standard symplectic space with capacity* c. *Let* $\{\phi_n : \mathbf{R}^{2n} \to \mathbf{R}^{2n}\}$ *be a sequence of diffeomorphisms that converges*[1] *to a diffeomorphism* $\phi : \mathbf{R}^{2n} \to \mathbf{R}^{2n}$. *Assume that each* ϕ_n *preserves the capacity of ellipsoids, i.e., for all* n, $c(\phi_n(E)) = c(E)$ *if* E *is an ellipsoid. Then* ϕ *preserves the capacity of ellipsoids also.*

The following is the nonlinear analogue to Theorem 7.7.10.

Theorem 7.7.17. *For the standard symplectic space* $(\mathbf{R}^{2n}, \omega_0)$ *with capacity* c, *let* $\phi : \mathbf{R}^{2n} \to \mathbf{R}^{2n}$ *be a diffeomorphism that preserves the capacity of ellipsoids. Then*

$$\phi^* \omega_0 = \pm \omega_0.$$

Proof. Let ϕ be a diffeomorphism that preserves the capacity of ellipsoids. For a fixed $x_0 \in \mathbf{R}^{2n}$, define a family of diffeomorphisms $\phi_t : \mathbf{R}^{2n} \to \mathbf{R}^{2n}$ by

$$\phi_t(x) = \frac{1}{t}\left[\phi(x_0 + tx) - \phi(x_0)\right].$$

In Exercise 7.25, we ask the reader to verify that ϕ_t preserves the symplectic capacity of ellipsoids for all t. Define $\Phi = \lim_{t \to 0} \phi_t$; in fact, $\Phi = (\phi_*)(x_0)$, and in particular, Φ is a linear map. By Lemma 7.7.16, Φ preserves the capacity of ellipsoids as well (for example by considering the sequence $\phi_{1/n}$), and so by Theorem 7.7.10, $\Phi^* \omega_0 = \pm \omega_0$. But this is the same as $\phi^* \omega_0 = \pm \omega_0$. By continuity of ϕ, the sign must be the same for all values of x_0. □

We are now in a position to state precisely the meaning of saying that the set of symplectic diffeomorphisms is "topologically closed."

Theorem 7.7.18. *Let* $\{\phi_n\}$ *be a sequence of symplectic diffeomorphisms of the standard symplectic space* $(\mathbf{R}^{2n}, \omega_0)$ *such that* $\{\phi_n\}$ *converges to a diffeomorphism* ϕ. *Then* ϕ *is also a symplectic diffeomorphism.*

Proof. Let c be a capacity on $(\mathbf{R}^{2n}, \omega_0)$. By Proposition 7.7.13, each of the maps ϕ_n preserves the capacity c. Hence by Theorem 7.7.17, $\phi^* \omega_0 = \pm \omega_0$. We therefore need to show that ϕ is not antisymplectic. Suppose to the contrary that $\phi^* \omega_0 = -\omega_0$, and consider the symplectic space

$$(\mathbf{R}^{2n} \times \mathbf{R}^{2n}, \omega_0 \oplus \omega_0);$$

see Example 7.2.10. For each n, define the diffeomorphism

$$\tilde{\phi}_n : \mathbf{R}^{2n} \times \mathbf{R}^{2n} \to \mathbf{R}^{2n} \times \mathbf{R}^{2n}$$

given by

[1] Convergence here and in the following is uniform convergence on compact sets.

$$\tilde{\phi}_n(x,z) = (\phi_n(x), z).$$

Note that

$$\tilde{\phi}_n^*(\omega_0 \oplus \omega_0) = (\phi_n^*\omega_0) \oplus \omega_0 = \omega_0 \oplus \omega_0,$$

and so the limit $\tilde{\phi} = \lim \tilde{\phi}_n$ given by

$$\tilde{\phi}(x,z) = (\phi(x), z)$$

satisfies

$$\tilde{\phi}^*(\omega_0 \oplus \omega_0) = \pm(\omega_0 \oplus \omega_0)$$

by Theorem 7.7.17. But this contradicts the assumption that $\tilde{\phi}^*(\omega_0 \oplus \omega_0) = (-\omega_0 \oplus \omega_0)$. Hence $\phi^*\omega_0 = \omega_0$. □

We close this section by noting that we have not stated the preceding sequence of results in their full generality. In fact, in the key Lemma 7.7.16, the conclusion is true even for a sequence of continuous maps (not even assuming an inverse!), as long as the limit map is continuous with a continuous inverse. This gives a sense of the strength of the capacity-preserving property. Readers interested in the statement and proof of the more general results are encouraged to consult the standard text [31] or the second chapter of Hofer and Zehnder's [23]. The latter text is an advanced book-length treatment of the existence of a symplectic capacity and its consequences.

7.8 For Further Reading

In contrast to the situation in contact geometry, there is a wealth of texts in symplectic geometry for a variety of audiences. For many years, the basic introductory text was Arnold's *Mathematical Methods of Classical Physics* [3], first published in 1978. Arnold notes in the preface to the 1989 edition, however, that, "The main part of this book was written 30 years ago." Indeed, the 1960s and 1970s might be called the classical period of symplectic geometry. Other representative texts of that period were R. Abraham and J. Marsden's *Foundations of Mechanics*, first published in 1978, and Guillemin and Sternberg's *Symplectic Techniques in Physics*, first published in 1984. Although it was published in 1987, one of the clearest presentations from a mathematician's perspective was Libermann and Marle's *Symplectic Geometry and Analytic Mechanics* [27].

It is not coincidental, as can be seen from the titles, that the basic texts of the classical period of symplectic geometry were heavily influenced by physics, and Hamiltonian systems in particular. Several authors have recounted the story of how this historical relationship spurred tremendous interest in symplectic geometry, but then experienced a lull due to the relative scarcity of physical Hamiltonian systems; see, for example, the short discussion in [27, p. 197] on completely integrable systems, or the paper "The Symplectization of Science" [21], which is easily accessible on the Internet.

The modern era of symplectic geometry opened in 1985 with Gromov's landmark paper, "Pseudoholomorphic Curves in Symplectic Manifolds" [22]. It is a testament to the power of that paper that new results were still being mined from it nearly two decades later. It is fair to say that the presence or absence of that paper in a text's bibliography determines whether the text is "classical" or "modern." In particular, that paper introduced the symplectic invariants of the type discussed in the previous section, and contained the first proof of the nonsqueezing theorem.

The impact of Gromov's paper is impossible to minimize. For at least two decades, a community of mathematicians led by Gromov, Eliashberg, Hofer, McDuff, and others, along with their respective schools, expanded the contours of symplectic topology. The best representative text of that period was McDuff and Salamon's *Introduction to Symplectic Topology* [31], first published in 1994 but significantly expanded in the second edition of 1998. It is now a standard reference aimed at an audience at the graduate level. The exposition on symplectic invariants in Sect. 7.7 is heavily indebted to their presentation.

7.9 Exercises

7.1. A mass attached to a spring has kinetic energy $T = \frac{1}{2}m\dot{x}^2$ and potential energy $U = \frac{1}{2}kx^2$, where m and k are physical constants.

(a) Write and solve the Euler–Lagrange equations of motion, as in Example 7.1.3.
(b) Write and solve the corresponding Hamilton's equations of motion as in Example 7.1.5.

7.2. The motion of a particle with mass m moving in the plane can be described using polar coordinates by means of a smooth curve $c : I \to \mathbf{R}^2$ with components $c(t) = (r(t), \theta(t))$, where I is an interval. In this case, the kinetic energy T associated to the system is given by

$$T(t) = \frac{1}{2}m\dot{r}^2 + \frac{1}{2}mr^2\dot{\theta}^2.$$

(a) Assuming that the particle moves without external forces, the potential energy U of the system is given by $U(t) = 0$ for all $t \in I$. Write, but do not solve, the Euler–Lagrange equations of motion.
(b) Write Hamilton's equations of motion for the system in part (a). Solve the system in the case that $m = 2$ and initial conditions are given by $c(0) = (1, 0)$, $\dot{c}(0) = (1, 1)$. Hint: Express the derivatives of r and p in terms of θ. The substitution $s = 1/r$ will come in handy.
(c) Write Hamilton's equations of motion for the system in the presence of a central gravitational force with potential energy given by $U(t) = \dfrac{-mM}{r}$, where M is a physical constant representing the mass of a second point object. Show that the motion of this system in the case $m = 2$ with the initial conditions in part

(b) must satisfy the following equation of an ellipse:

$$r(\theta) = \frac{1/M}{1 + (\mu/M)\cos(\theta + \theta_s)},$$

where $\mu = \sqrt{(1-M)^2 + 1}$ and $\theta_s = \cos^{-1}\left(\frac{1-M}{\mu}\right)$.

7.3. Suppose that $f : I \to \mathbf{R}$ is a convex function defined on an interval I, meaning that $f''(x) > 0$ for all x in I. Define further

$$I_1 = f'(I) = \{t \in \mathbf{R} \mid \text{There exists } s \in I \text{ such that } f'(s) = t\}.$$

Note that $f' : I \to I_1$ is a one-to-one and onto function with a differentiable inverse $(f')^{-1} : I_1 \to I$.

(a) Show that for a fixed $p \in I_1$, the value of x for which the function $D(x,p) = px - f(x)$ has a relative maximum satisfies $f'(x) = p$. (The function $D(x,p)$ represents the "vertical distance" between the curve $y = f(x)$ and the line $y = px$ for a given value of x.)
(b) Define the function $g : I_1 \to \mathbf{R}$ as

$$g(p) = D\left((f')^{-1}(p), p\right);$$

g is called the *Legendre transformation* of the function f. Compute the Legendre transformation of the following functions, identifying the domains of g:

(i) $f(x) = x^2$.
(ii) $f(x) = \frac{x^a}{a}$, where $a > 1$.
(iii) $f(x) = e^{ax}$, where $a > 0$.

(c) Let g be the Legendre transform of a convex function f. Show that g is convex.
(d) Show that if g is the Legendre transform of the convex function f, then the Legendre transform of g is f.

7.4. Suppose that f_1 and f_2 are convex functions with corresponding Legendre transforms g_1 and g_2 as in Exercise 7.3. Prove the following statements:

(a) If $f_2(x) = af_1(x)$ for $a > 0$, then $g_2(p) = ag_1\left(\frac{p}{a}\right)$.
(b) If $f_2(x) = f_1(ax)$ for $a > 0$, then $g_2(p) = g_1\left(\frac{p}{a}\right)$.
(c) If $f_2(x) = f_1(x) + c$ for $c \in \mathbf{R}$, then $g_2(p) = g_1(p) - c$.
(d) If $f_2(x) = f_1(x + c)$ for $c \in \mathbf{R}$, then $g_2(p) = g_1(p) - cp$.
(e) If $f_2(x) = f_1^{-1}(x)$, then $g_2(p) = -pg_1\left(\frac{1}{p}\right)$.

7.5. Provide the details of the proof of Proposition 7.2.3.

7.6. Suppose ω is a symplectic form on $\mathbf{R}^{2n}$ and $\phi : \mathbf{R}^{2n} \to \mathbf{R}^{2n}$ is a diffeomorphism. Show that $\phi^*\omega$ is also a symplectic form.

7.7. For each of the contact forms on $\mathbf{R}^3$ below (see Exercise 6.6), perform the symplectization procedure of Example 7.2.11 by writing the resulting symplectic form in coordinates.

(a) $\alpha_1 = dy - mdx$.
(b) $\alpha_2 = dz + xdy - ydx$.
(c) $\alpha_3 = (\cos z)dx + (\sin z)dy$.

7.8. For each of the symplectic forms constructed in Exercise 7.7, write the corresponding isomorphism Φ of Proposition 7.2.4. Use this calculation to write the Hamiltonian vector field X_f in coordinates for a smooth function f.

7.9. Let $(\mathbf{R}^{2n}, \omega_0)$ be the standard symplectic space with coordinate pairs (x_i, y_i). Let $\phi : \mathbf{R}^{2n} \to \mathbf{R}^{2n}$ be a diffeomorphism with component functions $\phi = (\phi_1^x, \phi_1^y, \ldots, \phi_n^x, \phi_n^y)$. Show that if ϕ is a symplectic diffeomorphism, then the component functions must satisfy the following partial differential equations (in block matrix form):

$$\sum_{k=1}^{n} \Phi_i^k J_0 \Phi_j^k = O \quad \text{when } i \neq j$$

$$\sum_{k=1}^{n} \Phi_i^k J_0 \Phi_j^k = J_0 \quad \text{when } i = j,$$

using the notation

$$\Phi_j^i = \begin{bmatrix} \frac{\partial \phi_i^x}{\partial x_j} & \frac{\partial \phi_i^x}{\partial y_j} \\ \frac{\partial \phi_i^y}{\partial x_j} & \frac{\partial \phi_i^y}{\partial y_j} \end{bmatrix}, \quad J_0 = \begin{bmatrix} 0 & -1 \\ 1 & 0 \end{bmatrix}, \quad \text{and} \quad O = \begin{bmatrix} 0 & 0 \\ 0 & 0 \end{bmatrix}.$$

Write each entry of the matrix equations for the case $n = 2$.

7.10. Prove Theorem 7.3.9.

7.11. Let $\omega_1 = d(e^t \alpha_1)$ be the symplectic form obtained by symplectizing the contact form $\alpha_1 = dy - mdx$ on $\mathbf{R}^3$ (see Exercise 7.7).

(a) Write the conditions on the components of a vector field X to guarantee that X is a symplectic vector field relative to ω_1.
(b) Write the component expression for the Poisson bracket $\{f, g\}$ of two smooth functions f and g relative to ω_1.

7.12. Prove Proposition 7.4.5.

7.13. The Hamiltonian function corresponding to the "Kepler problem" describing the motion of a particle moving under the influence of gravity due to a point mass, in rectangular coordinates, is given by the function $H : (\mathbf{R}^6, \omega) \to \mathbf{R}$,

$$H(x_1, p_1, x_2, p_2, x_3, p_3) = \frac{1}{2m}\left(p_1^2 + p_2^2 + p_3^2\right) + \left(\frac{G}{(x_1^2 + x_2^2 + x_3^2)^{1/2}}\right),$$

where m and G are physical constants and $(\mathbf{R}^6, \omega)$ is the symplectic space with coordinate pairs (x_i, p_i) and $\omega = \sum(dp_i \wedge dx_i)$.

(a) Compute the Hamiltonian vector field X_H of H relative to ω.
(b) Show that the functions $M_1 = x_2p_3 - p_2x_3$ and $M_2 = x_3p_1 - p_3x_1$ are first integrals of the Hamiltonian vector field X_H.
(c) Compute $M_3 = \{M_1, M_2\}$.
(d) Let

$$W_1 = \frac{1}{m}\left(p_2M_3 - p_3M_2\right) + \frac{Gx_1}{(x_1^2 + x_2^2 + x_3^2)^{1/2}}.$$

Show that W_1 is a first integral of the Hamiltonian vector field X_H.
(e) Compute $W_2 = \{W_1, M_3\}$ and $W_3 = \{W_2, M_1\}$.
(f) Show that $\{W_1, W_2\} = \dfrac{2H}{m}M_3$.
(g) Write a "multiplication table" for all possible Poisson brackets of the M_i and W_j.

7.14. Show that if ω_1 and ω_2 are two symplectic forms on $\mathbf{R}^{2n}$ such that for all smooth functions $f, g : \mathbf{R}^{2n} \to \mathbf{R}$, their associated Poisson brackets satisfy $\{f, g\}_1 = \{f, g\}_2$, then in fact $\omega_1 = \omega_2$. Hint: In any coordinate system, show that the components of X_f are uniquely determined by the components of ω. So if $X_f[g] = \tilde{X}_f[g]$ for all g, then $\omega_1 = \omega_2$.

7.15. Prove Proposition 7.5.2.

7.16. Let (U_1, ω_1) and (U_2, ω_2) be two symplectic spaces where $U_1 \subset \mathbf{R}^{2n}$ and $U_2 \subset \mathbf{R}^{2m}$ are domains, and let

$$(U_1 \times U_2, \omega = \omega_1 \oplus \omega_2)$$

be the product symplectic space as defined in Example 7.2.10. Show that for every $p \in U_2$, the subset $U_1 \times \{p\} \subset \mathbf{R}^{2n} \times \mathbf{R}^{2m}$ is a symplectic set in $U_1 \times U_2$.

7.17. Given a symplectic space (U, ω) with $U \subset \mathbf{R}^{2n}$, consider the product $U \times U$ with symplectic form $(-\omega) \oplus (\omega)$. Show that the set

$$\Delta = \{(x, x) \mid x \in U\} \subset \mathbf{R}^{2n} \times \mathbf{R}^{2n}$$

is a Lagrangian set in the product $U \times U$.

7.18. For each of the following functions $f : \mathbf{R}^2 \to \mathbf{R}$, find the Lagrangian sets corresponding to $df(\mathbf{R}^2) \subset T^*\mathbf{R}^2$ as described in Example 7.5.9.

(a) $f(x, y) = x^2 + y^2$.
(b) $f(x, y) = xy$.
(c) $f(x, y) = e^x \sin(xy)$ on $U = \{(x, y) \mid x^2 + y^2 > 0\}$.

7.19. Verify directly that the geometric set S described in Example 7.5.14 is coisotropic, without relying on Proposition 7.5.13.

7.20. Let $(\mathbf{R}^{2n-1}, \alpha)$ be a contact space, and let f, g be smooth functions on $\mathbf{R}^{2n-1}$. Define the *contact Poisson bracket* on $\mathbf{R}^{2n-1}$ relative to α by

$$\{f, g\}_\alpha = \alpha([X_f, X_g]),$$

where X_f and X_g are the contact gradients of f and g relative to α. (See Theorem 6.5.4.)

(a) Show that $X_{\{f,g\}_\alpha} = [X_f, X_g]$.
(b) Show that

$$\{f, \{g, h\}_\alpha\}_\alpha + \{g, \{h, f\}_\alpha\}_\alpha + \{h, \{f, g\}_\alpha\}_\alpha = 0.$$

(c) For a smooth function $f : \mathbf{R}^{2n-1} \to \mathbf{R}$, define the smooth function $F : \mathbf{R}^{2n} = \mathbf{R}^{2n-1} \times \mathbf{R} \to \mathbf{R}$ by $F(x, t) = e^t f(x)$. Show that $Y_F = X_f - \xi[f]\frac{\partial}{\partial t}$, where Y_F is the Hamiltonian vector field of F relative to the symplectic form $\omega = d(e^t\alpha)$, X_f is the contact gradient of f, and ξ is the Reeb vector field for α.
(d) For smooth functions f, g on $\mathbf{R}^{2n-1}$ and corresponding functions $F = e^t f$ and $G = e^t g$ on the symplectized space $(\mathbf{R}^{2n}, \omega)$ as in part (c), show that the associated (symplectic) Poisson bracket $\{F, G\}$ satisfies

$$\{F, G\} = e^t\{f, g\}_\alpha.$$

7.21. Let $(\mathbf{R}^4, \omega_0)$ be the standard symplectic space with coordinates (x_1, y_1, x_2, y_2), and let $X = \langle X^1, Y^1, X^2, Y^2 \rangle$ be a vector field on $\mathbf{R}^4$. Write the partial differential equations that the component functions of X must satisfy for X to be a Liouville vector field.

7.22. Show that the set of affine symplectomorphisms $A\mathrm{Sp}(\mathbf{R}^{2n})$ forms a group with the usual operation of composition of functions: (1) Id $\in A\mathrm{Sp}(\mathbf{R}^{2n})$, (2) if $\Phi, \Psi \in A\mathrm{Sp}(\mathbf{R}^{2n})$, then $\Phi\Psi \in A\mathrm{Sp}(\mathbf{R}^{2n})$, and (3) if $\Phi \in A\mathrm{Sp}(\mathbf{R}^{2n})$, then $\Phi^{-1} \in A\mathrm{Sp}(\mathbf{R}^{2n})$.

7.23. State the affine nonsqueezing theorem, i.e., the affine analogue of Theorem 7.7.3. Adjust the proof of Theorem 7.7.3 to prove the affine nonsqueezing theorem.

7.24. Prove Theorem 7.7.14, assuming that Theorem 7.7.15 is true. Hint: Follow the construction of the linear symplectic width.

7.25. Show that if $\phi : \mathbf{R}^{2n} \to \mathbf{R}^{2n}$ preserves the symplectic capacity of ellipsoids, then so does

$$\phi_t(x) = \frac{1}{t}\left[\phi(x_0 + tx) - \phi(x_0)\right],$$

where $x_0 \in \mathbf{R}^{2n}$ is fixed.
Hint: Write $\phi_t = \mu_{1/t} \circ T_{-\phi(x_0)} \circ \phi \circ T_{x_0} \circ \mu_t$, where $\mu_a(x) = ax$ for $a \in \mathbf{R}$ and $T_y(x) = x + y$ for $y \in \mathbf{R}^{2n}$.

7.26. Let c be a symplectic capacity on $\mathbf{R}^{2n}$. Show that if S contains a domain $U \subset \mathbf{R}^{2n}$, then $c(S) > 0$.

References

1. Abraham, R., Marsden, J.E., Ratiu, T.: Manifolds, Tensor Analysis and Applications, 2nd edn. Springer, New York (1988)
2. Anton, H., Rorres, C.: Elementary Linear Algebra with Applications, 9th edn. Wiley, New York (2005)
3. Arnold, V.I.: Mathematical Methods of Classical Mechanics, 2nd edn. Springer, New York (1989)
4. Audin, M., Lafontaine, F. (eds.): Holomorphic curves in symplectic geometry. In: Progress in Mathematics, vol. 117. Birkhäuser, Basel (1994)
5. Bachman, D.: A Geometric Approach to Differential Forms. Birkhäuser, Boston (2006)
6. Banyaga, A., McInerney, A.: On isomorphic classical diffeomorphism groups. III. Ann. Global Anal. Geom. **13**(2), 117–127 (1995)
7. Bates, L., Peschke, G.: A remarkable symplectic structure. J. Differential Geom. **32**(2), 533–538 (1990)
8. Bennequin, D.: Entrelacements et équations de Pfaff. Third Schnepfenried geometry conference, vol. 1, pp. 87–161 (Schnepfenried, 1982); Astérisque, vol. 107–108. Société mathématique de France, Paris (1983)
9. Blanchard, P., Devaney, R.L., Hall, G.R.: Differential Equations, 3rd edn. Thomson Brooks/Cole, Pacific Grove (2006)
10. Boeckx, E.: The case for curvature: The unit tangent bundle. In: Kowalski O., Musso E., Perrone D. (eds.) Complex, Contact and Symmetric Manifolds, vol. 234, p. 15. Birkhuser, Boston (2005)
11. Burke, W.L.: Applied Differential Geometry. Cambridge University Press, New York (1985)
12. do Carmo, M.: Differential Geometry of Curves and Surfaces. Prentice Hall, Englewood Cliffs (1976)
13. do Carmo, M.: Riemannian Geometry. Birkhäuser, Boston (1992)
14. do Carmo, M.: Differential Forms and Applications. Springer, New York (1994)
15. Dubrovin, B.A., Fomenko, A.T., Novikov, S.P.: Modern Geometry—Methods and Applications, Part I. Springer, New York (1984)
16. Folland, G.B.: Introduction to Partial Differential Equations, 2nd edn. Princeton University Press, Princeton (1995)
17. Geiges, H.: A brief history of contact geometry and topology. Expo. Math. **19**(1), 25–53 (2001)
18. Geiges, H.: Christiaan Huygens and contact geometry. Nieuw Arch. Wiskd (5). **6**(2), 117–123 (2005)
19. Geiges, H.: An Introduction to Contact Topology. Cambridge University Press, New York (2008)
20. Giaquinta, M., Hildebrandt, S.: Calculus of Variations II. Springer, New York (1996)

A. McInerney, *First Steps in Differential Geometry: Riemannian, Contact, Symplectic*, Undergraduate Texts in Mathematics, DOI 10.1007/978-1-4614-7732-7,

21. Gotay, M., Isenberg, J.A.: The symplectification of science. Gaz. Math. **54**, 59–79 (1992)
22. Gromov, M.: Pseudoholomorphic curves in symplectic manifolds. Invent. Math. **82**(2), 307–347 (1985)
23. Hofer, H., Zehnder, E.: Symplectic Invariants and Hamiltonian Dynamics. Birkhäuser, Boston (1994)
24. Kobayashi, S., Nomizu, K.: Foundations of Differential Geometry, vols. 1 and 2. Wiley, New York (1963, 1969)
25. Kühnel, W.: Differential Geometry: Curves–Surfaces–Manifolds, 2nd edn. American Mathematical Society, Providence (2006)
26. Lanczos, C.: The Variational Principles of Mechanics, 4th edn. Dover, New York (1986)
27. Libermann, P., Marle, C.-M.: Symplectic Geometry and Analytical Mechanics. D. Reidel, Boston (1987)
28. Lie, S.: Geometrie der Berührungstransformationen. B.G. Teubner, Leipzig (1896)
29. Lie, S., Engel, F.: Theorie der Transformationsgruppen, vol. 3. B.G. Teubner, Leipzig (1888–1893)
30. Marsden, J., Tromba, A.J.: Vector Calculus, 5th edn. W.H. Freeman and Co., New York (2004)
31. McDuff, D., Salamon, D.: Introduction to Symplectic Topology, 2nd edn. Clarendon Press, Oxford (1998)
32. Morgan, F.: Riemannian Geometry, A Beginner's Guide, 2nd edn. A K Peters, Ltd., Natick (1998)
33. O'Neill, B.: Elementary Differential Geometry, 2nd edn. Academic, New York (1997)
34. Peterson, P.: Riemannian Geometry, 2nd edn. Springer, New York (2006)
35. Rybicki, T.: Isomorphisms between groups of diffeomorphisms. Proc. Am. Math. Soc. **123**(1), 303–310 (1995)
36. Schwerdtfeger, H.: Geometry of Complex Numbers. Dover, New York (1979)
37. Spivak, M.: Calculus on Manifolds. Benjamin/Cummings, Reading (1965)
38. Spivak, M.: A Comprehensive Introduction to Differential Geometry, vol. 5, 2nd edn. Publish or Perish, Berkeley (1979)
39. Struik, D.: Lectures on Classical Differential Geometry, 2nd edn. Dover, New York (1988)
40. Warner, F.W.: Foundations of Differentiable Manifolds and Lie Groups. Springer, New York (1983)
41. Weintraub, S.H.: Differential Forms: A Complement to Vector Calculus. Academic, New York (1997)

Index

A. McInerney, *First Steps in Differential Geometry: Riemannian, Contact, Symplectic*, Undergraduate Texts in Mathematics, DOI 10.1007/978-1-4614-7732-7,

Printed in the United States
By Bookmasters